Santa Cruz Summer Workshops in Astronomy and Astrophysics

S.M. Faber
Editor

Nearly Normal Galaxies

From the Planck Time to the Present

The Eighth Santa Cruz Summer Workshop
in Astronomy and Astrophysics
July 21–August 1, 1986, Lick Observatory

With 133 Illustrations

Springer-Verlag
New York Berlin Heidelberg London Paris Tokyo

S. M. Faber
Department of Astronomy
University of California
Santa Cruz, CA 95064, USA

Library of Congress Cataloging-in-Publication Data
Santa Cruz Summer Workshop in Astronomy and Astrophysics
(8th : 1986)
Nearly normal galaxies.
(Santa Cruz summer workshops in astrophysics)
1. Galaxies—Congresses. 2. Astrophysics—
Congresses. I. Faber, Sandra M. II. Title.
III. Series.
QB856.S26 1986 523.1′12 87-9559

Softcover reprint of the hardcover 1st edition 1987

9 8 7 6 5 4 3 2 1
ISBN-13:978-1-4612-9145-9 e-ISBN-13978-1-4612-4762-3
DOI: 10.1007/978-1-4612-4762-3

ALBERT EDWARD WHITFORD

On October 22, 1985, Albert Whitford celebrated his eightieth birthday. Whitford is honored as one of the founding fathers of photoelectric photometry, the technique that transformed twentieth-century astronomy from a qualitative to a quantitative science. Together with collaborator and mentor Joel Stebbins, Whitford began, in the early 1930s, a systematic survey of stellar and galaxy photometry that initiated the modern era of this subject. Concepts encountered in this early work included interstellar extinction, color excess, effective wavelengths, galaxy surface-brightness profiles, and K-corrections. Techniques developed by Whitford and Stebbins to deal with these have become such a part of the lore of observational astronomy that their origins to today's students are obscure. Whitford's own contribution to this work lay principally in the perfection of the equipment, notably in low-noise amplification of the signal. Joining Stebbins as a junior collaborator and fresh Ph.D. (in physics from Wisconsin), Whitford exercised skills as an instrumentalist so outstanding that, within a short while, Stebbins came to regard him as indispensable.

World War II intervened, and Whitford spent five years in war work at the MIT Radiation Laboratory. After the war, he returned to Washburn Observatory at Wisconsin, where he shortly became Director and Full Professor. In 1958 he moved west to become Director of the Lick Observatory. There his technical acumen enabled him to take charge of the 120-inch telescope project when it was far behind schedule and get it rapidly back on track. Within little more than a year the telescope was in full operation and continues to this day as the mainstay of the optical astronomy program of the University of California.

At about this time Whitford wrote what is probably his most quoted paper setting forth a standard interstellar extinction curve, the so-called Whitford law. In the 1960s he was president of the American Astronomical Society and chaired the first National Academy of Sciences review on the progress and needs of observational astronomy. *Ground Based Astronomy: A Ten Year Program* was so well and wisely written that it became a model for NAS discipline reviews and has inspired two subsequent astronomical reviews in a pattern that now threatens to become established custom. Rereading the Whitford Report, one finds that its recommendations were closely followed and that its vision shaped the pattern for the technical advancement of observational astronomy from the 1960s onward.

In 1973, Albert Whitford became Emeritus Professor/Astronomer, but with retirement his scientific productivity if anything began to increase. In a profession renowned for the vigor of its older practitioners, Whiford's achievements after retirement inspire the greatest admiration. A major early impetus for the Stebbins-Whitford photoelectric photometry program was an interest in galaxy stellar populations and in quantifying comparisons between galaxy spectral-energy distributions and those of stars. In the 1970s Whitford realized that the nuclear bulge of the Galaxy provided a unique opportunity to study, star-by-star, a strong-lined, old stellar population like that found in elliptical galaxies and spiral bulges. Since then, he and his collaborators Jay Frogel, Mike Rich, and Don Terndrup have been collecting fundamental data – colors, magnitudes, and spectra – of bulge stars. As in all great enterprises, this one has not been without its surprises, turning up giant branch stars that are simultaneously metal-rich but rather blue. The bulge studies are clearly yielding unique data on old stars of fundamental importance to population syntheses of external galaxies.

In view of Whitford's age, it is not surprising that his recent collaborators are all younger than he, but it *is* significant that two of them – Rich and Terndrup – were graduate students. Albert Whitford seems to have had a special affinity for graduate students in his emeritus years. His office was and is a place students can go to find encouragement and a sympathetic ear. Several Lick students, including Alan Dressler, David Burstein, and Nick Suntzeff, remember Albert for the benign influence he has had on their graduate careers.

In recognition of his lifetime of achievement in the areas of photoelectric photometry and galactic stellar populations, the first session of this year's Santa Cruz Workshop is enthusiastically dedicated to Albert Edward Whitford.

S. M. Faber

July 1986

Preface

It is sometimes said that astronomy is the crossroads of physics. In the same spirit, it can forcefully be argued that galaxies are the crossroads of astronomy. Internal processes within galaxies involve all of the fundamental components of astrophysics: stellar evolution, star formation, low-density astrophysics, dynamics, hydrodynamics, and high-energy astrophysics. Indeed, one can hardly name an observational datum in any wavelength range on any kind of celestial object that does not provide a useful clue to galaxy formation and evolution.

Although internal processes in galaxies until recently occupied most of our attention, we now know that it is also vital to relate galaxies to their environment. How galaxies congregate in larger structures and are in turn influenced by them are crucial questions for galactic evolution. On a grander level we have also come to regard galaxies as the basic building blocks of the universe, the basic units whereby the large-scale structure of the universe is apprehended and quantified. On a grander level still, we also believe strongly that galaxies are the direct descendents of early density irregularities in the Big Bang. Galaxy properties are now viewed as providing a crucial constraint on the physics of the Big Bang and a vital link between the macroscopic and microscopic structure of the universe.

A picture of galactic evolution as the crossroads of astronomy and cosmology thus emerges very strongly – a melding of basic physics and astrophysics, an eclectic borrowing of diverse observational techniques, and the convergence of the very large with the very small. To understand galaxies in all their ramifications clearly entails a synthesis of the highest order.

Such a synthesis was the avowed goal of this summer's Santa Cruz Summer Workshop on galaxies. The extended, two-week format allowed participants to cover essentially all of the major issues of current interest in galaxy formation and evolution – from stellar evolution to cosmic strings and virtually everything in between. It was the first time, to my knowledge, that an organized meeting of such long duration was devoted to the subject, and the result was a more comprehensive, unified view than could be had from the usual two- or three-day conference.

This book, which summarizes the morning talks by the invited speakers, is therefore of special value. The fifty-odd papers contained herein provide a remarkably complete snapshot of the field of galactic evolution as it exists in 1986. Most of the papers are excellent reviews that provide both student and professional alike with a convenient jumping-off place into the literature. Because of the wide range of topics covered, the volume may also be useful as a collection of supplemental readings for graduate courses on galactic evolution. Special thanks are due to all the speakers for providing manuscripts of such uniformly high quality.

Readers will note that the present volume is the first of a new series. Since the last three Santa Cruz Workshops have produced proceedings in response to public demand, it seems that this is an idea that has taken permanent root. Accordingly, with the help of Springer-Verlag, we are intending to bring out a published summary of the morning sessions of future Summer Workshops as a regular activity.

In closing, it is worth noting the truly remarkable progress that has been made in understanding galactic evolution over the past decade, to which the current volume bears witness. A quick perusal of the Table of Contents reveals many papers whose very titles would have been incomprehensible a bare five years ago. Thanks to the injection of new ideas from particle physics and the stimulus of ever-grander observational programs, galactic evolution has recently experienced a rare quantum leap forward in understanding and insight.

One hopes fervently that this advance has not yet peaked but will continue in the years ahead with equal vigor. We look forward to the next Santa Cruz Workshop on galaxies, not too many years off, we hope, that will chronicle the next stage of our developing understanding.

S. M. Faber

Santa Cruz

December 1986

Acknowledgments

The Santa Cruz Astrophysics Workshops owe a great debt of gratitude to the several institutions and individuals who make continued workshops possible. The major share of extramural financial support for this year's workshop was provided by generous grants from the NSF and NASA. Important additional support came from the Institute for Geophysics and Planetary Physics (IGPP) at Lawrence Livermore National Laboratory and a local campus grant from the Natural Sciences Division.

We also gratefully acknowledge the continued close cooperation of the Lick Observatory. Although the workshops are financially and administratively independent of Lick, the Observatory provides invaluable assistance in lending facilities, personnel, and moral support. We also could not manage without the help of the astronomy graduate students, six of whom this year donated a great deal of their time and effort to the workshop.

Sandra Faber wishes especially to thank this year's Scientific Organizing Committee for their very able choice of topics, speakers, and organization of the morning sessions. The members of this committee are listed in the Table of Contents under the session they chaired.

Sandra M. Faber
Workshop Organizer

Sue Robinson
Workshop Coordinator

Pat Shand
Production Manager

Contents

1
2
3
4
5
6
7
8
9
10
11
12
13
14
15
16
17
18
19
20
21
22
23
24
25
26
27
28
29
30
31
32
33
34
35
36
37
38
39
40
41
42
43
44
45
46
47
48
49
50
51
52
53
54
55
56
57
58
59
60
61
62
63
64
65
66
67
68
69
70
71
72
73
74
75
76
77
78
79
80
81
82
83
84
85
86
87
88
89
90
91
92
93
94
95
96
97
98
99
100
101
102
103
104
105
106
107
108
109
110
111
112
113
114
115
116
117
118
119
120

1. Ray Carlberg
2. Brad Whitmore
3. Tony Tyson
4. Colin Norman
5. Art Wolfe
6. Lucio Buson
7. Mike Loewenstein
8. Sue Robinson
9. Bill Mathews
10. Albert E. Whitford
11. J. Jésus Gonzalez
12. Yehuda Hoffman
13. Erez Braun
14. Edmund Bertschinger
15. Doug Richstone
16. Bill Forman
17. Jim Gunn
18. Avishai Dekel
19. Josh Barnes
20. Simon White
21. Steve Kent
22. Gene Smith
23. Peter Martin
24. François Schweizer
25. Craig Hogan
26. Richard Kron
27. Paul Shapiro
28. Neta Bahcall
29. Marc Aaronson
30. Ofer Lahav
31. Sidney van den Bergh
32. Paul Schecter
33. Avery Meiksin
34. Pat McCarthy
35. Leonid Ozernoy
36. Paolo Salucci
37. Mike Dopita
38. Bruno Binggeli
39. Alexander Kashlinsky
40. Michael West
41. John Kormendy
42. Michael Strauss
43. Alar Toomre
44. Greg Bothun
45. George Blumenthal
46. Hyesung Kang
47. Richard Larson
48. Dongsu Ryu
49. Amos Yahil
50. Marc Davis
51. Ricardo Flores
52. Tod Lauer
53. Alan Dressler
54. Dave Carter
55. Jim Mark
56. Christine Jones
57. Albert Bosma
58. Ling Luan
59. Jo Ann Eder
60. Cedric Lacey
61. Jens Villumsen
62. Peter Quinn
63. Luis A. Aguilar
64. Ken Freeman
65. Steve Shectman
66. Gustavo Bruzual
67. Manolis Plionis
68. Gerry Gilmore
69. Rosie Wyse
70. David Koo
71. Stefano Kasertano
72. Jon Holtzman
73. Colin McGill
74. Jacqueline van Gorkom
75. Gregory Shields
76. Konrad Kuijken
77. Joel Primack
78. Mark Yates
79. Adrian Melott
80. Elaine Sadler
81. Ed Langer
82. Dave Burstein
83. Giuseppina Fabbiano
84. Unidentified
85. Renzo Sancisi
86. Michael Pierce
87. Marcia Rieke
88. Augustus Oemler
89. Merle Walker
90. Richard Elston
91. David Seckel
92. Gustavo Ponce
93. E. Athanassoula
94. Roger Davies
95. Sterl Phinney
96. David Band
97. J. Richard Bond
98. Scott Tremaine
99. Mike Fall
100. Carlos Frenk
101. Howard Bond
102. Marc Postman
103. Edward Rosenblatt
104. Jay Frogel
105. Robert Kirshner
106. Brent Tully
107. Carol Persson
108. Richard Shaw
109. Lloyd Robinson
110. Mark Henriksen
111. Trinh Thuan
112. John Tonry
113. David Spergel
114. Rick Pogge
115. Bruce Carney
116. Linda Stryker
117. Kap-Soo Oh
118. Sandra Faber
119. John Huchra
120. Timothy Beers

Participants

PARTICIPANT	INSTITUTION
AARONSON, Marc	Steward Observatory
AGUILAR, Luis A.	Smithsonian Center for Astrophysics
ALBRECHT, Andreas	Los Alamos National Labs
ATHANASSOULA, E.	Observatoire de Marseille, France
BAHCALL, Neta	Space Telescope Science Institute
BAND, David L.	Lawrence Livermore National Lab
BARDEEN, James M.	University of Washington
BARNES, Joshua E.	Institute for Advanced Study
BEERS, Timothy C.	California Institute of Technology
BERTSCHINGER, Edmund	University of California, Berkeley
BINGGELI, Bruno	University of Basel, Switzerland
BLUMENTHAL, George	University of California, Santa Cruz
BODENHEIMER, Peter	University of California, Santa Cruz
BOND, Howard E.	Space Telescope Science Institute
BOND, J. Richard	University of Toronto, Canada
BOSMA, Albert	Observatoire de Marseille, France
BOTHUN, Greg	California Institute of Technology
BRANDENBERGER, Robt.	University of Cambridge, England
BRAUN, Erez	Weizmann Institute of Science, Israel
BRUZUAL, Gustavo	Centro de Investigaciones de Astronomia Venezuela
BURKE, Bill	University of California, Santa Cruz
BURSTEIN, David	Arizona State University
BUSON, Lucio M.	Osservatorio Astronomico Padova, Italy

CARLBERG, Raymond	Johns Hopkins University
CARNEY, Bruce W.	University of North Carolina
CARTER, David	Mount Stromlo and Siding Spring Obs. Australia
CASERTANO, Stefano	Institute for Advanced Study
COURTEAU, Stephane	University of California, Santa Cruz
DAVIES, Roger L.	Kitt Peak National Observatory
DAVIS, Marc	University of California, Berkeley
DEKEL, Avishai	Weizmann Institute of Science, Israel
DJORGOVSKI, S. George	Smithsonian Center for Astrophysics
DOPITA, Michael A.	Mount Stromlo and Siding Spring Obs. Australia
DRAKE, Frank	University of California, Santa Cruz
DRESSLER, Alan	Mt Wilson and Las Campanas Obs.
EDER, Jo Ann	Yale University
ELSTON, Richard	Steward Observatory
FABBIANO, Giuseppina	Smithsonian Center for Astrophysics
FABER, Sandra M.	Lick Observatory
FALL, Michael	Space Telescope Science Institute
FLORES, Ricardo A.	Brandeis University
FORMAN, Bill	Smithsonian Center for Astrophysics
FREEMAN, Ken	Mount Stromlo and Siding Spring Obs. Australia
FREESE, Katherine	University of California, Santa Barbara
FRENK, Carlos S.	University of Durham, England
FRIEL, Eileen	University of California, Santa Cruz
FROGEL, Jay A.	Kitt Peak National Observatory
FULLER, George M.	Lawrence Livermore National Lab
GILMORE, Gerry	Institute of Astronomy, England
GONZALEZ, J. Jésus	University of California, Santa Cruz
GOODRICH, Bob	University of California, Santa Cruz

GRIEST, Kim	University of California, Santa Cruz
GUNN, James E.	Princeton University
HABER, Howard	University of California, Santa Cruz
HARTMANN, Dieter	University of California, Santa Cruz
HAYNES, Martha P.	Cornell University
HELLINGER, Douglas	University of California, Santa Cruz
HENRIKSEN, Mark	Space Telescope Science Institute
HO, Darwin D.-M.	Lawrence Livermore National Lab
HOFFMAN, Yehuda	University of Pennsylvania
HOGAN, Craig	Steward Observatory
HUCHRA, John	Smithsonian Center for Astrophysics
ILLINGWORTH, Garth	Space Telescope Science Institute
JONES, Christine	Smithsonian Center for Astrophysics
KANG, Hyesung	University of Texas
KASHLINSKY, Alexander	National Radio Astronomy Observatory
KENT, Stephen	Smithsonian Center for Astrophysics
KING, Ivan R.	University of California, Berkeley
KIRSHNER, Robert P.	Smithsonian Center for Astrophysics
KOO, David C.	Space Telescope Science Institute
KORMENDY, John	Dominion Astrophysical Observatory Canada
KORYCANSKY, Don	University of California, Santa Cruz
KRAFT, Robert P.	Lick Observatory
KRON, Richard	Yerkes Observatory
KUIJKEN, Konrad	Institute of Astronomy, England
LACEY, Cedric G.	Princeton University Observatory
LAHAV, Ofer	Institute of Astronomy, England
LARSON, Richard B.	Yale University
LAUER, Tod R.	Princeton University
LOEWENSTEIN, Michael	University of California, Santa Cruz
LUAN, Ling	University of California, Santa Cruz

MARK, James W-K.	Lawrence Livermore National Lab
MATHEWS, Grant J.	Lawrence Livermore National Lab
MATHEWS, William G.	University of California, Santa Cruz
MEIKSIN, Avery	University of California, Berkeley
MELOTT, Adrian L.	University of Kansas
MOREA, Cristina Dalle Ore	University of California, Santa Cruz
MURRAY, Stephen	University of California, Santa Cruz
McCARTHY, Patrick J.	University of California, Berkeley
McGILL, Colin	Institute for Advanced Study
NORIEGA-CRESPO, A.	University of California, Santa Cruz
NORMAN, Colin	Space Telescope Science Institute
OEMLER, Augustus	Yale University
OH, Kap-Soo	University of California, Santa Cruz
OSTERBROCK, Donald E.	Lick Observatory
OZERNOY, Leonid	Lebedev Physical Institute, USSR
PERSSON, Carol	California Institute of Technology
PHINNEY, E. Sterl	California Institute of Technology
PIERCE, Michael J.	Institute for Astronomy, Hawaii
PLIONIS, Manolis	University of Sussex, England
POGGE, Richard	University of California, Santa Cruz
PONCE, Gustavo A.	University of Texas at Austin
POSTMAN, Marc	Smithsonian Center for Astrophysics
PRIMACK, Joel	University of California, Santa Cruz
QUINN, Peter	Space Telescope Science Institute
REES, Martin	Institute of Astronomy, England
RICHSTONE, Douglas O.	University of Michigan
RIEKE, Marcia	Steward Observatory
ROBINSON, Lloyd B.	Lick Observatory
ROOD, Herbert J.	Institute for Advanced Study
RYU, Dongsu	University of Texas

SADLER, Elaine M.	Kitt Peak National Observatory
SALE, Ken	Lawrence Livermore National Lab
SALUCCI, Paolo	ISAS, Italy
SANCISI, Renzo	Kapteyn Laboratorium, Netherlands
SCHECHTER, Paul	Mt Wilson and Las Campanas Obs.
SCHWEIZER, Francois	Dept of Terrestrial Magnetism
SECKEL, David	University of California, Santa Cruz
SHAPIRO, Paul	University of Texas
SHAW, Dick	Lick Observatory
SHECTMAN, Stephen	Mount Wilson Observatory
SHER, Marc	University of California, Santa Cruz
SHIELDS, Gregory	University of Texas
SILK, Joseph	University of California, Berkeley
SMITH, Harding E.	University of California, San Diego
SPERGEL, David	Institute for Advanced Study
STEBBINS, Albert	University of California, Berkeley
STRAUSS, Michael	University of California, Berkeley
STRINGFELLOW, Guy S.	University of California, Santa Cruz
STRYKER, Linda L.	Arizona State University
SUNTZEFF, Nicholas	Cerro-Tololo Interamerican Observatory Chile
SWENSON, Fritz J.	University of California, Santa Cruz
TERNDRUP, Don	University of California, Santa Cruz
THUAN, Trinh X.	Leander McCormick Observatory
TOHLINE, Joel E.	Louisiana State University
TONRY, John	Massachusetts Institute of Technology
TOOMRE, Alar	Massachusetts Institute of Technology
TREMAINE, Scott	University of Toronto, Canada
TRIMBLE, Virginia	University of California, Irvine and University of Maryland
TULLY, Brent	Institute for Astronomy, Hawaii

TUROK, N.	Imperial College, England
TYSON, Tony	AT & T Bell Labs
van den BERGH, Sidney	Dominion Astrophysical Observatory
van GORKOM, Jacqueline	Princeton University Observatory
VEILLEUX, Sylvain	University of California, Santa Cruz
VILLUMSEN, Jens V.	California Institute of Technology
VITTORIO, Nicola	University of California, Berkeley
WALKER, Merle	Lick Observatory
WEST, Michael J.	Yale University
WHITE, Simon	Steward Observatory
WHITFORD, Albert E.	Lick Observatory
WHITMORE, Brad	Space Telescope Science Institute
WOLFE, Arthur M.	University of Pittsburgh
WOOSLEY, Stan	University of California, Santa Cruz
WYSE, Rosemary F.	Space Telescope Science Institute
YAHIL, Amos	State University of New York
YATES, Mark G.	Royal Observatory, Scotland

Session 1

Stellar Evolution in Galaxies

A Tribute to Albert Whitford
in His 80th Year

R. P. Kraft, Chair
July 21, 1986

The Stellar Population of Nuclear Bulge

JAY A. FROGEL

1. Preamble

Coincident with this workshop Albert Whitford and I submitted for publication a description of the results from the first phase of our study of stars in the Galactic nuclear bulge. I feel fortunate to have been able to work with Albert on this program for the past 6 years, and I am happy as part of the celebration of his 80th year to present in the proceedings of this workshop a slightly amended version of the summary section of our paper.

2. Introduction

The stellar content of the nuclear bulge of the Galaxy is relevant to two broad categories of problems - local and global. Amongst the former are: 1) what are the physical characteristics of old stars with a metal abundance considerably greater than solar ([Fe/H]>0.0); 2) how does the stellar content of the bulge vary as a function of radial distance from the centre; 3) what is the "nuclear bulge" and can it be distinguished from the inner parts of the galactic disk; 4) what can be learned about the early chemical and dynamical history of the inner parts of our Galaxy. By global problems I mean those concerned with the stellar content of other galaxies, particularly spheroidal systems such as E's, SO's, and the bulges of spirals.

The possibility that stars in Baade's Window could be relevant for tackling global problems was made evident by the work of MORGAN [1] and WHITFORD [2], who demonstrated that the strengths of atomic and molecular absorption features in the integrated light from Baade's Window are comparable to those seen in galaxy spectra.

2) The JHK colors and CO and H_2O indices of the bulge giants differ systematically from their solar neighborhood counterparts both as a function of spectral type and in terms of the dependence of one color or index upon another [23]. Bulge giants have markedly bluer mean colors than local giants of the same spectral type. Except for the reddest stars and the variables, all bulge M giants have J-K colors bluer than the mean for field giants of the same V-K. In the J-H, H-K plane all bulge M giants lie on the opposite side of the mean field giant line from that of globular cluster giants. Nearly all of the non-variable bulge M giants have CO indices significantly stronger than the mean value for field giants of the same color. These trends continue those seen in bulge K giants [22]. They are all consistent with the bulge M giants having a metallicity significantly in excess of solar.

3) Bulge M giants are between 1.5 and 2 magnitudes fainter than disk stars of same spectral class. In addition the giant branch they define in a color-magnitude diagram is intermediate to those of 47 Tuc and M67, in apparent contradiction to the other evidence which supports their SMR nature [24].

The progression of stars from three distinct populations in the J-H, H-K plane strongly points to the increasing importance of a blanketing agent in the stars of higher metallicity. We have attempted to examine the effects of blanketing on the infrared colors and indices with the use of narrow band filters at 1.02 and 2.20μm, two supposedly blanketing free regions of the spectrum. The H band appears to suffer the most severe blanketing and it is closely correlated with H_2O absorption. However, there is some amount of differential blanketing - with respect to solar abundance stars - at all wavelengths looked at. A change in the major source of opacity from H^- to the molecules present in the atmospheres of these SMR stars is proposed as a possible explanation for the blanketing and color differences between bulge and field stars.

We interpret the relative blueness of the bulge stars for their spectral type and in a color-magnitude diagram as arising from their SMR on the strengths of several molecular absorption bands; since they are from diatomic molecules they depend on the square of the metallicity. The net effect is to classify the M giants too late by as much as 4 spectral subclasses: the true effective temperatures of the stars are considerably hotter than implied by their late-type classification.

M giants contribute not more than 5-10% of the visual light from a spheroidal system, but they account for half of the bolometric luminosity and dominate the light longward of 1.0μm [3,4,5,6,7,8,9]. NASSAU and BLANCO and their collaborators [10,11,12,13] showed that the Galactic bulge also has large numbers of late M giants.

It appears that the stars which dominate the visual light in the central regions of spheroidal systems are super-metal- rich (SMR) in the sense that [Fe/H]>0.0 [14,15]. But generally, it has been solar neighborhood and Galactic globular cluster stars that have been used as the building blocks in stellar synthesis models. These models have often been less than satisfactory in reproducing the colors of galaxies, particularly in the ultraviolet and infrared spectral regions [16,17,18]. This paper is concerned with an attempt to resolve the problems in the infrared.

For a stellar synthesis model to be "successful" there are two rather obvious criteria which must be satisfied: 1) the stars in the model must have the correct physical characteristics (e.g. colors, luminosities, metallicities), and 2) the relative numbers of stars with different characteristics must be known. The similarity of optical spectral features and relatively large numbers of M giants suggest that stars in Baade's Window may be the appropriate ones to use in models.

3. The M Giants in Baade's Window

The observations that Whitford and I have made consist of photometry of about 200 M giants in Baade's Window in the JHKL and 10μm bands, limited R and I photographic photometry, and narrow band measurements of H_2O and CO absorption. About 150 of these stars constitute an unbiased sample of bulge M giants and LPVs drawn from the BLANCO [13], BLANCO, MCCARTHY, and BLANCO (hereafter BMB) [12], and LLOYD EVANS [19] surveys.

There are three important characteristics in which the Baade's Window giants differ from solar neighborhood and globular cluster stars which have been used as the basis for many stellar synthesis models:

1) Spectroscopic studies of Baade's Window K giants show that a substantial number are super metal rich (SMR), i.e. with [Fe/H]>0.0 [20,21]. Since M giants have K giants as precursors and giant branch track temperatures become cooler at higher metallicities, it is reasonable to conclude that the M giants will be SMR as well.

This would explain the relatively small number of large amplitude variables even amongst the M8-9 bulge giants [12].

With the exception of the variables, the dispersions of the M giants in one color or index at a given value of a second is consistent with that expected from the small observational uncertainties alone. This is surprising in view of the considerable spread in the metallicity of bulge K giants [20,21]. Perhaps the metallicity distribution amongst the M giants is weighted to have fewer stars of relatively low abundance. The advantage that SMR stars have in reaching the AGB domain [24,25] could promote such a bias among the latest of the M1-9 types that have a luminosity above the core He flash. But the bulge giants of lower luminosity types M1-5 likewise show a small dispersion. There may be another as yet unidentified parameter that influences the infrared colors of these stars. A similar situation exists with regards to giants in Galactic globular clusters: even with a range of a factor of 100 in metallicity, there is little apparent dispersion in some of their color-color relations [26]. In this case, however, the metallicities and temperatures of the globular cluster giants are probably such that they do not cross the threshold where the proportion of molecules in their atmospheres is sufficient to cause the blanketing that could distort these color relations.

The infrared photometry of the bulge M giants results in a well defined luminosity function well defined over a range of nearly three and a half magnitudes. There is a distinct cutoff for stars brighter than M_{bol} about -4.2. Fainter than this, the luminosity function is quite similar in shape to that for olddisk-giants [27]. Such an upper bound to the stellar luminosity function may provide a useful distance indicator when applied to similar populations in other galaxies.

Examination of the IRAS Point Source Catalogue shows that all but one of the 12μm sources in Baade's Window are LPVs previously identified in BMB's list of M giants [12] and/or LLOYD EVANS' [19] list of variables. The one exception is a luminous M giant from BMB with little or no thermal excess emission. It could well be a foreground star. Hence there are no hidden, dust-enshrouded objects in Baade's Window which would change its integrated infrared colors from those determined by observations of the spectroscopically identified sample of giants.

Since the LPVs are among the most luminous of the bulge giants and are the only stars in the bulge with obvious circumstellar shells, it is apparant why most of the bulge IRAS sources are LPVs.

4. Stellar Synthesis Models

Baade's Window giants have significantly different colors, indices, and magnitudes from M giants used previously in the construction of stellar synthesis models for galaxies. As an experiment Baade's Window M giants were substituted for the equivalent stars in some published models. Since the luminosity function of the Baade's Window M giants is fixed by the observations, only one free parameter had to be determined - the normalization factor needed to fit the M giant luminosity function onto that for earlier giants.

The 0.5 - 10µm colors of the model with Baade's Window M giants come close to those of typical elliptical galaxies. Differences are primarily at the shorter infrared wavelengths. (The contribution of the M giants to the optical light is <10% so that substitution of bulge M giants for whatever was used in any given model does not noticeably affect the optical colors). The problem is that the rest of the stars in the model - the non-M giants - are not of high enough metallicity as seems to be required by the mean metallicity derived for the bulge K giants [20,21]. If we adjust the magnitudes and colors of the non-M giants in the models by small amounts corresponding to reasonable estimates for the changes that would be induced by an increase in metallicity, the resulting model colors, with the bulge M giants, closely reproduce mean colors for a sample of E galaxies. We conclude that the Baade's Window M giants are the appropriate stars to use in modelling spheroidal stellar populations. It must be remembered, though, that there appears to be a range of more than a factor of ten in the metallicity of Baade's Window stars. In order to most accurately interpret the results of model fitting to the integrated light of a galaxy, the model should not consist only of stars of some mean value for the metallicity.

Finally, we have shown that the integrated K to 10µm energy distribution of a model elliptical galaxy containing M1-9 giants as are found in the bulge gives excellent agreement with the mean for a sample of elliptical galaxies measured by IMPEY, WYNN-WILLIAMS, and BECKLIN [30]. Their measurements indicate excess emission at 10µm in elliptical galaxies - about a factor of two above that expected from stellar photospheric emission alone. Our observations strongly point to the conclusion that for the majority of the ellipticals this excess emission is due to circumstellar dust around M giants. The M5-7 stars contribute over 80% of this excess whereas the red, luminous M8-9s contribute less than 20%. Considered as a group by themselves, the LPVs, which appeared to have the highest mass loss rate per star, contribute only about 20%. Since only half of the LPVs are IRAS sources, we find, in strong contrast to SOIFER et al's [31] result, that only a small fraction of the 10μm express emission in a spheroidal population can be attributed to typical IRAS bulge sources. Even with relatively small amounts of 10μm excess emission per star, their large relative numbers allow the M5-7s to dominate the integrated 10μm flux from elliptical galaxies. Integrated mass loss rates are consistent with previous estimates for similar stellar populations.

I am grateful to the Director, Michael Feast, and the staff of the South African Astronomical Observatory for providing the relaxed environment in which this text was prepared.

References

1. W.W. Morgan: Pub. A.S.P. 68, 509 (1956)
2. A.E. Whitford: Ap. J., 226, 777 (1978)
3. J. Stebbins, A.E. Whitford: Ap. J., 108, 413 (1948)
4. H.L. Johnson: Ap. J., 143, 187 (1966)
5. R.W. O'Connell: Ap. J., 206, 370 (1976)
6. M. Aaronson: Ph.D. Thesis, Harvard University (1977)
7. A.E. Whitford: Ap. J., 211, 527 (1977)
8. M. Aaronson, J.A. Frogel, S.E. Persson: Ap. J., 220, 442 (1978)
9. J.A. Frogel, S.E. Persson, M. Aaronson, K. Matthews: Ap. J., 220, 75 (1978)
10. J.J. Nassau, V.M. Blanco: Ap. J., 128, 46 (1958)
11. V.M. Blanco: in Galactic Structure, ed. A. Blaauw and M. Schmidt (Chicago: University of Chicago Press) p.242 (1965)
12. V.M. Blanco, M.F. McCarthy, B.M. Blanco: Astr. J., 89, 636 (BMB) (1984)
13. V.M. Blanco: Astr. J., 91, 290 (1986)
14. H. Spinrad, B.J. Taylor: Ap. J. Suppl., 22, 445 (1971)
15. S.M. Faber: Ap. J., 179, 731 (1973)
16. J.A. Frogel, S.E. Persson, J.G. Cohen: Ap. J., 240, 785 (1980)
17. J.G. Gunn, L.L. Stryker, B.M. Tinsley: Ap. J., 249, 48 (1981)
18. M. Aaronson, J.G. Cohen, J. Mould, M. Malkan: Ap. J., 223, 824 (1978)
19. T. Lloyd Evans: M.N.R.A.S., 174, 169 (1976)
20. R.M. Rich: Ph.D. Thesis, Calif. Inst. of Technology (1986)
21. A.E. Whitford, R.M. Rich: Ap. J., 274, 723 (1983)
22. J.A. Frogel, A.E. Whitford, R.M. Rich: Astr. J., 89, 1536 (FWR) (1984)
23. A.E. Whitford: in Spectral Evolution of Galaxies, ed. C. Chiosi and A. Renzini (Dordrecht: Reidel), p.157 (1986)
24. J.A. Frogel, A.E. Whitford: Ap. J. (Letters), 259, L7 (1982)
25. D.A. VandenBerg, P.G. Laskarides: Ap. J., in press (1986)
26. J.A. Frogel, S.E. Persson, J.G. Cohen: Ap. J. Suppl., 53, 713 (1983)
27. B.M. Tinsley, J.G. Gunn: Ap. J., 203, 52 (1976a)
28. D.M. Terndrup: Ph.D. Thesis, Univ. of Calif., Santa Cruz (1986)
29. D.M. Terndrup, R.M. Rich, A.E. Whitford: Pub. A.S.P., 96, 796 (1984)
30. C.D. Impey, G. Wynn-Williams, E.E. Becklin: Ap. J., in press (1986)
31. B.T. Soifer, W.L. Rice, J.R. Mould, F.C. Gillett, M. Rowan-Robinson, H.J. Habing: Ap. J., 304, 651 (1986)

Stellar Populations in Local Group Galaxies

L. L. STRYKER

1. Introduction

The word "population" is generally bound to the procedures of galaxy formation. Although we can find definite examples of what we call Population I and II, clearly there is a continuum of populations in our galaxy . Using these populations, we can follow the Galaxy's beginnings as a metal- poor sphere and its subsequent collapse to a flattened disk, now enriched in heavy elements. Kinematics bear this picture out; we understand what is meant by the terms Population I and II in our Galaxy. (We were reminded at the Baltimore Populations Meeting earlier this year that even kinematics evolve in time, which may affect our interpretations.) There is no particular reason to expect Population_i (our galaxy) = Population_j (any other Galaxy), unless we think that all galaxies (or all Sb galaxies) form and evolve in identical ways. This is especially in question when we are dealing with galaxies in the Local Group which display ranges in morphological type and mass, most likely in their formation timescales and procedures, in their star forming histories, in their differing histories and degrees of chemical enrichment, and in their differing mass fractions of gas and molecular clouds. Although it has been logical (and easiest) to use as templates, populations in our own galaxy, such as the globular clusters or solar neighborhood stars, when we try to understand the properties of other galaxies near us, why should we expect the systems of star clusters in say, the Magellanic Clouds or in M31, to have the same properties as ours or as each other's? Does "Population II" refer to a particular thing in the Large Magellanic Cloud, or do we mean only that an old population that we see in the Large Cloud resembles that which we know as Population II in the Milky Way? Just making a simple color comparison of our globular clusters with the clusters seen in the LMC, as was done in earlier decades, shows for the LMC a very dissimilar distribution of integrated colors, where we find blue as well as red globulars. And we now know that there is a completely different distribution in ages for Cloud clusters in contrast to the age distribution for Milky Way clusters. We have known this for some time now, so you might think that a lesson could be learned here. What is defined as "old" for one galaxy may differ from what is called "old" in another.

This talk is highly selective in that it discusses a few of the more interesting recent papers under the subject of populations and luminosity and mass functions in the Local Group. During the Populations Meeting at Yale in 1977, the following was recorded for posterity: "We now know so much that we begin to see how much we don't know . . ."; ". . . can we tolerate any longer an IMF that is uniform and constant?"; "Why do we use a single parameter "Z" for all abundances greater than Helium?"; "The averaged IMF is roughly universal."; "What is the age of an 'old' population?"; "... reconsider the classical interpretation of galactic formation and enrichment in terms of a one- to-one relationship between chemical composition and age ..."; "Populations will be different in different galaxies" (Populations are bivariate functions, i.e., the distribution of its component stars with respect to age and composition.) ; "... increasing recognition of the complexity of population structures ...". At the Populations Meeting at the Space Telescope Science Institute in 1986, the following remarks were overheard: "Galaxies are different from each other."; "Zorro (Renzini) destroyed everything!" (i.e., our dependence on Yale isochrones); "We should use the Magellanic Cloud clusters as templates for extra-galactic evolution and population synthesis."; "Is the IMF bi-modal?" ; "What is the low- mass cutoff?" ; "Are there galaxies which contain only old stars?"; "Are there single- burst galaxies?"; "We have learned a great deal... Now we know a lot more about what the things are that we don't know." At least we seem to be worrying about the same problems! The suggestion was made in Baltimore that we test our population- synthesis techniques on Magellanic Cloud clusters which have good, detailed color- magnitude diagrams. This is the best way to test the procedures used, to check the uniqueness of the solutions, and to establish whether these solutions have any resemblance to reality.

To learn more about the "universality" of old populations, BURSTEIN et al. [1] studied integrated light spectral indices of elliptical galaxies (including M32) and of globular clusters in the Milky Way and in M31. In brief, although the populations appear to be similar on the surface and show quite overlapping behavior in most indices, plots in Hß and CN show large differences for the two systems. This is in contrast to the usual idea that the "old stellar populations" found in globular clusters and in elliptical galaxies form a simple, one- parameter continuum (in metallicity). We might have taken some hints in this direction from our observations of clusters in the LMC, but perhaps we think to ourselves that the LMC is an Irregular galaxy and therefore maybe its clusters are unusual. It does not occur to us that maybe populations in our very own familiar galaxy, the Milky Way, may be the unusual ones.

2. M32

M32 is by now a very well- studied object (FABER [2], O'CONNELL [3], PRITCHET [4], WU et al. [5], FROGEL et al. [6]). We are probably zeroing in on the population makeup of M32. For M32, BURSTEIN et al. find a 50% stronger Hß index than that seen in metal- rich Galactic globulars. This finding would require 70% of the light to come from a metal- poor component and thus, considering the over- all constraints on the line- strengths of ellipticals, rules out such a component, agreeing with O'CONNELL and GUNN et al. [7]. Hß findings combined with UV results show that one cannot add A stars to the metal- rich base to get a match with Hß; but a 25% contribution of late F stars does the trick, as suggested previously by O'CONNELL and BRUZUAL [8]. Does this mean a younger age for M32 or could these be stars above the turnoff?

2.1 Constraints on the populations in M32.

Using well- understood, reddening- independent, blue (3400 Å to 4500 Å) line- strength ratios at high resolution, ROSE [9] re-investigated population synthesis findings of other workers who reported seeing an excess of some 20% in the blue integrated light of elliptical galaxies over their models. Some of these workers concluded that the excess was due to a hot, young component or to evolved blue stragglers; others ascribed the excess to a metal-poor component or to a younger overall system age. We are all aware that population synthesis techniques suffer from possible non- uniqueness problems and from the inherent ambiguity of distinguishing metallicity characteristics from age ones. Plus we often put into the synthesis only single- valued metallicity and age. And we find that although the single- burst, old (10-15 Gyr), metal- rich population fits well in the red, it fails at wavelengths shortward of 4300Å. In brief, ROSE reports that "all models requiring hot stars to provide the 'excess' blue light in elliptical galaxies can be ruled out." ROSE's indices show that M32 has 80% of its light at 4000Å coming from main sequence and turnoff stars and 20% from giants. Only a very small population contributing about 2 to 4 % could be attributed to a hot, young star , or blue straggler contribution (A3-A4 stars). O stars are ruled out and a 20% contribution of stars even only as early as A3 stars is ruled out at the 15 sigma level. For a metal- poor component, ROSE used M5 and found only an 8% contribution. According to his indices, the blue light is coming from dwarf stars, presumably from a strong intermediate- age main sequence. Ten other giant elliptical and S0s showed the same 80/20 proportions plus no noticeable variations in the hot contribution. NGC 205, a dwarf elliptical near M31, on the other hand, is dominated by a strong, young main- sequence population. Galactic metal- rich globulars have dwarf/giant ratios of about 50/50. Basically, it is the weak SrII/FeI index and the strong $\lambda 3886/\lambda 3859$ feature, both seen in M32, that provide this result, along with the CaII H + He/CaII K index. Also, ROSE finds that the mean abundance of M32 at 4000 Å is somewhat metal poor with respect to the sun. He finds that M32 is not an extension of metal rich globular clusters. The dwarf enrichment in M32 points to a large intermediate- age population (previously suggested by O'CONNELL); ROSE rules out any unusual IMF, horizontal branch morphology, giant branch morphology (AGB tip), sampling area, or the possible coverage of a wider range in metallicity in M32. Fifty percent of the light at 4000 Å comes from stars born 5 to 10 Gyr after the first star- forming epoch when apparently significant star formation was occurring, and then there was a large decline in star production 10 to 15 Gyr after that.

3. The Magellanic Clouds

To come a little closer to home, just as we thought we were getting a good feeling for the chemical evolution of the Magellanic Clouds, along comes a cluster to mess things up. In SMITH AND STRYKER [10], we see a rough updated version of the chemical histories of the LMC and SMC. In the Galaxy, there was rapid initial enrichment, followed by the slow, steady rise in abundances. We see a steady rise in the LMC but a very long, uneventful story for the SMC, where abundances have taken off only recently. Some of the evidence for the SMC's flat era comes from BUTLER et al. [11], and from the MOULD-Da COSTA group [12,13, 14]. Much additional work still needs to be done, but apparently in the SMC, at least in the outer regions, there was little chemical activity for perhaps up to 10 Gyr, then a substantial increase took place in a relatively short period. The finding for the field near NGC 121, was unusual in that very little is seen of the ubiquitous intermediate- age population as reported earlier in STRYKER [15] for most other Cloud fields. Now, back to that perverse cluster, MATEO et al. [16] report that ESO 121-SC03, at 10° out, in the LMC, has [Fe/H] = -0.9 ± 0.2, but an age of 8 or 10 Gyr, depending on the adopted distance modulus. That puts it off the LMC line in the chemical history diagram [10]. This reported age is not found in any other cluster in the LMC but the lack may be due to selection effects or to destruction of clusters in that age bracket which are nearer to the main body of the LMC. It may well be that the clusters we see now are only the remains of a much different distribution in ages. But it is also interesting, and should be enlightening to our ideas of star formation, that stars and clusters are able to form in environments where there is essentially no detectable neutral hydrogen or CO. However, M. DOPITA informs us from the floor that a mapping of H_2 in the LMC shows that CO is not a tracer of hydrogen, in spite of what we all learned in graduate school.

3.1 Low- mass Luminosity Functions

On the low- mass end of the stellar scale (absolute magnitudes $-2.5 < Mv < +4$), we have luminosity function studies of a few field areas in the Magellanic Clouds from several workers. There is very good agreement among the studies for various locations in the LMC, and with the solar neighborhood luminosity function, that one begins to see some evidence for a "universal" mass function. More will be said on this below. Given the new work on the local luminosity function by SCALO [17], it is not obvious that the same conclusions would be reached con- cerning the change in slope seen in the luminosity function at $Mv = +3$ in the Large Magellanic Cloud compared to an erstwhile change at $Mv = +4$ for the Galaxy [18,19] -- as no clear- cut slope change is seen now in SCALO's adopted luminosity function in the Galaxy. On the basis of a difference in slope change, the conclusion was reached that a strong burst (or strong enhancement) of star formation occurred in the LMC about 3 to 5 Gyr ago. There was always a nagging, residual worry that perhaps the interpretation should have been that the LMC was merely deficient in producing stars with masses less than about 1 or 2 Mo. Fortunately, this star- formation enhancement finding does not depend only on the main-sequence luminosity function results. We can look to asymptotic giant branch luminosity functions (FROGEL AND BLANCO [20]) and sub-giant branch populations (HARDY AND DURAND [21]) to give another indication of an enhanced population arising from the predicted era. Two age groups (3-5 Gyr) and (10^8), are distinct, the latest star forming strength apparently down by a factor of about 10 [18]. Although cluster ages and the carbon stars and extended giant branches seen in the Clouds may point to a more continuous star- formation rate, it is generally agreed that there does exist a significant enhancement in intermediate- age populations relative to old in the Clouds. This is observed in other nearby galaxies as well, where strong intermediate- age populations are seen in M33, IC 1613 , and, as inferred from population synthesis and blue line indices mentioned above, in M32. The Clouds have very distinguishable Populations, but we cannot call them Population I, II, or III. Namely, there is (a) the old ($\geq$ 10 Gyr): RR Lyraes and the handful of old clusters, SEARLE et al. [22] (SWB) type VII; (b) the intermediate- age (1-2 to 8 Gyr): the red clusters and bright AGB stars, SWB IV-VI; and (c) the young (< 1 Gyr): cepheids, supergiants, and blue clusters, SWB I-III. See FROGEL [23] for further discussion.

The answer to the question of the reality of the possible low- mass deficiency in the Clouds is very likely to come from the Hubble Space Telescope which will make possible deeper color- magnitude diagrams and luminosity functions (to Mv = +6 or +7). This will give us a critical region in the luminosity function to observe whether the slope rises, dips, or remains flat.

4. Dwarf Spheroidals

Many color-magnitude diagrams now exist which include the main- sequence turnoff for fields and clusters in both Clouds and for some fields in nearby dwarf spheroidal galaxies: Draco, Ursa Minor, Carina, Fornax, Sculptor (ZINN [24]; also AARONSON, this volume). I will leave details to Marc Aaronson, but I will show one diagram. This is a new, deep color- magnitude diagram by CARNEY AND SEITZER [25] for the Draco dwarf spheroidal galaxy. They conclude from a comparison of Draco's luminosity function and that of M92, that although Draco has a population younger than that of M92 it is super- imposed upon one equally old. In fact, they point out from isochrone fits to the turnoff region, that the oldest population stopped its significant star formation only about 8 billion years ago. Draco, too, may have experienced two epochs of star formation, with the younger epoch supplying the red horizontal branch; Draco displays 4 billion old blue stragglers as well.

5. High-mass Initial Mass Functions

Color- magnitude diagrams for the upper portions of fields in galaxies of the Local Group such as IC 1613 and M33 are now appearing in journals. Luminosity functions for several of these nearby galaxies have been produced. We usually assume a universal power- law initial mass function (IMF), independent of position within a galaxy or among galaxies and unchanging with time or metal abundance. This is the simplest thing to do. But the adoption of any particular IMF in models of galaxy evolution has great impact on what answers come out. As SCALO [17] has recently pointed out, the IMF is a conditional probability and may be based on several galactic properties (chemical composition, gas density, turbulent velocity, ISM, . . .) some of which are known to be functions of position and time. For galaxy evolution work we are not interested in local fluctuations, but in an IMF averaged over large spatial scales ($\geq$ 1 kpc). The problems in attaining good, discriminating luminosity functions in the Local Group are that there are few galaxies and hence few stars so that statistical reliability suffers; distances are not well- established, even to the nearby Magellanic Clouds; because of differing distances, there may not be much overlap in the magnitude range covered; the star forming procedures have probably been different in each case (so how does the present luminosity function really relate to the IMF?) For the brightest stars one can avoid the ever- present dichotomy of SFR-IMF and take the SFR as being constant over the last 10^7 years.

FREEDMAN [26] recently examined these ideas for the brightest stars in ten nearby (distance moduli < 28) galaxies including M33 and NGC 6822 which exhibit quite a range of environments in chemistry, kinematics, present- day activity in star formation, and masses. There is no direct information about the low- mass end of the IMF nor about what the IMF was like in the distant past. After dealing with the usual problems of foreground contamination, crowding, incompleteness, and sample size, which we have all come to know and love, and all of which tend to flatten the luminosity function or even to cause a turn down, FREEDMAN attained luminosity functions complete to 19 - 20 mag covering the magnitude range $-9 < M_V < -5$. She found for all ten galaxies a similarity of slope in the luminosity functions with very little variation from one galaxy to the next. The largest deviations from universality occurred at the very brightest end where only a small numbers of objects are counted; many of the youngest stars were not included because of clumpiness in the image (only "stellar" objects are counted by the automated reduction techniques). To examine the proposals by TERLEVICH AND MELNICK [27] and others that the slope of the IMF is a function of metallicity, and by GARMANY et al. [28] that the IMF slope shows a positional gradient within galaxies, FREEDMAN studied the radial behavior of the luminosity function in M33, since an abun- dance gradient has been found there by several wor- kers. The log ([OIII]/Hß) varies by a factor of ten over M33, so there is a tem- perature gradient as well. From her study, FREEDMAN finds no evidence of any statistically significant change in slope with radius in M33. The variations in IMF cannot be large (if they exist) and no great sensitivity to chemical composition is seen. Her conclusion is that the upper end of the IMF is evidently a universal function. SCALO [17] concurs with this. At the very brightest end, small- number statistics hide any conclusions.

To go from a luminosity function to an actual mass function is non- trivial. To be considered are bolometric corrections, effective temperature scale calibrations, theoretical tracks, mass loss, opacities, the roles of convection and overshoot, and mass- luminosity relations. The luminosity function itself must be known to good precision since the mass-luminosity function relation for high- mass stars goes as approximately the 5th power. And the IMF may well be bi-modal, that is, not a monotonic function of stellar mass. The double- peaked function does seem to answer a number of problems concerning the evolution, stellar content and amount of unseen matter in the solar neighborhood. As LARSON [29] has said recently, galaxy evolution models including a bi- modal IMF, and decreasing star- formation rate with most of the mass in remnants, work much better than the usual models in Spiral galaxies to reproduce the observed colors, mass-to- light ratios, and gas contents; and in Ellipticals in allowing for the rise in metallicity and mass-to-light ratio as a function of total mass.

6. Wrench in the Works

The last comment is this: we may want to consider the argument found in a little- known preprint by WILLSON et al. [30] concerning the consequences of mass- loss on stars on the main sequence. The authors hypothesize that main- sequence stars in the pulsation instability strip could experience mass loss at rates in excess of 10^{-9} Mo/year and end up evolving down- ward on the main sequence. This selective mass loss (stars with about 1-3 Mo) could cause some star clusters to have turnoffs indicating older ages than they really have, i.e., bringing globular ages more in line with the somewhat younger age being proposed recently for the age of the universe. It could explain blue stragglers as the very slowly rotating stars left behind when faster rotaters evolve downward; blue stragglers can then have any mass that allows them to remain on the main sequence, and it could drastically alter the luminosity function (and hence mass function) for stars on the main sequence. Clusters with metallicities less than a tenth of the solar value and with ages more than a few times 10^9 years will appear to be 14-18 Gyr years old. The metallicity, not the age, determines the maximum mass on the main sequence for these populations. These authors write, " . . . it is possible that no stars in our galaxy are older than 7-10 Gyr old." All of astrophysics and cosmology would be profoundly affected if their hypothesis turns out to be correct.

In conclusion, it appears that although we have learned quite a lot about stellar populations in the Local Group, I think that what we have learned most is that we have much more to learn.

References

1. D. Burstein, S. Faber, C. M. Gaskell, N. Krumm: Astrophys.J. 287, 586 (1984)
2. S. M. Faber: Astrophys.J. 179, 731 (1973)
3. R. W. O'Connell: Astrophys.J. 236, 430 (1980)
4. C. J. Pritchet: Astrophys.J.Suppl. 35, 397 (1977)
5. C. C. Wu, S. M. Faber, J. S. Gallagher, M. Peck, B. M. Tinsley: Astrophys.J. 237, 290 (1980)
6. J. A. Frogel, S. E. Persson, M. Aaronson, K. Matthews: Astrophys.J. 220, 75 (1978)
7. J. Gunn, L. Stryker, B. Tinsley: Astrophys.J. 249, 48 (1981)
8. G. Bruzual A.: Astrophys.J. 273, 105 (1983)
9. J. A. Rose: Astron.J. 90, 1927 (1986)
10. H. A. Smith, L. L. Stryker: Astron.J. 92, 328 (1986)
11. D. Butler, P. Demarque, H. Smith: Astrophys.J. 257, 592 (1982)
12. R. M. Rich, G. S. Da Costa, J. R. Mould: Astrophys.J. 286, 517 (1984)
13. G. S. Da Costa, J. R. Mould, M. D. Crawford: Astrophys.J. 297, 582 (1985)
14. L. L. Stryker, G. S. Da Costa, J. R. Mould: Astrophys.J. 298, 544 (1985)
15. L. L. Stryker: in IAU Symposium No. 108, Structure and Evolution of the Magellanic Clouds, ed. by S. van den Bergh and K. S. de Boer (Reidel, Dordrecht 1984) p.79
16. M. Mateo, P. Hodge, R. Schommer: Astrophys.J. in press (1986)
17. J. Scalo: Fund.Cosmic Phys. 11, 1 (1986)
18. H. R. Butcher: Astrophys.J. 216, 372 (1977)
19. L. L. Stryker, H. R. Butcher: in IAU Colloq. No. 68, Astrophysical Parameters for Globular Clusters, ed. by A. G. Davis Philip and D. S. Hayes (Davis, Schenectady 1981) p.255
20. J. A. Frogel, V. M. Blanco: Astrophys.J.Lett. 274, L57 (1983)
21. E. Hardy, D. Durand: Astrophys.J. 279, 567 (1984)
22. L. Searle, A. Wilkinson, W. Bagnuolo: Astrophys.J. 239, 803 (1980)
23. J. A. Frogel: Publ.Astron.Soc.Pac. 96, 856 (1984)
24. R. J. Zinn: Mem.Soc.Astron.Ital. 56, 223 (1985)
25. B. W. Carney, P. O. Seitzer: Astron.J. 92, 23 (1986)
26. W. L. Freedman: Astrophys.J. 299, 74 (1985)
27. R. Terlevich, J. Melnick: ESO preprint 264. (1983)
28. C. D. Garmany, P. S. Conti, C. Chiosi: Astrophys.J. 263, 777 (1982)
29. R. B. Larson: Mon.Not.R.Astron.Soc. 218, 409 (1986)
30. L. A. Willson, G. Bowen, C. J. Struck-Marcell: preprint (1986)

Star Formation in Colliding and Merging Galaxies

FRANÇOIS SCHWEIZER

1. INTRODUCTION

Our awareness of the importance of star formation in colliding and merging galaxies has increased dramatically since the Infrared Astronomical Satellite (IRAS) began its sky survey in 1983. As TOOMRE and TOOMRE [1] had guessed already, the violent mechanical agitation of close tidal encounters seems to bring deep into galaxies gas located previously in their outskirts or even in neighbors, and seems to trigger prolific star formation. It is becoming increasingly evident that such processes may yield glimpses of galaxy formation itself. Observations of present-day collisions and mergers give information about (1) the spatial distribution of star formation in such events, (2) global processes that govern the formation of molecular gas and -- ultimately -- of stars, (3) star formation rates and efficiencies, and (4) the initial mass function. This paper reviews recent progress in these areas and proposes that collisions and mergers may trigger the formation not only of stars, but also of globular clusters.

2. STARBURST GALAXIES

The triggering of star formation in galaxies by gravitational interactions occurs with very different degrees of intensity. In this review, the emphasis is on the most spectacular forms of star formation, often called "starbursts." Yet one ought to keep in mind that for every violent collision there must be at least several, if not many, less violent tidal interactions and milder star-forming events. Such events may be, e.g., the presumed density wave set up in M51 by the passage of its companion, and the low-level star formation apparently taking place in polar rings around S0 galaxies.

2.1 Spatial Distribution of Star Formation

Two trends have recently come together to give us a better picture of the spatial distribution of star formation in colliding galaxies. One trend started with the discovery by IR observers in the 1970's that the central regions of some galaxies emit several times more energy in the 10-100 μm range than they do in the visible and near IR [2]. It became clear that M82 and NGC 253 are experiencing starbursts which extend beyond the nucleus to ~0.5 kpc radius. The main IR emitter seems to be dust heated to 30-50 K by large numbers of massive young stars [3,4]. What remained then unclear was the cause of this phenomenon. In a second trend, since the early observations by Zwicky and Arp a vast body of evidence has accumulated that among galaxies with strong central activity, interacting galaxies are overrepresented. The nuclear equivalent width of Hα in a sample of strongly interacting galaxies is 5-6x higher than in a control sample of relatively isolated galaxies [5]. And at 10 μm, a sample of strongly interacting galaxies contains a population of extremely luminous nuclei that has no counterpart in the control sample. The average 10 μm luminosity of these nuclei is about 20x that of the nuclei in non-interacting galaxies [6].

It is now established that the majority of interacting galaxies with IR-bright centers show resolvable IR emission beyond the nuclei [6,7]. Occasionally, starbursts extend even across an entire disk, as, e.g., in the Sb galaxy NGC 3310, which is thought to be the merged remnant of a gas-rich disk and a smaller companion [8,9]. Extended emission argues directly for young stars embedded in dust being the main source of energy, because even a central point source of 10^{10} $L_\odot$ can heat dust grains to the observed temperatures only out to a few parsecs. Therefore, any resolvable IR emission cannot stem from a central monster. This conclusion is supported by high-resolution radio observations, which show that the same galaxies are also resolved in the radio continuum, with diameters of typically 1-2 kpc and spectral indices around -0.7 [10]. This value is characteristic of nonthermal emission in supernova remnants. In general, then, starbursts in colliding and merging galaxies extend into the disk or main body, although they do tend to be strongly peaked toward the center.

2.2 Molecular Gas

Star formation has long been known to be associated closely with the presence of molecular clouds. What has come as a surprise in starburst galaxies are the sometimes tremendous amounts of molecular gas and the very high luminosities in individual molecular lines.

The main molecules detected so far have been H_2, CO, OH, and H_2O, the latter two only because their emission is amplified in masers. The 2.6 mm emission of CO is now detected routinely with radio telescopes. On the other hand, the detection of the quadrupole emission lines S(1) and S(0) of rotationally-vibrationally excited H_2 has so far been restricted to a few extreme starbursts, as in Arp 220 and NGC 6240. The luminosity in the S(1) line of NGC 6240 is 4×10^8 $L_\odot$, or the equivalent of 2×10^7 Orion Molecular Clouds [11]. The masses of H_2 derived directly from the 2.1 μm emission line are deceptively small, about 10^4-10^5 $M_\odot$, because they refer only to the hot, shock-excited H_2 at ~2000 K temperature. However, the total H_2 masses inferred from CO observations of some 40 interacting galaxies are 4×10^9- 4×10^{10} $M_\odot$, which is 2-20 times the molecular gas content of the Milky Way [12,13]. The upper limit is about 4x more than is observed in the most molecule-rich spirals such as M51 and NGC 6946. Among different galaxies, the ratio of molecular hydrogen to atomic hydrogen varies greatly, $M(H_2)/M(H\ I) = 0.01$-10, with the highest values observed in colliding systems [14].

Both the origin and the excitation of the molecular gas in colliding galaxies seem to be related to the presence of shocks of galactic dimensions. Although excitation of H_2 by UV fluorescence is possible, observations of H_2 line ratios in colliding systems clearly favor shock excitation [15,16]. The galactic shocks must be strongly supersonic since galaxies collide with velocities of typically 100-300 km/s, whereas the sound velocity in the interstellar medium is about 6-10 km/s. Detailed analyses by various authors suggest that as such a shock propagates through the gas, most pre-existing H_2 molecules dissociate, but dust grains survive. These grains then allow the H_2 to form again in the high-density postshock region, where the gas can actually become fully molecular [17]. Such gas at high densities cools very fast, thus forming ideal conditions for fragmentation and rapid star formation.

2.3 Star Formation Rates

Models of evolving stellar populations show that the UBV colors of the vast majority of galaxies can be explained if SFRs decline monotonically with time and at different rates for different Hubble types [18]. For a small fraction of galaxies, however, the observed colors can be reproduced only if bursts of star formation are assumed to have occurred recently. By comparing normal galaxies and Arp galaxies in the (U-B,B-V)-diagram and overlaying grids of evolutionary models, LARSON and TINSLEY [19] have found Arp objects with bursts as short as $2x10^7$ yr and involving up to about 5% of the total mass; the strongest bursts invariably seem to involve tidal interactions. A problem with this method is how to account for extinction, which in objects like Arp 220 and NGC 6240 reaches A_V = 10-15 mag [11].

With the availability of 12-100 μm fluxes from IRAS, the estimation of star formation rates has become more straightforward. The energy distributions of starburst galaxies from the visual through the IR to radio wavelengths look all remarkably similar; they peak near 100 μm and are typical of thermal radiation from dust heated to ~50 K [7]. Because of the large optical depth of the dust surrounding the young stars, photons get converted from UV to IR wavelengths nearly completely, and the integrated 10 μm - 1 mm flux is a good measure of the total radiation from young stars. In strong starburst galaxies, the integrated luminosities reach 10^{12} $L_\odot$, or about 100 times the luminosity of normal spirals and still 10 times the luminosity of M82 and NGC 253. These high luminosities translate into high SFRs and into gas conversion rates of about 100 $M_\odot$/yr, as compared to the canonical 1 $M_\odot$/yr in normal spirals.

Extremely fast star formation cannot last long without exhausting the gas supply. From detailed models of NGC 6240, RIEKE et al. [11] conclude that the time constant for exponential decline is 20-40 Myr and the age of the burst is only 60-100 Myr. The SFR per unit volume is about the same as that in M82, but the star-forming volume is 10-20 times larger. An interesting consequence of intense star formation in a volume of several kiloparsec dimension is the predicted formation of a supernova-driven, supersonic wind [20]. Evidence for the existence of such a wind in M82 has been accumulating for some time, and bipolar outflows of 10-100 $M_\odot$/yr have been identified tentatively in Arp 220 and NGC 6240 [21]. If the quoted mass-loss rates are correct, previous estimates of burst durations may have to be revised downward since rates of gas depletion due to winds and to astration seem to be comparable.

Of as great interest as the peak star formation rates are the mean star formation rates in colliding galaxies. About 3% of all galaxies currently have infrared-to-blue luminosity ratios $L_{IR}/L_B > 1$, and most of them are strongly interacting. Their mean SFR is about 3x normal, suggesting that ~10% of the total massive star formation in present-day galaxies is being triggered by collisions [22].

2.4 Star Formation Efficiency

A quantity of fundamental interest is the star formation efficiency (SFE), which is the fraction of gas that gets converted into stars during a starburst. This quantity is difficult to determine, of course, since it involves modeling the time history of such a burst.

Recently, the ratio between IR luminosity and CO luminosity, L_{IR}/L_{CO}, has been introduced as a measure of the star formation efficiency [13,14]. The basis for this choice is the observed strong correlation between L_{IR} and L_{CO}. The claim is that L_{IR} measures the current star formation rate and L_{CO} measures the molecular mass, whence L_{IR}/L_{CO} would be some momentary measure of the SFE. This argument has been criticized on grounds that L_{CO} may measure not only the mass of gas, but also the exciting flux. Whatever the exact interpretation, it is interesting that L_{IR}/L_{CO} correlates well with the dust temperature: Galaxies with low L_{IR}/L_{CO} -- such as M31, M33, and the vast majority of normal spirals -- have low mean dust temperatures ($\lesssim$30 K), whereas colliding galaxies with high L_{IR}/L_{CO} have high mean dust temperatures ($\gtrsim$50 K). Typically, the latter systems have L_{IR}/L_{CO} enhanced by a factor of 3-5 over normal, with a few extreme starburst galaxies like NGC 3310 and Arp 220 reaching a factor 10 enhancement.

The true SFEs of NGC 6240 and Arp 220 have been computed from detailed models [11]. The resulting values are around 50%, suggesting that about half the total gas will have been converted into stars by the end of the starburst. But this number is, of course, very dependent on the assumed initial mass function and on some estimate of the total gas mass.

An important result emerging from this difficult business of SFEs is that the local processes of star formation seem to be only moderately more efficient in colliding galaxies than in normal spirals. What really seems to make a big difference is the large-scale dynamics: Collisions lead to the formation of shocks of galactic dimensions, which in turn form 10-100 times more molecular gas than normal.

2.5 Initial Mass Function

Does the present-day IMF in colliding galaxies differ from that in normal galaxies? The question is of interest because there has long been evidence that the IMF may vary among different Hubble types [18], and because it has recently been proposed again that the IMF may be bimodal [23]. The reason for this suggestion is that at particle densities in excess of $\sim 10^4$ cm^{-3} and in a strong radiation field, the lower mass limit for star formation may increase dramatically due to the coupling of dust and gas temperatures. Such conditions seem to occur in giant molecular clouds and may be reached rather easily in the shock fronts of colliding galaxies.

Yet observationally the situation remains unclear. On the one hand, there is little doubt that most of the luminosity emerging from luminous IRAS galaxies is due to massive young stars. The main piece of evidence is the presence of strong CO absorption bands around 2.3 μm, which are signatures of red giants and supergiants. This statement simply reflects the fact that in a young population with a Salpeter IMF most of the luminosity stems from high-mass stars; it does <u>not</u> imply that the IMF is particularly enriched in high-mass stars. On the other hand, there have been claims that in some starburst galaxies, there is a strong excess of high-mass stars in the IMF relative to a Salpeter function [4,11]. In these cases, the apparent paucity of low-mass stars is invariably a consequence of total-mass constraints imposed on the models. Although these constraints certainly represent the best of our knowledge, I find them shaky enough in each case to conclude, for the moment at least, that we lack solid observational evidènce for a significantly different IMF in colliding galaxies.

3. MERGERS AND GLOBULAR CLUSTER FORMATION

In ending this review, I would like to suggest that colliding and merging galaxies form not only stars, but also globular clusters. The distinction between plain star formation and the inclusion of globular-cluster formation is important because VAN DEN BERGH [24] and others have objected to the hypothesis that ellipticals form through mergers of disk galaxies on grounds that the number of globular clusters per unit luminosity seems to be larger in ellipticals than in spirals. Therefore, the argument goes, an elliptical cannot be the sum of two spirals.

It seems to me that this objection is based on two unfounded premises: (1) Globular clusters formed all long ago, and none formed since; and (2) the total number of globular clusters involved in a galaxy merger remains constant. Certainly, the first premise is wrong. We know that both young and intermediate-age globular clusters exist in the Magellanic Clouds, in M33, and in M31. It just happens that our Milky Way does not seem to possess any. So at least four out of five nearby galaxies do know how to make globular clusters long after the Big Bang. It is unknown at present whether the second premise is correct or not. However, it would seem strange indeed if colliding galaxies like Arp 220 and NGC 6240 produced vast amounts of H_2 and stars, but did not also produce some globular clusters. What better environment is there to produce massive clusters than the highly crunched gas in such systems?

Although with different scenarios in mind, various authors have pointed out that 10^6 $M_\odot$ globular clusters may have formed in the compressed gas behind strong shocks [25,26]. Already in 1977, SEARLE [27] outlined an attractive picture of globular-cluster formation in relatively isolated fragments of gas during the collapse of a protogalaxy. It seems to me only natural that the same mechanism should also apply to the fragmented gas in merging galaxies, criss-crossed as it is by shock waves. Therefore, given enough time for the young populations to age and fade, I would predict that remnants of merged spirals must have more globular clusters per unit luminosity than the spirals had originally. This seems to be exactly what is observed in ellipticals.

This picture of globular-cluster formation explains the presence of radial abundance gradients in globular-cluster systems just as naturally as the galactic collapse picture does. Further, given the strong central concentration of starbursts, the similarly strong central concentration of globular clusters comes as no surprise. Finally, the fact that some centrally located cluster galaxies have unusually many globulars per unit luminosity may simply mean that they formed from an above-average number of relatively early, and therefore relatively gas-rich, mergers.

I gratefully acknowledge partial support from the National Science Foundation through grant AST 83-18845.

REFERENCES

1. A. Toomre, J. Toomre: Astrophys. J. 178, 623 (1972).
2. G.H. Rieke, M.J. Lebofsky: Ann. Rev. Astron. Astrophys. 17, 477 (1979).
3. C.M. Telesco, D.A. Harper: Astrophys. J. 235, 392 (1980).
4. G.H. Rieke, M.J. Lebofsky, R.I. Thompson, F.J. Low, A.T. Tokunaga: Astrophys. J. 238, 24 (1980).
5. H.A. Bushouse: Astron. J. 91, 255 (1986).
6. R.M. Cutri, C.W. McAlary: Astrophys. J. 296, 90 (1986).
7. R.D. Joseph, G.S. Wright: Mon. Not. Roy. Astron. Soc. 214, 87 (1985).
8. C.M. Telesco, I. Gatley: Astrophys. J. 284, 557 (1984).
9. F. Schweizer: In Internal Kinematics and Dynamics of Galaxies, IAU Symp. No. 100, ed. by E. Athanassoula (Reidel, Dordrecht 1983), p. 319.

10. J.J. Condon, M.A. Condon, G. Gisler, J.J. Puschell: Astrophys. J. 252, 102 (1982).
11. G.H. Rieke et al.: Astrophys. J. 290, 116 (1985).
12. J.S. Young, J. Kenney, S.D. Lord, F.P. Schloerb: Astrophys. J. Lett. 287, L65 (1984).
13. D.B. Sanders et al.: Astrophys. J. Lett. 305, L45 (1986).
14. J.S. Young, F.P. Schloerb, J. Kenney, S.D. Lord: Astrophys. J. 304, 443 (1986).
15. J.M. Shull, S. Beckwith: Ann. Rev. Astron. Astrophys. 20, 163 (1982).
16. G.S. Wright et al.: Paper presented at conference on Star Formation in Galaxies (Caltech, Pasadena, June 1986).
17. D. Hollenbach, C.F. McKee: Astrophys. J. Lett. 241, L47 (1980).
18. L. Searle, W.L.W. Sargent, W.G. Bagnuolo: Astrophys. J. 179, 427 (1973).
19. R.B. Larson, B.M. Tinsley: Astrophys. J. 219, 46 (1978).
20. R.A. Chevalier, A.W. Clegg: Nature 317, 44 (1985).
21. T.M. Heckman et al.: Paper presented at conference on Star Formation in Galaxies (Caltech, Pasadena, June 1986).
22. C.J. Lonsdale, S.E. Persson, K. Matthews: Astrophys. J. 287, 95 (1984).
23. R.B. Larson: Mon. Not. Roy. Astron. Soc. 218, 409 (1986).
24. S. van den Bergh: Publ. Astron. Soc. Pacific 94, 459 (1982).
25. J.E. Gunn: In Globular Clusters, ed. by D. Hanes and B. Madore (Cambridge University Press, Cambridge 1980), p. 301.
26. W.H. McCrea: In Progress in Cosmology, ed. by A.W. Wolfendale (Reidel, Dordrecht 1982), p. 239.
27. L. Searle: In The Evolution of Galaxies and Stellar Populations, ed. by B.M. Tinsley and R.B. Larson (Yale University Observatory, New Haven 1977), p. 219.

Star Formation in Disks and Spheroids

RICHARD B. LARSON

1. INTRODUCTION

A major goal of studies of the stellar populations in galactic disks and spheroids is to understand how these systems formed. In particular, we wish to understand how the observed stars formed, and how the star formation rate and initial mass function (IMF) may have varied with time and location. It is also important to understand the formation of star clusters, since they serve as crucial population tracers. If the dark matter in galactic halos is in faint stars or stellar remnants, its properties, too, need to be understood in terms of star formation processes.

2. DISK STAR FORMATION

Disk star formation is much better understood than spheroid star formation because it is still occurring in many locations and has received extensive study. It appears possible to understand the origin of the disk stars in our Galaxy by processes like those occurring in nearby star forming regions, where low-mass stars are presently forming in small dark clouds like the Taurus clouds and more massive stars in the dense cores of giant molecular clouds like the Orion cloud. On larger scales, the spatial distribution and the rate of star formation in galactic disks have been extensively studied by observations at optical, infrared, and radio wavelengths. Some important qualitative results have been established, as will be noted below, although quantitative conclusions remain uncertain owing to the uncertainty in the stellar IMF that must be assumed to interpret the data [1, 2].

In general, the present rate of star formation per unit mass in galactic disks is higher in galaxies of later Hubble type, and also in galaxies with higher fractional gas contents. However, these broad correlations include a great deal of scatter, and when examined closely star formation proves to be a rather irregular process; it has a patchy spatial distribution and probably also fluctuates with time, especially in the irregular galaxies, whose star forming properties seem to be dominated by random scatter [3]. Moreover, strong bursts of star formation can occur in galaxies, often apparently triggered by interactions or accretion processes. The rate of formation of massive stars inferred for some "starburst" galaxies is so high that it is not easily reconciled with a conventional IMF, and this has led some authors to suggest that the IMF in starbursts strongly favors massive stars [1].

The irregular behavior of star formation and the sensitivity of the star formation rate to external perturbations suggests that some sort of instability or threshold effect may trigger star formation in galactic disks. Gravitational instabilities in gas disks can play such a role, and may serve both to generate spiral structure [4, 5] and to initiate or organize star formation [6, 7]. The stability of a rotating gas disk or layer is governed by the parameter Q which depends on the surface density and velocity dispersion of the gas and on the epicyclic frequency [8]. In a marginally stable disk, a small increase in the local surface density or a decrease in the velocity dispersion may suffice to trigger instability and hence star formation; small random fluctuations in these parameters could lead to patchy or sporadic star formation, while large-scale perturbations could cause star formation in a more organized spiral pattern. A sudden large increase in surface density caused by a merger or accretion event might well result in a strong burst of star formation. In irregular galaxies, the whole disk might be just above or just below the threshhold of instability, causing the galaxy to be either active or inactive in forming stars, and possibly accounting for the large variations in the star formation rate observed in irregular galaxies [3].

If gravitational instabilities in disks are an important mechanism of star formation, one might expect star formation to be particularly rapid during early stages of galactic evolution when the gas surface density is high and galaxy interactions and accretion events are frequent. As gas is consumed by star formation and the gas surface density decreases, disks become more stable, so star formation processes and spiral structure driven by gravitational instability should decay strongly with time. If the rapid star formation occurring in young disks has an IMF that favors massive stars, as has been suggested for starburst galaxies, the resulting rapid formation and evolution of massive stars can generate large numbers of stellar remnants that contribute importantly to the total mass present in galactic disks [9].

3. SPHEROID STAR FORMATION

Understanding the origin of the stars in galactic spheroids is a more challenging problem because we presently see little, if any, star formation taking place in spheroids. Indeed, spheroids appear to consist mostly of very old stars; the halo of our Galaxy has an age comparable to the Hubble time, and most spheroids are believed to have similar ages, although there is debate as to whether they also contain populations as young as perhaps 5 Gyr. In any case, most of the star formation in spheroids probably took place relatively early, i.e. within the first several Gyr of the history of the universe.

Nearly all of the star formation that we observe at present, even in peculiar galaxies, occurs in disks or disturbed disks. At least some elliptical galaxies must have been formed or built up by mergers of disk galaxies [10], since some elliptical galaxies have peculiarities indicative of recent or ongoing mergers or accretion of disks [11]. Thus the conclusion cannot reasonably be avoided that at least some of the stars in spheroids were formed in disks. Theoretically, it has been argued that fragmentation and star formation do not occur in a collapsing gas cloud until it has attained a flattened configuration such as a disk [12]; fragmentation can then be understood as resulting from the gravitational instability of such a disk [8]. Star formation in proto-elliptical galaxies almost certainly occurred in dense, gravitationally bound subsystems [13], and in order to form globular clusters they probably had to have masses several orders of magnitude larger than those of the clusters themselves (see below). These systems may well have been flattened and may have resembled, or evolved into, small disk galaxies. Thus a plausible hypothesis is that all of the stars in spheroids actually formed in disks.

There are, of course, significant differences between the present stellar populations in galactic spheroids and disks. Most prominent of these is that globular clusters are much more abundant in spheroids than in disks. Spheroids also generally have higher mass-to-light ratios than disks, especially if the dark matter in galactic halos is basically stellar and forms part of the spheroid population. In our Galaxy, there are differences in chemical abundance ratios between halo and disk stars, notably a higher oxygen-to-iron ratio in halo stars. However, these differences may not be fundamental, because the oldest disk population in our Galaxy is in fact similar to the halo; the old disk contains globular clusters [14], and has a high oxygen-to-iron ratio like the halo [15]. If the star formation rate in the galactic disk was higher at earlier times, as in plausible models [16, 9], the mass-to-light ratio of the old disk is also larger due to the contribution of stellar remnants [1]. Thus spheroid star formation may not have differed essentially from early disk star formation.

4. GLOBULAR CLUSTERS AS FOSSILS

It has long been realized that globular clusters are key tracers of the early development of galaxies, so it is of great interest to understand the origin of these clusters. The well-studied globular clusters in our Galaxy have ages that are equal within the uncertainties to the Hubble time, so they must have formed within the first few Gyr of the history of the universe. Globular clusters may moreover be the only identifiable fossils surviving from this early era, making it particularly important to learn how to read this fossil record.

Some salient properties of globular clusters that may cast light on the earliest phases of galactic evolution are the following: (1) In general, globular clusters have a more extended spatial distribution and a flatter density profile than the background stars in galactic spheroids [17]; in the brightest elliptical galaxies the globular cluster density distribution is approximated by $\rho \propto r^{-n}$ where $n \sim 2.5$, compared with $n \sim 3.0$ for the background stars. However, the globular cluster density distribution is not as shallow as that usually assumed for the dark matter in galactic halos, which has $n \sim 2$. (2) The globular clusters are bluer and thus presumably more metal-poor than the background stars at the same distance from the centers of elliptical galaxies [18]. (3) With few exceptions, globular clusters are internally quite homogeneous in metallicity; all of the stars in each cluster have the same metallicity within the measuring errors [19].

Properties (1) and (2) are consistent with the globular clusters having formed before the bulk of the stars in elliptical galaxies [17]. In any picture of galaxy formation involving progressive condensation and enrichment of the gas, a population formed earlier will have both a lower metallicity and a shallower density profile [20]. If the dark matter in galactic halos is basically stellar and derives from an early "Population

III", property (1) also suggests that the globular clusters formed after Population III and thus represent an intermediate population between Population III and the bulk of the stars in spheroids.

Property (3), the internal chemical homogeneity of globulars, is an important constraint on the formation process because such homogeneity is not easy to bring about. Thorough mixing of the pre-cluster gas must occur subsequent to its chemical enrichment but before the observed stars form, and this requires that the gas be retained in a bound system long enough for it to be homogenized by turbulence generated by continuing energy input from massive stars [21]. Thus, the pre-cluster gas must already have undergone considerable processing before forming globular clusters; the globular clusters cannot be primordial objects. In order to retain and adequately mix this gas, the pre-existing bound systems probably had to have masses much larger than those of the clusters themselves, and some of them may well have been sizeable galaxies.

Observations of nearby regions of star formation in our Galaxy also suggest that bound clusters form only as small parts of much larger systems. Young clusters generally represent only a small fraction of the stellar mass, and an even smaller fraction of the gas mass, in a region of star formation. For example, surrounding the Trapezium in Orion is an exceptionally dense grouping of young stars that will probably become a bound open cluster [22], but its mass is only a few hundred $M_\odot$, compared with a total of several thousand $M_\odot$ in young stars and a few times 10^5 $M_\odot$ in gas in the Orion star forming complex. Thus in this region only about 10^{-3} of the total mass seems likely to form a bound cluster. If the same low efficiency applies to the formation of globular clusters, they must have formed in systems having total masses of at least 10^8 $M_\odot$. It may be relevant that the smallest galaxy known to contain globular clusters, the Fornax dwarf spheroidal galaxy, has a present mass of at least 2×10^7 $M_\odot$, and it may well have had a substantially larger initial mass.

It is important to note that the smaller systems from which spheroids formed could not have been just like typical present-day irregular or spiral galaxies, which are too loose in structure and have too low a rate of star formation; moreover they do not form clusters as massive as typical globular clusters. In order to form stars sufficiently rapidly, the first bound systems must have been more compact than present-day irregular or spiral galaxies, since faster star formation requires a higher mean density [23]. Also, they probably did not develop as high a degree of flattening and symmetry as typical present spiral galaxies. If observed today, such a system would probably be called a blue compact or starburst galaxy. An example may be provided by the nearby starburst galaxy M82 [24], in which massive stars are forming at a high rate in a compact nuclear gas disk having a diameter of about 1 kpc and a mass of about 10^8 $M_\odot$ [25]. M82 also contains some extremely luminous stellar knots with luminosities exceeding 10^8 $L_\odot$ [26], and they may represent, or contain, young globular clusters.

Since most of the stars in elliptical galaxies apparently formed after the globular clusters, the formation of globular clusters must have stopped or become less likely during the later stages of spheroid star formation. A similar development evidently occurred during the early evolution of the disk of our Galaxy: globular clusters formed during an initial period of a few Gyr or less [14], but subsequently only much smaller open clusters have formed. A possible reason for this is that the typical masses of star-forming clouds and hence of the clusters formed in them decreased with time as the gas settled into a thinner disk. This might be expected since the typical size of condensations formed by gravitational instability in a gas disk is proportional to the scale height of the disk [8]. There is, in fact, some evidence that the luminosities of the disk globular clusters are positively correlated with distance from the galactic plane, since those more than 500 pc from the plane are more luminous on the average by about 1.3 mag than those closer to the plane [14].

The above considerations suggest that the globular clusters and the field stars in elliptical galaxies formed at different stages during the evolution of pre-existing disk systems. If so, differences in the globular cluster frequency among elliptical galaxies [17] could reflect differences in the stage at which disk formation was truncated by mergers. For example M87, which has an unusually high abundance of globular clusters, may have formed from subsystems that merged at a particularly early stage and never developed very mature disks; globular clusters may nevertheless have continued to form in starbursts caused by the mergers [13, 27].

If the IMF in starburst systems strongly favors massive stars, and if a similar IMF applied during the early stages of spheroid star formation, large numbers of massive stars may have formed along with or before the globular clusters and collapsed into dark remnants that now occupy galactic halos [28]. The early formation of mostly massive stars, which might be expected because of the relatively high gas temperature at early times [9], is not incompatible with the simultaneous formation of low-mass stars in globular clusters because low-mass stars can still form in small, exceptionally dense cloud cores [1]. Again, this suggests that globular clusters are exceptional objects that form with low efficiency under extreme conditions.

5. GALAXY FORMATION: A SKETCH

To tie together the ideas that have been discussed, I suggest the following qualitative picture of spheroid and disk formation. Initially, when the smallest bound pre-galactic condensations began to form, their collapse was probably little impeded by angular momentum and very high densities were attained, leading to intense bursts of star formation. Because the gas temperature was initially high, only massive stars formed and they soon collapsed into dark remnants. As time went on, the gas collected into progressively larger and more flattened star-forming systems, which eventually may have resembled present compact dwarf galaxies. Globular clusters containing low-mass stars began to form in the densest cloud cores, while elsewhere mostly massive stars continued to form. Most of these early dwarf systems were soon disrupted and their contents were dispersed by collisions or gas loss that left them unbound. The globular clusters and the dark remnants thus released now populate galactic halos, while a few surviving original systems are still observed as dwarf spheroidal galaxies. Because of the early predominance of massive stars in the IMF, the stars formed at early times have a high oxygen-to-iron ratio, and because most of the mass that formed stars is now in stellar remnants, galactic halos and dwarf spheroidal galaxies have large mass-to-light ratios.

Some of the collisions between star-forming systems led not to dispersal but to aggregation into larger systems. Gaseous dissipation played a role in allowing them to become progressively more condensed, and accumulation of gas at their centers and in surrounding disks led to the development of dense cores and extended flat disks. The metallicity of the gas in these disks increased steadily with time, while the gas temperature generally decreased, causing star formation to favor progressively less massive stars. The formation of globular clusters became progressively less likely and eventually ceased, possibly because the gas layers became too thin to form sufficiently massive clouds. Collisions between developing disk galaxies continued in the densest environments, forming increasingly condensed spheroids containing increasingly metal rich stars. In this way the visible parts of elliptical galaxies and spiral bulges were formed; because they have benefitted from some dissipation, they are more spatially concentrated than the extended halos containing debris from earlier stages of galaxy formation.

If merging continued until the remaining gas was used up or became too hot to form stars, an elliptical galaxy would result. If mergers ceased early and if sufficient cool gas remained and continued to collect in a disk, a spiral or S0 galaxy would be formed, a spiral galaxy resulting from a more protracted period of disk formation [29]. Continuing accretion of much smaller galaxies by large spirals could play a role in building thick disks and in replenishing interstellar gas in spirals, but such events have by now become fairly rare. Evidently most of the visible stars in the universe formed after major merging events ceased to be important, since these stars are now in disks. However, most of the dark matter and most of the globular clusters must have been in existence before the merging stopped, since they are now located in galactic halos.

At present we see only occasional mergers and starbursts, and they are almost certainly just a pale reflection of what happened at earlier times [10, 13]. However, the action may not be all over yet, even for a rather normal galaxy like our own with a disk that has obviously escaped major damage since early in its history. In the not too distant future, our Galaxy will probably accrete the Magellanic Clouds, which will add significant new material and will cause some disruption of the disk. Eventually M31 and its companions, which are presently moving toward us, will very likely also collide and merge with our Galaxy, and the debris will form a giant elliptical galaxy. Any astronomers of the distant future who study this elliptical galaxy, the graveyard of the entire Local Group, will clearly have a major challenge in trying to unravel its long and exceedingly complex history. Will they suspect that a small bit of it involved a gathering of astronomers who met to discuss the properties and origins of "nearly normal galaxies" like the rather diverse dozen or so that they called their "Local Group"?

REFERENCES

1. R. B. Larson: In Stellar Populations, ed. by A. Renzini and M. Tosi (Cambridge Univ. Press, 1987), in press.
2. R. C. Kennicutt: In Stellar Populations, ed. by A. Renzini and M. Tosi (Cambridge Univ. Press, 1987), in press.
3. D. A. Hunter, J. S. Gallagher: Publ. Astron. Soc. Pacific 98, 5 (1986).
4. P. Goldreich, D. Lynden-Bell: Mon. Not. Roy. Astron. Soc. 130, 125 (1965).
5. A. Toomre: In The Structure and Evolution of Normal Galaxies, ed. by S. M. Fall and D. Lynden-Bell (Cambridge Univ. Press, 1981), p. 111.
6. B. G. Elmegreen, D. M. Elmegreen: Mon. Not. Roy. Astron. Soc. 203, 31 (1983).
7. R. B. Larson: Highlights of Astronomy 6, 191 (1983).
8. R. B. Larson: Mon. Not. Roy. Astron. Soc. 214, 379 (1985).
9. R. B. Larson: Mon. Not. Roy. Astron. Soc. 218, 409 (1986).
10. A. Toomre: In The Evolution of Galaxies and Stellar Populations, ed. by B. M. Tinsley and R. B. Larson (Yale University Observatory, New Haven, 1977), p. 401.
11. F. Schweizer: In Internal Kinematics and Dynamics of Galaxies, IAU Symp. No. 100, ed. by E. Athanassoula (Reidel, Dordrecht, 1983), p. 319.
12. J. E. Tohline: Astrophys. J. 235, 866; 239, 417 (1980).
13. B. M. Tinsley, R. B. Larson: Mon. Not. Roy. Astron. Soc. 186, 503 (1979).
14. R. J. Zinn: Astrophys. J. 293, 424 (1985).
15. R. E. S. Clegg, D. L. Lambert, J. Tomkin: Astrophys. J. 250, 262 (1981).
16. M. Schmidt: Astrophys. J. 137, 758 (1963).
17. W. E. Harris: Astron. J. 91, 822 (1986).
18. W. E. Harris: Publ. Astron. Soc. Pacific 95, 21 (1983).
19. L. Searle, R. Zinn: Astrophys. J. 225, 357 (1978).
20. R. B. Larson: Mon. Not. Roy. Astron. Soc 145, 405 (1969).
21. L. Searle: In The Evolution of Galaxies and Stellar Populations, ed. by B. M. Tinsley and R. B. Larson (Yale University Observatory, New Haven, 1977), p. 219.
22. G. H. Herbig, D. M. Terndrup: Astrophys. J. 307, in press (1986).

23. R. B. Larson: In The Evolution of Galaxies and Stellar Populations, ed. by B. M. Tinsley and R. B. Larson (Yale University Observatory, New Haven, 1977), p. 97.

24. R. B. Larson: Comments Astrophys. 6, 139 (1976).

25. G. H. Rieke, M. J. Lebofsky, R. I. Thompson, F. J. Low, A. T. Tokunaga: Astrophys. J. 238, 24 (1980).

26. S. van den Bergh: Astron. Astrophys. 12, 474 (1971).

27. F. Schweizer: This conference.

28. R. B. Larson: Comments Astrophys., 11, in press (1986).

29. R. B. Larson, B. M. Tinsley, C. N. Caldwell: Astrophys. J. 237, 692 (1980).

Star Formation in Disks: IRAS Results

CAROL J. PERSSON

The importance of the far infrared emission to the study of stellar populations in galaxies is that in the local universe a comparable amount of energy emerges in the far infrared as in the uv-optical region [1-2]. IRAS has now made available to us far infrared fluxes for over 20,000 galaxies.

Interpretation of the IRAS far infrared data for galaxies is limited by the fact that the flux is observed only after absorption and re-radiation of uv-optical photons by dust grains. Therefore, in optically thin situations, we must make assumptions about the physical properties of the grains and their geometrical distribution with respect to the heating sources in order to interpret the resulting far infrared spectral energy distribution. Also, IRAS sensitivity is restricted to flux in the 40-120 micron range, for most galaxies. Fortunately, this wavelength band usually straddles the peak of the spectral energy distribution [1], but substantial emission does arise at both longer and shorter wavelengths.

Despite these concerns, IRAS far infrared fluxes are potentially powerful as a measure of the total massive star formation rate in galaxies, especially if the flux arises in relatively optically thick situations so that the dust grain properties can be ignored to first order. The IRAS far infrared flux of a star forming region can then be converted to a star formation rate (SFR) if an initial mass function (IMF) for the heating sources is assumed, and a correction for the flux missed by IRAS is made. The uncertainty in this correction factor is on the order of a factor of two.

1. IRAS Detection Rates and Completeness for Galaxies

IRAS detection rates for UGC galaxies [3] have been investigated by PERSSON and RICE [4]. As shown originally by de JONG et al. [5], IRAS detects few ellipticals and S0s, and the detection rate increases towards later types. At 60 microns IRAS detects essentially all spirals brighter than a B magnitude of 13, but by 15th magnitude only about 20% of late-type galaxies are detected [4]. Thus the IRAS survey is complete for late-type galaxies only to a B magnitude of about 13: beyond this level IRAS detects only galaxies with increasingly larger than average infrared-to-blue luminosity ratio (IR/B).

2. The Far-Infrared Luminosity Function

Various groups have constructed far infrared luminosity functions from the IRAS data base [2,6-7]. These are all in excellent agreement [2]. SOIFER et al.'s luminosity function is particularly useful since an attempt has been made to cast it in terms of bolometric luminosity, and to compare it to the bolometric luminosity functions of other galaxy types. At moderate luminosities (10^{10-11} L_o for H_o=75 km/s/Mpc), the space density of 60 micron-selected galaxies is much like that of the starburst galaxy population. At very high luminosities (>$2x10^{11}$ L_o), IRAS galaxies are the dominant population [2].

The far infrared luminosity function is, as expected, dominated by galaxies that are relatively luminous in the far infrared. The IR/B ratio of an L^*_{IR} galaxy (a galaxy at the break in the far infrared luminosity function) is five to six times larger than the the median IR/B of optically selected galaxies [4]. Thus the far infrared energy density of the local universe is dominated by a minority of the optically selected galaxies.

3. Origin of the Far-Infrared Flux

At the low luminosity end of the far infrared luminosity function we find 'normal' spiral and irregular galaxies. There is now general agreement that this emission is composed of roughly equal contributions from two principal sources: (i) thermal emission from warm dust associated with the young O and B star population, and (ii) a lower color temperature component from the diffuse neutral medium [5,8-14]. This second component is probably thermal emission from grains heated by the general interstellar radiation field. PERSSON and HELOU [11] require much of this diffuse far infrared emission to originate with non-ionizing stars.

This scenario is supported by the far infrared maps of RICE et al. [15], who have coadded IRAS data for all galaxies large enough to show structure. The images are dominated by smooth, cool, extended emission. For the very largest nearby galaxies, a few of the giant HII-region complexes can be resolved from this extended disk emission, and spiral arm structure is discernible in a few cases.

An alternative viewpoint is provided by BURSTEIN and LEBOFSKY [17], who have proposed that the far infrared flux of most normal galaxies is dominated by emission from the inner (500-1000 pc) regions. This is based on the result that there is an inclination dependence of the detection rate of UGC galaxies by IRAS. However, the maps of RICE et al. [15] do not support the conclusion that the far infrared flux is sharply concentrated to the nuclear regions. Also, there is no dependence of the 60/100 micron color temperature of UGC galaxies on inclination [4], which might be expected if edge-on disks are optically thick in the far infrared.

As we move up the far infrared luminosity function we find that increasing far infrared luminosity is loosely correlated with IR/B ratio and with 60/100 micron color temperature [2,5]. This is consistent with an increasing contribution from the warmer of the two emission components seen in low luminosity galaxies: that associated with the massive star forming regions. Moreover, the increasing IR/B ratio is indicative of a move towards more optically thick conditions, such as seen in dusty nuclear starbursts like that in M82 [17]. This interpretation of the moderate far infrared luminosity galaxies as starburst galaxies is consistent with the similarity of their space density to that of known starburst galaxies [2]. It is further borne out by optical spectroscopy, which shows a predominance of strong HII-region-like spectra among 60 micron-selected galaxy samples [18-21].

A minority of IRAS-selected galaxies in the mid far infrared luminosity range have active nuclei [18]. There is evidence that Seyfert galaxies are characterized by large 25/60 micron ratio [22-23]. DePoy [24] has found evidence for a broad Brackett line in the extremely luminous Arp 220, which may indicate that this galaxy, and others like it, is powered by a buried quasar.

An alternative model for the far infrared luminous galaxies is that, rather than an active nucleus or a starburst, they are powered directly by excitation in the collision of two nuclear disks [25]. Shu [26] has stressed the relative inefficiency of this model.

Whether or not the shock model is viable, a relatively large fraction of moderate and high luminosity IRAS galaxies do seem to show some evidence for morphological peculiarity [28] which might implicate interactions and mergers in the far infrared flux. Well-known, merger cases show clear evidence for enhanced far infrared luminosity [27-30], but among the less strongly interacting pairs far infrared luminosity is less strongly enhanced [30-31], and selection effects plague objective studies of far infrared luminosity as a function of impact parameter, relative velocity, etc. The presence of a bar also seems to enhance the probability that a galaxy will be infrared-bright [32-34].

4. Star Formation Rates

In low luminosity galaxies, star formation rates cannot be derived very reliably from the far infrared data because much of the flux originates in optically thin conditions from a mixture of heating sources [11]. For the more luminous galaxies that are thought to be powered by starbursts, it is likely that optically thick conditions pertain and the far infrared luminosity can be converted with some reliability to an SFR if an IMF is assumed. For example, for a Miller-Scalo [35] IMF extending from 0.1 to 60 M_o, the far infrared luminosity of a galaxy with L^*_{IR} given by Soifer et al.'s luminosity function ($2x10^{10}$ for H=75) converts to an SFR of 10-15 M_o/yr

[36], assuming that all the flux of the young population emerges in the far infrared. This large rate, implying a gas depletion timescale of about 10^9 yr for a typical giant spiral, has led several people to the conclusion that starbursts, especially the very luminous ones, are likely to form only massive stars [cf 17, 37-39]. However, the space densities of IRAS galaxies are such that a typical L^*_{opt} galaxy need not go through more than one high IR/B phase of duration 10^8 yr during a Hubble time [2, 28], so a solar neighborhood-like IMF during the starburst may not be incompatible with the gas depletion timescales. It should also be pointed out that a short depletion timescale is just what proponents of the merger model of E galaxy formation would like to find.

Acknowledgements: Space does not permit me to reference all the excellent contributions which have been made to this field over the last three years, nor does it allow proper recognition of all the people with whom I have had stimulating discussions about the IRAS galaxy data. I am particularly grateful to G. Bothun, S.E. Persson and W.L. Rice.

References:

1. Telesco, C.M. and Harper, D.A. 1980, Ap.J., **235**, 392
2. Soifer, B.T. et al. 1986,[1]
3. Nilson, P. 1973, Uppsala General Catalogue of Galaxies
4. Persson, C. J. and Rice, W.L. 1986,[1]
5. de Jong, T. et al. 1984, Ap.J.Lett., **278**, L67
6. Lawrence, A. et al. 1986, M.N.R.A.S., **219**, 687
7. Rieke, G.H., and Lebofsky, M.J. 1986, Ap.J., **304**, 326
8. Cox, P. and Mezger, P.G. 1986,[1]
9. Perault, M. et al. 1986,[1]
10. Sodroski, T.J. et al. 1986,[1]
11. Persson, C.J., and Helou, G. 1986,[1]
12. Helou,G. 1986, Ap.J.Lett, submitted
13. de Jong, T. and Brink, K. 1986,[1]
14. Rowan-Robinson, M. 1986,[1]
15. Rice et al. 1986, in preparation
16. Burstein, D. and Lebofsky, M.J. 1985, Ap.J. **301**, 683
17. Rieke, G.H. et al. 1980, Ap.J., **238**, 24
18. Persson, S.E. et al. 1984, B.A.A.S., **16**, 471
19. Elston, R., et al. 1985, Ap. J., **296**, 106
20. Allen, D. A. et al. 1985, M.N.R.A.S., **213**, 67P
21. Lonsdale, C.J. 1986, in Spectral Evolution of Galaxies, ed C. Chiosi and A. Renzini, p 91
22. Miley, G.K. et al 1985, Ap.J.Lett., **293**, L11
23. de Grjp, M.H.K. et al 1985, Nature, **314**, 240

24. de Poy, D. L. 1986,[1]
25. Harwit, M. et al 1986, Ap.J., submitted
26. Shu, F. 1986,[1]
27. Gerhz, R.D. et al 1983, Ap. J., **267**, 551
28. Lonsdale,C.J., et al 1984, Ap.J., **287**, 95
29. Joseph, R.D. and Wright, G.S. 1985, M.N.R.A.S., **214**, 87
30. Cutri, R. M., and McAlary, C.W. 1985, Ap.J., **296**, 90
31. Kennicutt, R.C. et al 1986,[1]
32. Hawarden, T.G. et al 1985, in Light on Dark Matter, ed. F.P. Israel (Dordrecht: Reidel), p 455
33. Devereux, N.A. 1986, [1]
34. Puxley, P.J. et al 1986, [1]
35. Miller, G.E. and Scalo, J.M. 1979, Ap.J.Suppl., **41**, 513
36. Telesco, C.M. and Gatley, I. 1984, Ap.J., **284**, 557
37. Augarde, R. and Lequeux, J. 1985, Astron.Ap., **147**, 273
38. Wyse, R. and Silk, J. 1986,[1]
39. Larson, R.B. 1986, in Stellar Populations in Galaxies. eds A. Renzini and M. Tosi, in press

[1] in Star Formation in Galaxies, ed. C.J. Persson, in press

The Metallicity Distribution of the Extreme Halo Population

Timothy C. Beers

1. A Survey for Extremely Metal – Poor Stars

The first stars to form in the Galaxy were likely to have been extremely deficient in their content of heavy metals relative to stars formed in more recent generations. If this initial population included objects less massive than $0.8 M_{\odot}$, then stars of nearly primordial metal abundance should still be observable today. By identifying a large and unbiased sample of extremely metal-poor stars in the Galactic halo, we can address several long-standing questions concerning the early Galaxy. For example, did the Galaxy form from pristine primordial material, or has it suffered an initial enhancement prior to the formation of the presently observed stars? What was the Initial Mass Function of the Galaxy like? What are the kinematic properties of the extremely low metallicity halo of the Galaxy, and how do they differ from those of the globular clusters and other halo tracers?

Beers, Preston, and Shectman [1] (hereafter BPS) describe a new objective prism survey optiized for detecting stars of the lowest metal abundance. To date, the survey covers roughly 1940 square degrees in the South Galactic Hemisphere. Visual scans of these plates were made to identify about 1800 candidates with weak or absent lines of CaII H and K. Digital spectroscopy (1 Å resolution) of the first 450 candidates was used to identify a subset of 134 stars exhibiting CaII K strengths consistent with metallicities in the range $[Fe/H] \leq -2.0$. The metal-poor stars span the range of spectral types from F0 to G5, and are a mix of stars near the halo main-sequence turnoff and stars on the red horizontal branch, giant branch, and asymptotic giant branch.

2. The Metallicity Distribution

Grids of CaII K strength as a function of $B - V$ color were used to assign individual [Fe/H] values to the stars in BPS [2]. The one-sigma scatter in these in the abundances is roughly 0.3 dex. Metallicities for the most extreme stars in the BPS sample must be regarded as tentative at present, owing to our ignorance of the ratio [Ca/Fe] for stars below $[Fe/H] = -3.0$. If any systematic offset is present, however, it is likely to operate in the sense that the sub -3.0 stars are even more metal-poor than their assigned values.

Figure 1 is a stem-and-leaf diagram of the [Fe/H] values for stars in the BPS sample. The 'stems' to the left of the double bar are the [Fe/H] values to two significant figures; the 'leaves' to the right of the double bar are the corresponding third significant figure for each star. For example, one can immediately read off that the values of [Fe/H] in the range $-2.0 \geq [Fe/H] > -2.1$ are $-2.00, -2.04, -2.05, -2.05$ and -2.09, respectively. The numbers to the far right are the cumulative number of stars less than the listed metallicity stem to the left. As seen in the Figure, among the 120 stars with $[Fe/H] \leq -2.0$, there are 50 stars below $[Fe/H] = -2.5$ and 10 stars with $[Fe/H] \leq -3.0$. It is interesting to compare these numbers with the predictions of the simple 'closed box' model for Galactic nucleosynthesis [3], [4]. If the Galaxy began forming stars from gas of essentally zero initial metal abundance, a sample of 120 stars below $[Fe/H] = -2.0$ should include 38 below $[Fe/H] = -2.5$ and 12 below $[Fe/H] = -3.0$. Although such a comparison is still uncertain due to small number statistics, it does indicate that there is no deficit of the most metal-poor stars compared to the simple model.

Because stars in the BPS sample were originally selected on the basis of the apparent weakness of the CaII H and K lines in the objective prism spectra, we expect that the completeness of the survey will be a function of temperature, in particular for the coolest stars. Over the range of spectral types included in the survey, effective temperature is well correlated with the $B - V$ color or with the reddening– independent Balmer line index H'. Although we cannot, at present, precisely estimate the temperature for which completeness effects become significant, inspection of the original survey plates suggests that the liklihood of selection does not vary appreciably among stars with CaII K indices $K \leq 4$Å. As can be seen in Figure 8 of BPS, stars with $K \approx 4$Å and $[Fe/H] \approx -2.0$ have a Balmer line index $H' = 1.7$, which corresponds to a $B - V \approx 0.50$. The hot stars ($H' > 1.7$) outnumber the cool stars ($H' < 1.7$) by a factor of 2 to 1. Figure 2 presents histograms of the metallicities for these two sub-samples, parameterized in terms of $z/z_{\odot} = 10^{[Fe/H]}$.

The cooler stars in the BPS sample have CaII K equivalent widths in the range 2.5Å $\leq K \leq 7$Å. Because the CaII K line of the cool metal-poor stars is rather strong, it is likely that many will be omitted during the original selection as discussed above. This effect is clearly seen in Figure 2a as a depletion of stars with $z/z_{\odot} > 0.004$ ($[Fe/H] > -2.4$). Note, however, that for $z/z_{\odot} < 0.004$, the distribution of metallicities is flat, as predicted for the simple model. We expect that the hotter stars (Figure 2b) should not suffer incompleteness at the metal-rich end. Although the overall trend of Figure 2b is still flat, there is some indication of a bump in the range $0.002 \leq z/z_{\odot} \leq 0.004$, and a depletion beyond $z/z_{\odot} = 0.002$. One likely explanation for these features at the low metallicity end is the contamination of the *intrinsic* CaII K feature of a substantial fraction of these stars by *interstellar* CaII K. Based on a comparison of the assigned metallicities as a function of the K index at a given H', the estimate of [Fe/H] for a star can be raised from 0.2 – 0.6 dex for the lowest metallicity stars simply due to the presence of the interstellar line. If the [Fe/H] measures for as few as one-third of the stars in the region $0.002 \leq z/z_{\odot} \leq 0.004$ are affected to this degree, the bump in the region of Figure 2b is easily smoothed out.

[Fe/H]			Total < [Fe/H]
-1.6	6		133
-1.7			
-1.8	2 5 6		132
-1.9	0 1 5 5 7 8 9 9		128
-2.0	0 4 5 5 9		120
-2.1	0 0 0 2 2 3 3 3 3 4 5 5 6 7 7 8 8 9 9		115
-2.2	0 0 0 0 1 3 3 4 4 5 6 7 8 8 9 9 9 9		96
-2.3	0 1 2 3 3 3 4 4 4 5 6 6 7 9		78
-2.4	0 1 2 2 4 5 5 5 5 6 7 8 9 9		64
-2.5	0 0 1 2 2 2 2 2 3 3 3 6 6 6 7 7 8 9		50
-2.6	2 3 4 4 4 5 6 6 8 9		32
-2.7	1 5 9 9		22
-2.8	0 3 5 7		18
-2.9	2 4 5 9		14
-3.0			
-3.1	3 4 8	22948-66, 22948-93, 22897-8	10
-3.2	1 1	22891-200, 22891-209	7
-3.3	4	22888-31	5
-3.4	2	22885-96	4
-3.5	9	22968-14	3
-3.6			
-3.7			
-3.8	0 8	22876-32, 22881-39	2

Figure 1: Stem-and-leaf diagram of the metallicities ([Fe/H]) for stars in the list of BPS.

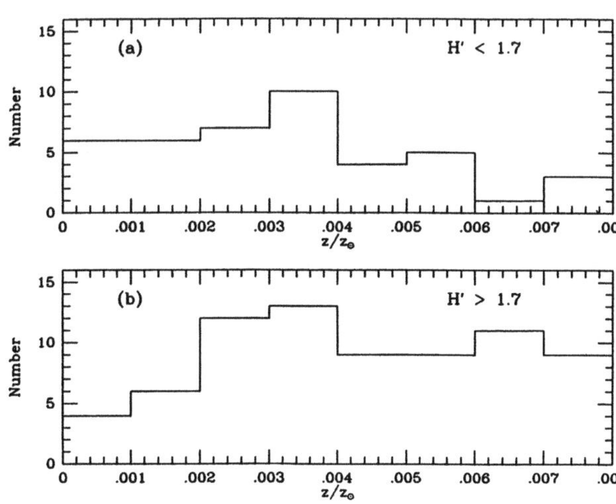

Figure 2a: Distribution of metallicities parameterized by $z/z_{\odot}$ for the (cool) stars satisfying $H' < 1.7$Å.

Figure 2b: Distribution of metallicities parameterized by $z/z_{\odot}$ for the (hot) stars satisfying $H' > 1.7$Å.

3. Conclusions

The first results from an objective prism survey for extremely metal-poor stars indicate that there is *no apparent dearth* of stars in the metallicity range [Fe/H] ≤ -2.6, as had been previously found [5]. Rather, the evidence seems to favor a metallicity distribution which remains constant down to the lowest measured [Fe/H] . It should be kept in mind that this data represent only the first quarter of the original survey candidates. A more detailed analysis of the metallicity distribution and kinematics of the extreme halo population will be deferred until more of the data can be included.

References:

1. T.C. Beers, G.W. Preston, S.A. Shectman: Astronomical Journal 90 , 2089 (1985)
2. J. Norris: Astrophysical Journal (Suppl.) 61, 667 (1986)
3. F.D.A. Hartwick: Astrophysical Journal 209, 418 (1976)
4. L. Searle and R. Zinn: Astrophysical Journal 225, 357 (1978)
5. H.E. Bond: Astrophysical Journal 248, 606 (1981)

Session 2

Small Objects

J. Huchra, Chair
July 22, 1986

Systematic Properties of Extragalactic Globular Clusters

DAVID BURSTEIN

1. Introduction

The globular cluster system of our own Galaxy has been used to determine the Sun's distance to the center of the Galaxy, to understand the basics of stellar evolution, to deduce the formation and early evolution of the Galaxy and to determine the total mass of the Galaxy. Galactic globular clusters are generally acknowledged to be the oldest stellar systems in our Galaxy and hence place a strong constraint on estimates of the age of the universe.

The above comments are an incomplete list that emphasizes the importance of Galactic globular clusters to modern astrophysics. Indeed, globular clusters have played such an important role that our handling of other data can be subtly biased by the known properties of Galactic globular clusters.

The purpose of this review is to explore three of the more implicit assumptions that we have inherited from our knowledge of Galactic globular clusters: 1) On average, the chemical abundance properties of stars are similar in globular clusters of similar overall metal abundance in different galaxies. 2) The globular cluster systems in giant spirals like the Milky Way and M31 are the oldest stellar systems in their respective parent galaxies. 3) The globular cluster systems in giant spirals are only formed during the early stages of formation of the parent galaxy.

Each of these assumptions is 'conservative', in that it simply extends the properties of a known archetype, the Galactic globular cluster system, to unknown but apparently similar systems. However, these assumptions were born of a time when little was known about the systematics of globular cluster systems in other galaxies.

2. 'Known' Globular Cluster Systems of Other Galaxies

This review will use the term 'known' to denote those globular clusters that have spectroscopic observations either of their integrated light, or of their individual stars. Spectroscopy of enough globular clusters in a single galaxy to infer the systematic properties of the globular cluster system is currently available for six galaxies other than our own: the Large Magellanic Cloud (LMC), the Small Magellanic Cloud (SMC), the Fornax dwarf elliptical, M31, M87 and NGC 5128 (=Cen A).

On the basis of the visible spectroscopy and broad-band colors, what are the distinguishing characteristics of the globular cluster systems of these galaxies, and how do these characteristics compare to that of the Galactic globular cluster system?

The Milky Way: In integrated visible and infrared light, the spectra of Galactic globular clusters arrange themselves in a nearly one-parameter sequence (cf. [1-3]). (While such is not the case for the integrated ultraviolet spectra [4], little is known about the ultraviolet spectra of extragalactic globular clusters.) The ages of Galactic globular clusters appear to be very old (>10 billion years) and similar for both metal-poor and metal-rich clusters (e.g., [5]).

The LMC and the SMC: It is well-known that the globular clusters in these two galaxies have a wide range in age, with perhaps an increase of metallicity with time for the globular clusters in the LMC [6-8]. The radial velocities of the LMC globular clusters show disk-like structures that are different for old and young clusters [9].

The Fornax Dwarf Elliptical: The five globular clusters in this galaxy are very metal-poor ([Fe/H]<-1.8), with ages comparable to that of the oldest metal-poor clusters in the Milky Way [10].

M31: While M31 and the Milky Way are apparently similar in structure and size, the globular cluster system of M31 differs from the Galactic globular cluster system in two key respects: a) In addition to a general correlation that exists for the broad-band colors and metal-line strengths among the M31 clusters, there also appears to be an intrinsic dispersion in the Balmer line strengths at a given metallicity [11-13]. Such a dispersion is hard to detect in the integrated spectra of Galactic globular clusters with known differences in horizontal branch morphology [1,11]. b) Both Balmer line strengths and CN strengths are enhanced in M31 clusters relative to galactic clusters of similar color and Fe-line strength [11,14].

NGC 5128: The brightest NGC 5128 globular clusters are redder than the reddest known Galactic globular clusters [15], and this globular cluster system appears to have a significantly different mean radial velocity from that of the parent galaxy [16].

M87: The huge globular cluster population of M87 contains many very luminous globular clusters, some of which appear to be much redder than galactic globular clusters (e.g., [17]). The origin of this abundant globular cluster population is ambiguous, as the position of M87 is believed to be coincident with the center of the Virgo cluster.

It is evident from these brief sketches that the known globular cluster systems of galaxies have few systematic properties in common either among themselves, or with the Galactic globular cluster system. It is in this context that the three 'conservative' assumptions should be reconsidered.

3. Does Similar Metallicity Imply Similar Chemical Abundance?

A spectroscopic comparison of the M31 globular cluster system with that of our Galaxy gives a clear negative answer to this question [11]. In fact, the M31 globular clusters have very weird visible spectra by the standards of galactic stars. How weird?

The study by our group in 1984 [11] showed that significant differences in stellar populations exist between the brightest M31 globular clusters, taken as a group, and a cross section of Galactic globular clusters. Additional differences also clearly separated the stellar populations of elliptical galaxies from these two populations.

Both the Balmer lines and the CN feature at 4150Å are stronger in many of the M31 clusters compared to Galactic globular clusters at a given Mg_2 line strength, Fe line strength, (B-V) or (J-K) color (cf. figures in [11,14]). For the purposes of the present more general discussion, it is worth re-emphasizing how strange the M31 globular cluster giant stars must be compared to galactic stars [14]:

Compare CN4150 line strengths for the elliptical galaxy M32, the M31 globular cluster V87 and the Galactic globular cluster NGC 6356. These three objects are chosen for this comparison as they otherwise have similar integrated colors and Mg_2 line strengths. The value of CN4150 = 0.023 mag for the nucleus of M32, 0.114 mag for V87 and 0.015 mag for NGC 6356 (one sigma errors of 0.015 mag).

The CN feature in integrated light is produced only by giant stars, which contribute at most only 40% of the light at 4150Å in old, metal-rich systems [18]. The difference of 0.1 mag in CN4150 strength between V87 and the other two systems thus requires the K giants in V87 to have CN4150 strengths 0.2-0.3 mag stronger than the K giants in M32 and NGC 6356.

Direct observations of K giants in M71, which is somewhat more metal-poor than NGC 6356 [11], imply CN4150 strengths in the range 0.25-0.35 mag for the NGC 6356 K giants. This in turn implies CN4150 strengths equal to 0.55-0.65 mag for the V87 K giants! As all of the metal-rich M31 globular clusters observed so far show enhanced CN4150, one infers that such 'CN-enormous' stars must be the dominant kind of K giant star (in the color range B-V =0.9-1.2) in most metal-rich M31 globulars.

In contrast, only two K giants in this color range in our own Galaxy are known to possess CN4150 bands that are of comparable strength: the star ROA 253 (B-V = 1.1) in Omega Centauri [19], and a star in the open cluster NGC 2158, #4305 [20]. The spectrum of NGC 2158#4305 is reproduced from JANES and HUI in Fig. 1 (below), along with a spectrum of a normal K giant of similar color in the same cluster (next page).

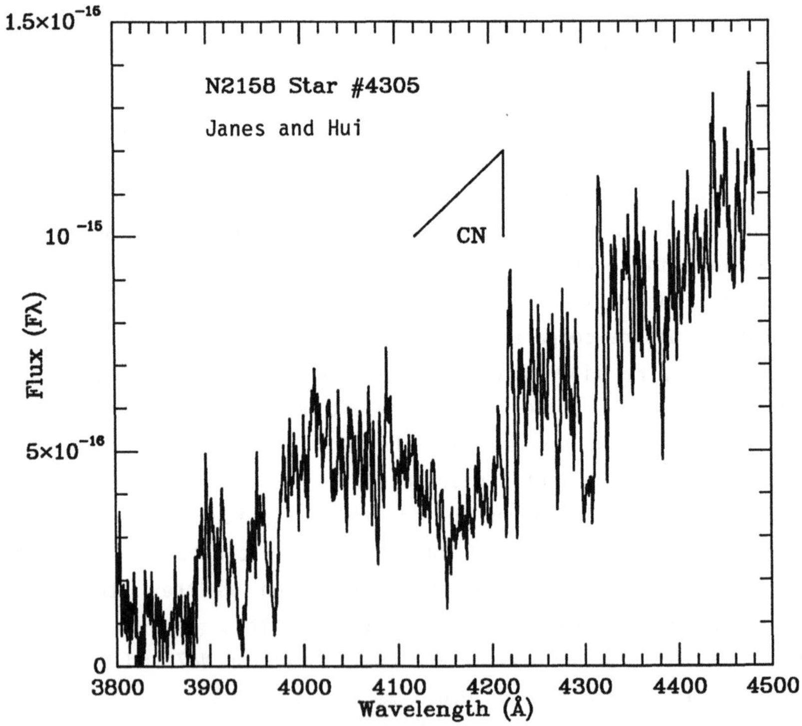

Figure 1a: The spectrum of the 'CN-enormous' K giant star in NGC 2158, #4305 (= #1199 in the number system of [21]). In addition to the prominent CN4150 feature, also note the very low level of continuum in the CN3800 band, indicating saturation of that feature. Copies of the spectra in a) and b) are kindly provided by JANES and HUI [20].

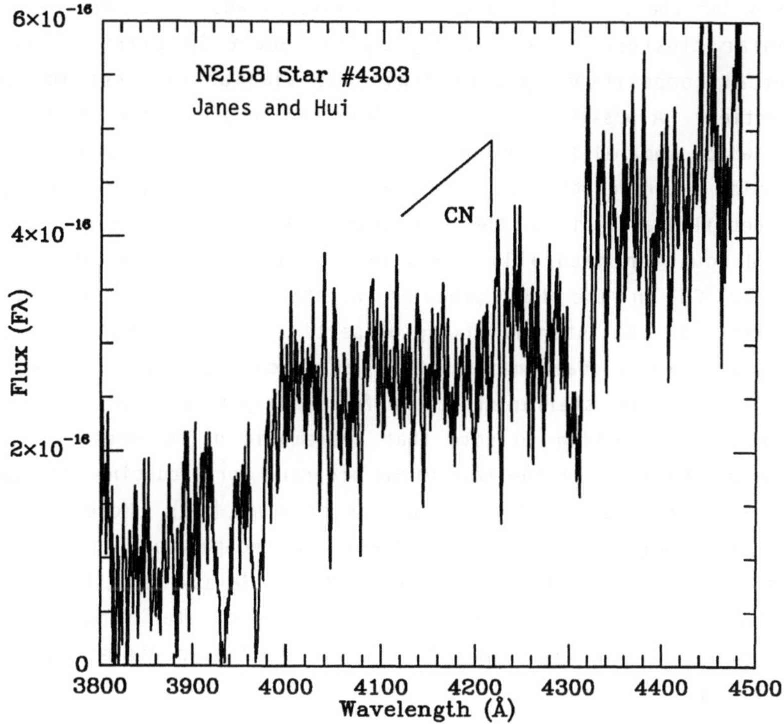

Figure 1b: A spectrum of the NGC 2158 normal-CN K giant, star #4301 (= star #1223 in the numbering system of [21]), as obtained from [20]. Note the lack of saturation of the CN3800 feature, as well as the much weaker CN4150 band.

The cluster NGC 2158 has a measured abundance of [Fe/H] = -0.5 and an approximate age of 3 billion years [21]. As estimated from these graphs, the CN4150 strength of the normal K giant star, #4301, is 0.15-0.20 mag, while that of #4305 is approximately 0.60 mag! The unreddened B-V colors of both stars are near 1.1 mag.

The existence of both ROA 253 and NGC 2158#4305 demonstrates that such CN-enormous stars do exist at some level in our own galaxy. However, the frequency of such stars here must be very small. As a rough estimate, assume that spectra of sufficient wavelength resolution are known of about 2000 galactic K giants. If so, the frequency of such 'CN-enormous' stars must be less than 0.1%. In contrast, the frequency of such CN-enormous stars must be >95% in the metal-rich M31 globular clusters.

How can the stars in M31 globular clusters and the stars in galactic globular clusters be apparently so different in their surface CNO chemical composition, yet otherwise so similar in other integrated properties? A possible clue comes from the investigation of ROA 253 by [19], which implied internal mixing as the source of the CN enhancement. A similar interpretation is also plausible for NGC 2158#4305, although a more detailed analysis of CNO abundance in the NGC 2158 stars is needed. This discussion naturally leads one to ask two questions: Is the enhanced CN in the M31 globular clusters caused by a systematic difference in the internal mixing properties of the stars? If so, what is (are) the global mechanism(s) that can produce systematically different internal stellar structure from galaxy to galaxy?

It is interesting to note that systematic differences in internal mixing has been one of the mechanisms proposed for producing the 'second-parameter' problem in Galactic globular clusters [22]. Moreover, it has also been suggested that the Galactic second-parameter problem is a function of distance from the center of our Galaxy (e.g,[23]). Is it possible that, contained within the processes of star formation, there are physical mechanisms that can produce systematic differences in the internal mixing of stars on a galaxy-wide scale?

4. Are All Globular Clusters (in Giant Spirals) Very Old?

In the context of the above discussion, one must now ask what direct evidence exists that all globular clusters in giant spirals and elliptical galaxies are very old? The only direct evidence this reviewer can find is that Galactic globular clusters are very old.

One must then also ask, what direct evidence exists that some globular clusters in giant spirals and ellipticals are relatively young? This reviewer knows of no direct evidence to the specific question. While the Magellanic Clouds and NGC 55 [24] do contain young globular clusters, none of these galaxies is a giant spiral.

It is apparent that there is little direct evidence for either position regarding the ages of extragalactic globular clusters, and what direct evidence that does exist is unconvincing.

Little indirect evidence exists as well. The dynamic anomaly of the globular cluster system in NGC 5128 cited in Sec. 2 could be used to infer a dynamical origin for these clusters that is different from that of the central galaxy. However, this anomaly is still based on only 30 clusters, and must await further verification.

It has been suggested by [11] that the brighter M31 globular clusters could be signficantly younger than Galactic globular clusters. The brightest clusters in NGC 5128 are also much brighter than Galactic globular clusters [15]). Younger ages could easily account for this increased brightness, but, of course, do not prove that these clusters are young.

Measurements of the relative frequency of globular clusters in proportion to the 'bulge' luminosity (the 'S' parameter) in spirals and ellipticals have been cited as support for old ages of extragalactic globular clusters [25,26]. As pointed out by VAN DEN BERGH [26], these arguments are tied to our understanding of both globular cluster formation and galaxy formation. Unfortunately, neither of these latter physical processes is understood at the current time.

Given the lack of clear evidence, together with the observed dissimilarities of the globular cluster systems of different galaxies, it would appear to be prudent to replace the underlying assumptions about the ages of globular clusters with a more agnostic position: The ages of globular clusters in other galaxies, and their relationship to the parent galaxy, should be directly determined for each galaxy.

5. Speculations

The lack of any convincing evidence on the ages of globular clusters in giant ellipticals and spirals, coupled with the known existence of young globular clusters in smaller galaxies, leads one to speculate on the consequences of not assuming that the globular clusters in giant galaxies must all be old.

First and foremost, how are more recent globular clusters formed? Since we strongly suspect that some globular clusters are formed in the initial collapse of a galaxy (e.g, our own globular cluster system), it is reasonable to assume that a rapidly changing gravitational field and/or very strong shocks could produce globular clusters. In the present day, such physical conditions are reproduced in interacting galaxies. Thus, one is led to consider the possibility that globular clusters could be formed in galaxy-galaxy interactions.

If globular clusters can be formed in interactions, then the luminosity function of globular clusters in a galaxy could evolve with time, as new bright clusters are added and older clusters fade in brightness. The brightest globular clusters in a given galaxy will depend on the history of interactions, which, in turn, could plausibly depend on the environment of the galaxy. The radial distribution of metallicities of the globular clusters would not be necessarily related to the overall structure of the parent galaxy.

Ironically, in this context the 'S' parameter can be used to support the merger hypothesis for the origin of ellipticals, as realized by VAN DEN BERGH [26]: the higher value of S for ellipticals would be the result of globular clusters formed in the interaction.

The purpose of these speculations is to draw attention to the tenuousness of conclusions that are drawn regarding the properties of globular clusters in other galaxies. Similar, plausible chains of argument have been made in favor of old ages for globular clusters in giant galaxies [25,26]. If one chooses not to make the three basic assumptions given in the Introduction, it is clear that our present knowledge of globular cluster systems can be used to yield contradictory conclusions about their formation processes and about the origins of their parent galaxies.

6. Future Possibilites for Obtaining Direct Evidence

New data pertaining to the problem of the ages of extragalactic globular clusters can be assembled in at least three different ways:

(a) Integrated Stellar Population Studies: High signal-to-noise integrated ground-based spectra of the bright globular clusters in NGC 5128, M87 and nearby galaxies should be able to detect differences in age between clusters less than one billion years old and those greater than ten billion years old. Ultraviolet spectra could extend this discrimination to smaller differences in age, perhaps between 3 billion years old and 10 billion years old [27]. However, as emphasized recently by O'CONNELL [28], the techniques for interpreting integrated spectra in this manner need to improve, and the ability to observe fainter globular clusters is required (e.g., the Ten-Meter class of telescopes and the Hubble Space Telescope). Note that the use of broad-band colors alone leads to confusion in separating age effects from metallicity effects in integrated light [28].

(b) Direct tests of interactions as a means of forming globular clusters: A number of nearby groups contain interacting galaxies. These include the M81-82 group and the Centaurus chain, which includes NGC 5128. High quality spectra should be taken of globular cluster candidates of the galaxies in these groups to search for young globular clusters. If the galaxy-galaxy interaction is very recent, it is possible that the globular clusters that are formed will be very young.

(c) The M31 Globular Cluster System: A much more thorough investigation of the M31 globular cluster system is needed, one which will include a large fraction of the M31 globular clusters that are as faint as typical Galactic globular cluster ($m_V > 16$). Are the fainter M31 globular clusters as spectroscopically-peculiar as their brighter counterparts? Is the velocity field of the fainter M31 globular clusters systematically different from that of the brighter clusters? If the two different populations are both spectroscopically and kinematically different, this would be evidence in support of at least two epochs of globular cluster formation in this galaxy.

7. Concluding Remarks

One of the purposes of any workshop is to offer the participants an opportunity to reexamine the basic assumptions that often underly our work. That has been the purpose of the present review. Globular clusters justifiably occupy an important position in our current understanding of the formation of our own Galaxy. Yet, this review has shown that there are significant differences among the systematic properties of globular cluster systems in different galaxies. These differences are enough to warrant more direct proof of the role played by globular clusters in the formation and evolution of other galaxies.

It is a pleasure to thank Ken Janes for supplying me with the spectra shown in Fig. 1, in advance of publication. The hospitality of the University of California, Santa Cruz, and of the Santa Cruz Workshop staff is gratefully acknowledged. This review benefitted from discussions with John Huchra and Sandra Faber.

References:

1. R. Zinn and M.J. West: Ap. J. Suppl. 55, 45 (1984)
2. J.A. Frogel, J.G. Cohen and S.E. Persson: Ap. J. 275, 773 (1983)
3. J.E. Hesser and S.J. Shawl: P.A.S.P. 97, 465 (1985)
4. T.S. van Albada, K.S. de Boer and R.J. Dickens: M.N.R.A.S. 195, 591 (1981)
5. D.A. VandenBerg: Ap. J. Suppl. 51, 29 (1983)
6. L. Searle, A. Wilkinson and W.G. Bagnuolo: Ap. J. 239, 803 (1980)
7. R.A.W. Elson and S.M. Fall: Ap. J. 299, 211 (1985)
8. J.G. Cohen: Ap. J. 258, 143 (1983)
9. K.C. Freeman, G. Illingworth and A. Oemler, Jr.: Ap. J. 272, 488 (1983)

10. R. Buonanno, C.E. Corsi, F. Fusi Pecci, E. Hardy, and R. Zinn: Astron. Ap. 152, 65 (1985)
11. D. Burstein, S.M. Faber, C.M. Gaskell and N. Krumm: Ap. J. 287, 586 (1984)
12. L. Searle: as presented at STScI conference on Stellar Populations, May 1986
13. J. Huchra and J. Stauffer: private communication (1986)
14. D. Burstein: P.A.S.P. 97, 89 (1985)
15. J.A. Frogel: Ap. J. 278, 119 (1984)
16. J.E. Hesser, H.C. Harris, and G.L.H. Harris: Ap. J. Lett. 303, L51 (1986)
17. S.E. Strom, J.C. Forte, W.E. Harris, K.M. Strom, D.C. Wells, and M.G. Smith: Ap. J. 245, 416 (1981)
18. J.A. Rose: Astron. J. 90, 1927 (1985)
19. M.S. Bessel and J. Norris: Ap. J. 208, 369 (1976)
20. K.A. Janes and X. Hui: private communication (1986).
21. C.A. Christian, J.N. Heasley and K.A. Janes: Ap. J. 299, 683 (1985)
22. G.E. Langer: P.A.S.P. 97, 382 (1985)
23. R. Zinn: Ap. J. 241, 602 (1980)
24. G.S. Da Costa and J.A. Graham: Ap. J. 261, 70 (1982)
25. W.E. Harris: Ap. J. 251, 497 (1981)
26. S. van den Bergh: P.A.S.P. 96, 329 (1984)
27. D. Burstein, F. Bertola, L.M. Buson, S.M. Faber, and T.R. Lauer: In preparation (1987)
28. R.W. O'Connell: preprint, to appear in the Proceedings of the STScI conference on Stellar Populations (1986)

Stellar Populations in Dwarf Spheroidals

MARC AARONSON

1. INTRODUCTION

Be they ever so humble, dwarf spheroidals have nevertheless become the focus of considerable recent attention. There are plentiful reasons for this, not the least of which is that spheroidals are the most numerous type of galaxy in the Local Group, and thus very probably in the Universe as a whole. The simple structure of these systems makes them ideal test sites for closed box models of galaxy collapse and chemical evolution. At the same time, the extreme characteristics of the spheroidals, and particularly their low luminosities and very small surface brightnesses, begs the question of how star formation even proceeds in such galaxies to begin with.

The widespread availability of CCD's has sparked renewed interest in the study of the nearby spheroidals. These objects are fully resolved and can be examined star-by-star, avoiding the frustrations inherent in conventional stellar synthesis work. Many spheroidals have now been found to possess an intermediate age stellar population. Evidence also continues to mount indicating that some of these systems contain a lot of dark matter, which if true may set interesting contraints both on galaxy formation models and on the nature of the non-luminous material itself. The "classical" view of spheroidals as the simple, low mass tail of larger-size ellipticals is gradually giving away to a picture in which low surface brightness objects in general -- both spheroidals and dwarf irregulars -- have a similar ancestry that is distinct from more massive galaxies. In many respects, the spheroidals seem to well fit the theme of this Workshop, as they are (not) nearly (as) normal as was once thought.

This discussion of the dwarf spheroidals will be brief, since several rather lengthy reviews of the subject have recently appeared [1,2]. The topics covered below encompass structural parameters, stellar content (including carbon and variable stars), ages, abundances, mass-to-light ratios, and some thoughts on the origins of the spheroidals. Most of our knowledge about these objects comes from the ten examples that are very close at hand -- the seven systems in the Milky Way's outer halo (Fornax, Leo I and II, Sculptor, Carina, Draco, and Ursa Minor), and the three Andromeda dwarfs. It is perhaps illuminating to remember that possibly only the brightest of these, Fornax, would have been within the completion limits of the very extensive Virgo cluster dwarf galaxy survey recently completed by SANDAGE and co-workers [3].

2. STRUCTURAL PROPERTIES

In total magnitude, the spheroidals range from Fornax ($M_V \sim -12.6$) down to Ursa Minor ($M_V \sim -8.5$). Though many of these estimates remain poorly determined, the spheroidals clearly overlap in luminosity with galactic globulars. The former are of course much more extended in size, ranging from ~ 0.5 to 3 kpc in estimated tidal radius. It remains an interesting but open question as to whether Ursa Minor defines the low luminosity end of galaxies, or if still smaller objects exist. The Milky Way spheroidals lie at distances from ~ 70 to 220 kpc; there is a rough correlation present between M_V and distance, which is perhaps a hint that even in our own halo the census of such objects is incomplete. The tidal to core radii of the spheroidals is small, with values in the range of 3 to 6. This means that both King models and exponential disks provide equally good fits to the available star counts [4]. The central surface brightnesses range from ~ 23 to 26 magnitude per square arc second [5], making some of these objects difficult to locate casually on Palomar Sky Survey plates. The central surface brightnesses are positively correlated with spheroidal luminosity, opposite to the sense of the behaviour found for larger ellipticals. This and other core parameter relations have been carefully investigated by KORMENDY [6]. Following an earlier suggestion [7], his results seem to strongly favor the notion that early type galaxies come in two varieties: high surface brightness ellipticals, whose low mass tail is defined by objects such as M32; and low surface brightness ellipticals, whose high mass tail is composed of objects such as NGC 185 and 205.

The core parameter relations of dwarf spirals and irregulars are similar to the low surface brightness ellipticals [6], though nuclei appear to be more common among the brighter examples of the latter. It wculd be extremely valuable, though very difficult, to accurately measure structure parameters for the very large, low surface brightness dwarfs that have been discovered in Virgo [3]. While presumably cousins to the spheroidals, there are nonetheless no nearby field analogues of these objects, so precisely how they relate to the two track scheme of early-type galaxies outlined above remains unclear. One possibility is that these huge dE's are irregulars which have been puffed up after a stripping episode.

3. STELLAR CONTENT

The upper part of the HR diagram in spheroidals superficially resembles that of classical Population II systems, with the presence of a generally steep giant branch and a well-developed horizontal branch. There is no sign of dust, gas, or any other evidence for very recent star formation in these systems. However, in all cases except Ursa Minor, a red rather than blue horizontal branch is seen, an effect that is also prevalent among the outer halo globulars. One system, Fornax, has globulars of its own, whose color-magnitude diagrams [8] show a blue horizontal branch, implying some similarity with galactic clusters.

There are also a number of well-known peculiarities associated with the variable stars in these systems. RR Lyraes have been definitely identified in five of the halo spheroidals to date (and are undoubtably present in all seven). In Draco, Leo II, and Carina, the mean periods of the $\langle P_{ab} \rangle$ variables lie between the two Oosterhoff classes. One or more anomalous Cepheids have also been located in each of the halo spheroidals, excepting Carina, where a number of potential candidates may well exist (see [9]). The predominantly red horizontal branches and the oddities of the variables are now generally accepted as being a result of the spread in spheroidal ages and abundances discussed below. In this regard, the spheroidals can serve as valuable templates to match against advanced stages of stellar evolution theory.

Only a few irregular and long period variables have been found, though in general the searches for such objects are not complete. The first W Virginis star to be located in a spheroidal was recently reported [10] in Fornax. The association of this object with one of the Fornax globulars, rather than with the general field star population, is in line with the notion that such variables are very old.

Carbon stars have been identified in all seven spheroidals. Except in Draco and Ursa Minor, the bolometric luminosities of these objects reach well above the first giant branch tip, indicating the presence of an intermediate-age population. A luminous carbon giant has also been discovered in the Andromeda II dwarf [11], and it seems likely that these stars and the intermediate-age stellar content associated with them are general features of spheroidals. Indeed, the well-known presence of O stars in NGC 185 and 205, combined with the evidence of an extended star-forming period in many other dE's [12], suggests that some continuing star formation is probably the norm for low surface brightness, early-type galaxies.

Low luminosity C stars that fall below the tip of the giant branch, reaching down to $M_V = -1.1$ for one Draco object [1,13], are present as well. Dredge-up theory is unable to account for such objects, and it is thought instead that their origin is connected with mass-transfer binaries. The census of low-luminosity C stars in the spheroidals is possibly quite incomplete. The general absence of similar stars in galactic globulars may be related to the survival difficulty of binaries in the stronger tidal fields present.

Contrary to earlier expectations, the fractional carbon star number does not appear to be correlated with mean spheroidal abundance, a result that can nevertheless be understood within the context of dredge-up theory (see [1]). The fractional C star light can, however, be used to roughly gauge the ratio of intermediate-age to old stars. The results of the test [1] imply that in Fornax and both Leo systems, roughly 20% of the population is intermediate age, while in Carina this is true for ~ 70% of the population. There is also a rough correlation between bolometric magnitude of the most luminous C star (a measure of the most recent epoch of star formation), and total magnitude of the parent spheroidal. This suggests that the more massive systems have indeed been able to retain gas the longest to form successive generations of stars.

4. AGES and ABUNDANCES

CCD's have now been used to obtain deep CM diagrams and reach the main sequence turnoff in four spheroidals: Carina [14], Sculptor [15], Ursa Minor [16], and Draco [17]. In Carina, the turnoff occurs at only ~ 7 Gyr; the Carina luminosity function is consistent with the fractional carbon star light in suggesting little evidence for the presence of a very old population. The discovery of RR Lyraes in Carina [9] does indicate that some old stars are there, but a scaling of the RR Lyrae number with that in Draco again implies only ~ 15% of the stellar content is really ancient. In Sculptor the turnoff appears to set in at ~ 13 Gyr, i.e. 2-3 Gyr less than for galactic globulars. The CM diagram of Ursa Minor is well fit by an M92 ridge line. This is also true for Draco, though there may be a hint of a younger population in the luminosity function.

Blue stragglers, however, have been found in all four of the above systems. The question still remains open as to whether these objects are more massive single stars with ~ 3 Gyr ages, or less massive binaries with ~ 10 Gyr ages. The number of blue stragglers seems adequate to account for the anomalous Cepheids, if the former evolve into the latter, as is generally supposed.

The aforementioned turnoff ages are in rough accord with the luminous extent of the AGB. Indeed, in Fornax a clump of stars has been found below the horizontal branch which is likely to be the top of a 3 Gyr old main sequence, just the right age to account for the very luminous Fornax C stars [8]. The existence of differing mean ages between spheroidals, and of a real age spread within some individual dwarfs, appears now to be firmly established.

From either various morphological features in the CM diagram, or individual giant star spectrophotometry, a reasonable estimate of the mean abundance is today available for all seven spheroidals. The resulting abundance-absolute magnitude relation is strikingly good -- even better, perhaps, than should be expected [1,2]. The spheroidals extrapolate naturally onto the [Fe/H], M_V relation that is seen for more luminous galaxies, be the latter either early or late-type, so it is difficult to make any kind of common heredity argument on this basis alone.

A real abundance spread over an order of magnitude seems convincingly demonstrated in Fornax [8]. Somewhat smaller spreads, ~ 0.6 dex in [Fe/H], are probably also present in Draco and Sculptor [17,18]. It remains unclear if any spheroidals possess as small an abundance dispersion as is seen in the vast majority of galactic globulars. Again, the spheroidal ages and abundances can be generally understood in the context of the more massive systems being able to retain star-forming gas the longest. How they manage to do so may be problematic, since even for Fornax the escape velocity (assuming no dark matter) is smaller than stellar mass loss speeds. Prolonged star formation may be assisted by the presence of non-luminous material, the latter being the subject we turn to next.

5. VELOCITY DISPERSIONS and MASSES

Over the past few years, several groups have been independently pursuing measurement of the dynamical masses of spheroidals via determination of the velocity dispersion from spectra of individual giants. This sort of work puts considerable demands on present-day technology, as velocities for metal-poor, weak-featured stars having $V > 17$ mag must be observed with accuracies < 5 km s^{-1}, and preferably 1-2 km s^{-1}. Data of variable quality and extent is now available for five spheroidals: Fornax, Sculptor, Carina, Draco, and Ursa Minor (see [1] for a summary). There appears to be a general trend emerging of increasing central mass-to-light ratio M/L with decreasing spheroidal luminosity. It is in fact intriguing that the total masses of these five systems come out all roughly the same. A similar change of M/L with M_V is also seen in a sample of very low luminosity dI's recently mapped in H I [19].

The author, along with Ed Olszewski, has over the last several years devoted much attention to the Draco and Ursa Minor systems. It is worthwhile to give an update of these results here, as good progress has been made since the most recent reporting of this work [1]. In particular, velocities of four new stars have been measured in Draco, and three new ones in Ursa Minor.

In the past we have been limited to the small number of bright giants ($V < 17.5$) identified just in the dwarf centers by the existing proper motion studies. Using the "MX" spectrograph on the Steward Observatory 2.3 m, we have begun undertaking an areally complete survey of all bright giants in Draco and Ursa Minor. The MX is a multi-object fiber-fed device, with fibers that can be repositioned in real time; it allows us to collect > 20 spectra at once in 30 min for stars down to 18th magnitude, from which $\pm$ 30 km s^{-1} accuracy velocities can be extracted. This is more than adequate to weed out non-members, as the systemic velocities of Draco and Ursa Minor are so high. We shall eventually have a much larger sample to draw upon for 1-2 km s^{-1} velocity measurement with the MMT Echelle. The first efforts here have already borne fruit with the seven new stars we recently observed. Both Draco and especially Ursa Minor are elongated, and it may one day be possible to say something about $\sigma(r)$ and v/σ in these systems. The latter quantity might be of particular interest in further delineating the relation between dSph and dI types. Some progress along these lines has already been made for the Fornax dwarf [20].

At the present time, we have in Draco a total of 15 stars; after two velocity variables are eliminated, the remainder (which includes 11 K giants and 2 C stars) gives a dispersion of 11 ± 2 km s^{-1}. Of this pruned sample, one star has been measured at 4 epochs, five stars at 3 epochs, and one star at 2 epochs. In Ursa Minor the total now stands at 13 stars. After discarding three variables, the remainder (all K giants) yield a dispersion of 10 ± 2 km s^{-1}. Again, in the reduced sample, the two stars that are the extreme velocity members have been observed at 4 epochs, while one other star has been measured at 3 epochs, and two others at 2 epochs.

As both we elsewhere [21], and Scott Tremaine at this Workshop have emphasized, the dispersions found in Draco and Ursa Minor cannot be readily accounted for by either atmospheric motions or duplicity effects. It is also worthwhile to point out that a KS test shows there to be no statistically significant difference between the velocity histograms and gaussians. Having said all this, it is nonetheless difficult for many to accept the implied King model M/L values: 60 for Draco, and 80 for Ursa Minor. We note, however, that the lowest luminosity dI system examined in H I, LGS3, has an M/L ratio of 27 [19]; a one magnitude fading of this object would put its total luminosity and M/L in the same regime as Draco and Ursa Minor. While some of the dI mass-to-light ratios may be biased toward large values because of gas stirring by young, massive stars, it seems difficult to make such a case for LGS3. In this object, the most luminous stars are red, not blue, and it has been argued [22] that no star formation has occurred in it for at least 10^8 yr, and perhaps for much longer than 10^9 yr.

The presence of nonluminous material in the spheroidals is important from a cosmological standpoint, as various forms of this stuff might then be ruled out [23]. On the other hand, instead of involving some exotic particle, it may be possible to put the dark matter in the form of low mass stars. A related alternative [24] invokes stellar remnants left over from an IMF that is biased by low abundance toward high mass stars. However, it remains unclear if this latter scheme can avoid difficulties with preventing enrichment and/or complete disruption of the system by the very large number of supernovae that must have at one time been going off.

6. WHERE DO SPHEROIDALS COME FROM?

There are two broad answers to this question, involving either tidal disruption or isolated collapse. Tidal models once appeared nice in that they could roughly explain the coplanar distribution of halo spheroidals and at the same time avoid the problem of how stars form in such low density environments. On the other hand, our present day knowledge of spheroidal stellar content, and in particular the various abundance and age spreads, and the tightness of the [Fe/H], M_V relation, do not seem easily accounted for by tidal creation in anything but a very arbitrary manner. Tidal models also do not appear to readily account for any dark matter.

The present author therefore favors a picture in which the formation and evolution of spheroidals occurred under more isolated circumstances. It has long been appreciated that if the "simple model" of chemical evolution is at all a good description of these systems, they must have lost an appreciable amount of their initial mass to keep the mean abundance from growing too large. Supernovae driven winds are usually assumed to be the mechanism behind this gas loss [25]. It has been argued by DEKEL and SILK [26] that such a picture cannot reproduce the observed [Fe/H], M_V and M_V, surface brightness relations unless a dark halo is also present. By normalizing to the observed relations, these authors predict $M/L \sim L^{-.37}$, which is remarkably enough in reasonably good accord with the trends seen for both dSph and dI systems [1]. In their view, the question of whether dI's turn into dSph's is really beside the point. Rather, both types come from one-sigma peaks in the density fluctuation spectrum that gives rise to galaxy formation. A further prediction of their biasing scheme is that low surface brightness dwarfs ought to be less clustered than more massive galaxies.

7. THINGS TO LOOK FORWARD TO

It would be nice to have new, modern star counts in the spheroidals from which better estimates of their absolute magnitudes and structure parameters could be derived. There is obviously also considerable work still to be done on the abundances and especially the kinematics of these systems, though truly dramatic progress here may await the coming generation of large ground-based telescopes. The Hubble Space Telescope will hopefully provide us with deep color-magnitude diagrams and luminosity functions for all the spheroidals, from which we can expect to derive much better limits on the stellar age distributions. Perhaps the most intriguing question, though one that may prove difficult to address, is the issue of just how many spheroidals there really are floating about in the Universe.

It is a pleasure to acknowledge Ed Olszewski for the use of our unpublished velocity dispersion data. Preparation of this article was partially supported with funds from NSF grant AST83-16629.

REFERENCES

1. M. Aaronson: In Stellar Populations, ed. by C. Norman, A. Renzini, and M. Tosi (Cambridge University Press, Cambridge 1986), in press.
2. R. Zinn: Mem. Soc. Astron. Ital. 56, 223 (1985).
3. A. Sandage, B. Binggeli: A. J. 89, 919 (1984).
4. S. M. Faber, D. N. C. Lin: Ap. J. (Letters) 266, L17 (1983).
5. J. Kormendy: In IAU Symposium No. 117, Dark Matter in the Universe, ed. by J. Kormendy and G. R. Knapp (Dordrecht, Reidel 1986), in press.
6. J. Kormendy: Ap. J. 295, 73 (1985).
7. A. Wirth, J. S. Gallagher: Ap. J. 282, 85 (1984).
8. R. Buonanno, C. E. Corsi, F. Fusi Pecci, E. Hardy, R. Zinn: Astr. Ap. 152, 65 (1985).
9. A. Saha, D. G. Monet, P. Seitzer: A. J. 92, 302 (1986).
10. R. M. Light, T. E. Armandroff, R. Zinn: A. J. 92, 43 (1986).
11. M. Aaronson, G. Gordon, J. Mould, E. Olszewski, and N. Suntzeff: Ap. J. (Letters) 296, L7 (1985).
12. N. Caldwell: A. J. 88, 804 (1983).
13. M. Azzopardi, J. Lequeux, B. E. Westerlund: Astr. Ap. 161, 232 (1986).
14. J. Mould, M. Aaronson: Ap. J. 273, 530 (1983).
15. G. S. DaCosta: Ap. J. 285, 483 (1984).
16. E. W. Olszewski, M. Aaronson: A. J. 90, 2221 (1985).
17. B. W. Carney, P. Seitzer: A. J. 92, 23 (1986).
18. T. E. Armandroff, G. S. Da Costa: (1986), in preparation.
19. W. L. W. Sargent, K. Y. Lo: In Star Forming Dwarf Galaxies and Related Objects, ed. by D. Kunth and T. X. Thuan, in press.

20. G. Paltoglou, K. C. Freeman: (1986), in preparation.

21. M. Aaronson, E. W. Olszewski: In IAU Symposium No. 117, Dark Matter Matter in the Universe, ed. J. Kormendy and G. R. Knapp (Dordrecht, Reidel, 1986), in press.

22. C. A. Christian, R. B. Tully: A. J. 88, 934 (1983).

23. D. N. C. Lin, S. M. Faber: Ap. J. (Letters) 266, L21 (1983).

24. J. Melnick, R. Terlevich: The Observatory 106, 69 (1986).

25. G. H. Smith: Pub. A.S.P. 97, 1058 (1985).

26. A. Dekel, J. Silk: Ap. J. 303, 39 (1986).

Blue Compact Dwarfs: Extreme Dwarf Irregular Galaxies

TRINH X. THUAN

1. Introduction

In Hubble's classification scheme, dwarf irregular galaxies (dIs) are low luminosity systems ($M_B \gtrsim -16$) which are at the end of the galaxy morphological sequence and which lack both a dominating nucleus and rotational symmetry [1]. We concentrate here on a class of dIs which are so extreme in these properties that they appear almost stellar in appearance, with no obvious underlying galaxy, at least on the Palomar Sky Survey prints (PSS), the blue compact dwarf galaxies (BCDs). Our motivation is twofold: (1) it is easier to discuss the physical characteristics of dIs by focussing on their extreme manifestations; and (2) the properties of dIs have been well reviewed elsewhere [2]. The present review of BCDs is intended to be complementary and will attempt to contrast the properties of BCDs with those of dIs.

2. The Stellar Populations

2.1 The Young Stellar Population and the Ionized Gas

Sargent and Searle [3] first demonstrated spectroscopically that BCDs have a large young ionizing stellar population of OB stars. The optical spectra of BCDs are similar to those of HII regions in the Galaxy and the outer regions of nearby galaxies such as M101: they show weak, nearly flat and almost featureless continua on which are superposed very strong emission lines covering a narrow range of ionization. The [ArIV]λ4715 line (ionization potential = 40.8 eV) is sometimes seen in the hottest objects (see for example the spectrum of IZw36 in [4] while the [OI]λ6300 line (ionization potential = 13.6eV) is weakly seen. A typical BCD optical spectrum is shown in Fig. 1. A small ($\lesssim$ 10 percent) fraction of BCDs show broad (FWHM ~ 20Å) HeIIλ4686 and NIIIλ4640 emission implying the presence of large numbers (10^2 - 10^4) of Wolf-Rayet stars of late WN subtypes [5].

The measured electronic temperatures determined from the fluxes of the [OIII]$\lambda\lambda$4959, 5007 doublet and that of the [OIII]λ4363 line range from 10000 to 20000 K. These temperatures, high as compared to those of galactic HII regions, are due to the low metallicity of BCDs relative to the solar neighborhood. The BCD metallicity ranges from ~ 1/3 to ~ 1/30 of the solar metallicity. The abundance distribution peaks at ~ 1/10 of the solar value and drops off sharply on the low abundance

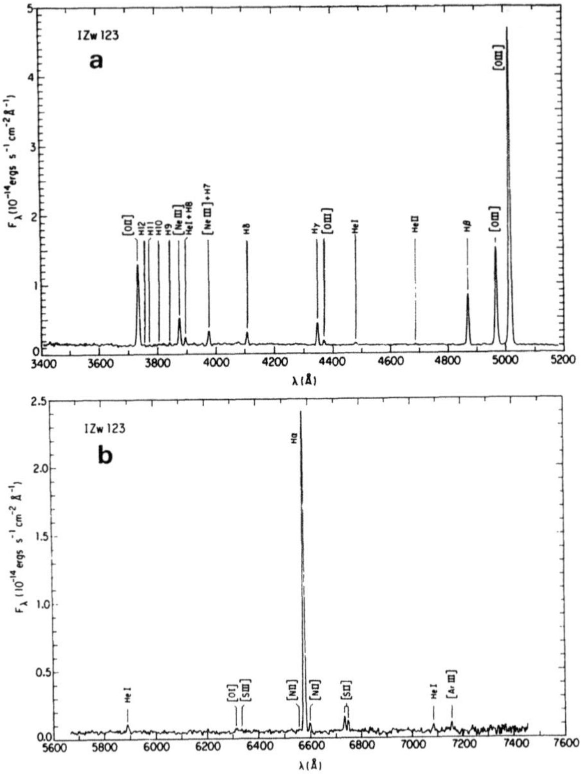

Fig. 1. The blue and red optical spectra of the BCD IZw123 (M_B = - 15.6, V_H = 652 km s^{-1} [17]) through a circular 6".1 diameter entrance aperture. Both spectra have a half-width resolution of 6.7 Å.

side [6]. Abundance analysis of a large sample of dIs and BCDs confirms that from the chemical evolution point of view, BCDs do not form a distinct class of objects but are the low luminosity and metallicity end of dIs [7]. In order not to exceed the observed low metallicity, BCDs cannot have made stars at the present rate for more than ~ 10^7 years, i.e., star formation in these galaxies cannot be continuous but must proceed by bursts. The most metal-deficient BCD known is IZw18 = MKN 116 ($Z \sim Z_\odot/30$) which, by a stroke of beginner's luck, was also one of the first discovered [3]. Since 1970, hundreds of other BCDs have been observed with none being as metal-deficient as IZw18. The lack of very metal-poor BCDs is probably due to the fact that star formation need only proceed for ~ 4 x 10^6 years to produce as much metals as is observed in IZw18, and the chance of catching a BCD just at ~ 4 x 10^6 years after the beginning of the first burst of star formation is very small [6].

The advent of the International Ultraviolet Explorer (IUE) satellite permitted to probe directly the young stellar population in BCDs. The ultraviolet spectra demonstrate unambiguously the presence of hot massive stars in BCDs: (1) they all show a continuously rising continuum toward the blue, characteristic of the UV spectra of OB stars. Because the stellar continuum is much more important in the UV than in the optical relative to the nebular emission, the ultraviolet spectra do not show strong nebular emission, contrary to the optical spectra. (2) They exhibit P-Cygni profiles with blue-shifted absorption wings in the lines of SiIVλ1405 and CIVλ1550, characteristic of mass loss in radiatively driven stellar winds in OB stars. A typical BCD ultraviolet spectrum is shown in Fig. 2. A detailed recent review of the ultraviolet properties of BCD spectra is given in [8].

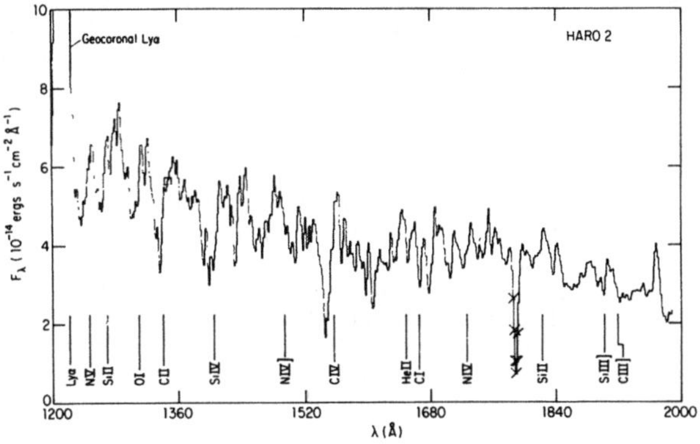

Fig. 2. The ultraviolet spectrum of the BCD Haro 2 = MKN33 (M_B = - 18.8, V_H = 1454 km s^{-1} [15,17]) taken with the short-wavelength camera of the IUE satellite through a 10" x 20" aperture. It has a resolution of 7 Å. The location of the prominent ultraviolet lines are marked. Only the P-Cygni profiles of the SiIVλ1405 and CIVλ1550 lines are certain.

The optical and ultraviolet data can put constraints on the star formation history of BCDs as characterized by the Initial Mass Function (IMF), the star formation rate (SFR), the age or duration of the burst of star formation and the number of bursts before the present one. The IMF can be parameterized by its slope x defined such that dn(M)/dM α $M^{-(1+x)}$ and the lower and upper stellar mass cutoffs m_L and m_U. Studies of a few BCDs [4,9] find that the data are consistant with x ~ 1.5 (close to the Salpeter value of 1.35 but significantly flatter than the solar neighborhood of ~ 2 for M > 1.8 $M_\odot$). Whether this enhancement of the IMF at the high mass end is related to the metal deficiency of BCDs or not is not known. Values of m_U ~ $100M_\odot$ fit the data, although curiously, for the BCD IZw36 = MKN209, a lower mass cut-off at ~ 4 $M_\odot$ is needed to fit the large equivalent width of the Hβ line, implying perhaps bimodal star formation for high and low mass stars [4]. The BCD IZw18 which has a smaller Hβ equivalent width has a smaller m_L ($\lesssim 2M_\odot$). Durations of the bursts are typically between ~ 10^6 and ~ 10^7 years. Star formation rates are very high, varying from ~ 0.01 to ~ 1 $M_\odot$ yr^{-1}, or normalizing to unit area, from ~ 6 x 10^{-9} to ~ 8 x 10^{-8} $M_\odot$ yr^{-1} pc^{-2}, generally higher than the values found for dIs of 3 x 10^{-10} to 5 x 10^{-9} with an average of 2 x 10^{-9} $M_\odot$ yr^{-1} pc^{-2}. For comparison Kennicutt [10] finds SFR between 5 x 10^{-10} and 6 x 10^{-8} with an average of 5 x 10^{-9} $M_\odot$ yr^{-1} pc^{-2} for spiral galaxies and the global SFR of the Milky Way and Magellanic Clouds is 4 x 10^{-9} $M_\odot$ yr^{-1} pc^{-2} [2]. The observed metallicity in BCDs put strong constraints on the possible number of bursts in the past lifetimes of these objects. In all the cases which have been studied, the number of bursts cannot exceed a few (≤ 5), and in several cases, e.g. IZw18 or IZw36, the present burst is sufficient to explain the observed metallicity.

Is there a contradiction between our picture of massive star formation in BCDs which proceeds by bursts lasting ~ 10^6 - 10^7 years separated by long periods of inactivity lasting several billion years, and that which Hunter and Gallagher [2] describe for dIs: 'galaxies which have roughly (to factors of 3) constant SFRs over a fair fraction of a Hubble time'? These two pictures can only be reconciled if the many centers of star formation in dIs which locally undergo bursts in comparable intensity to those in BCDs add up to a global rate of star formation which is regulated by an unknown mechanism to be roughly constant.

2.2 The Older Stellar Population

Ever since their discovery, the question has arisen whether BCDs are truly young systems, where stars are being formed for the first time now, or old galaxies with an older underlying stellar population on which are superposed the young stars formed in the current burst of star formation [11]. Thuan [12,13] attacked the issue by obtaining near-infrared measurements of BCDs and attributed the near-infrared fluxes to an older stellar population of K and M giants. He concluded that there are no truly young galaxies in his sample, although the near-infrared observations obtained through small (~ 8") apertures centered on the star-forming regions could not unambiguously distinguish between main-sequence, giant or supergiant light.

Perhaps the best technique to look for an older stellar population in BCDs is to obtain very deep CCD pictures. Such a program was carried out by Loose and Thuan [14] who could classify the BCDs into four main classes of objects according to their morphology. The first type, the iE galaxies (the details of the classification scheme are found in [14]), is by far the most common, ~ 70% of the BCDs being of this type. The high-surface-brightness star forming regions which are located near but not at the center of the galaxy show a clumpy complex irregular (i) structure. They are superposed on a more extended low-surface-brightness non-star-forming component (invisible on the PSS prints) with regular elliptical (E) outer isophotes. The BCD IZw36 [4] is a good representative of this class (Fig. 3a). The second type, the nE galaxies, includes ~ 20% of the BCDs. Instead of an irregular central structure, nE galaxies have a regular single high-surface-brightness star-forming region located precisely at the nucleus (n) and with elliptical isophotes. The non-star-forming low-surface-brightness component is similar to that in the iE galaxies. A good representative of this class is the BCD Haro 2 = MKN33 (Fig. 3b) [15] which has remarkable properties: its light profile is not distinguishable from that of a low-luminosity elliptical galaxy, but yet it has so much star formation at its center that it was picked up by both Haro and Markarian in their objective prism surveys. As expected, the extended low-surface brightness component is redder and thus older than the star-forming region, but more unexpectedly, both optical and near-infrared colors were not red enough to be consistent with a dominant population of K and M giants. The oldest stars consistent with the observed colors are F6 main-sequence stars. Thus, Haro 2 is a relatively young elliptical galaxy, the oldest stars it contains being made less ~ 4 billion years ago. It is clear that the nE BCDs form a very interesting class of objects with extreme examples of low-luminosity star-forming young elliptical galaxies.

The remaining ~ 10% of the BCDs fall into the last two types, the iI and iO galaxies. The iI galaxies are similar to the iE galaxies, except that the outermost isophotes of the low-surface-brightness non star-forming component are irregular (I) instead of elliptical. IZw18 (Fig. 3c) is a good example of an iI BCD. The iO galaxies are extremely rare ($\lesssim$ 5%). They are galaxies which do not show more faint outer structure on CCD pictures than on the PSS plates (O stands for zero outer component). Thus they appear to lack an underlying older stellar population and are the best candidates for truly young galaxies. The BCD POX 186 falls in this category [16].

To summarize, more than 95% of the BCDs show an extended underlying older stellar population on CCD pictures. Even BCDs such as IZw18 or IZw36 whose observed metallicities do not require previous star formation show a more extended fainter outer component. This would argue for some sort of bimodal star formation where the massive star formation responsible for the metals and the high-surface-brightness component is decoupled from the less massive star formation responsible for the low-surface-brightness component. This adds more weight to the idea of bimodal star formation in some BCDs, which was already suggested by the high lower-mass cut-off (~ $4M_\odot$) of the IMF in the BCD IZw36.

What is the appearance of a BCD in between bursts? After the death of the massive stars, only the low-surface brightness component remains and the BCD takes the appearance of a low-surface-brightness (LSB) dI such as the one shown in Fig. 3d. This is confirmed by the fact that LSB dIs and BCDs have very similar near-infrared colors and HI properties [13]. By contrast, dwarf ellipticals (dE), at least in the Virgo cluster, (as the one shown in Fig. 3e) show bluer near-infrared colors than LSB dIs or BCDs (dEs are probably more metal rich, $Z_\odot/3 \lesssim Z \lesssim Z_\odot$) and therefore it is very unlikely that a LSB dI could evolve to a dE by being stripped of its gas [13].

3. Gas and Dust

Gas is needed to feed star formation. BCDs contain ~ 10^8 $M_\odot$ in neutral gas, ~ 15% of their total mass (as computed from the width of the hydrogen line) [17]. Aperture synthesis of a few isolated (not undergoing interactions or mergers with other galaxies) BCDs such as IZw18 [18] or IZw36 [4] show the following general features. The HI distribution shows a core-halo structure, with the core containing 30-50% of the total HI mass, while the halo is diffuse and contains several less massive HI clumps. The star-forming region is associated with the HI core, although is slightly (by ~ 1 Kpc) offset with respect to the HI peak surface density. The HI gas is more extended (by a factor of ~ 5 in size) than the optical distribution, measured to an isophotal level of 25 mag arcsec^{-2}. While there are systematic motions (contraction, expansion or rotation) within the most massive HI clump where the star formation is occurring, the amount of random motions is also significant. The HI clumps are moving disorderly with respect to each other and undergo collisions which are probably responsible for the bursts of star formation.

BCDs also possess a 'missing mass' problem. Assuming virial equilibrium, a total mass M_T can be obtained from the width of the HI line. The fractional neutral hydrogen gas mass $M(HI)/M_T$ is only ~ 0.2. What constitutes the remaining ~ 80% of M_T ~ 5 x 10^8 $M_\odot$? It cannot be the young stars whose total mass is only ~ 10^4 $M_\odot$. The older stellar populaa

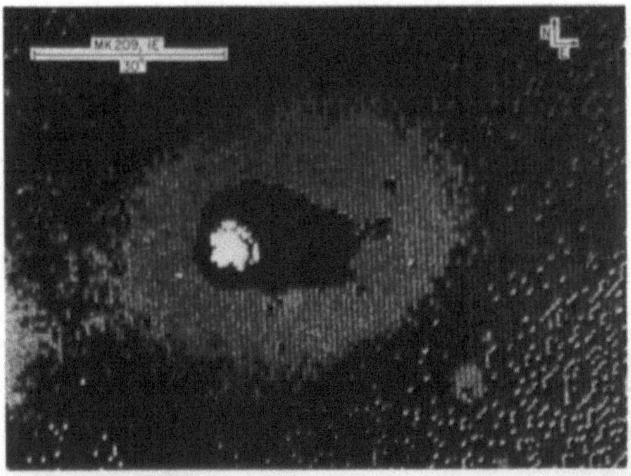

b

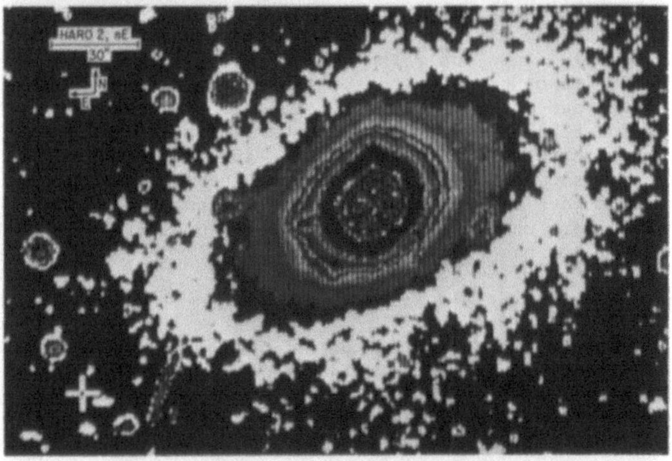

c

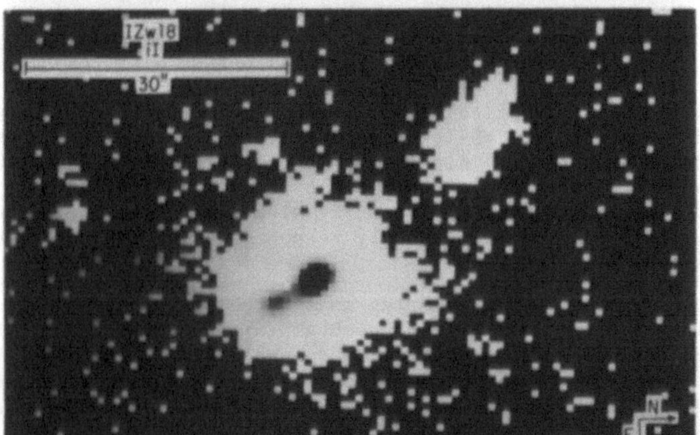

d

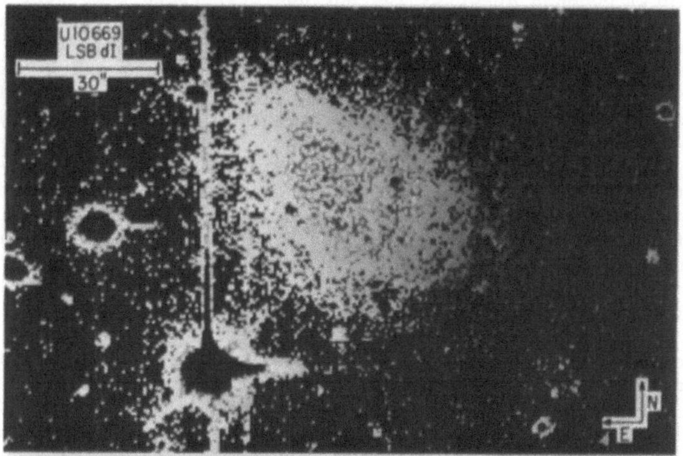

e

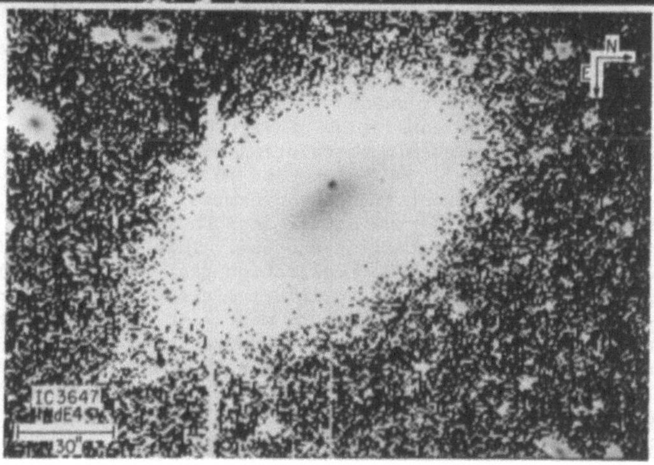

Fig. 3. Three of the four main morphological types of BCDs [14] are shown in the first three panels. The iE type (MKN 209 = IZw36, M_B = - 13.9, [4], fig. 3a) has an off-centered irregular star-forming region superposed on an elliptical low-surface-brightness outer component. The nE type (Haro 2, M_B = - 18.8, [15], fig. 3b) has a regular star-forming region centered precisely on the elliptical low-surface-brightness outer component. The iI type (IZw18 = MKN116, M_B = - 14.3, [18], fig. 3c) is the same as the iE type, except that the low-surface-brightness component is irregular. All three pictures have been taken with a CCD camera (RCA chip) attached to the cassegrain focus of the No. 1 0.9 m telescope at KPNO through a B filter. On the Palomar Sky survey prints, the low-surface-brightness outer component is invisible. Notice that IZw18 (fig. 3c) shows a low surface brightness blob to the NW of the main galaxy which is not associated with any present star formation. Since both MKN209 and IZw18 do not need any previous massive star formation before the present burst to account for the observed metallicity, the presence of a low-surface-brightness outer component not coexistent spatially with the star-forming region may imply a bimodal mode of star formation for high and low mass stars. The fourth type of BCD (not shown here) does not

Fig. 3 (cont.) possess an outer low-surface-brightness component on CCD pictures. Figure 3d shows an R CCD picture of the low-surface-brightness dwarf irregular U10669 (M_B = - 15.5, V_H = 443 km s^{-1}, [22]) taken at the prime-focus of the KPNO 4-m telescope. This is the appearance that a BCD takes after the death of the young stars: only the outer low-surface-brightness component remains. Finally, Figure 3e shows a B CCD picture obtained with the KPNO No. 1 0.9m telescope of the dwarf elliptical in the Virgo cluster IC3647 = VCC1857 [23] for comparison. The near-infrared colors for the Virgo dEs are bluer than those of the LSB dIs, making an evolutionary connection between those two classes of dwarfs very unlikely [13]. Notice that there is a bar-shaped high-surface brightness region at the center of the IC3647, suggesting perhaps recent ($\lesssim$ 5 billion years) star formation [13].

tion has a mass of the order of the HI mass and so probably cannot account for the entirety of the missing mass [15]. Stringent upper limits exist for the amount of CO molecular gas in BCDs [19]. In Zw36 [4], the CO upper limit implies that, if molecular hydrogen were to account for the virial mass, the value of the CO/H_2 abundance ratio would have to be smaller than the galactic value by a factor of ~ 1000. This may not be implausible if C/O α O/H, i.e., CO/H_2 α $(O/H)^3$, IZw36 being 10 times more metal-deficient than the solar neighborhood. Thus it appears that exotic nonbaryonic dark matter is not needed to explain the virial mass in BCDs. This is consistent with the conclusion of Vigroux et al. [7] who found that, independently of any chemical evolution model, the 'chemical' total mass to blue light ratio M_T/L_B of ~ 1 needed to account for heavy element enrichment is not too different from the dynamical M_T/L_B (~ 2-4), implying that nearly all ($\gtrsim$ 25-50%) the matter in BCDs is of baryonic nature and has participated to the astration process. This conclusion is strengthened even further with the realization that the dynamical masses of BCDs calculated assuming virial equilibrium are probably upper limits. The optical and HI components of BCDs are most likely not self-gravitating (K. Freeman, this workshop). The many supernovae resulting from the death of the massive stars in the star-forming regions inject energy in the neutral gas creating systematic motions. The outer parts of the BCD HI haloes are probably still infalling into the central regions. Both effects increase the width of the HI line and hence the virial mass.

The optical intrinsic extinction E(B-V) due to dust, as derived from the Balmer decrement and corrected for galactic extinction, is typically ~ 0.1-0.2. The optical extinction is slightly smaller than that derived from radio-continuum observations, but larger than that derived from spectral synthesis of ultraviolet spectra. The extinction in the ultraviolet is negligible (corresponding to E(B-V) ~ 0) [8]. Observations at different wavelengths reveal different regions in the BCDs, with the ultraviolet-emitting regions being the least dusty. In the BCD IZw36 [4], the dust-to-gas mass ratio is overabundant by ~ 3 compared to the solar neighborhood, assuming that ratio to be proportional to metallicity and that there is only atomic HI and no molecular gas. If H_2 exists in the amount required by the virial mass, the dust-to-gas mass ratio is 'normal', which may be an argument in favor of the existence of H_2 in BCDs. The dust acts as an efficient frequency converter, absorbing ultraviolet photons and re-emitting far-infrared photons. Many BCDs have been detected by the IRAS satellite and there is a tight correlation between the 60 and 100 μm luminosities and the radio-continuum 6cm luminosity [20]. Since a significant fraction of the radio continuum radiation at 6 cm is of thermal origin, and is probably associated with the regions of star formation [21], this tight correlation implies that most of the far-infrared emission comes from HII regions rather than from giant molecular clouds or diffuse neutral hydrogen clouds in BCDs [20].

4. Conclusion

From the preceding discussion, it is clear that BCDs play a very important role in galaxy formation and evolution studies. They are nearby examples of relatively 'young', chemically unevolved galaxies. Understanding such questions as the triggering mechanisms of the starbursts or the shape of the IMF (Is it enriched in high-mass stars? Is it bimodal?) in BCDs can yield insights into the star formation processes in the protogalactic phases of normal galaxies.

Acknowledgements

I thank the hospitality of the Service d'Astrophysique of the Centre d'Etudes Nucleaires de Saclay in France where this review was written. I acknowledge financial support from NASA grant NAG 5-257, the Air Force Office for Scientific Research grant AFOSR 85-0125 and the Commissariat à l'Energie Atomique in France.

References

1. E. P. Hubble: Ap. J. 64, 321 (1926).
2. D. A. Hunter and J. S. Gallagher: Pub. A.S.P. 98, 5 (1986).
3. W. L. W. Sargent and L. Searle: Ap. J. Letters 162, L155 (1970).
4. F. Viallefond and T. X. Thuan: Ap. J. 269, 444 (1983).
5. D. Kunth and H. Schild: Astr. Ap. in press (1986).
6. D. Kunth and W. L. W. Sargent: Ap. J. 300, 496 (1986).
7. L. Vigroux, G. Stasinska and G. Comte: Astr. Ap. in press (1986).
8. T. X. Thuan: in Star-Forming Dwarf Galaxies and Related Objects, ed by D. Kunth, T. X. Thuan and T. T. Van (Editions Frontieres, Paris 1986), p. 105.
9. R. W. O'Connell, T. X. Thuan and S. J. Goldstein: Ap. J. Letters 226, L11 (1978).
10. R. C. Kennicutt: Ap. J. 272, 54 (1983).
11. L. Searle, W. L. W. Sargent and W. G. Bagnuolo: Ap. J. 179, 427 (1973).
12. T. X. Thuan: Ap. J. 268, 667 (1983).
13. T. X. Thuan: Ap. J. 299, 881 (1985).
14. H. H. Loose and T. X. Thuan: In Star-Forming Dwarf Galaxies and Related Objects, ed. by D. Kunth, T. X. Thuan and T. T. Van (Editions Frontieres, Paris 1986), p. 73.
15. H. H. Loose and T. X. Thuan: Ap. J. October 1 (1986).
16. D. Kunth, J. M. Martin, S. Maurogordato and L. Vigroux: In Star-Forming Dwarf Galaxies and Related Objects, ed. by D. Kunth, T. X. Thuan and T. T. Van (Editions Frontieres, Paris 1986) p. 89.
17. T. X. Thuan and G. E. Martin: Ap. J. 247, 823 (1981).
18. J. Lequeux and F. Viallefond: Astr. Ap. 91, 269 (1980).
19. F. Combes: In Star-Forming Dwarf Galaxies and Related Objects, ed. by D. Kunth, T. X. Thuan and T. T. Van (Editions Frontieres, Paris 1986), p. 307.
20. D. Kunth and F. Sevre: In Star-Forming Dwarf Galaxies and Related Objects, ed. by D. Kunth, T. X. Thuan and T. T. Van (Editions Frontieres, Paris 1986) p. 331.
21. U. Klein, R. Wielebinski and T. X. Thuan: Astr. Ap. 141, 241 (1984).
22. T. X. Thuan and P. O. Seitzer: Ap. J., 231, 327 (1979).
23. B. Binggeli, A. Sandage and G. A. Tammann: A. J., 90, 1681 (1985).

Dwarf Galaxies and Dark Matter

SCOTT TREMAINE

Abstract

The nearby dwarf spheroidal galaxies help us to probe the nature and extent of dark matter in at least two ways. First, if they are regarded as test particles orbiting our Galaxy, their Galactocentric radial velocities and distances can be used to constrain the distribution of dark matter in the Galactic halo. Although this problem has been investigated several times in the past [1,2,3,4], there are at least two motivations for another look: (i) A number of accurate new radial velocities for dwarf spheroidals and distant globular clusters have become available in the past few years; the present study [5] is based on 10 galaxies and globular clusters—all those with distances between 50 and 200 kpc and velocity errors of less than $25\,\mathrm{km\,s^{-1}}$; (ii) We have devised a new method of statistical analysis based on Bayes' theorem, which directly yields confidence intervals for the mass of the Galaxy once the eccentricity distribution of the orbits is specified. Assuming an isotropic velocity ellipsoid, we find that the mass of the Galaxy is $\lesssim 5 \times 10^{11}\,\mathrm{M_\odot}$ at the 95% confidence level. If we assume that the Galaxy has a flat rotation curve out to several hundred kpc, then for isotropic orbits, the circular speed is $\lesssim 160\,\mathrm{km\,s^{-1}}$ at the 95% confidence level. These results suggest that the Galaxy's massive dark halo extends to $\lesssim 50\,\mathrm{kpc}$; a more massive halo is only consistent with the data if the orbital velocities of the satellites are mostly tangential. We have investigated the suggestion of LYNDEN-BELL et al. [2] that predominantly tangential velocities might arise because satellites on elongated orbits have been tidally disrupted, and find that tidal disruption in an initially isotropic satellite population with a power-law density distribution is unlikely to have an important effect on our results.

The internal dynamics of dwarf spheroidal galaxies also provide evidence for dark matter. The line-of-sight velocity dispersions of $\approx 10\,\mathrm{km\,s^{-1}}$ in Draco and Ursa Minor suggest that most of the mass in these systems is in some invisible form [6,7,8]. One alternative possibility is that the dispersions reflect the orbital motion of binary stars rather than the motion of the stars through the galaxy. We have investigated this hypothesis using Monte Carlo simulations of a population of binary stars, similar to the simulations described by AARONSON and OLSZEWSKI [6]. Although neither the fraction of binary stars nor their semi-major axis distribution is known, it turns out that these parameters are not needed: there is a strong, largely model-independent, correlation between the velocity dispersion due to orbital motion and the fraction of stars which show velocity variations over a fixed interval of time. A crude rule of thumb is that for plausible stellar masses, the dispersion is roughly $10f\,\mathrm{km\,s^{-1}}$, where f is the fraction of stars showing velocity changes $\geq 4\,\mathrm{km\,s^{-1}}$ in a one year interval. AARONSON and OLSZEWSKI [6] find that $f \lesssim 40\%$ in Draco and Ursa Minor; this result implies that binaries contribute at most $4\,\mathrm{km\,s^{-1}}$ to the observed dispersion of $10\,\mathrm{km\,s^{-1}}$. Thus it appears that binarism is not a viable alternative to dark matter in these galaxies.

A related possibility is that the stars in these galaxies may be clumped into "hard" subsystems or clumps whose internal dispersion is much larger than the orbital velocity dispersion. However, it is straightforward to show that these clumps would either evaporate (through star-star encounters) or merge on a timescale much shorter than the Hubble time. Thus, so far I have been unable to find an acceptable dynamical explanation for the large velocity dispersions in Draco and Ursa Minor which does not imply that large quantities of dark matter are present.

References

1. F. D. A. Hartwick, W. L. W. Sargent: *Astrophys. J.* **221**, 512 (1978).
2. D. Lynden-Bell, R. D. Cannon, P. J. Godwin: *Mon. Not. Roy. Astron. Soc.* **204**, 87P (1983).
3. R. C. Peterson: *Astrophys. J.* **297**, 309 (1985).
4. E. W. Olszewski, R. C. Peterson, M. Aaronson: *Astrophys. J. Lett.* **302**, L45 (1986).
5. B. Little, S. Tremaine: In preparation (1986).
6. M. Aaronson: *Astrophys. J. Lett.* **266**, L11 (1983).
7. M. Aaronson, E. Olszewski. "The Search for Dark Matter in Draco and Ursa Minor: A Three Year Progress Report", in J. Kormendy and G. R. Knapp, eds. *Dark Matter in the Universe.* Reidel, Dordrecht 1986
8. J. Kormendy, "Dark Matter in Dwarf Galaxies", in J. Kormendy and G. R. Knapp, eds. *Dark Matter in the Universe.* Reidel, Dordrecht 1986

Session 3

Galactic Structure and Dynamics: Observations

J. van Gorkum, Chair
July 23, 1986

Photometry and Mass Modeling of Spiral Galaxies

S. KENT

1. Introduction

Observations of extended rotation curves in spiral galaxies show that the cumulative mass distribution in a galaxy increases roughly linearly with radius, unlike the light distribution which rises much more slowly [1]. This discrepancy is taken as evidence for a dark, extended halo distinct from the visible stellar disk. The question then arises as to what are the relative contributions of the luminous and dark components to the total mass of a galaxy and how are their relative distributions related to galaxy formation. This paper examines several aspects of galaxy photometry and how photometric and kinematic data can be combined to model the mass distribution in spiral galaxies.

2. Photometric accuracy

The evidence for dark matter relies upon joint analyses of photometric and kinematic observations of galaxies. In principle, the measurement of the luminosity profile in a galaxy is quite straightforward: one simply takes a 2-dimensional image of a galaxy and computes the average intensity, either from a single cut along the major axis or by averaging over some appropriately oriented elliptical contours. Aside from the nuisance caused by non-axisymmetric structure (such as spiral arms), there is the problem that often one is trying to measure light intensities which are only a few percent of the night sky. Small errors in the night sky correction can make large errors in the derived properties of the dark halo. Consequently, independent checks on photometry are useful.

Figure 1 shows the luminosity profile for a galaxy with one of the best HI rotation curves, NGC 3198 [2]. Two observed profiles are drawn, one from photographic photometry by WEVERS [3] and the other from CCD photometry by KENT [4]. Also plotted is the luminosity profile which

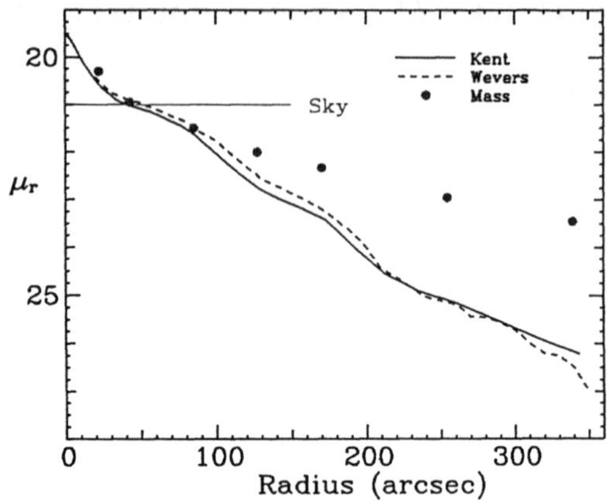

Figure 1: Observed luminosity profile of NGC 3198 compared with mass density required to produce the observed rotation curve.

would be required to produce the observed rotation curve assuming a constant mass/light ratio [5]. The two observed profiles agree reasonably well (within 0.3 mag) and are clearly incompatible with the constant M/L ratio curve. The point to be made is that reasonably accurate photometry at few percent of the sky background is quite feasible and that the evidence for dark matter in spiral galaxies is not an artifact of poor photometry.

3. Three-dimensional Structure of Spiral Galaxies and Bulge/Disk Decompositions

Normally one idealizes the structure of a spiral galaxy as the sum of an infinitely thin disk plus a spheroidal bulge. In practice, disks have some finite thickness and the distinction between bulge and disk, especially in late-type galaxies, is not alway obvious. For mass modeling, it would be more desirable to have the 3-dimensional luminosity profile and not worry about bulge or disk components. One would like to take the 2-dimensional image of a galaxy and, under some assumption such as axisymmetry, reconstruct the full 3-dimensional structure. Unfortunately, RYBICKI [6] has shown that except for galaxies viewed edge-on, this is not possible. To demonstrate this fact, consider the 3-dimensional Fourier transform of the density profile of a galaxy. If one views the galaxy in projection, then Rybicki shows that the 2-dimensional Fourier transform of the projected image corresponds to a planar slice through the 3-dimensional Fourier space. If the galaxy is axisymmetric, then so is its transform, so one can rotate this plane about the symmetry axis in Fourier space to recover the 3-dimensional transform. If the galaxy is viewed edge-on, the symmetry axis lies in this plane and the full 3-dimensional transform is recovered. If the galaxy is viewed at some other inclination, there is a cone about the symmetry axis which is not filled in, the size of the cone increasing as the viewing angle approaches face-on. Consequently, except for systems viewed edge-on, one cannot recover the full 3-dimensional structure of a galaxy even in principle. Of course, for edge-on systems internal extinction is a problem.

For this reason one is led to characterize spiral galaxies in terms of bulge and disk components. Typically, the decomposition into components is accomplished by taking the major axis luminosity profile (or equivalent), adopting a priori forms for the bulge disk luminosity profiles, and fitting them to the observed profile by adjusting various scale parameters. Usually one takes a de Vaucouleurs law for the bulge and an exponential law for the disk:

Bulge: $\log I = \log (I_e) - 3.33 [(r/r_e)^{1/4} - 1]$

Disk: $\log I = \log (I_0) - 1.087 [r/h]$.

If the bulge or disk components of a galaxy deviate significantly from these forms, then the decomposition can become meaningless. In fact, deviations in both components are seen often. FREEMAN [7] already noted that disks display non-exponential behavior - his type 2 disks (of which M31 is an example) have profiles that are fainter in the center compared with an exponential fit to the outer disk profile. The opposite behavior is also seen: in M33, the luminosity profile shows an excess of light in the center characteristic of a bulge, yet inspection of the optical image shows that aside from a semi-stellar nucleus, this galaxy essentially has no bulge (Fig. 2a). BOROSON [8] and KENT [9] likewise found non-exponential behavior quite common. Significant variations in the shapes of bulge profiles are also seen. Figure 2b show the minor-axis profiles for three high inclination galaxies with prominent bulge components. Only in NGC 4594 does the bulge follow a de Vaucouleurs law. FRANKSTON and SCHILD [10] showed that the bulge in NGC 4565 is actually exponential.

For these reasons, it is not possible to arrive at a sensible bulge/disk decomposition for many galaxies. However, some decomposition is needed for mass modeling. KENT [11] has developed an alternative

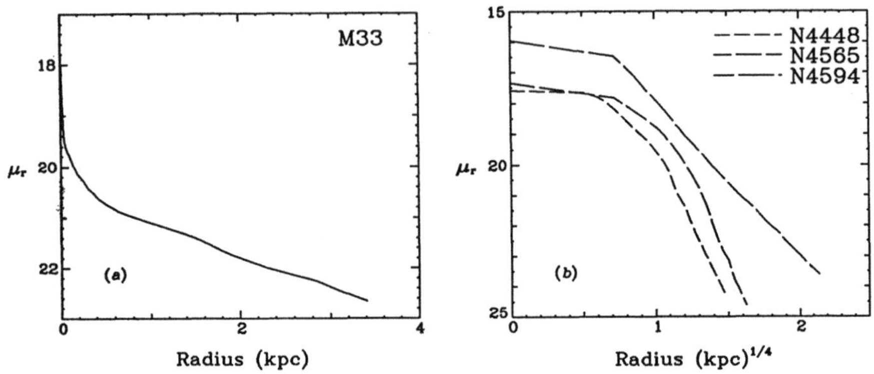

Figure 2: (a) Luminosity profile of M33 demonstrating non-exponential behavior of disk. (b) Minor axis profiles of three galaxies showing differences in bulge profile shape.

decomposition scheme for highly inclined galaxies that does not require a priori knowledge of the luminosity profile shapes. One only assumes that the bulge and disk individually have projected images in which the isophotes have constant flattening but which can be different for the two components. In descriptive terms, one takes a minor-axis cut of a galaxy to provide a first estimate of the bulge profile and a major-axis cut to be the disk profile. The bulge profile is scaled in radius by the bulge flattening and subtracted from the major-axis cut to arrive at an improved disk profile. This disk profile is then scaled in radius by the disk flattening and subtracted from the minor-axis cut to arrive at a new bulge profile. The process is repeated until convergence is achieved. Figure 3 shows the results of applying this procedure to NGC 4448. This decomposition procedure has been used in the work described in the next section.

4. Mass Modeling with Rotation Curves

The number of galaxies with both photometry and rotation curve data is increasing steadily [3,11,12,13,14,15]. With these data it is possible to construct mass models for a galaxy in which one can try to account explicitly for the relative distribution of luminous and dark matter. There are two approaches one can use: invert the rotation curve to get the expected mass distribution, or use the photometry to compute the expected rotation curve (say, assuming a constant stellar M/L ratio.) In

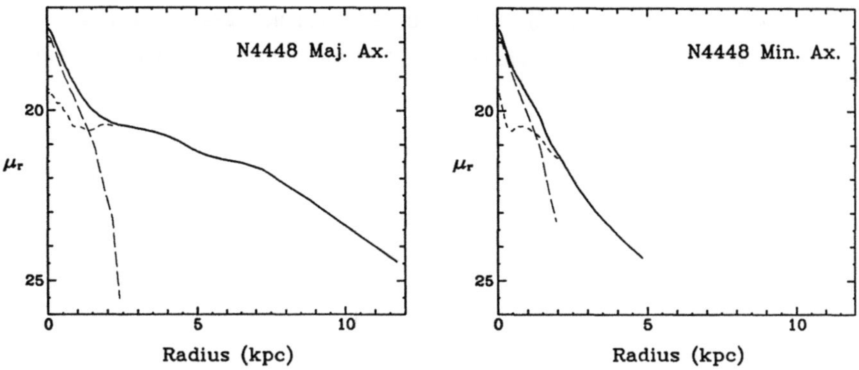

Figure 3: Non-parametric bulge/disk decomposition of NGC 4448.

either case, a shortfall in either the computed mass distribution or rotation curve is attributed to a dark halo. The second method is generally preferred since the photometry is usually better determined than the rotation curve. The problem one has is that neither the stellar M/L ratio nor the dark matter distribution is known well a priori, and to proceed one must make some further assumptions. Three approaches which have been used include the following:

(a) In the maximum-disk method, an upper limit to the disk M/L ratio is estimated by fitting the inner part of a rotation curve with a constant M/L ratio disk (and bulge if present) and ignoring the halo initially. Usually the inner parts of rotation curves can be fit well out to a radius where the disk contribution turns over. The additional mass needed to reproduce the outer part of a rotation curve provides a lower limit to the halo mass. If interpreted formally, such a model produces a halo with a hollow center. One can then lower the disk M/L ratio to produce a halo with a more reasonable density profile [2].

(b) One can assume a form for the dark matter distribution such as an isothermal sphere. By modeling the rotation curve as the sum of contributions from a bulge, disk, and halo, one can make a simultaneous solution for the bulge and disk M/L ratios and two scale parameters for the halo [15].

(c) If disks are stable against various perturbations (e.g. bar instabilities) there is a minimum halo mass needed to provide stability. In this way one can constrain the stellar M/L ratio between certain limits [16].

Figure 4 shows examples of four galaxies with photometry and rotation curves and with models fit to each. NGC 801 and NGC 2998 have only optical rotation curves [12] while UGC 2259 and NGC 2841 have HI rotation

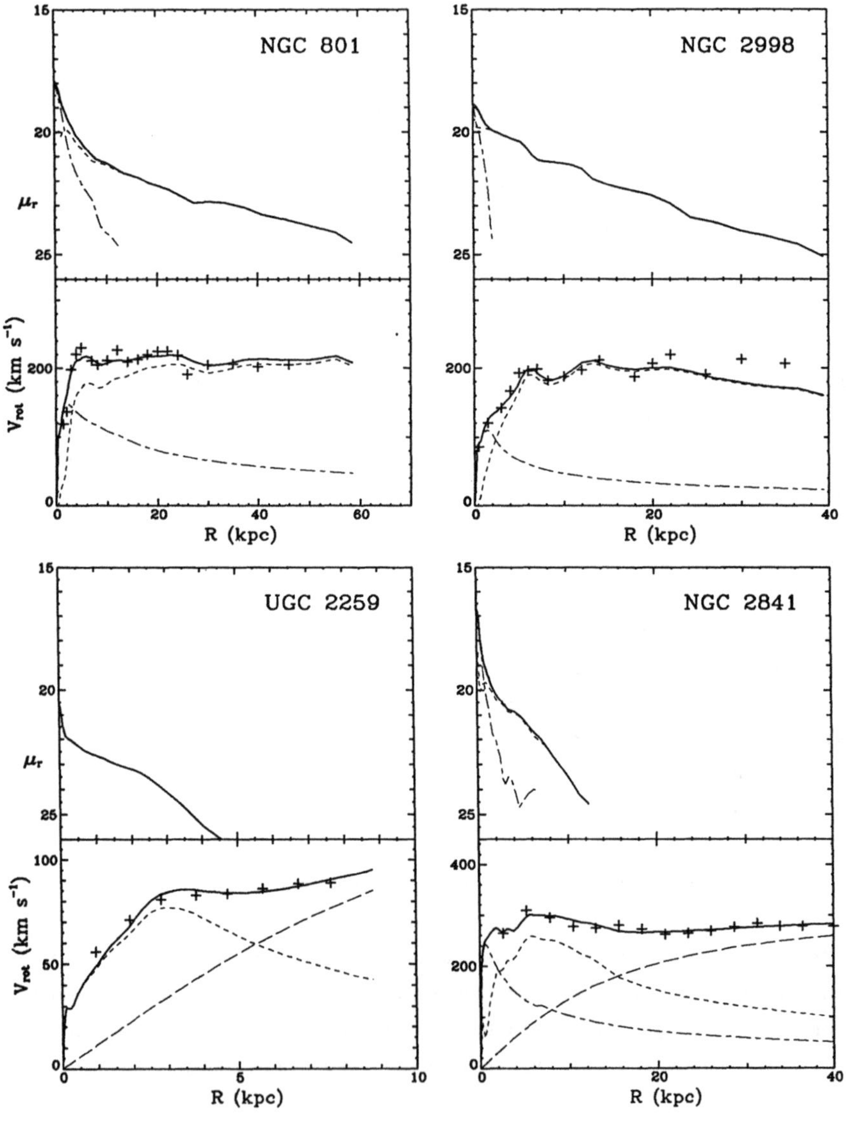

Figure 4: Luminosity profiles and rotation curves for four galaxies. In both panels, the individual bulge and disk contributions are shown. In the bottom panels, the crosses are the observed points and the solid curve is the model fit.

curves [13, 17]. The first two galaxies have been fit with maximum-disk models while the other two are fit with a disk plus halo model. If a bulge is present, separate M/L ratios have been determined for the bulge and disk. These galaxies illustrate several points:

(a) Compared with the extent of the photometry, optical rotation curves do not extend as far as HI rotation curves and seldom provide much information on the halo component. In a study of 37 galaxies with optical rotation curves, KENT [9] found about 1/3 could be fit with models containing no halo.

(b) Maximum-disk solutions usually (though not always) provide a good fit to the inner parts of rotation curves. Evidence for halos is indicated only in the outer parts. HI rotation curves are needed to delineate the halo contribution properly.

(c) The peak velocity from the maximum-disk solution is usually close to the asymptotic rotation velocity produced by the dark halo. If the true disk M/L ratio is close to this maximum value, then one is left with a "disk-halo" conspiracy. Low-luminosity galaxies may be an exception to this behavior.

(d) Dark halos do not dominate at any single characteristic radius: UGC 2259 shows evidence for a halo inside 5 kpc while NGC 801 shows no evidence for a halo out to 40 kpc. While it has been speculated that the ratio of dark to luminous matter inside a Holmberg radius is of order unity in all spiral galaxies, there is not yet enough data to support or refute this conjecture.

The disk M/L ratios, corrected for internal extinction, range from 1.6 to 3.2 (red bandpass).

The question of whether the true disk M/L ratios are close to their maximum-disk values is still an open question. The relatively good fit one gets with the maximum-disk solution suggests that stellar matter does dominate the inner parts of galaxies. BURSTEIN and RUBIN [18] have grouped the rotation curves of 60 galaxies into three mass types based on their shapes and find little correlation between shape and morphological type. On this basis they suggest that the mass and light distribution are not well correlated and that dark and luminous matter are well-mixed at all radii. However, if one tries to quantify the shapes of velocity and luminosity profiles, then a correlation between them is found. Figure 5 shows the relation between a velocity concentration index $C_V = 5\log[r_{25}/r(v_{25}/2)]$ and a luminosity concentration index

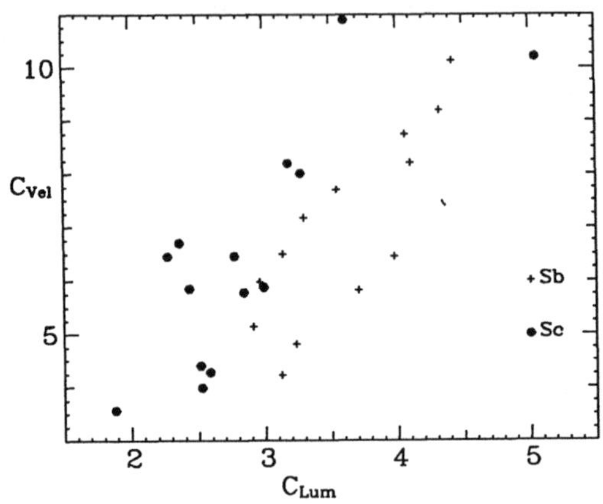

Figure 5: Correlation between luminosity and velocity concentration indices.

C_L=5log[r(80%)/r(20%)] for a set of galaxies studied by KENT [9]. Here, r_{25} is the de Vaucouleurs radius, $r(v_{25}/2)$ is the radius at which the rotation curve has risen to 1/2 the velocity at r_{25}, and r(80%) and r(20%) are the radii containing the specified fraction of the total light. The essential result from Fig. 5 is that galaxies with a low luminosity concentration have slowly rising rotation curves while the converse is true for high concentration systems. It is also significant that although Sb's and Sc's overlap in both luminosity and velocity concentration, they are offset from one another in the diagram.

An independent estimate of the mass density in disks can be made by studying the vertical structure of disks. If a population of stars has an exponential scale height h and velocity dispersion σ, then the surface density is given by $\mu=\sigma^2/2\pi Gh$. VAN DER KRUIT and SEARLE [19] showed that h is about 300-400 pc from a study of edge-on spirals, independent of position within a galaxy. By measuring the velocity dispersion in face-on systems, VAN DER KRUIT and FREEMAN [20] found the total mass in spiral disks is about 1/3 the total mass out to the optical edge of the disks. Although not all the relevant physical parameters can be measured for any single galaxy, this method of measuring disk masses has the advantage that it does not suffer from the ambiguities inherent in fitting mass models to rotation curves.

REFERENCES:

1. S. M. Faber, J. S. Gallagher: Ann. Rev. Astron. Astrophys. 17, 135 (1979)
2. T. S. van Albada, J. N. Bahcall, K. Begeman, R. Sancisi: Astrophys. J. 295, 305 (1985)
3. B. M. H. R. Wevers: Ph. D. thesis, Groningen University (1984)
4. S. M. Kent: In preparation (1986)
5. J. Binney: Preprint (1985)
6. G. Rybicki: Preprint (1986)
7. K. Freeman: Astrophys. J. 160, 811 (1970)
8. T. Boroson: Astrophys. J. Suppl. 46, 177 (1981)
9. S. M. Kent: Astrophys. J. Suppl. 59, 115 (1985)
10. M. Frankston, R. Schild: Astron. J. 81, 500 (1976)
11. S. M. Kent: Astron. J. 91, 1301 (1986)
12. V. C. Rubin, D. Burstein, W. K. Ford, N. Thonnard: Astrophys. J. 289, 81 (1985)
13. A. Bosma: Astron. J. 86, 1791 (1981)
14. A. Bosma, P. C. van der Kruit: Astron. Astrophys. 79, 281 (1979)
15. C. Carignan, K. C. Freeman: Astrophys. J. 294, 494 (1985)
16. E. Athanassoula, A. Bosma: In Nearly Normal Galaxies: from the Planck Time to the Present, ed. by S. Faber (Springer-Verlag, New York, 1986)
17. C. Carignan, R. Sancisi, T. van Albada: In Dark Matter in the Universe, I. A. U. Symposium No. 117 (1985)
18. D. Burstein, V. C. Rubin: Astrophys. J. 297, 423 (1985)
19. P. C. van der Kruit, L. Searle: Astron. Astrophys. 95, 105 (1981)
20. P. C. van der Kruit, K. C. Freeman: Astrophys. J. 303, 556 (1986)

The Stellar Population at the Galactic Center

MARCIA J. RIEKE

I. Introduction

The center of our own galaxy may be the site of very recent star formation [1], and hence its study is of considerable interest in relation to star formation in other galactic nuclei but also as an example of star formation in a bulge environment. The star formation process in the deep gravitational potential well at the center of a galaxy may well differ from that process in the less extreme circumstances found in a disk.

Because of the complicating factor of 30 magnitudes of extinction at visible wavelengths, the study of the Galactic Center has been done at infrared and radio wavelengths. The availability of area-format detectors in the infrared recently has permitted surveying areas of the Galactic Center that are comparable in size to those studied in other galaxies. Preliminary results from a 2um survey covering an area of 400 pc^2 (5'x5') centered on the Galactic Center are presented here. That a large amount of star formation has occurred recently is confirmed.

II. Observations

Maps at K(2.2μm) and H(1.6μm) were obtained using the Steward Observatory 1.54-meter telescope and infrared camera using a 32x32 HgCdTe hybrid CCD fabricated by Rockwell International. This system [2] was equipped with relay optics yielding 1."2 per pixel. The camera was mounted on an offset guider which was used to position each frame accurately with respect to neighboring frames. The entire area mapped required 9x9 frames of 32x32 pixels each. Note that a block of four frames from the southeastern corner are missing because of the lack of an offset guide star.

The integration times were 30 seconds per frame at K and 60 seconds at H. This yielded 5-sigma detections at K=13.5 and at H=13.8. The K map is shown in Figure 1 where individual frames were flat-fielded and contoured and then combined into a large map by manually positioning the frames. The contour intervals were chosen to permit faint structure to be seen, and consequently, sources brighter than K=10 appear as "doughnuts".

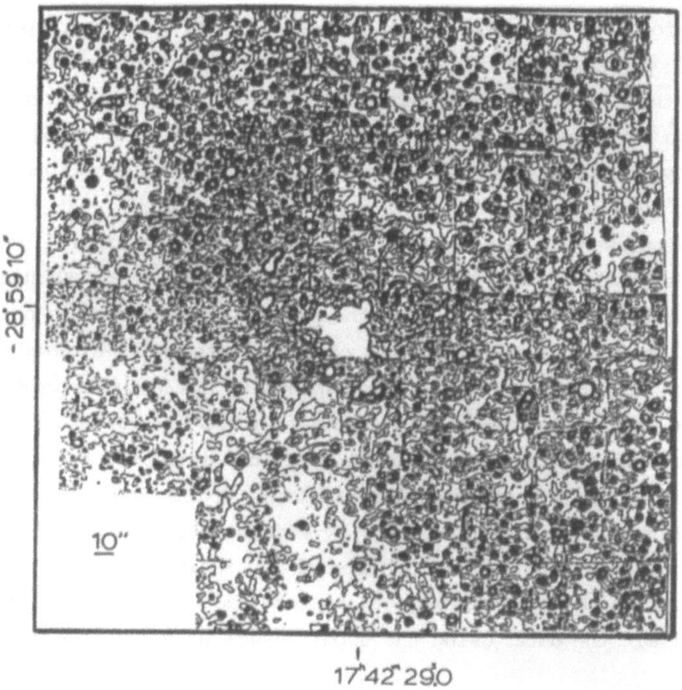

Figure 1: A region 5'x5' at the Galactic Center imaged at 2.2um with 1".2 pixels. The contours are linear in steps of 2.7 mJy.

A preliminary list of objects detected at K was derived by selecting all sources by hand to yield a list complete to K=12. The difference between this limit and the 5-sigma detection limit may be the result of crowding. These sources were processed by a photometry program which yielded fluxes in a 1".5 round aperture and positions. The list of positions was used to center an aperture on the H-frames, and 1.6-μm fluxes were derived. Before fluxes were derived from the H-frames, the centering of these frames relative to the K-frames was checked and small (<0".7) shifts were applied as necessary. Because of extreme crowding, the central frame was not processed and is not included in the statistics which follow.

III. Data

The influence of heavy extinction along the line-of-sight must be understood before source counts and statistics can be discussed. An average value of about 30 magnitudes for A_V to the Center itself has been accepted for some time (eg. [3,4]). The extinction at K corresponding to A_V=30 is 3 magnitudes and at H is 5 magnitudes [5] so the K map is more complete than the H map. The existence of areas near the Center where the extinction may be significantly larger has also been known [6], but is frequently ignored. These regions correspond to molecular clouds lying along the line-of-sight.

Figure 2 presents a histogram of the raw H-K colors for all objects detected at both H and K. The unobscured colors for A0 to M9 stars are indicated, and it is obvious that the objects in this map are heavily reddened. The average H-K color for objects in the well-studied region at the Center is H-K=2.3 [1] which corresponds to A_V=30 and a dereddened H-K=0.4. The average value for all sources in Fig. 2 is H-K=2.2 which for an average dereddened $H-K_o$=0.3 yields A_V=30. This intrinsic H-K will be used in the following discussion because it is the average H-K for M giants in Baade's Window [7,8]. The most heavily reddened sources detected at both wavelengths have A=60 for this choice of intrinsic color. This is only a lower limit on the highest extinction since the relative sensitivities at H and K limit the data. Using the upper limits at H for some objects

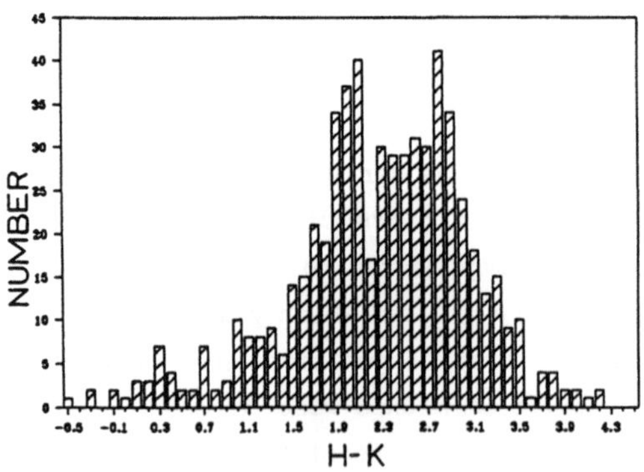

Figure 2: The distribution of observed H-K colors for Galactic Center sources.

detected at K only yields A_V>75 magnitudes. Note that the infrared extinction law has been derived using objects with large amounts of extinction and including objects lying along nearly this same line-of-sight [5] so that even though the differential extinction between H and K is small, the uncertainty in A_V is mainly from the uncertainty in the intrinsic color rather than uncertainties in the extinction law. The entire range of H-K color for stars is -0.1 to 0.5 which leads to an uncertainty of about ± 4 magnitudes in A_V or ± 0.4 magnitudes in dereddened K magnitude.

That the extinction is non-uniform is apparant from Fig. 1. The large region east and south of the Center with a much lower source density is highly obscured as is a region starting about 45" northwest of the Center and extending to the west. Other smaller areas with reduced source counts may also be regions of extinction higher than the average A_V of 30.

IV. Discussion

Figure 3 shows the distribution of dereddened magnitudes, K_0. Objects must have also been detected at H to be included in this figure; a total of 603 sources were detected at both wavelengths. The first questions to be answered are whether this concentration of objects is unusual and what fraction of them actually lie at the 7 Kpc distance of the Galactic Center and what fraction of them are foreground objects. These issues are central to whether or not these objects are likely to have been formed as part of a starburst at the Galactic Center.

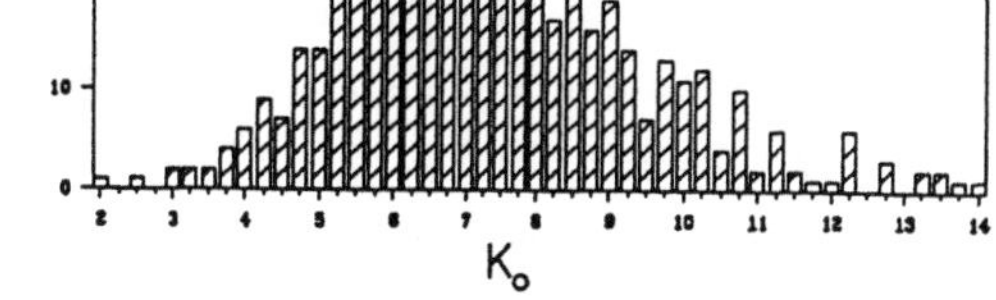

Figure 3: The distribution of dereddened K magnitudes.

If these stars are predominately members of a starburst, they are probably members of the same burst as those sources in the central region which have been studied spectroscopically and have been confirmed as being young (of order 20 million years) and as having massive ($>10M_\odot$) progenitors. These sources are shown as solid dots in Fig. 1. This group of stars has a brightest bolometric magnitude of -8 and K_o of about 3.

If these stars are older objects, they may be members of the asymptotic giant branch and have characteristics similar to the population seen in Baade's Window. A comparison of the Galactic Center population with the Baade's Window population can be made using the luminosity function for Baade's Window [7,8]. Bolometric magnitudes for Galactic Center stars have been estimated crudely in the following manner. A relationship between K_o and M_{BOL} was derived from the data in [7,8] assuming a distance of 7 Kpc for the Galactic Center. This relationship was then applied to K_o from Fig. 3. The resulting distribution of bolometric magnitudes is shown in Figure 4 which also shows the distribution for Baade's Window stars normalized to the same number of objects brighter than an observed K magnitude of 11 where each sample is complete.

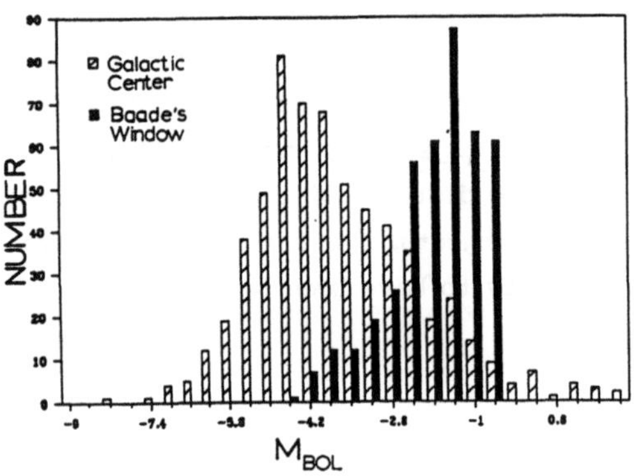

Figure 4: Distribution of bolometric magnitudes in Baade's window and the Galactic Center.

The Galactic Center objects appear to represent a stellar population 3 magnitudes brighter than that seen in Baade's Window. This result would be expected if there has been recent star formation at the Galactic Center. The other possibility is that this group of stars are foreground objects of probably the same population as seen in Baade's Window. If so, these objects must lie 4 times closer to compensate for their greater brightness. To estimate whether this is probable, the surface brightness at 2.4um from ODA et al. [9] for a Galactic longitude corresponding to 5.4 Kpc from the center (38°) is converted to an equivalent number of stars per square degree. This number of stars is computed using the luminosity function from Baade's Window and hence is an overestimate of the number of stars present brighter than the completeness limit for this sample. The number of stars needed is less than 6000/sq. deg brighter than K_o=7.5. The areal density in the Galactic Center map corresponds to 57000/sq. deg. brighter than K_o=7.5 which implies that these stars cannot be interpreted as just foreground bulge stars; a genuine excess is present. Also note that ODA et al. [9] show that the extinction peaks strongly at the Center, and A_V=30 is very unlikely for a region at 38° longitude.

V. Conclusions

New mapping of a large region at the Galactic Center has revealed a population of bright, heavily reddened stars. These objects have an average A_V=30, similar to the well-studied sources right at the Center. Their average bolometric magnitude is -4.6 if they lie at the 7 Kpc distance of the Galactic Center, and implies recent star formation. Interpreting them as a foreground population is not consistent with the observed surface brightness profile of the Galactic plane.

The assistance of G. H. Rieke and E. F. Montgomery in obtaining and analyzing these data is greatly appreciated. This work has been supported by the National Science Foundation.

References:

1. M. J. Lebofsky, G. H. Rieke, A. T. Tokunaga: Ap. J. 263, 736 (1982)
2. M. J. Lebofsky, E. F. Montgomery, W. F. Kailey: in Second Infrared Detector Technology Workshop, ed. C. R. McCreight, (NASA Ames, Moffett Field 1985)
3. E. E. Becklin, K. Matthews, G. Neugebauer, S. P. Willner: Ap. J. 220, 831 (1978)
4. G. H. Rieke, C. M. Telesco, D. A. Harper: Ap. J. 220, 556 (1978)
5. G. H. Rieke, M. J. Lebofsky: Ap. J. 288, 618 (1985)
6. M. J. Lebofsky: A. J. 84, 324 (1979)
7. J. A. Frogel, this conference.
8. J. A. Frogel, A. E. Whitford: Ap. J. submitted
9. N. Oda, T. Maihara, T. Sugiyama, H. Okuda: A. and. A. 72, 309 (1979)

Hot Gas Evolution in Nearly Normal Elliptical Galaxies

MICHAEL LOEWENSTEIN AND WILLIAM G. MATHEWS

The interstellar medium has long been a familiar feature of nearly normal spiral galaxies while elliptical galaxies, until recently, had been regarded as largely gas free – neither optical nor radio observations indicated significant gas within ellipticals. This mistaken impression has now been corrected by X-ray observations which reveal the presence of $5 \times 10^8 - 5 \times 10^{10}$ $M_\odot$ of hot ($T \sim 10^7$) K gas distributed throughout and beyond the optically bright parts of elliptical galaxies (FORMAN *et al.* [1], BIERMANN and KRONBERG [2]; NULSEN, STEWART, and FABIAN [3]; FORMAN, JONES, and TUCKER [4]; TRINCHIERI and FABBIANO [5]; CANIZARES, FABBIANO, and TRINCHIERI [6]). This amount of gas is comparable to or greater than that in typical gas-rich spiral galaxies and is consistent with the amount of gas expected to have been ejected from stars since the galaxies formed. For example, the current mass loss rate resulting from stellar evolution in an old population of stars, $\alpha_* = (dM_*/dt)/M_* \approx 2.2 \times 10^{-20}$ s^{-1}, when combined with the Hubble time, indicates that a fraction $\alpha_* t_H \approx 0.01$ of the stellar mass has been converted to gas; this is comparable to but somewhat less than the observed fraction.

The X-ray luminosities of ellipticals span a range $L_x \approx 10^{39} - 10^{42}$ ergs s^{-1} and correlate with optical luminosities $L_x \propto L_B^{1.64}$, although the scatter around this correlation is large and observationally real. Ellipticals having low optical luminosities or unusually small X-ray thermal emission from the hot gas can still exhibit a significant X-ray flux from X-ray binaries. Nevertheless, it is generally agreed that most of the emission from ellipticals having moderate to large L_x is thermal emission from hot gas (see for example [4, 6]). The distribution of X-ray surface brightness Σ_x, resolved in several elliptical members of the Virgo cluster, is found to be proportional to the stellar surface brightness $\Sigma_* \propto R^{-2}$ ([7]), but this may not be a universal property of all ellipticals. Ellipticals embedded in high pressure cluster gas, such as the three galaxies in Virgo, have X-ray luminosities that do not differ substantially from those of more isolated ellipticals.

We do not attempt here a comprehensive review of the theoretical implications of these X-ray observations nor do we wish to provide detailed models of the X-ray emission from any particular elliptical galaxy. Instead we are interested in the evolution of the hot gas component of these galaxies over the age of the universe. Specifically, we wish to illustrate how the observed luminosities L_x from isolated ellipticals not immersed in cluster gas can be understood as a result of stellar mass loss. In accomplishing this, we also demonstrate that massive dark halos are essential for retention of this gas and that the heating rate of the gas by type 1 supernovae must not be excessive [$\lesssim 0.0005$ type 1 SN yr^{-1} $(10^{10}\ L_{\odot B})^{-1}$]. When our model galaxies are evolved to the present time, we find general agreement with the $L_x \propto L_B^{1.6}$ correlation and approximate agreement with the $\Sigma_x \propto \Sigma_*$ correlation (SARAZIN [8, 9]). Neither large amounts of residual gas left over from galaxy formation nor continued gaseous infall from the intergalactic medium are required to understand the observed X-ray fluxes. After accumulating gas within the galactic potential over a Hubble time, the gas flows at present are very slow, usually (but not always) moving inward as the gas cools at the center into optically emitting clouds at $T \sim 10^4$ K. Except in this interesting central region, the gas flow is very subsonic and can be approximated by hydrostatic models. It now appears that in most cases strong galactic winds are not possible in the presence of massive halos; winds produce steep X-ray surface brightness profiles ($\Sigma_x \propto R^{-3.6}$) unlike those observed.

Before reporting the results and interpretation of our gas dynamical calculations, we discuss the equations used. The computed models of the evolution of hot gas described here are an extention of the earlier results of MATHEWS and LOEWENSTEIN [10].

I. Equations

The equations for galactic flows are the equation of continuity

$$\frac{\partial \rho}{\partial t} + \frac{1}{r^2}\frac{\partial (r^2 \rho u)}{\partial r} = \alpha \rho_*, \tag{1}$$

equation of motion

$$\rho\left(\frac{\partial u}{\partial t} + u\frac{\partial u}{\partial r}\right) = -\frac{\partial P}{\partial r} - \rho\frac{\partial(\phi_* + \phi_h)}{\partial r} - \alpha\rho_* u, \tag{2}$$

the equation for thermal energy

$$\rho\frac{d\varepsilon}{dt} - \frac{P}{\rho}\frac{d\rho}{dt} = -\frac{\rho^2\Lambda}{M^2} + \alpha\rho_*\left[\frac{3kT_o}{M} - \frac{5P}{2\rho} + \frac{u^2}{2}\right], \tag{3}$$

and the equation of state $\varepsilon = 3P/2\rho$. These are the same equations derived by MATHEWS and BAKER [11] from the Boltzmann equation. Here $\alpha = \alpha_* + \alpha_{sn}$ is the sum of the specific mass loss rates from stars and supernova although $\alpha \approx \alpha_*$ in practice. Both α_* and α_{sn} are time-dependent. The gravitational potential in the equation of motion is a superposition of potentials for a King-type stellar mass distribution $\rho_* = \rho_{*o}/(1 + (r/r_c)^2)^{3/2}$ and a quasi-isothermal mass distribution for the dark halo $\rho_h = \rho_{ho}/(1 + (r/r_{ch})^2)$, where subscripts o and c refer to central and core values. Both distributions vanish beyond r_t. We ignore the self-gravity of the hot gas and assume that the potentials ϕ_* and ϕ_h do not change during the calculation.

The temperature $T_o(r,t) = (\alpha_* T_*(r) + \alpha_{sn} T_{sn})/\alpha$ that appears in the energy equation is the characteristic temperature of new gas introduced into the galactic flow. The term $\alpha_* T_*/\alpha$ represents the heat given to the gas by the relative motion of the galactic stars which have a local temperature given by $6kT_*(r)/M \approx \phi(r)$; $T_*(r)$ has been computed by solving the equation of stellar hydrodynamics. Optically thin radiative cooling is described by the cooling rate coefficient $\Lambda(T)$ [12].

We have determined the time variation of α_* and α_{sn} by adopting the model for stellar evolution of an old stellar population developed by RENZINI and BUZZONI [13] using a Salpeter initial mass distribution (see also MACDONALD and BAILEY [4]). The evolution of type 1 supernovae, $\alpha_{sn}(t)$, involves some assumptions about the frequencies of binaries and their initial separations as a function of stellar mass (IBEN and TUTUKOV [15]) and can be normalized to the (somewhat uncertain) current value. We consider two possible current values of α_{sn} corresponding to supernova rates $\nu_{sn1} = 1/500$ and $\nu_{sn2} = 1/2000$ in units of supernovae yr^{-1} $(10^{10} L_\odot)^{-1}$. These are similar to rates determined by TAMMANN [16]. Both $\alpha_{sn}(t)$ and $T_o(r,t)$ are reasonably constant with time, while $\alpha_*(t)$ decreases as $\sim t^{1.1}$.

We are not concerned here with the earliest stages of galactic evolution lasting up to times comparable to the gravitational free fall time $t \lesssim 10^8$ yrs and during which stars of mass $M \gtrsim 9 M_\odot$ evolve off the main sequence. This is a period of great uncertainty which may involve numerous supernovae resulting from massive star evolution and a time-dependent galactic potential. Our approach here is to ignore these complications and to begin the solution of equations (1) - (3) at $t \approx 10^8$ yrs with the assumption that the interstellar gas content at that time is negligible and that the stellar component of the galaxy has completely formed.

We consider two model galaxies A and B with properties given in Table 1. Model B is slightly smaller, denser, and less massive. The dark halos in both galaxies are constrained to contain 90 percent of the total mass but are dominated by the stellar potential at their centers, $\gamma = (\rho_{*o}/\rho_{ho}) > 1$ (BURSTEIN and RUBIN [17]; BLUMENTHAL *et al.* [18]). The model galaxies are designed so that the characteristic gas temperature $0.6 T_o$ (see [11]) does not exceed twice the equivalent temperature of the isothermal dark halo, i.e. $\gamma\beta^2 \gtrsim 0.2$, where $\beta = r_{ch}/r_{c*}$. This satisfies an approximate global virial condition for binding the hot gas to the total galactic potential.

Table 1

Properties of the Model Galaxies

model	r_{c*} (*kpc*)	ρ_{0*} (10^{-21} $gm\ cm^{-3}$)	M_{t*} (10^{11} $M_\odot$)	L_B (10^{44} $ergs\ s^{-1}$)	r_{ch} (*kpc*)	ρ_{0h} (10^{-23} $gm\ cm^{-3}$)	r_t (*kpc*)	M_t (10^{12} $M_\odot$)	$\gamma\beta^2$
A:	0.50	7.20	8.29	5.39	5.0	1.75	100.0	8.31	0.243
B:	0.25	21.8	3.14	2.04	2.5	5.30	50.0	3.14	0.243

II. Four Evolutionary Models

We are interested in the influence of the massive halos and supernova rates on the evolution of hot gas in ellipticals and how galaxies of different mass respond to these uncertain parameters. To this end we compare solutions of a grid of four cases, A1, A2, B1, and B2, where the letter denotes the galactic model (as in Table 1) and the number, the assumed current supernova rate, ν_{sn1} or ν_{sn2}. Equations (1) - (3) are solved using an explicit Lagrangian numerical procedure. The spatial zones slowly enlarge outwards to accommodate both the central cooling region ($\sim$ 100 pc) and the distant parts of the flow affected by the massive halo ($\sim$ 100 kpc). The innermost zone is dropped from the calculation when the strong central cooling has proceeded to $\sim 10^4$ K and the gas is in supersonic freefall; for this reason, occasional rezoning is required. Approximately $\sim 10^5$ time steps are necessary to compute the flow from the epoch of galaxy formation to their present age taken to be $t_H = 15 \times 10^9$ yrs.

We adopt model A2 as a standard reference and the results of its evolution are shown in Figs. 1-3. The first of these figures illustrates the glacially subsonic inward flow of the gas toward the rapidly cooling region near the center. Apart from the residual outward travelling wave at $r \sim 10^5$ pc at the earliest time shown in Fig. 2, there is very little hydrodynamic activity in the hot gas.

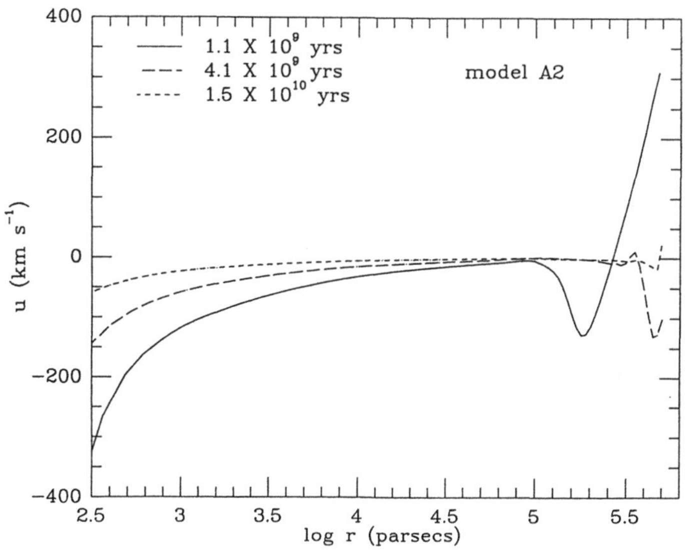

Figure 1: Velocity vs. radius at three times for model A2 illustrating the slowly weakening inflow resulting from a time-increasing energy source term.

The secular deceleration of the inflow is brought about by the increased heating by supernovae ($d(\alpha_{sn}/\alpha)/dt > 0$). Figures 2 and 3 show that the density and temperature vary little during the entire evolution and that the temperature is remarkably isothermal throughout the galaxy. However, for $r > r_t$, ρ_* vanishes and the source terms in equations (1) - (3) no longer apply; this accounts for the more nearly adiabatic flow in Fig. 3 for $r > r_t$. The slow decrease in gas density in the galactic flow shown in Fig. 2 is responsible for a decrease in the X-ray luminosity by over an order of magnitude as the galaxy evolves.

During the evolution of the hot interstellar medium, the total mass of hot gas typically increases with time, but its growth is mediated by an even larger mass of gas which cools into the galactic core. For example, the ratio of the mass of cooled gas to that of the residual hot gas after 15×10^9 yrs is 1.5, 5.0, 1.6, and 5.3 respectively for models A1, A2, B1, and B2. The stagnation radius, $r_{stag}(t)$, which separates the central inflow from the outflow beyond, becomes very large ($\sim$ 500 kpc) for model A2, but steadily decreases (to $\sim$ 7 kpc) in model A1 where most of the gas eventually flows outward, although very subsonically. Over a Hubble time, the X-ray luminosity in the 0.5 - 4.5 keV band, L_x, usually decreases by factors of 20 - 40, a luminosity evolution which may have observational consequences. After $t = 5 \times 10^9$ yrs, models A1, A2, and B2 inhabit the region in the (L_x, L_B)-plane occupied by the observed galaxies indicating that the assumptions we have used are likely to be reasonable. This is illustrated in Fig. 4.

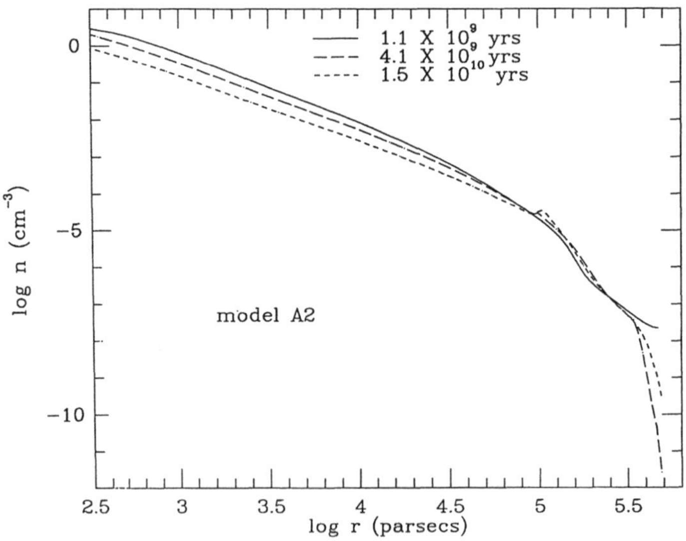

Figure 2: Number density vs. radius at three times for model A2 illustrating the slowly decreasing dentral density resulting from a time-decreasing mass source term. Note that the density distribution becomes less steep with time.

While models A1, A2, and B2 behave in a similar manner, the development of model B1 becomes remarkably aberrant at later times. For the first eight billion years after the calculation was begun, the evolution of B1 and B2 are similar. But the higher supernova rate in model B1 eventually raises the central temperature sufficiently to reverse the direction of the flow. Figure 5 shows that in its subsequent evolution model B1 develops a wind which begins at the center and propagates outwards. The wind decreases the gas density near the center and consequently the final soft X-ray luminosity at $t = t_H$ is only $L_x = 3.4 \times 10^{37}$ ergs s^{-1}, four orders of magnitude less than that of B2 for which the supernova rate is only four times lower.

We were pleased when model B1 turned into a wind since this illustrates the importance of allowing for non-steady gas flow throughout the calculation and pointed to meaningful astronomical restrictions on the assumed parameters. The final thermal emission of the gas in model B1 is so low that its X-ray flux and spectrum would be dominated by X-ray binaries not the hot gas. While such galaxies may exist, it appears from the observations that the X-ray emission from most massive ellipticals arises from the hot gas. We conclude that the supernova rate (including the efficiency of the supernova heating process) must not exceed ν_{sn1} for galaxies in the mass range of model B. Finally, we note that the general heating trend which dominates model B1 after eight billion years also operates more slowly in an incipient manner in the other models. We expect therefore that the three well behaved models – A1, A2, and B2 – would eventually develop into strong wind flows if our calculations were carried billions of years into the future.

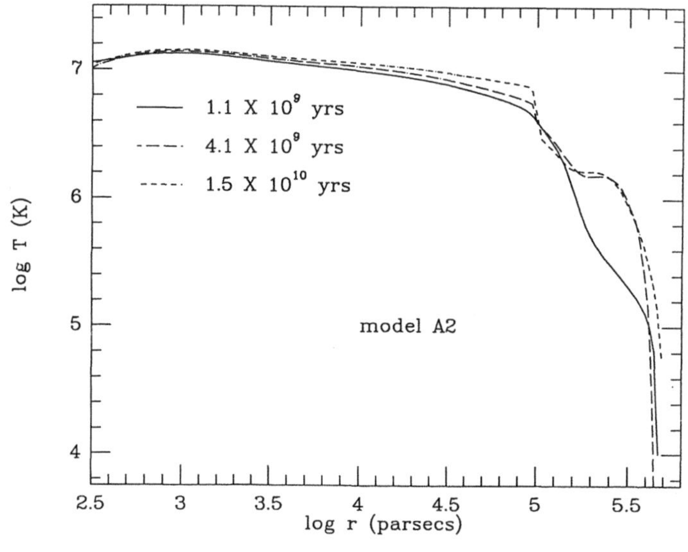

Figure 3: Temperature vs. radius at three times for model A2 illustrating the very slight temperature rise in the nearly isothermal part of the flow as the energy source term increases with time. As this occurs, compression becomes increasingly less important in establishing the roughly steady thermal balance.

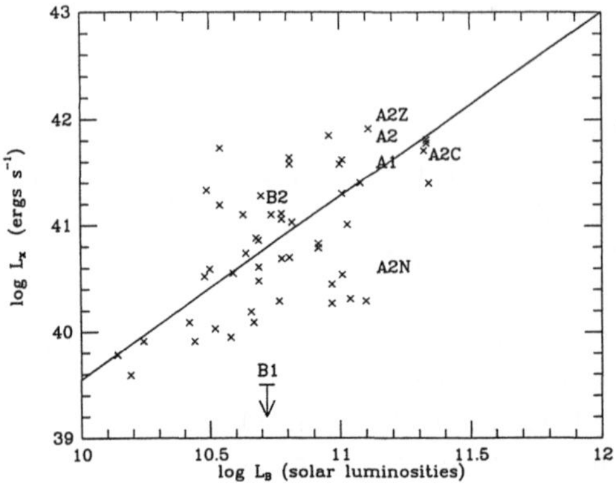

Figure 4. The positions of our calculated models at $t = 1.5 \times 10^{10}$ years in the (L_B, L_x)-plane. For comparison, the observations from Canizares *et al.* (1986) and the best fit to their data are also shown. Model B1 is off the scale of the plot at log $L_x = 37.5$. Model A2C is shifted 0.2 dex to the right for clarity.

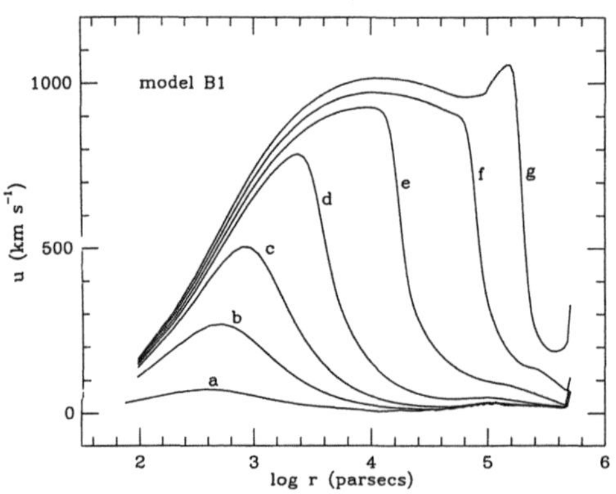

Figure 5. The later stages of the evolution of the velocity in model B1 illustrating the "unloading" of the gas as the energy input from supernovae becomes too great to prevent the development of a supersonic galactic wind. Times shown, in units of 10^9 years, are: (a) 9.0, (b) 10.0, (c) 11.0, (d) 12.0, (e) 13.0, (f) 14.0, and (g) 15.0.

III. Effects of Dark Halo, Conductivity, and Abundances

In addition to the four basic models described above, we computed the evolution of several variations on the A2 model designed to examine the specific influence of massive halos, thermal conductivity, and abundance on the overall gas flow.

Model A2N is identical to A2 in every respect except the massive dark halo is omitted. It is of considerable interest therefore that model A2N developed a wind flow very similar to that of B1 with an X-ray luminosity after $t_H = 15 \times 10^9$ that is 17 times lower than that of A2. Although this X-ray luminosity is well below values expected from very massive elliptical galaxies corresponding to model A parameters, L_x is not reduced by orders of magnitude as occurs with model B1. An examination of model A2N reveals that as the wind grows from the inside out, it is impeded by the inertia of gas that had been ejected at earlier epochs. In fact the flow in A2N remains unsteady from the inception of the central wind to the end of the calculation at $t = t_H$. The X-ray surface brightness Σ_x of A2N at t_H decreases much faster with projected radius than that of models A1 and A2 (see Fig. 6) and is much steeper than the stellar brightness Σ_*. Our results with model A2N underscore the important role of massive halos in retaining gas ejected from stars in elliptical galaxies.

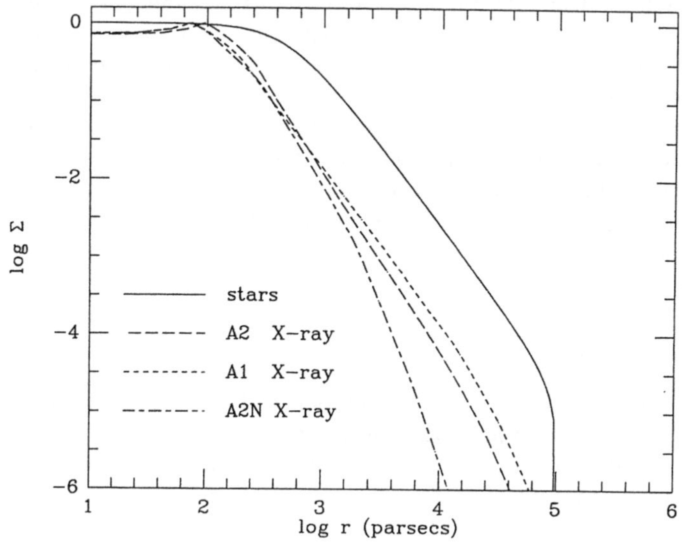

Figure 6: Comparison of relative X-ray surface brightness distributions in models A2, A1, and A2N. The surface brightness distributions of the stellar light is also shown. The X-ray slope for model A1 is nearly the same as the slope for the stars.

Another variant of model A2, A2C, is a repeat of the A2 calculation with an additional thermal conduction term

$$\frac{1}{r^2}\frac{\partial}{\partial r}\left(5.0 \times 10^7 T^{5/2} r^2 \frac{\partial T}{\partial r}\right)$$

included in equation (3). Although thermal conduction can only proceed along the direction of the local magnetic field and is therefore restricted to flux tubes, an argument can be advanced that the field in the hot gas is radially oriented; this arrangement would be expected as a combined result of buoyant motion of supernova remnants and field reconnection. In any case, the addition of conductive energy transport to model A2 has a very minor effect on its observable properties; for example, the final X-ray luminosity is still 5×10^{41} ergs s^{-1}. The most significant effect of conduction is an outward transport of thermal energy which heats the tenuous gas beyond r_t producing a hot wind in this (unobservably faint) region. The X-ray surface brightness profile of model A2C is somewhat steepened compared to that of A2. The effect of conductivity on cluster flows (BERTSCHINGER and MEIKSIN [19]) could be more important than hot gas flows in individual galaxies since the gas temperature is $\sim$ 10 times greater in the cluster medium and the conduction term varies as $T^{7/2}$.

For both models A1 and A2 the inflow of cold gas into the galactic center is large, comparable to the stellar core mass. Although we have not allowed for the distortion of the central stellar mass and optical surface brightness distributions or the increase in central stellar velocity dispersion due to the inflow of substantial amounts of gas (or hypothetical stars formed from the gas, see JURA [20] or SARAZIN and O'CONNELL [21]) into the galactic cores, such effects could be observable and should be included in future calculations. It is also interesting to explore aspects of the problem that might reduce the amount of inflowing gas. One possibility is to consider the effect of metal abundance on the cooling rate and the net inflow of gas.

IV. Discussion and Summary

Of the many parameters which describe the evolution and present day behavior of hot galactic flows, two – the supernova rate and the fraction of mass that is nonvisible – are of considerable astronomical interest. Fortunately, the evolution of the hot gas is sensitive to both. Consider for example a series of hot gas flow calculations with a fixed supernova rate but with a variable amount of dark mass (all having the same spatial distribution). All solutions with sufficiently large amounts of dark matter must eventually become pure inflow; however, models with no dark matter develop into pure winds since supernova heating usually exceeds the rate that thermal energy can be radiated. For intermediate values of the dark matter fraction, we expect a sequence of partial wind solutions. Conversely, for models having a fixed mass distribution, strong winds must always occur if the supernova rate is sufficiently high and total inflow must develop if the supernova rate is not large enough to offset radiative cooling. While our preliminary calculations do not sample every combination of these two parameters, a pattern of flow solutions similar to that shown in Fig. 7 has emerged. Models A1, A2, and A2N shown in Fig. 7 illustrate final flows after 15 billion years for three points on the parameter plane.

The soft X-ray luminosity varies strongly across the plane of Fig. 7 from $\sim 10^{42}$ ergs s^{-1} at the lower right to $\sim 10^{40} - 10^{41}$ ergs s^{-1} in regions corresponding to partial winds, and drops to very low values, $\ll 10^{40}$ ergs s^{-1}, at the upper left where strong winds analogous to that of model B1 occur. The stagnation radius varies in a sense opposite to L_x, decreasing from large values at the lower right to zero at the upper left.

The lines that separate the four regions in Fig. 7 are very poorly established at present, but the overall trend is reliable and of great interest since the evolutionary gas flow solutions are sensitive to parameters in an astronomically interesting range. Evidently, most observed galaxies occupy the "pure inflow" or "subsonic partial wind" regions of the plane, corresponding to the presence of significant massive halos and modest supernova rates. However, it must be emphasized that all the models in Fig. 7 are compared at the same time since galaxy formation. The morphological distribution of galactic flows in Fig. 7 evolves with time, so identical galaxies examined at different times (*i.e.* having a range of ages) could experience different flow morphologies. In particular, when considered at later times, the region of pure outflow must enlarge toward the lower right in diagrams analogous to Fig. 7 since α_{sn}/α increases with time even though α_{sn} may be fairly constant.

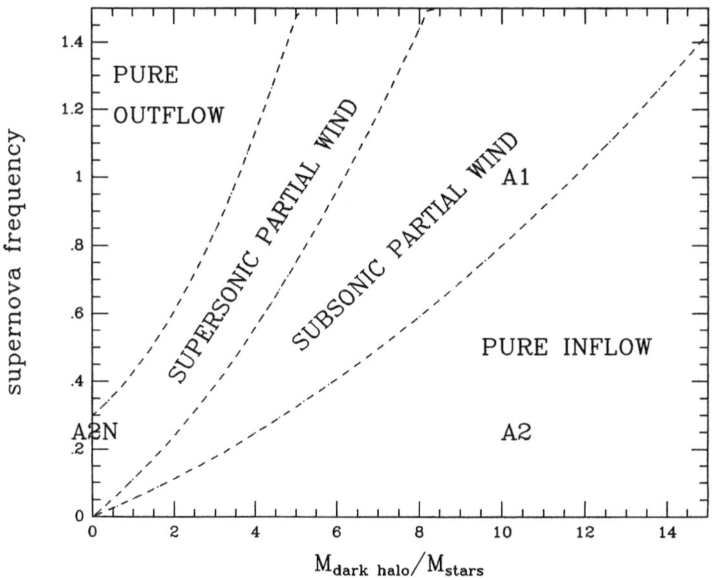

Figure 7: Schematic representation of the nature of solutions after 15 billion years as a function of the mass of the dark halo and the present supernova frequency. The supernova frequency α_{sn} is in units of 0.002 supernovae yr^{-1} $(10^{10} L_\odot)^{-1}$. Models A1, A2, and A2N are plotted.

Another matter of interest is the radial distribution of surface brightness Σ_x in those relatively few galaxies that can be spatially resolved in X-ray emission. In three ellipticals, all members of the Virgo cluster, TRINCHIERI, FABBIANO, and CANIZARES [7] found that Σ_x is proportional to the stellar surface brightness Σ_* across most of the projected image. This proportionality can most easily be understood if the flow velocity in the hot gas is so slow that the gas temperature is governed by the source terms in equation (3), $\rho^2 \approx Mk\Lambda(T)^{-1}\alpha(3T_o - 5T)\rho_*$, which leads to proportional emissivities, $\rho^2 \propto \rho_*$, and surface brightnesses if the temperature is uniform (see also SARAZIN 1986b). However, $\Sigma_x \propto \Sigma_*$ also follows if either (i) the flow is in an exact steady state, $\dot{M}$ = constant, or (ii) the gas is heated primarily by compression, the temperature gradient is small, and $\rho\partial u/\partial r \ll u\partial\rho/\partial r$, as usually occurs. However, the hot gas flows we have calculated are not in steady state (since the source terms change on time scales comparable to the inflow time) so on-the-spot reradiation of supernova energy or possibility (ii) above are more likely candidates to account for $\Sigma_x \propto \Sigma_*$. In either case, the gas temperature must be remarkably uniform in the flow; any small radial gradients will interfere with $\Sigma_x \propto \Sigma_*$ and make this proportionality somewhat sensitive to the particular X-ray band pass used. We find that $\Sigma_x \propto \Sigma_*$ is quite well achieved in model A1, but Σ_x drops slightly faster than Σ_* in model A2. It is unlikely, however, that $\Sigma_x \propto \Sigma_*$ maintains for all ellipticals since this would imply $L_x \propto L_*$ if the mass to light ratios and gas temperatures were similar for all galaxies; this is shallower than the observed distribution of galaxies in the (L_x, L_*)–plane (Figure 4). We note finally that the three galaxies for which $\Sigma_x \propto \Sigma_*$ is best established are immersed in the high pressure environment of the extensive hot ($T \sim 3 \times 10^7$ K) gas surrounding M87 in the Virgo cluster, which undoubtedly affects the evolution and distribution of the galactic hot gas.

Most massive elliptical galaxies appear to have substantial cooling flows in their central regions (see FABIAN, NULSEN, and CANIZARES [22] for a review of cooling flows). Our evolutionary calculations indicate that the X-ray luminosity is very low for galaxies having strong winds for which the central cooling region is small or absent. A possible exception to this would be those galaxies which are in a transient stage of development into a strong wind from the inside out, like model B1, a process which may take billions of years as the inertia of the previously produced and slower moving gas in the flow is overcome. In any case, the widespread incidence of cool, optically emitting gas near the centers of elliptical galaxies (DEMOULIN-ULRICH, BUTCHER, and BOKSENBERG [23]; PHILLIPS *et al.* [24]) is consistent with the presence of cooling flows that become thermally unstable (MATHEWS and BREGMAN [25]). Whether the cooled gas at the centers of these flows forms into stars is a controversial issue (SARAZIN and O'CONNELL [21]; WHITE and SARAZIN [26, 27]; SILK *et al.* [28]), but in any case the total amount of mass flowing into the center may be large – comparable to the stellar core mass in many of our models. While our main interest here has been in the overall global evolution of galactic hot gas flows, it is clear that future studies must confront the issue of the large central mass inflows we have encountered. Either the parameters of the models need to be adjusted to reduce the inflow, or the mass inflows – in the form of either gas or stars – must be shown to be consistent with the observed light distributions at the centers of ellipticals which are by no means identical in all galaxies (LAUER [29]).

Our goal in these initial evolutionary calculations has been to understand the soft X-ray luminosities from an evolutionary perspective, and we have been largely successful in achieving this goal. A significant component of dark matter is necessary to retain the hot gas, a conclusion consistent with studies based on hydrostatic hot gas containment in ellipticals [4, 7, 30, 31] A simple Salpeter power law for the initial stellar mass function provides enough hot gas to account for the X-ray luminosities; neither a period of extended infall of primeval gas nor innoculation of new gas by merging with gas-rich dwarfs appears to dominate the gas budget in early type galaxies. While merging with dwarfs almost certainly occurs in some ellipticals (SILK and NORMAN [32]) continued diffuse infall seems unlikely due to the apparent absence of cold intergalactic gas and the peculiar motions of individual galaxies which would defocus and disorganize a hypothetical continued inflow of intergalactic gas. Some of the scatter in the observed luminosities in Figure 4 could be due to these perturbations, but they could also be a result of modest variations in composition, age, frequency of binaries or their mean separation [influencing $\alpha_{sn}(t)$] or in the amount or distribution of dark matter. Ellipticals in clusters containing substantial intracluster gas might be expected to have large L_x if isotropic compression dominates, or small L_x if stripping has robbed some of their galactic gas. Since so many factors can contribute to the scatter in Figure 4, it is not surprising that correlations of L_x/L_B with other galactic parameters have been difficult to establish.

We wish to acknowledge the National Science Foundation (Grant AST 83-12971), the California Space Institute (Grant CSS 65-85), and a UCSC Faculty Research Grant for financial assistance in supporting our work. We are also grateful for computing time granted us under the IBM - UCSC Jointly Defined Effort sponsored by IBM Academic Information Systems and administered by Harwood Kolsky.

References

1. Forman, W., Schwarz, J., Jones, C., Liller, W., and Fabian, A. C. 1979, *Ap. J. (Letters)*, **234**, L27.

2. Biermann, P., and Kronberg, P. P. 1982 *Ap. J. (Letters)*, **268**, L69.

3. Nulsen, P. E. J., Stewart, G. C., and Fabian, A. C. 1984, *M. N. R. A. S.*, 208, 185.

4. Forman, W., Jones, C., and Tucker, W. 1985, *Ap. J.*, **293**, 102.

5. Trinchieri, G., and Fabbiano, G. 1985, *Ap. J.*, **296**, 447.

6. Canizares, C. R., Fabbiano, G., and Trinchieri, G. 1986, Ap. J., submitted.

7. Trinchieri, G., Fabbiano, G., and Canizares, C. R. 1986, *Ap. J.*, submitted.

8. Sarazin, C. L. 1986a, *Rev. Mod. Phys.*, **58**, 1.

9. ________. 1986b, in IAU Symposium 117, *Dark Matter in the Universe*, eds. Kormendy, S., and Knapp, G. (Dordrecht: Reidel), in press.

10. Mathews, W. G., and Loewenstein, M. 1986, *Ap. J. (Letters)*, **306**, L7.

11. Mathews, W. G., and Baker, J. 1971, *Ap. J.*, **170**, 241.

12. Raymond, J. C., Cox, D. P., and Smith, B. W. 1976, *Ap. J.*, **204**, 290.

13. Renzini, A., and Buzzoni, A. 1985, in *The Spectral Evolution of Galaxies*, eds. Chiosi, C., and Renzini, A. (Dordrecht:Reidel), in press.

14. MacDonald, J., and Bailey, M. E. 1981, *M. N. R. A. S.*, **197**, 995.

15. Iben, I., Jr., and Tutukov, A. V. 1984, *Ap. J. Suppl.*, **54**, 335.

16. Tammann, G. A. 1974, in *Supernovae and Supernova Remnants*, ed. Cosmovici, C. B. (Dordrecht: Reidel).

17. Burstein, D., and Rubin, V. C. 1985, *Ap. J.*, **297**, 423.

18. Blumenthal, G. R., Faber, S. M., Flores, R., and Primack, J. R. 1986, preprint.

19. Bertschinger, E., and Meiksin, A. 1986, *Ap. J. (Letters)* **306** L1.

20. Jura, M. 1977, *Ap. J.*, **212**, 634.

21. Sarazin, C. L., and O'Connell, R. W. 1983, *Ap. J.*, **268**, 552.

22. Fabian, A. C., Nulsen, P. E. J., and Canizares, C. R. 1984, *Nature,* **310**, 733.

23. Demoulin-Ulrich, M.-H., Butcher, H. R., and Boksenberg, A. 1984, *Ap. J.*, **285**, 527.

24. Phillips, M. M., Jenkins, C. R., Dopita, M. A., Sadler, E. M., and Binette, L. 1986, *Astron. J*, **91**, 1062.

25. Mathews, W. G., and Bregman, J. N. 1978, *Ap. J.*, **224**, 308.

26. White III, R. E., and Sarazin, C. L. 1986a, *Ap. J. Suppl.*,, submitted.

27. ________. 1986b, *Ap. J. Suppl.*, submitted.

28. Silk, J., Djorgovski, S., Wyse, R. F. G., and Bruzual A., G. 1986, *Ap. J.*, **307** 415.

29. Lauer, T. R. 1985, *Ap. J.*, **292**, 104.

30. Fabian, A. C., Arnaud, K. A., Nulsen, P. E. J., and Mushotzky, R. F. 1986a, *Ap. J.*, in press.

31. Fabian, A. C., Thomas P. A., Fall, S. M., and White, R. E. III 1986b, *M. N. R. A. S.*, **221**, 1049.

32. Silk, J., and Norman, C., 1979, *Ap. J.*, **234**, 86.

Hot Coronae Around Early Type Galaxies

CHRISTINE JONES

The study of gas in early type galaxies began at least as early as 1957 when SANDAGE [1] estimated that during the course of normal stellar evolution, the stars of a bright elliptical galaxy would shed $\sim 5\times10^9\ M_\odot$ of gas. Over the last 25 years, limits on various forms of gas in early type galaxies became increasingly stringent. Models for a galactic wind, like that of MATHEWS and BAKER [2], were developed to explain the absence of gas. However NORMAN and SILK [3] and SARAZIN [4] argued that galaxies with massive halos could maintain gaseous coronae over cosmological times and could overcome the heating from supernovae.

X-ray images have shown that bright early type galaxies are not free of gas but contain a hot gaseous corona [5,6,7,8,9]. Isointensity contours of the x-ray emission from two elliptical galaxies in the Perseus-Pisces filament are shown in Figure 1a [10]. The appearance and properties of coronae can be affected by their environment. Figure 1b shows the ram pressure effects of the M87/Virgo medium on the corona around M86 (NGC 4406). M86 travels toward us, with a velocity through

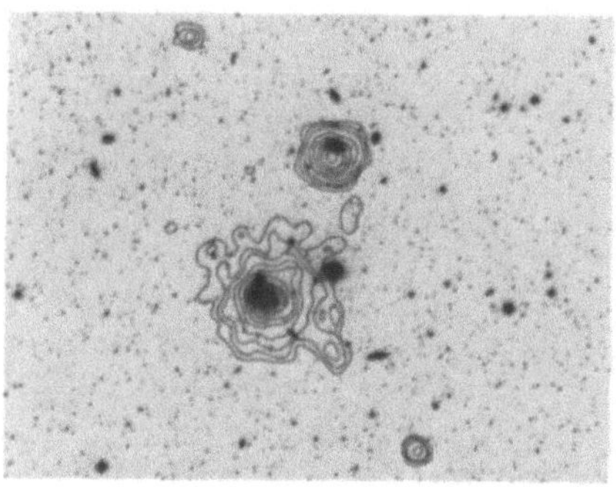

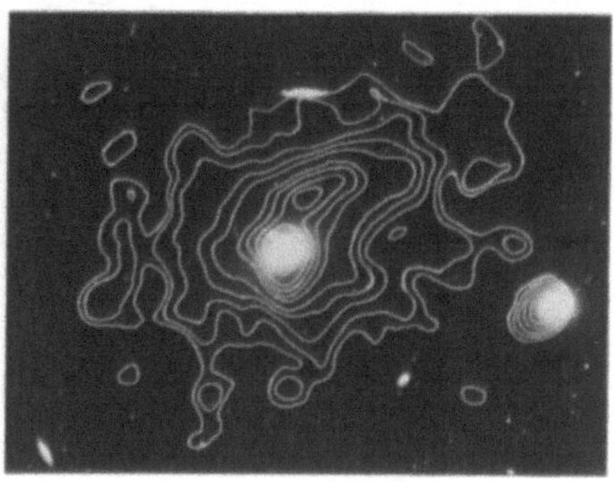

Figure 1a (left) and 1b (right). X-ray isointensity contours of NGC 499 (north) and NGC 507 and of M86 superposed on optical photographs.

Virgo of nearly 1500 km/sec. Its encounter with the denser medium in the cluster core apparently has stripped some of its corona and produced a plume of gas not yet incorporated into the cluster medium [5,11].

Observations such as these have shown the presence of hot gas around early type galaxies in good agreement with simple predictions. Even more important, the existence of extensive hot coronae around early type galaxies implies the presence of a dark halo around the parent galaxy [9,12,38].

Several surveys have been undertaken to determine the x-ray properties of early type galaxies [13,9,14,15,16,17]. Other studies have focussed on the detailed properties of a few galaxies [5,8,10,18,19,20,21,22]. Galaxies included in these surveys are generally either in the Virgo cluster or in low velocity dispersion groups. Central cluster galaxies are not included.

Figure 2a shows a plot of the x-ray luminosity against absolute blue magnitude for 150 early type galaxies observed with Einstein [17]. Although this larger sample allows better comparisons to be made among galaxies of different morphological types or in different environments, the primary conclusions are those reached by FORMAN, JONES, and TUCKER [9] based on a sample of 55 galaxies. There is considerable scatter in the correlation of x-ray luminosity with optical magnitude. This scatter is not due primarily to differences in x-ray emission from one galaxy type to another. As shown in Figure 2b, the elliptical galaxies are well mixed with SO galaxies. However, as noted by Forman, Jones, and Tucker, galaxies with some cool interstellar medium (Cen A, Sombrero and other Sa galaxies) tend to fall below the general correlation. Some dispersion also may arise from differences in the galaxy environment and

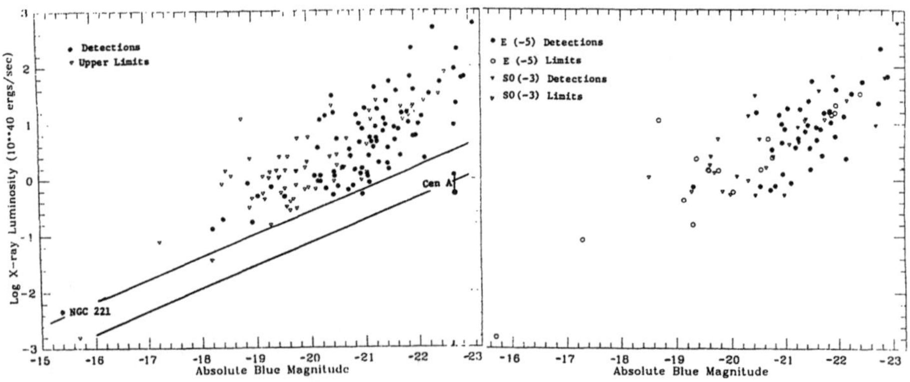

Figure 2a and 2b. Distribution of X-ray luminosity vs. optical luminosity for ~150 early type galaxies and for E and S0 galaxies only. Estimates of galactic source contribution are shown as solid lines (2a).

the effects of that environment on the galaxy as evidenced by M86 in Virgo (Figure 1b). For the low luminosity galaxies, some scatter may arise from different contributions of nuclear or galactic sources, and some dispersion arises from uncertainties in the x-ray flux and errors in absolute quantities arising from distance uncertainties.

The spatial extents observed for the galaxies as well as the spectral properties of the x-ray emission provide strong evidence in favor of a thermal mechanism for production of the x-rays [9]. For ten galaxies whose x-ray luminosities range from 2×10^{40} to 6×10^{42} erg/sec, the emission is constrained to have gas temperatures of about 10^7 K [9,20,17]. A population of galactic bulge or globular cluster x-ray sources whose integrated spectrum would be significantly harder cannot account for the x-ray emission. Figure 2a shows two estimates of the galactic x-ray source component. One used the diffuse emission from Cen A as a limit which agrees with the value derived for the bulge of our galaxy [9] and the other larger estimate is based on extrapolating the properties of the bulge of M31 [15,16].

Although thermal emission from a hot corona is the dominant source of emission for luminous, early type galaxies, for galaxies below some absolute magnitude, the contribution to the x-ray flux from hot gas must become unimportant since the galaxy will not be massive enough to bind hot gas. (Much cooler gas would not be visible in the x-ray band). For an M/L of 100, 10^7 K gas will not be bound in galaxies fainter than -18. (Note in Figure 2 that the x-ray emission from NGC 221 (M_B = -15.5) is produced by a nuclear point source.)

Surface brightness profiles have been obtained for about twenty galaxies [9,10,20,17] and are all similar in form. At large radii, the surface brightness varies approximately as r^{-2}. TRINCHIERI et al. [20] noted the similarity between the x-ray and optical surface brightness. Since the x-ray emission is generated by bremsstrahlung and line radiation, for the Einstein energy band, the density profile is fairly well constrained from the surface brightness profile. Integrating the density profiles to the radius of detected x-ray emission, one finds total gas masses of 5×10^8 to 5×10^{10} $M_\odot$ [9]. Thus the fraction of mass in presently observed gas is less than ~7% of the mass in stars (with M/L = 6 for stellar material). The observed hot gas therefore contributes little to the total galaxy mass.

The gas mass observed is equal within the uncertainties to that predicted by the accumulation of gas lost by present generation stellar systems [9]. However much larger amounts of gas are predicted to be produced by young stellar systems in a relatively short interval at early stages of galaxy evolution [2,23]. Thus the gas observed in galactic coronae may be either mass accumulated using present epoch mass loss rates, if most of the early epoch gas is lost from the galaxy, or if the true value for present epoch mass loss is near its lowest estimate, most of the x-ray emitting gas would be that remaining from the earliest stages of galaxy evolution.

The observed gas mass compared to the sweeping time of a galactic wind requires that the gas be in hydrostatic equilibrium unless the mass replenishment rate from stars is much higher than expected. Also, since the gas density in the corona decreases as $r^{-1.5}$, the bulk of the gas mass lies at large radii. For example, 65% of the observed gas mass lies between the largest detected radius and half that value. Thus partial winds over a significant fraction of the corona would require unreasonably large replenishment rates.

While the outer regions are in hydrostatic equilibrium, the high central gas density in the cores implies short radiative cooling times [9,20]. In the absence of heating, accretion flows similar to those in the central regions of clusters may occur. In a study of eighteen early type galaxies, THOMAS et al. [22] found that all had mass flow rates in the range $\approx$0.03 to 1 $M_{\odot}$/yr. As with clusters of galaxies, heat conduction may reduce this rate [24,25,26]. Such cooling flows could be important for the initiation of nuclear activity seen in the form of radio emission or for the production of optical emission lines which are found in about half of the early galaxies [27,28,29].

In addition to determining the gas parameters of the hot coronae and its effects on the galaxy, the x-ray emitting gas can be used to derive the total galaxy mass. The general absence of a cool interstellar medium at large radii has hampered the determinations of the mass distributions for individual early type galaxies. Only for a few early type galaxies can mass determinations from polar rings [30], shells [31], or HI observations [32] be attempted. The constituents of the x-ray coronae are ideal test particles to trace the total mass distributions since they obey well-understood laws and comprise a small fraction of the total galaxy mass. The fact that hot coronae are a general property of early-type galaxies allows (with adequate observations) this technique to be used on a morphologically diverse set of galaxies in different environments.

The use of hot gas to trace the gravitational potential was first applied to M87 by MATHEWS [33] and BAHCALL and SARAZIN [34]. FABRICANT, GORENSTEIN, and LECAR [35,36] used the Einstein images to determine more accurately the underlying mass around M87. A similar analysis was performed on NGC 4696 in the Centaurus cluster [37]. This technique also has been used to estimate the total mass in halos around the early type galaxies NGC 4472, NGC 499, and NGC 507 which are not central cluster members [9,10].

The method relies on the gas being in hydrostatic equilibrium. With the ideal gas law one solves for the total mass inside a radius

$$M(r) = -\left(\frac{kT}{G\mu m_p}\right)\left(\frac{d\ \log\rho}{d\ \log r} + \frac{d\ \log T}{d\ \log r}\right) r$$

$$= 3\times10^{12}\ (T/10^7\ K)\left(\frac{d\ \log\rho}{d\ \log\ r} + \frac{d\ \log T}{d\ \log r}\right)(r/100\ kpc).$$

Thus if one measures the gradients of temperature, T, and density, ρ, and the temperature at a radius r, one can determine the total mass within that radius. The minimum mass corresponds to the largest allowed increase in gas temperature with increasing radius and the lowest allowed temperature at radius r. The x-ray emission in systems from individual early type galaxies to rich clusters provides a powerful tool for measuring the distribution of total mass.

While reliable density profiles have been determined for many systems, limits on the temperature gradient have been obtained only for the five galaxies noted above. The most conservative (minimum) mass-to-light ratios for these five systems (calculated only to the radius for which the temperature is measured) exceed 25.

NORMAN and SILK [3] argued that coronae would be roughly isothermal outside the central core since the conduction time scale is shorter than the cooling timescale. The isothermal assumption which is consistent with the available data, leads to galaxy masses of up to 10^{13} $M_\odot$ with mass-to-light ratios as large as $\sim$90 $M_\odot/L_\odot$ [9,10]. Detailed studies of NGC 4472 by THOMAS [21] support the large halo mass derived for this galaxy.

FABIAN and collaborators [12] relaxed the isothermal assumption and derived the minimum total galaxy mass by assuming that the temperature gradient was no steeper than that allowed by a convectively stable atmosphere. Any steeper profile would be unstable and return to the stable configuration. Their derived masses implied average mass-to-light ratios $\gtrsim$ 60 for plausible parameters.

MATHEWS and LOWENSTEIN [38] have extended the work of MATHEWS and BAKER [2] to explore the effects of dark halos. They confirm the earlier results - galaxies with no dark halos produce supernova driven galactic winds and no observable x-ray emission. On the other hand, the presence of dark halos affects the behavior of the winds leading to approximately hydrostatic atmcspheres with isothermal temperatures and x-ray properties similar to those observed. Thus while there remain considerable uncertainties in mass measurements with the current x-ray data, the evidence from x-ray, polar ring and shell observations as well as theoretical models continues to increase in support of massive halos around individual early-type galaxies.

Studies of the hot coronae around early type galaxies have only just begun. The present observations are sufficient to measure the density distribution and integrated temperature of the hot gas. These observations require the existence of massive halos, demonstrate that early type galaxies do not presently drive hot galactic winds and suggest that modest cooling flows occur in the galaxy cores.

Observations with ROSAT and AXAF will permit the full potential of the x-ray studies of early type galaxies to be realized. Figure 3 shows a comparison of the areas of the Einstein, ROSAT and AXAF telescopes. Although the ROSAT energy range does not extend beyond 2.5 keV, it is sufficient to measure the temperature of 10^7 K gas. With ROSAT one can determine the radial temperature distribution as well as the density distribution and thus the distribution of the total galaxy mass. The full compliment of instruments planned for AXAF will permit detailed spatial spectroscopy on these objects. For a 10^7 K gas with solar abundances, 60% of its radiation is in line emission. By measuring elemental abundances with good spatial resolution from one galaxy to another, one can study the iron enrichment. The increased sensitivity of AXAF will permit one to trace the gas to fainter flux levels and will allow the detection of coronae at redshifts ≥ 1.

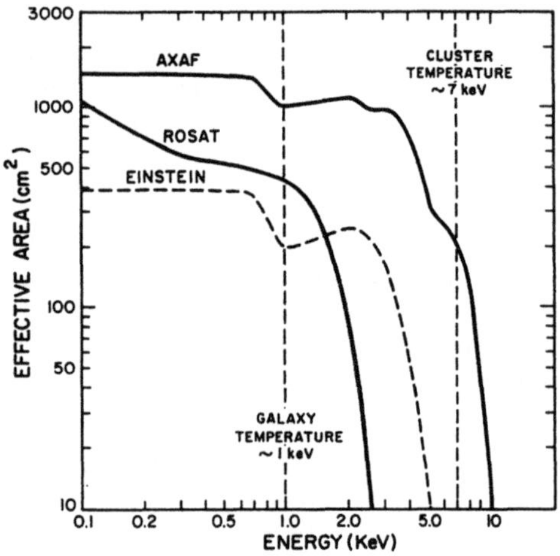

Figure 3. Effective areas of Einstein, ROSAT and AXAF.

The discovery of hot coronae around early type galaxies has opened a rich field of research. Problems related to the origin of gas, its iron enrichment and evolution over time in various environments will be addressable as will problems of the total galaxy mass. More important, future observations will allow us to determine the shape of the underlying potential. Thus, we will be able to provide important information on the formation of galaxies such as the relative amount of dissipation undergone by the baryonic matter comprising the stars and the material defining the dark halos.

I am grateful to Bill Forman, Wallace Tucker, and Richard Burg for their assistance and use of unpublished work and to Karen Modestino for the preparation of this manuscript. This work was supported by NASA grant NAS8-30751.

References

1. Sandage, A., 1957, Ap.J., 12.5, 422.
2. Mathews, W., and Baker, J., 1971, Ap.J., 170, 241.

3. Norman, C., and Silk, J., 1979, Ap.J., 233, L1.
4. Sarazin, C., 1979, Astrophysical Letters, 20, 93.
5. Forman, W., Schwarz, J., Jones, C., Liller, W., and Fabian, A., 1979, Ap.J., 234, L37.
6. Bechtold, J., Forman, W., Giacconi, R., Jones, C., Schwarz, J., Tucker, W., and Van Speybroeck, L., 1983, Ap.J., 265, 26.
7. Biermann, P., and Kronberg, P., 1983, Ap.J., 268, L69.
8. Nulsen, P., Stewart, G., and Fabian, A., 1984, MNRAS, 208, 185.
9. Forman, W., Jones, C., and Tucker, W., 1985, Ap.J., 293, 102.
10. Jones, C., Sullivan, W., and Bothun, G., 1986, preprint.
11. Fabian, A., Schwarz, J., and Forman, W., 1980, MNRAS, 192, 135.
12. Fabian, A., Thomas, P., Fall, M., and White, R., 1986, preprint.
13. Long, K., and Van Speybroeck, L., 1983, in Accretion-Driven Stellar X-ray Sources, ed. W. Lewin and E. van den Heuval (Cambridge University Press), p.117.
14. Stanger, V., and Schwarz, J., 1985, preprint.
15. Trinchieri, G., and Fabbiano, G., 1985, Ap.J., 296, 447.
16. Canizares, C., Fabbiano,G., and Trinchieri,G., 1986, preprint.
17. Forman, W., Jones, C., Burg, R., and Tucker,W., 1986, preprint.
18. Feigelson, E., Schreier, E., Delvaille, J., Giacconi, R., Grindlay, J., and Lightman, A., 1981, Ap.J., 251, 31.
19. Warwick, R., and Stanger, V., 1986, preprint.
20. Trinchieri, G., Fabbiano, G., and Canizares, C., 1986, preprint.
21. Thomas, P., 1986, preprint.
22. Thomas, P., Fabian, A., Arnaud, K., Forman, W., and Jones, C., 1986, preprint.
23. Faber, S., and Gallagher, J., 1976, Ap.J., 204, 365.
24. Tucker, W., and Rosner, R., 1983, Ap.J., 267, 547.
25. Bertschinger, E., and Meiksin, A., 1986, Ap.J., 301, L1.
26. Rosner, R., Tucker, W., and Najita, J., 1986, preprint.
27. Phillips, M., Jenkins, C., Dopita, M., Sadler, E., and Binetti, L., 1986, A.J.
28. Caldwell, N., 1984, PASP, 96, 287.
29. DeMoulin-Ulrich, M., Butcher, H., and Boksenberg, A., 1984, Ap.J., 285, 527.
30. Schweizer, F., Whitmore, B., and Rubin. V., 1983, A.J., 88, 909.
31. Hernquist, L., and Quinn, P., 1986, preprint.
32. Van Gorkum, J. H., Knapp, G. R., Raimond, E., Faber, S. M., and Gallagher, J., 1986, preprint.
33. Mathews, W., 1978, Ap.J., 219, 413.
34. Bahcall, J., and Sarazin, C., 1977, Ap.J., 213, 699.
35. Fabricant, D., Gorenstein, P., and Lecar, M., 1980, Ap.J., 241, 552.
36. Fabricant, D., and Gorenstein, P., 1983, Ap.J., 267, 535.
37. Matilsky, T., Jones, C., and Forman, W., 1985, Ap.J., 291, 621.
38. Mathews, W., and Lowenstein, M., 1986, Ap.J., 306, L7.

The Relative Masses of the Milky Way's Components, the Ostriker-Caldwell Approach, and Differential Rotation Beyond the Solar Circle

P. L. SCHECHTER

The determination of the relative masses of the disks, bulges and dark halos of galaxies has received considerable attention in the present proceedings. The attention is well deserved, since any attempt to understand the formation of galaxies must take account of the relative proportions of dark and luminous matter.

Yet there is a fundamental problem in attempting such a determination in an external galaxy: one cannot simultaneously observe both the velocity dispersion perpendicular to a galactic disk and the scale height of that disk. One is forced either to the assumption of a constant mass-to-light ratio [1,2], or to a statistical treatment based on observations of velocity dispersions in face-on galaxies and scale heights in edge-on galaxies [3,4].

The Milky Way, by contrast, offers a unique opportunity in that both the scale height of its disk and the velocity dispersion perpendicular to it can be measured, with relative ease at the Sun's position and perhaps elsewhere as well. Using the best values for such measurements one can then obtain the contribution of the disk to the rotation curve of the Milky Way, with the deficit accounted for by the bulge and the dark halo.

A model for the distribution of mass within the Galaxy includes scale lengths, characteristic densities, and shapes of the mass distributions for each of its components. Among the various approaches for determining the parameters which characterize the components, I find the one employed by OSTRIKER and CALDWELL [5,6,7] especially illuminating. What follows is, for the most part, a review of their work, emphasizing the strengths of their approach and pointing to those places where new data would considerably improve the accuracy of the derived parameters.

A key feature of the Ostriker-Caldwell (henceforth abbreviated O/C) approach is the attempt to determine the parameters of their model from kinematic data rather than photometric data when at all possible. Ostriker and Caldwell thus avoid the assumption that the mass-to-light ratio is constant throughout the components of interest. This principle is ultimately compromised in their adoption of a photometric core radius for the bulge (indeed they use the core radius for the bulge of M31); the kinematic data do not suffice to constrain it.

A second important feature of the O/C approach is the computation of uncertainties in the model parameters, an example worthy of emulation. However the value of Ostriker and Caldwell's quoted uncertainties is somewhat diminished by their failure to indicate the covariances among the parameters. Without covariances it is impossible to estimate proper uncertainties for the various interesting quantities which are functions of the model parameters.

A third feature of the O/C approach, one that is especially useful to observers, is the systematic attempt to indicate which observations constrain which model parameters. Thus Caldwell and Ostriker find, not surprisingly, that the central density of the disk is most tightly constrained by the observed acceleration perpendicular to the Galactic plane, and next most tightly constrained by the observed value of Oort's "A" parameter [6].

The basic elements of the O/C approach involve first, an assumed set of density profiles for the three mass components of their model (which they call model "outputs"), and second, a set of observational constraints (model "inputs").

One might choose to quibble with Ostriker and Caldwell's choice of density profiles, but the O/C approach is quite flexible and capable of incorporating different profiles. They adopt a Hubble law for the surface density of the bulge (which they call the "spheroid"), introduce a gap into their exponential disk, and cut off their halo (which they call a "corona") at a radius r_{tot}.

The parameters in the O/C model are tabulated below. Adopting their convention, ρ indicates a space density and Σ indicates a surface mass density. The distance to the Galactic center must be included as a free parameter because many of the constraints are evaluated at the Sun's position or multiples thereof.

Ostriker-Caldwell Parameters

Disk:	Σ_d, r_d, r_g	$\Sigma(r) = \Sigma_d[\exp(-r/r_d) - \exp(-r/r_g)]$
Spheroid:	ρ_s, r_s	$\Sigma(r) = \Sigma_s(1 + r/r_s)^{-2}$
Corona:	ρ_c, r_c, r_{tot}	$\rho(r) = \rho_c[1 + (r/r_c)^2]^{-1}$
Scale:	R_0	

The available data cannot support all 9 parameters in the O/C model. The corona cutoff radius, r_{tot}, is poorly constrained by the available data; Ostriker and Caldwell therefore present solutions for a range of reasonable values for r_{tot}. They also constrain the disk scale length, r_d to be only slightly larger than the gap scale length, r_g. This has the effect of producing a surface density profile of the form $r \cdot \exp(-r/r_d)$.

In each of their three papers Ostriker and Caldwell attempt to use the most reliable observational constraints available. In the four years between their first and third papers new observational constraints were added (e.g. σ_{bul}, the stellar velocity dispersion in the inner part of the bulge) while others (e.g. the escape velocity at the Sun's position) were dropped. The constraints used in the last of their papers, grouped by the units in which the quantities are measured, are tabulated below.

Caldwell-Ostriker Constraints

$(\text{length})(\text{time})^{-1}$:	σ_{bul}, $\Delta V_{0.3}$, $\Delta V_{0.5}$, $\Delta V_{0.7}$, $\Delta V_{1.6}$, V_{glob}, V_{MC}
$(\text{time})^{-1}$:	A, B
$(\text{length})(\text{time})^{-2}$:	$K_z(z)$
(length):	R_0, r_s

Ostriker and Caldwell take pains to emphasize that observations of the differential rotation of the Milky Way are exactly that, and no more. Any constant angular velocity could be added to the model without changing the predicted values of $\Delta V_{0.3}$, et cetera. The angular velocity is not, however, unconstrained. It is equal to A - B, and is further constrained by the circular velocities V_{glob} and V_{MC}, which are inferred, in the case of V_{glob}, from the velocity dispersion of distant globular clusters and nearby dwarf galaxies, and in the case of V_{MC}, from the orbit of the Magellanic clouds.

These last two are among the more uncertain of the adopted constraints (see, for example, TREMAINE's contribution to the present proceedings), yet they are the leading constraints for the corona parameters, ρ_c and r_c. Oort's B is next most important in constraining the corona. This serves to confirm what may already have been obvious, that a) proper motions for those distant globular clusters and dwarf galaxies, b) a larger sample of distant stars from which to compute a radial velocity dispersion, and c) a better determination of Oort's B would all be exceedingly valuable in constraining the properties of the corona.

In their third paper [7], Ostriker and Caldwell emphasize the importance of improved measurements of differential rotation exterior to the solar circle for determining the degree of coronal dominance interior to the solar circle. The acceleration due to the disk, $K_z(z)$, must also play an important role in deciding this question.

I would not want it thought that I was emphasizing the virtues of the Milky Way as a nearly normal galaxy without following my own advice. Marc Aaronson, Victor Blanco, Kem Cook and I have been studying differential rotation beyond the solar circle using carbon stars. They are not the best of standard candles, with an rms scatter at K of the order of 0.75 mag, but carbon stars have the advantages of being abundant, easy to find, easy to observe, and neither too young nor too old. We have searched for carbon stars in particularly transparent parts of the Galactic plane, using the Curtis Schmidt telescope. Followup JHK photometry has then been obtained at Cerro Tololo, and velocities have been measured from spectra obtained at Las Campanas.

We borrow a trick from radio astronomy to obtain the apparent magnitude a typical carbon star would have were one to be found at the center of the Galaxy, independent of the value of R_0. The locations of the extrema and zero crossings in the velocity-distance relations along different lines of sight occur at known geometric multiples of the distance to the galactic center. We obtain our typical magnitude by observing carbon stars near these extrema and zero crossings.

The details of our study will be reported elsewhere. Our principal results are that we observe carbon stars at distances of up to $2R_0$ from the galactic center, and that the quantity $d\omega/d(1/R)$, which is equal to $2AR_0$ when evaluated at the Sun's position, is constant, to within the uncertainties, between R_0 and $2R_0$, with a value, averaged over that interval, of 232 +/- 12 km/s.

We expect that when our results are included in the O/C model, the degree of coronal dominance interior to the solar circle will be smaller than was estimated in its most recent incarnation [7]. With the incorporation of new results on $K_z(z)$ and Oort's B, it is reasonable to expect a decisive determination of the relative contribution of the corona to the mass of the Milky Way, at least interior to the last point at which circular velocities can be measured.

This work was supported, in part, by the National Science Foundation under grant AST83-18504.

References:

1. C. Carignan, K. C. Freeman: Astrophys. J. 294,494 (1986)
2. S. M. Kent: Astron. J. 91,1301 (1986)
3. P. C. van der Kruit, L. Searle: Astron. Astrophys. 110,61 (1982)
4. P. C. van der Kruit, K. C. Freeman: Astrophys. J. 303,556 (1986)
5. J. P. Ostriker, J. A. R. Caldwell: "The Mass and Light Distribution of the Galaxy: A Three Component Model" in W. Burton, ed.: The Large-Scale Characteristics of the Galaxy, pp. 441-450. Reidel, Dordrecht 1979
6. J. A. R. Caldwell, J. P. Ostriker: Astrophys. J. 251,61 (1981).
7. J. P. Ostriker, J. A. .R. Caldwell: "A Model for the Galaxy with Rising Rotation Curve", in W. L. H. Shuter, ed.: Kinematics, Dynamics and Structure of the Milky Way, pp. 249-257. Reidel, Dordrecht 1983

Mass Models for Disk and Halo Components in Spiral Galaxies

E. ATHANASSOULA AND A. BOSMA

1. Introduction

Following KALNAJS [1], several authors have derived mass models for spiral galaxies using both photometric and kinematical data. Constant mass-to-light ratios for bulge and disk are assumed, a rotation curve is calculated from the light distribution, and possible differences with the observed rotation data are attributed to a dark halo component. Most authors ([2] - [5]) have concentrated on 'maximum disk' solutions, in which the mass-to-light ratios of bulge and disk are taken as high as possible without overshooting the observed rotation curve. However, other types of models have not been excluded.

We have recently undertaken an analysis of the rotation curves of a large number of galaxies for which both photometric and kinematic data of reasonable quality are available in the literature (ATHANASSOULA et al. [6], hereafter ABP). We have applied additional dynamical constraints, based on spiral structure theory, to the modelling. We will describe here briefly our method and present the main results.

2. Method

The following steps are employed in our mass model procedures :

a. Bulge/disk decomposition. We assume the bulge to be spherical and the disk to be infinitesimally thin. The bulge is given a finite radius, and the disk luminosity profile is extrapolated inwards exponentially.

A different decomposition scheme has been employed by Kent [7]. When we analyze his sample with our method we find good agreement in the resulting mass-to-light ratios.

b. We compute the rotation curves of the bulge and the disk, assuming a constant mass-to-light ratio for each, and compare them with the observed rotation curve. From this comparison we derive a halo rotation curve.

c. Assuming spherical symmetry, we can now evaluate a number of halo parameters as follows :

- directly from the numerical values of the halo rotation curve. We derive the total halo mass up to an outer radius, chosen here to be the optical radius R_{25}, $MH(R_{25})$, and concentration indices $MH(R_{25}/2)/MH(R_{25})$ and $MH(R_{25}/3)/MH(R_{25})$
- fitting a power law to the halo densities
- fitting an isothermal sphere to determine a core radius, a central density and a velocity dispersion.

Thus in principle, except for the freedom of M/L ratios, we have complete mass models. The M/L can take any value between a maximum, for which the rotation curve of the luminous material reaches as high as possible without overshooting the observed rotation curve, and a minimum infinitesimal value, for which all the mass in the galaxy lies in the halo. The first solution is well known as 'maximum disk', while the second is the 'minimum disk'. These two extremes bracket a wide margin which we would like to narrow.

For this we use constraints from spiral structure theory. Swing amplification theory (TOOMRE [8]) shows clearly that too much halo inhibits spiral structure and that the amount necessary to do so is a function of the multiplicity m of the spiral structure (1 armed, 2 armed etc.). Thus for reasonably symmetric two armed galaxies we have an upper and a lower limit of the disk M/L. The requirement that the disk should not be too massive, so as to prevent asymmetries, sets the upper limit. The lower limit is set if the disk is not massive enough to allow for two armed spirals. The first solution we will call maximum disk with $m = 1$ inhibited, or for short 'no $m = 1$', while the second one maximum disk with $m = 2$ inhibited or for short 'no $m = 2$'. Thus for every galaxy in our sample we have calculated four solutions, the maximum disk, the 'no $m = 1$', the 'no $m = 2$', and the minimum disk.

In ABP we used mainly HI rotation curves, which extend on average twice as far out as optical data. If halos are present, we adopt as maximum disk solution that solution which would leave a halo with non-hollow core. This more or less automatically guarantees stability against $m = 1$ perturbations in the majority of cases. For part of the sample we used optical data, and ran into some difficulties : several galaxies can be fitted without the need for a halo (cf. [1] and [7]). Most of these 'truly maximum disk' models are found to be unstable against m = 1 perturbations. Since these can only be cured by a halo (the bulge is in most cases insufficient, cf. SELLWOOD [10] for the BAHCALL et al. [11] model of our Galaxy), we obtain 'no m = 1' models by forcing the presence of a halo with a non-hollow core. For Kent's sample we then found M/L values for the disk on average 35% lower than the maximum possible. However, due to the short range in radii of the optical rotation data, the resulting halo curves are not well determined in most cases, so that the derived halo parameters are quite uncertain.

3. Incorporating the gas

For Sc galaxies the gas fraction becomes high enough to contribute substantially to the rotation curve. For a number of these we have made models with and without the gas included explicitly. For the gas we take the observed HI surface density. It can be argued that the molecular gas scales roughly with the blue luminosity (e.g. YOUNG [12]), so that its contribution to the surface density is already subsumed in the M/L value for the disk.

The main effect of the gas is in the outer parts of the disk, so that keeping the same mass-to-light ratios gives equally good fits. Only the corresponding halo rotation curve is lower, and rises less steep than in the case gas is not taken into account. Thus the inclusion of the gas leads to smaller halo masses within the optical radius, larger halo core radii, and lower halo concentrations.

4. Checking the spiral structure extent

For the 'no m = 1' models of the galaxies in the ABP sample, we have calculated the extent over which the amplification of the m = 2 component can take place. In general, the observed spiral structure can extend up to roughly twice this range (thereby assuming that the spiral can end at the outer Lindblad resonance). For several Sc galaxies indeed this whole range is needed.

One galaxy only did not conform to this prescription: NGC 4736. This is a very complex galaxy, having two distinct surface brightness zones in the main disk, and an outer ring around it. The two zones may well have different mass-to-light ratios, or the disk mass distribution might have been rearranged substantially during its evolution leading to the formation of the ring.

5. Gas content of the resulting models

For our sample we calculated for those galaxies for which CO-data are available an estimate of the global gas fraction in the disk. We evaluated $M_{gas} = M_{HI}(1+f)$, where f represents the contribution of molecular hydrogen. We estimate f from CO-data using several conversion factors for CO to H_2 masses, reflecting the considerable uncertainties and disagreements concerning this issue. Nevertheless, we can, in this way, roughly derive the ratio M_{gas}/M_{lum} (out to R_{25}) for a number of galaxies. The major result is that the gas fractions come out to be reasonable for the 'no m = 1' models, but very high (60% for some blue galaxies) for the 'no m = 2' models. This suggests that solutions with the disk M/L just high enough to allow a bit of m = 2 amplification are not very realistic with respect to their gas content, at least for blue galaxies. In fact, if one adopts the global star formation rates given by KENNICUTT [13], the average Sc galaxy would have had a very short lifetime if its mean past star formation rate equals the present rate. No such difficulties occur for the 'no m = 1' models : the gas fractions for blue galaxies are reasonable (about 30%) and the disk could have been built up in a Hubble time at a rate of star formation equal or higher than the present one.

6. Mass-to-light ratios

The disk mass-to-light ratios have been compared with the galactic evolution models of LARSON and TINSLEY [14] and LARSON [15]. The conventional models of Larson and Tinsley do not fit the data for the blue galaxies at all. Larson's models, based on a bimodal IMF, produce too high M/L values, since he includes also the dark matter within the optical radius. A fit to our 'no m = 1' data with his simple two population model yields mass-to-light ratios for the blue and red components which are higher than those expected for a young blue population with continuous star formation (by a factor of 2.2), and for a red population formed according to the conventional single burst model (by a factor of 1.4). The fit to our 'no m = 2' results is a factor of 2 lower in the M/L values, and hence at the red end lower than a standard Larson and Tinsley model.

The M/L values found by VAN DER KRUIT and FREEMAN [16, 17] based on considerations of the vertical scaleheights and velocity dispersions of stars and gas agree very well with the range for our 'no m = 1' models, but not with the range of the 'no m = 2' models.

7. Halo parameters

The mean halo to disk mass ratio within the optical radius for our 'no $m = 1$' models shows the trend with B-V color expected by TINSLEY [18] : bluer galaxies have relatively more dark material within the optical radius than do redder galaxies.

The halo parameters in our models are closely coupled to the disk parameters. There is a good correlation between V_{max}(disk) and the velocity dispersion of the halo σ (halo). Tully-Fisher relations hold separately for the disk and the halo, i.e. both V_{max}(disk) and σ (halo) correlate with M_B. Note however that, if we include low luminosity galaxies whose rotation curves are still rising at the optical radius, the slopes of the two are noticeably different.

It is thus clear that a sort of conspiracy exists between disk and halo. Another way of studying this is to define a disk central volume density by dividing the disk central surface density by a characteristic scaleheight. Since the ratio scaleheight to scalelength in several edge-on galaxies is reasonably constant (VAN DER KRUIT and SEARLE [19]), we may divide the disk surface density by the optical radius. The resulting disk volume density is found to correlate fairly well with the central density of the halo.

Variations in central concentration of the halo as function of morphological type are relatively difficult to establish. To study them we derived for each morphological type bin the mean values of quantities such as the ratio halo core radius/optical radius, the concentration index $MH(R_{25}/2)/MH(R_{25})$, or the slope of the power fitted to the halo density profile. These behave in a regular fashion, showing that the halos of early type disk galaxies tend to be more concentrated than the halos of late type disk galaxies.

These trends in the halo parameters are in good agreement with the idea of halo contraction due to baryonic infall during the formation of the bulge and the disk, as discussed by several authors elsewhere in this volume.

8. Conclusions

Our main conclusion is that the spiral structure constraints limit disk mass-to-light ratios to a range of 0.3 dex, and that the maximum disk with $m = 1$ disturbances inhibited seems to produce the most reasonable models in terms of gas fraction and mass-to-light ratios. The resulting halo parameters are closely coupled to the disk parameters, in agreement with the idea that infall of the disk forming material contracts the halo.

References:

1. A.J. Kalnajs: in, E. Athanassoula, ed.: Internal Kinematics and Dynamics of Galaxies, IAU Symp. 100, p.87, Reidel Dordrecht (1983)
2. C. Carignan, K.C. Freeman: Astrophys. J. 294, 494 (1985)
3. T.S. van Albada, J.N. Bahcall, K. Begeman, R. Sancisi: Astrophys. J. 295, 305 (1985)
4. J.N. Bahcall, S. Casertano: Astrophys. J. 293, L7 (1985)
5. T.S. van Albada, R. Sancisi: Dark Matter in Spiral Galaxies. Preprint 1986
6. E. Athanassoula, A. Bosma, S. Papaioannou: Halo Parameters of Spiral Galaxies, Preprint 1986
7. S.M. Kent: Astron. J. 91, 1301 (1986)
8. A. Toomre: Theory of Spiral Structure, in S.M. Fall, D. Lynden-Bell, eds.: The Structure and Evolution of Normal Galaxies, pp.111-140. Cambridge Univ. Press, Cambridge, U.K.
9. K.C. Freeman: Astrophys. J. 160, 811 (1970)
10. J.A. Sellwood: M.N.R.A.S. 217, 127 (1986)
11. J.N. Bahcall, M. Schmidt, R.M. Soneira: Astrophys. J. 258, L23(1982)
12. J.S. Young: in, E. Athanassoula, ed.: Internal Kinematics and Dynamics of Galaxies, IAU Symp. 100, p.49. Reidel Dordrecht (1983)
13. R.C. Kennicutt: Astrophys. J. 272, 54 (1983)
14. R.B. Larson, B.M. Tinsley: Astrophys. J. 219, 46 (1978)
15. R.B. Larson: M.N.R.A.S. 218, 409 (1986)
16. P.C. van der Kruit, K.C. Freeman: Astrophys. J. 278, 81 (1984)
17. P.C. van der Kruit, K.C. Freeman: Astrophys. J. 303, 556 (1986)
18. B.M. Tinsley: M.N.R.A.S. 194, 63 (1981)
19. P.C. van der Kruit, L.T. Searle: Astron. Astrophys. 110, 61 (1982)

Session 4

Galactic Structure and Dynamics: Theory

K. Freeman, Chair
July 24, 1986

The Structure and Evolution of Disk Galaxies

R. G. CARLBERG

ABSTRACT. The structure and evolution of self-gravitating disks are dominated by the spiral waves that provide the principal source of dynamical relaxation. This review emphasizes the quasilinear theory of shearing waves as providing a substantial basis for understanding the properties of the waves occurring in large N-body experiments. The nonlinear consequences of spirals, heating and angular momentum transfer, can be reliably measured in the N-body experiments, and are comparable to estimates made using quasilinear theory. The heating time scale is of order 10 orbital periods, and the angular momentum transfer time is of order 100 orbital periods. Many of the important aspects of disk galaxies are directly modelled in these N-body experiments. Finally, a speculative origin of SO galaxies is suggested.

1. INTRODUCTION

The low velocity dispersion disks of galaxies are richly responsive to disturbances. As a result, most disk galaxies exhibit dramatic spiral patterns. The spirals are more than a pretty veneer on the sturdy frame of the galaxy; the spirals are the agent of considerable structural change in the disk, transporting angular momentum and heating the disk. Moreover, spirals compress the gas and are associated with the formation of stars. In the past few years the body of analytic description of disk galaxies laid down has been applied to understanding the spiral activity observed in ever higher quality N-body experiments. A coherent understanding of spiralling and evolution in N-body models is emerging. In brief, N-body spiralling is largely the propagation of shearing wavelets that originate in the slightly lumpy particle distribution. The typical numbers of particles, of order 10^5, induce potential fluctuations directly comparable to those induced by molecular clouds. Some caution is required in applying all the results of N-body experiments to the disks that we observe, since the N-body models do not yet fully describe the 10^{11} stars present.

2. LINEAR THEORY

The basis for understanding spirals is the linear theory of shearing waves in disks, an impressive body of literature delightfully illustrated in the wide ranging review of Toomre (1981). The basic process is conceptually simple: a small amplitude leading wave is sheared and amplified into a strong trailing spiral that then phase mixes away to nothing (see "dust to ashes", Fig. 8 of Toomre 1981). The amplification depends primarily on the gravitational liveliness of the cool disk, but is partly a natural conspiracy between the epicyclic frequency and the shear frequency, that enhances compressed regions as they shear. The amplification from leading to trailing depends on the responsiveness of the disk, and is characterized by two main dimensionless numbers, Q and X.

The stars in a disk must have a minimum random velocity to free from simple radial instabilities. The parameter Q is defined for a stellar disk as the ratio of the radial velocity dispersion to the minimum for axisymmetric stability,

$$Q = \frac{\sigma_u \kappa}{3.36 G\mu}, \tag{1}$$

(Toomre 1964), where σ_u is the radial velocity dispersion, κ is the epicylic frequency and μ is the local surface density of the disk. A similar expression with π replacing the factor of 3.36 applies for a pure gas disk (Goldreich and Lynden-Bell 1965). Q values less than 1 are physically implausible, since the disk is locally unstable to a short wavelength radial instability that quickly acts to heat the disk up to Q of 1 or greater. The linear response to an imposed perturbation potential depends on Q^{-2} Hot disks provide very little response density to help the wave grow.

A characteristic length of gravitational instabilities is,

$$\lambda_c = \frac{4\pi^2 G\mu}{\kappa^2}. \tag{2}$$

The dimensionless ratio X is simply λ_θ/λ_c, where λ_θ is the azimuthal wavelength, $= 2\pi r/m$, where m is number of arms in the spiral.

To follow the amplitude of a shearing wave requires the equations of continuity and momentum conservation, and Poisson's equation for the gravitational field. The maximum amplifications of a wave as it shears from leading to trailing in several disk models are summarized by Toomre (1981, Fig. 7). The most strongly amplified waves have an X value near to 2, with a weak dependence on the shape of the rotation curve. Q's much greater than 2.5 nearly completely snuff out the amplification. Consequently, stellar populations with Q's much greater than 2.5 or so might as well be in the halo as far as disk stability is concerned. To be free of spiral patterns, S0 galaxies must have fairly large Q values. How this comes about is discussed below. To summarize, a disk with spirals will have a mean Q value in the neighbourhood of 1.5 to 2.0, as seen in N-body disks and several galaxies, including our own (Toomre 1974, Sellwood and Carlberg 1984, Carlberg and Sellwood 1985, van der Kruit and Freeman 1985).

If the surface density of the disk is lowered the characteristic wavelength becomes shorter, leading to a pattern with more spiral arms around the disk. A simple relation can be given for the preferred number of arms in a disk. A simple model disk is the self similar disk imbedded in a halo that gives a circular velocity constant at V_0. In this case $\kappa = \sqrt{2}V_0/r$ and the surface density is $\mu = fV_0^2/(2\pi G r)$, where f is the fractional mass in the disk. Taking $X_{peak} = 2$, the most strongly amplified patterns have $m = 1/f$. The applicability of this simple expression has been noted in various N-body experiments (Sellwood and Carlberg 1984, Carlberg and Freedman 1985), one example of which is shown in Figure 1. The three experiments shown have identical initial particle distributions, so that the relative amplification of different m components is discernible. The amplitude of the potential fluctuations scales linearly

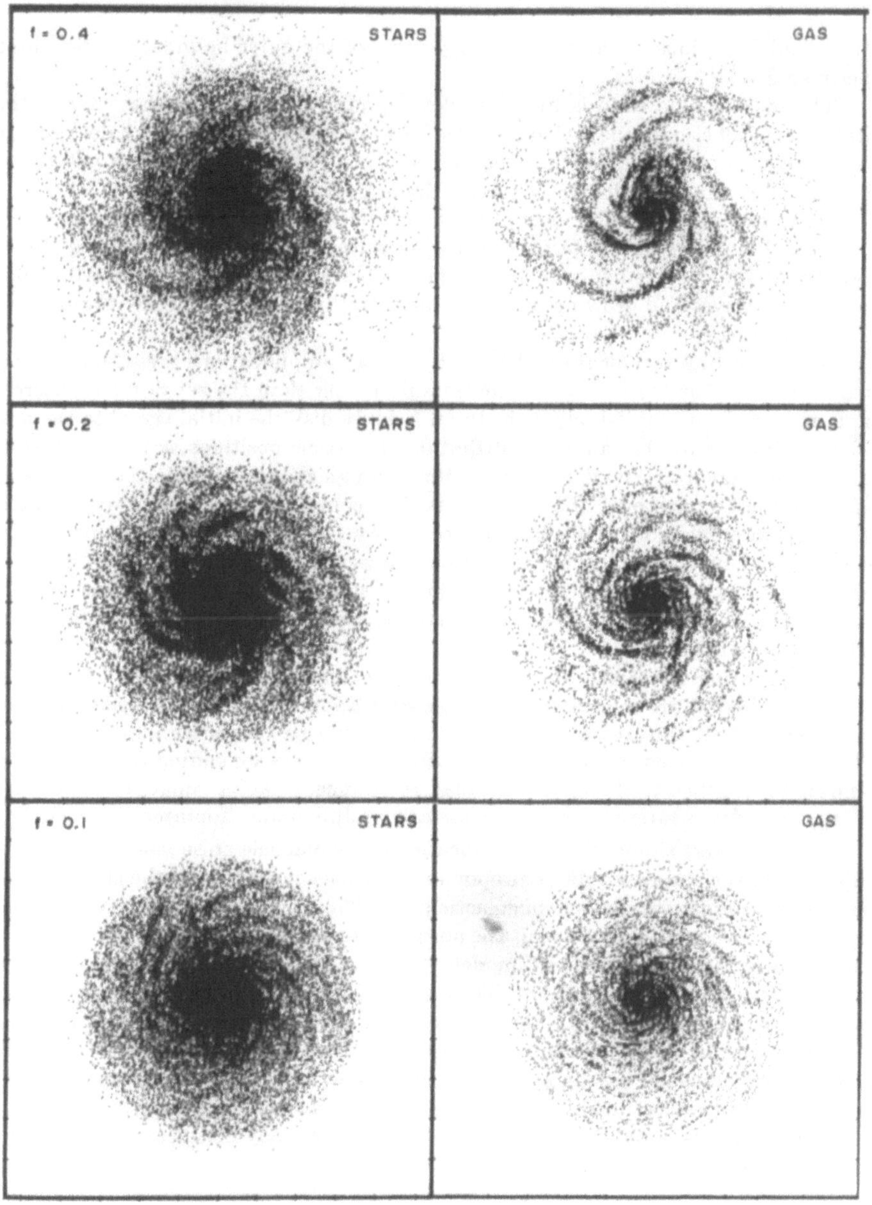

Figure 1: The distribution of star and gas particles in three experiments, varying only the fractional mass in the disk of the galaxy. The experiments are shown at an age of 5 rotation periods, measured at the half mass radius. All experiments start with the same initial particle positions, leading to a similarity in the pictures. The gas particles are 1/4 of all particles, and are collided with a simple sticky particle code, with a coefficient of restitution of 0.9. (Reproduced from Carlberg and Freedman, 1985).

with f, and is obvious in the relative raggedness of the outer boundary of the three experiments.

The disk to halo ratio is the main determinant of the overall character of the spiral arms, if all else is equal. Companions, bars, and possibly distorted halos provide quite a lot of input signal to the amplifier at $m = 1$ and $m = 2$, and would need relatively little amplification to dominate the appearance of the disk.

To a first approximation most of the behaviour observed in large N-body experiments can be understood in terms of shearing waves. Sellwood and Carlberg (1984) showed that the observed spirals were transient features, starting from small amplitude leading waves and growing into prominent trailing patterns, nearly complete in accord with theory.

An interesting technical point is that "swing" amplifications are usually calculated assuming that the disk was a continuum, rather than the set of many discrete points that make up an N-body disk. In an N-body disk the initial signal to the amplifier is provided by the random distribution of particle positions, scaling as $1/\sqrt{N}$. Kalnajs and Toomre (private communication) recognized that this initial perturbing signal does not vanish once the wave starts, but continues to drive the wave a little bit, giving amplifications typically a factor of two larger than in its absence, in the relevant range of Q and X. This correction brings the swing amplification theory in a shearing sheet within a few percent of the measurements of shearing sheet N-body experiments, and probably repairs the factor of two discrepancy noted in Sellwood and Carlberg (1984).

These shearing waves can be viewed equally well as the overlapping stellar wakes that accumulated around particles (Julian and Toomre 1967). The wakes are not small scale, local features and can quite reasonably be viewed as global patterns. The $m = 2$ component of particle at r_0 will have a wake extending from the inner Lindblad resonance to the outer Lindblad resonance, a radius range from 0.3 r_0 to 1.7 r_0, for a flat rotation curve, as illustrated in Julian and Toomre (1967).

If spirals were simply linear amplification of the particle noise in a disk, then the amplitude of the spirals would be proportional to the square root of the effective N of the disk, which may vary widely in galactic disks. This is open to direct test in N-body experiments, where it is found that the spiral patterns are only weakly dependent on N for N larger than 10^4 for the SC models. This is a contradiction of the prediction of linear theory. The difference is not due to some simple numerical problem in N-body code. The same code running a Mestel-Toomre-Zang constant circular velocity model finds the expected $1/\sqrt{N}$ decline of the spiral intensity (Sellwood 1986). The MTZ model is very unusual, having a rotation curve that is absolutely constant to $r = 0$. Evidently more realistic disk models having a more accessible central region behave differently. That is, a purely linear description of an N-body disk seems to miss some slight feedback that helps maintain spiral structure.

3. NONLINEAR SPIRALS AND EVOLUTION

The "anti-spiral" theorem of Lynden-Bell and Ostriker (1967) proves that in linear theory a dissipationless disk cannot have a constant amplitude pattern with any spirality. It appears that disks evade this difficulty twice over, first the spirals are transient, and second the disks evolve. There is a considerable body of literature on secular changes, most calculations springing from the papers of Kato (1971) and Lynden-Bell and Kalnajs (1972). The overall mechanics are reasonably straightforward, but the analytical estimates have differred on the magnitude of the changes

The most obvious nonlinear effect of transient spiral activity is the increase in velocity dispersion in the disk. Left to itself, a disk heats to a Q of 2 in a few rotation periods and the spiral patterns diminish to near invisibility. A source of cooling is required to maintain spirality. Hohl (1970) returned 10% of the particles to circular orbits every rotation to keep the average velocity dispersion low enough to maintain strong spiralling. A more complicated, but effectively similar, tack was taken by Miller, Prendergast, and Quirk (1970).

An effective method of cooling is to increase the surface density of the disk. The dynamical temperature of the disk is given by Q, a ratio of velocity dispersion to surface density. Sellwood and Carlberg (1984) found that an accretion of a few percent of the disk mass per rotation was enough to keep the spiralling going. Accretion in this case simply means that low Q material is added to the disk, and the material may originate as nearby as hot gas surrounding the disk, or as remote as cosmological accretion.

Simply relying on the gas component to provide a low velocity dispersion is not enough. The experiments of Carlberg and Freedman (1985) show that continuous spiralling is maintained in the gas component, however once the disk stars heat beyond a Q of 2.5, the apparent number of arms in the spiralling changes, as if all the disk stars were in the halo.

The direct connection between stars and gas with low velocity dispersion and the presence of spiralling provides a nice explanation of the age velocity dispersion relation of the Population I disk. The older stars have been subject to relatively more spiral waves, and have higher velocity dispersions. The measured age velocity dispersion relation has been characterized as $\sigma \propto t^{1/2}$ by Wielen (1974). A reanalysis by Carlberg, Dawson, VandenBerg and Hsu (1985) found a much shallower rise for the old stars, more in keeping with the $t^{1/4}$ or slower expected from spiral heating (Carlberg and Sellwood 1985). The bonus feature of spiral heating is that it turns on only when it is needed, and then slowly becomes less effective as the stars heat up, so that the disk has an equilibrium value of Q that it maintains if the rate of disk cooling remains constant.

In spite of their dramatic appearance the spirals have a remarkably low degree of nonlinearity. Quasilinear analysis relies on the validity of linear perturbation theory. Although spirals are a fairly strong density perturbation, 20-40% or so in single component models, the potential perturbations are only a few percent. Although the nonlinear effects in disks have important consequences, a reliable description based on linear analysis, and its extensions to higher order, appears to be in substantial agreement with the more easily measured nonlinear properties.

Lynden-Bell and Kalnajs (1972) showed that spiral waves transport angular momentum from the inner Lindblad resonance to the outer resonance, allowing the galaxy to increase its central concentration, somewhat analogous to the initial evolution of a globular cluster. In disks the interactions are wave-particle scattering, instead of particle-particle scattering of star clusters. In a disk where the rate of change of the spiral pattern is comparable to the epicyclic frequency the resonances are broadened to cover virtually the entire range of the spiral wave. The radial range of good coupling between the wave and the particles therefore extends over the entire range of the wave.

The rate of change of the surface density is slow, but significant. The relative rates of surface density change and heating are roughly in the ratio of the energy in random motion to circular motion. The rate of shortening of the scale length r_D of the exponential fitted over the middle range of the disk can be reliably estimated over a short time. For the models in Carlberg and Sellwood (1984) the rate is $d(\ln r_D)/dt = -0.0055$ per rotation period at the half mass radius. This change was measured over 10 rotation periods. One can extrapolate this value to estimate that the scale length will roughly halve in a Hubble time. Such an extrapolation is somewhat dangerous because the rate of transport depends on the details of the spiralling (for instance the amplitude squared) over the course of time, that may well change somewhat from amplitude observed over the short time interval. Given that the disk structure not change too much, in particular that accretion doesn't change the disk mass by a large factor, this estimate should be reliable to a factor of two or so. Working in favour of the reliability of this estimate is the requirement that the spirals always work hard enough to maintain the mean Q of the disk near some equilibrium in the range 1.5 to 2.0, dependent on the rate of cooling. The heating is stable, and angular momentum transport is simply another side of the heating coin. One concludes that the structural change in disks over their lifetime is bound to be significant. Whether there is a dynamically preferred form for the surface density is unknown.

Although the intimate connection of spiralling and angular momentum transport has been recognized for some time, the analytical estimates of the rate of transport vary by several orders of magnitude. Carlberg and Sellwood (1985) showed that the ratio of the heating rate of surface density change is $m(\sigma_r/v_\phi)$, where v_ϕ is the rotation speed of the stars. The transport rate is typically 2% of the heating rate. Since the heating time scale is 5 or so rotation periods the surface density change time scale is 250 rotation periods. The time scale for change measured in the N-body experiment reported above is 180 rotations. Both of these are nearly identical to Bertin's (1983) estimate of 214 rotation periods for time scale for angular momentum transport. However Bertin (1983) choses quite a long time scale for rotation, 0.25 Gyr, such as is appropriate to the very outermost disk. At the half mass radius the dynamical time scale is much shorter, and galaxies should show significant surface density evolution in a Hubble time.

4. NONLINEAR EFFECTS WITH GAS

The gas in the disk provides a source of dissipation in the disk. The disk stars are on epicyclic orbits conveniently thought of as being driven harmonic oscillators, and the gas orbits as a damped harmonic oscillator. The stars respond either exactly in phase or out of phase with the perturbing field. However the damping in the gas causes a phase shift, proportional to the rate of damping, between the stellar response and the gas response. As a result the gas spiral tends to lead the star spiral inside of corotation and trail it outside corotation (Norman 1984, Carlberg and Freedman 1985). Consequently there is a net torque between the stars and the gas that transfers angular momentum out of the stars and into the gas. Carlberg and Freedman (1985) found that the rate of loss of angular momentum was $d(\ln h)/dt = -0.0065$) from the gas to the stars. This loss of angular momentum makes for an additional transport of gas inward, on top of the usual outward transfer of angular momentum. Hence, the gas drifts inwards, on a time scale equal to the Hubble time. In as much as disk galaxies contain gradients of abundance of metals, the gradient may be partly due to the inward transport of metal enriched gas.

5. A NONLINEAR BAR INSTABILITY

The importance of nonlinear effects can be seen in the bar instability of disks with nearly flat rotation curves, the ELN models from Efstathiou, Lake and Negroponte (1981). The continuing improvement of the Sellwood (1981) polar grid code and improved methods of analysis now allow very slowly growing modes to be discovered and isolated. Sellwood (1981) has shown that analytically known rapidly growing spiral modes are easily detected and measured with an error of only a few percent.

Linear modes that grow slowly are hard to detect and measure. The difficulty stems from the particle nature of the disk. The disk continues to support the swing amplified particle noise that rattles around in the disk and jiggles any real mode around its normal smooth growth. Even a stable disk has plenty of action present from the swing amplified particle noise. Given a long enough train of data the small but coherent mode signals could, in principle, be filtered out. However the slight nonlinear effects present then begin to destroy the original careful distribution of the disk particles, and new, rapidly growing, modes often appear. Nevertheless Sellwood has advanced his methods so that modes with growth rates of only a few percent of the pattern speed can be examined.

The slowly growing modes of interest occur when the rotation curve of the disk is flat into a very small core radius. Global mode bar instabilities can occur in a disk when the swing amplifier has a feedback loop through the centre that returns some leading signal with the appropriate phase. The applicability of this simple idea was hinted at in Bardeen (1975) and has been impressively documented by Toomre (1981). There are therefore two ways of killing a bar instability. One is to squash the amplifier, either by adding halo to push the two armed pattern out of the amplifier's bandpass, or by increasing the Q of the amplifying region. There is very little observational room to increase Q in the outer parts of most disks. Suppressing the bar through the addition of some modest halo, comparable to the mass of the disk, is one of the classic arguments for the presence of dark halos surrounding disks (Ostriker and Peebles 1973).

Rather than kill the amplifier one can live more dangerously and try leaving the amplifier on, but simply cutting off the feedback (Toomre 1981). One way is to heat the centre of the disk up (Athanassoula and Sellwood 1986). Another way to cut the feedback is to insert an inner Lindblad resonance, by decreasing the core radius of the rotation curve. A small core forces $\Omega - \kappa/2 = (1 - 1/\sqrt{2})V_0/r$ up cutting off all the lower pattern speeds. The pattern speeds of modes are quantized by the necessity of having a phase match between waves that travel radially through the disk and waves that propagate around a circle. The mode frequencies are not free to move much, so the the signal can only leak a little through the resonance and the mode growth plummets as the barrier is raised.

"Cutting the feedback wire" works well to suppress linear instabilities. Toomre (1982) has analyzed the growth rates of modes in a cold ELN disk, finding that a core radius of 0.1 of the exponential scale length cuts the growth rate of the mode to nearly zero even with very little halo. For a V_m of only 0.75, in the ELN notation, for soft core models, is wildly unstable to the immediate formation of a bar. Sellwood's improved code confirms Toomre's (1982) analysis down to core radii of 0.2, beyond which the modes are growing too slowly to be easy to find. It is therefore possible to say that this small core disk has no significant linear instabilities. Furthermore, this result gives considerable confidence in the N-body code.

The quiet start N-body experiments were specifically designed to eliminate as much as possible of the usual swing amplified noise present in the disk. What if this stable ELN disk is subject to some particle noise? A dramatic bar develops. The bar appears to develop from a swing amplification event that becomes "stuck", and organizes itself into a bar. A further piece of evidence for the fundamentally nonlinear nature of the bar is that if we restrict the code to have only $m = 2$ fourier components in the azimuthal direction, then the bar does not form. It seems that this bar is the result of a finite amplitude instability.

A particularly interesting feature of this bar is that it contains a substantial "inner Lindblad resonance", (in the symmetrized potential) whereas the usual N-body bar grows from a spiral mode and has only a weak, or no, inner resonance. The resonance is due to the strong central potential provided by the halo. Orbits within this bar have a significant probability of being semi-stochastic, about 25%. Bars formed from unstable linear global modes and in softer central potentials appear to have a lower stochastic fraction (Sparke and Sellwood 1986, Binney 1982). Such a bar may be of use in moving gas into the central region of the galaxy.

6. A SPECULATION ON THE ORIGIN OF S0 GALAXIES

It is a simple consequence of shearing wavelets that the preferred number of arms in a spiral pattern is roughly equal to the inverse of f, the fractional mass in the disk. In the inner parts of a galaxy, where the bulk of their mass lies, the bulge to disk ratio, B/D, is a fairly good estimator of $f, = 1/(1 + B/D)$, where B and D have been appropriately weighted with M/L ratios. To a substantial degree it does seem that the appearance of the arms does correlate with the bulge to disk ratio, as in the Hubble sequence. There are many cautions in applying this relation, since the dark matter must be included, and one should allow for the population structure of the disk, since the old disk stars may have such a large velocity dispersion that they do not effectively participate in spiralling.

Star formation appears to be correlated with the spiral arms. Some further evidence for this possibility comes from the work of Wyse (1986), who finds that the surface density of CO correlates with surface density of HI and the local rotation rate Ω as $\mu_{CO} \propto \Omega(r)\mu_{HI}^2$. This would be expected if spirals increase the collision rate in the arms. Therefore, take for the moment the assumption that the rate of star formation increases with the number of spiral arms, given an equal gas content. This allows a simple explanation for the correlation of gas content with Hubble type: large bulge galaxies have relatively more spiral arms, and use up their allotment of gas proportionately quicker. Once the gas is gone, the disk continues to support spirals for a few rotation periods, until the last few stars to form heat sufficiently that no visible spirals are left. An extension of this model would take all disks to be initially similar, so that S0 galaxies simply evolved more quickly, and are now finished.

The rapid evolution of gas to stars should drive a similarly rapid structural evolution. Recalling that the scale length of an exponential decreases, with a characteristic time scale of 200 rotation periods, it is intriguing to note that S0 galaxies have relatively higher central surface brightnesses, and scale lengths usually a factor of two smaller than gas rich disks (Boroson 1981). At high redshifts, proto S0 disks should look like most other disks, except with higher star formation rates. Stripping of gas from disks must play some role, although Sancisi (1981) suggests that stripping mostly removes the extended low column density gas beyond the Holmberg radius.

This entire picture leaves aside the question as to why galaxies that are dominated by their spheroidal population should be found in densely clustered environments.

ACKNOWLEDGEMENTS

Much of this work has been done in collaboration with Jerry Sellwood and many of the views presented developed in our joint work. Alar Toomre is a constant source of enthusiastic guidance and constructive criticism. The ideas about S0's arose partly in discussions with Colin Norman. I am particularly grateful to Johns Hopkins University and the Space Telescope Science Institute for their support.

REFERENCES

Athanassoula, E., and Sellwood, J. A. 1986, *Astr. Ap.*, in press.

Bertin, G. 1984, *Astr. Ap.*, **127**, 145.

Bardeen, J. M. 1975, in *Dynamics of Stellar Systems*, I. A. U. Symp. No. 69, ed. A. Hayli, (Dordrecht: Reidel), p. 297.

Binney, J. 1982, *M. N. R. A. S.*, **201**, 1.

Carlberg, R. G., and Sellwood, J. A. 1985, *Ap. J.*, **292**, 79.

Carlberg, R. G., and Freedman, W. L. 1985, *Ap. J.*, **298**, 486.

Carlberg, R. G., Dawson, P. C., Hsu, T., and VandenBerg, D. A. 1985, *Ap. J.*, **294**, 674.

Efstathiou, G., Lake, G., and Negroponte, J. 1982, *M. N. R. A. S.*, **199**, 1069.

Goldreich, P., and Lynden-Bell, D. 1965, *M. N. R. A. S.*, **130**, 124.

Hohl, F. 1971, *Ap. J.*, **168**, 343.

Julian, W. H., and Toomre, A. 1966, *Ap. J.*, **146**, 810.

Kalnajs, A. J., and Toomre, A. 1985, private communication.

Kato, S. 1971, *Pub. Astr. Soc. Japan*, **23**, 467.

Lynden-Bell, D., and Kalnajs, A. J. 1972, *M. N. R. A. S.*, **157**, 1.

Lynden-Bell, D., and Ostriker, J. P. 1967, *M. N. R. A. S.*, **136**, 293.

Miller, R. H., Prendergast, K. H., and Quirk, W. J. 1970, *Ap. J.*, **161**, 903.

Norman, C. A. 1984, in *Formation and Evolution of Galaxies and Large Scale Structures in the Universe*, ed. J. Audouze and J. Tran Thanh Van (Dordrecht: Reidel), p. 327

Ostriker, J. P., and Peebles, P. J. E. 1973, *Ap. J.*, **186**, 467.

Sancisi, R. 1983, in Internal Kinematics and Dynamics of Galaxies, I. A. U. Symp. 100, ed E. Athanassoula, (Dordrecht: Reidel)

Sellwood, J. A. 1981, *J. Comp. Phys.* **50**, 337.

Sellwood, J. A., and Carlberg, R. G. 1984, *Ap. J.*, **282**, 61.

Sellwood, J. A. 1986, in preparation.

Sparke., L. S., and Sellwood, J. A. 1986, preprint

Toomre, A. 1964, *Ap. J.*, **139**, 1217.

Toomre, A. 1974, in *Highlights of Astronomy*,**3** p. 457.

Toomre, A. 1981, in *Normal Galaxies*, ed S. M. Fall and D. Lynden-Bell (Cambridge: Cambridge University Press), p. 111

van der Kruit, P. C., and Freeman, K. C. 1984, *Ap. J.*, **278**, 81.

Wielen, R. 1974, in *Highlights of Astronomy*,**3**, p. 395.

Wyse, R. F. G. 1986, submitted to Ap. J.

The Effects of Satellite Accretion on Disk Galaxies

P. J. QUINN

ABSTRACT. Over a Hubble time, the disk of a spiral galaxy and its system of satellites can be severely altered by the action of dynamical friction. Satellites sink rapidly to the center of the disk when their orbits bring them within the optical radius. Those satellites in direct orbits with orbital planes close to that of the disk sink the fastest. The resultant satellite population does not however have as much projected anisotropy as that observed by Holmberg (1969). The orbital energy of the devoured satellites is deposited into z and radial motions of the disk stars. Self-gravitating simulations indicate that the z heating produces a noticeable thickening of the disk even for satellite masses of only a few percent of the mass of the disk.

1. Why Worry About Disk-Satellite Interactions?

Satellite systems are a somewhat neglected component of normal galaxies. Little is known of their systematic properties or their longterm evolution and interactions with parent galaxies. The following is a list of observational facts and suspicions that has prompted us (PJQ, Jeremy Goodman, Lars Hernquist and Jens Villumsen) to investigate the dynamical evolution of disk-satellite systems in more detail.

Tremaine (1980) has computed the amount of material that could be accreted by a galaxy over a Hubble time due to relatively slow dynamical friction on an isothermal halo. Folding together the equation for a decaying circular orbit, the two point correlation function and the luminosity function, he concluded:

$$M_{acc.} = 4 \times 10^8 \Big[H_0 t ln\Lambda \frac{100km/s}{\sigma}\Big]^{0.6} (\frac{M}{L})^{1.6} h^{-2.6} M_\odot \simeq 6 \times 10^9 M_\odot$$

where $\sigma \sim 140km/s$, $\frac{M}{L} \sim 6$, $h \sim 1$, $ln\Lambda = 3$, and $t =$ Hubble time. This is about 10% of the mass of the Milky Way disk and comparable to the bulge mass. We would then at least expect the bulge to be significantly modified by the accreted matter.

Disks typically contain about 1% of their total kinetic energy in motions perpendicular to the disk (z direction). Hence a satellite with a mass of 1% of the disk's mass could double the disk's thickness if it surrendered all of its orbital energy to z motions of disk stars. Thin disks may be fragile.

The total HI content of a satellite similar to the LMC is comparable to the HI content of the Milky Way. Hence by accreting a LMC-like satellite we could significantly modify the gas content and gas kinematics of the disk.

Interacting systems are well known for being sights of star formation and nuclear activity (Keel et. al. 1985). Satellites may well be responsible for triggering activity of this type especially if they were carrying massive objects (Gaskel 1985). The rather dramatic spiral response induced by a near encounter with the disk (Toomre 1980) could also lead to a disk-wide burst of star formation.

Holmberg (1969) observed that satellites of disk galaxies avoided the disk plane. Is this due to increased friction near the disk? Bothun and Sullivan (1977) observed that ellipticals have ten times fewer satellites than spirals. Are ellipticals made with fewer satellites or do their satellite systems evolve more rapidly than similar systems around spirals? Einasto, Saar and Kaasik (1974) have noted that elliptical satellites are found preferentially closer to their parent galaxies than are spiral or irregular companions. What might this be telling us about the way different types of satellites are formed or the ways in which the parent galaxy affects the evolution of the satellites?

In order to address some of these issues Quinn and Goodman (1986) conducted a large number of experiments on disk-satellite interactions using a restricted N-body code. The aims of their study were (1)to determine how quickly a satellite sank into the disk given its mass and orbital parameters; (2)to uncover the essential dynamical mechanisms responsible for the sinking,and (3)to determine whether the observations of Holmberg (1969) could be the result of satellite orbit decay due to friction on the disk.

2. Friction in Disk-Satellite Systems

Given that disk galaxies are surrounded by massive, isothermal halos we can derive the sinking rate for satellites on initially circular orbits at large radius.

$$\frac{dln r_s}{dt} = \frac{2.5}{10^{10} yrs}\left[\frac{M_s}{10^9 M_\odot}\right]\left[\frac{V_c}{220 km/s}\right]^{-1}\left[\frac{r_s}{10kpc}\right]^{-2}\frac{ln\Lambda}{3}$$

where M_s is the mass of the satellite, V_c is its circular velocity and r_s is its orbital radius. Hence the inward drift rate due to the halo alone is quite slow. The change in radius is only $5kpc$ over a Hubble time for $M_s \sim 10^9 M_\odot$, $V_c \sim 220km/s$ and $r_s \sim 50kpc$. The Quinn and Goodman experiments did not include friction due to halo. Their conclusion was that the disk began to dominate the friction at a radius $\sim 10\alpha^{-1}$ where α^{-1} is the exponential scale length of the disk.

The interaction between the disk and satellite is in general quite complex involving several competing effects (see Quinn and Goodman 1986 for details). In the case of direct circular orbits in the disk plane, the dynamics can be divided into two clear regimes.

(A): $\Delta R > \frac{GM_s}{\Delta V^2}$. Here ΔR is the distance between the satellite and the edge of the disk and ΔV is the difference in velocity between the satellite and the edge of the disk. In this case the stars near the edge of the disk suffer only small perturbations due to the presence of the satellite. Lynden-Bell and Kalnajs (1972) have shown that in this perturbative case a net torque on the satellite arises only from stars on resonant orbits. In the case where the satellite had a mass of $0.04M_D$ (M_D is the disk mass) and was initially on a circular orbit at $8\alpha^{-1}$ ($R_D \sim 4\alpha^{-1}$), the outward drift rate was $\sim 1\alpha^{-1}$ per 10^{10} years. Even a small disk density in the vicinity of the satellite led to its eventual sinking. Hence the perturbative results may not in general be relevant to disk galaxies.

(B): $\Delta R < \frac{GM_s}{\Delta V_s^2}$. Now the disk stars are strongly scattered by the satellite. Figure 1 shows the evolution of a restricted N-body calculation (no self-gravity). Stars upstream of the satellite receive a positive torque from the satellite and move out in radius behind the satellite ($t = 36.3$). Downstream stars are retarded and lose angular momentum to the satellite causing them to sink inward. The excess density behind the satellite and downstream "cavity" results in a slowing of the satellite and sinking into the disk. Once in the disk, the density gradient of the disk always ensures that there are more stars taking angular momentum than receiving it from the satellite (horseshoe orbits, Goldreich and Tremaine 1982) which leads to rapid sinking.

3. A Summary of Sinking Simulations

The following summarises the main results of the Quinn and Goodman simulations. The results apply to a satellite with a mass $\sim 0.04M_D \sim 2\times10^9 M_\odot$ and a restricted N-body disk.

(a) Once the satellite comes within the disk the sinking is rapid $\lesssim$ few $T_{rot.}$ for direct orbits.

(b) Retrograde orbits sink slower ($\simeq$ 2 to 3 times) than direct orbits due to the larger ΔV at a given r_s.

(c) The sinking rate decreases with the inclination of the orbit to the disk plane due to the decrease in the time spent in the disk and the larger relative velocity of disk and satellite.

(d) It is important to include the center-of-mass motion of the disk when the satellite is outside of the disk. The sinking rate is 2 to 3 times slower when the center-of-mass motion is included in agreement with White(1983).

(e) Inclined satellites sink to the disk plane before they sink to the center of the disk. This results in a preferential z heating in the outer parts of the disk and a radial heating near the center.

(f) Although satellites near the disk plane are preferentially removed by friction, the projected spatial distribution of satellites does not show the strong anisotropy reported by Holmberg. Even a family of purely polar, circular orbits will contribute a significant density to the disk plane when projected on the sky.

4. Self-Gravitating Simulations

The restricted N-body calculations of Quinn and Goodman did not include the self-gravity of the disk. In order to investigate the effects of self-gravity on sinking rates and to deal with the z heating of the disk in a self-consistent manner, Hernquist, Quinn and Villumsen have conducted some initial self-gravitating experiments. The principle aim of this new study is to determine how fragile disks are and what constraints can be placed on the amount of accreted material by insisting disks remain thin for a Hubble time.

Figure 2 shows the evolution of a direct, coplanar encounter of a $0.04M_D$ satellite and an exponential disk (32768 particles, three dimensional, Bahcall and Soneria (1980) model). The disk evolves in a similar manner to that shown in Figure 1,but self-gravity has increased the rate of sinking by a factor of approximately 2.

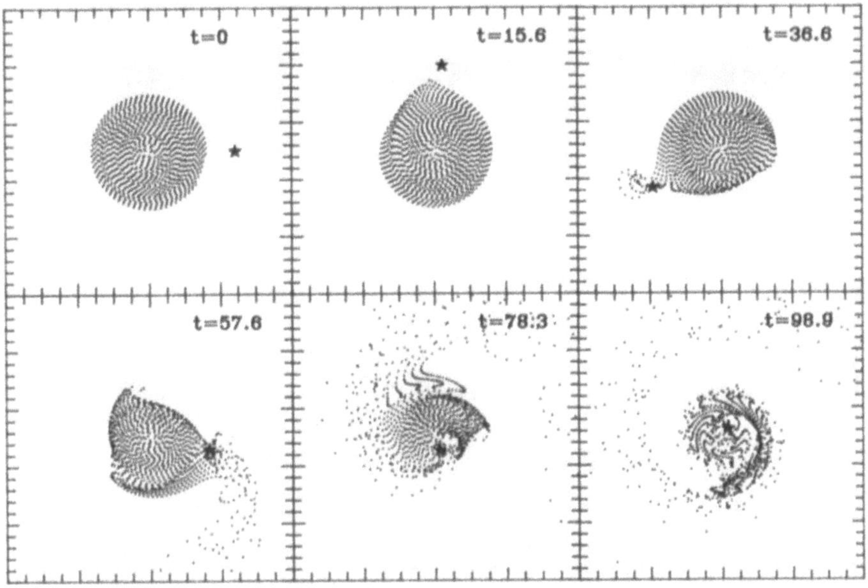

Figure 1. Evolution of a restricted N-body disk with an exponential surface mass density. $M_s = 0.04$; direct, coplanar encounter and the time units are 10^7 yrs. and the boxes are $20\alpha^{-1}$ on edge.

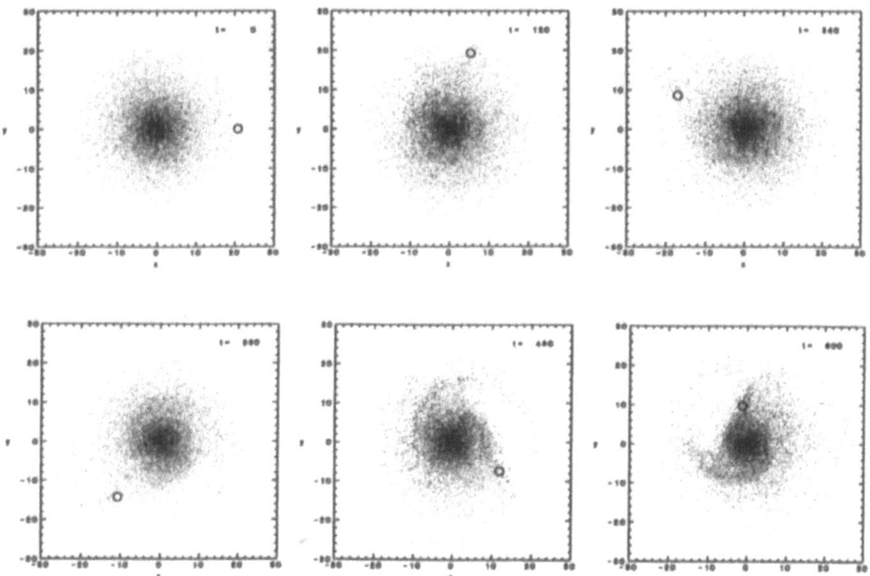

Figure 2. Evolution of a self-gravitating disk. M_S=0.04; direct, coplanar encounter and the time units are 10^6 yrs and the boxes are $17\alpha^{-1}$ on edge.

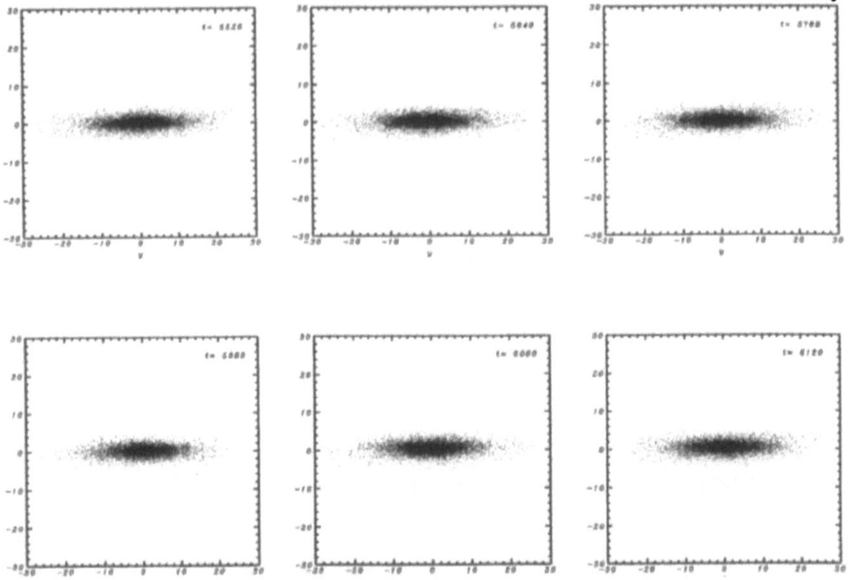

Figure 3. Evolution of a self-gravitating disk and an inclined (30°),$M_s = 0.04$ satellite on a direct orbit seen from in the disk plane. Times and dimensions are as in Figure 2.

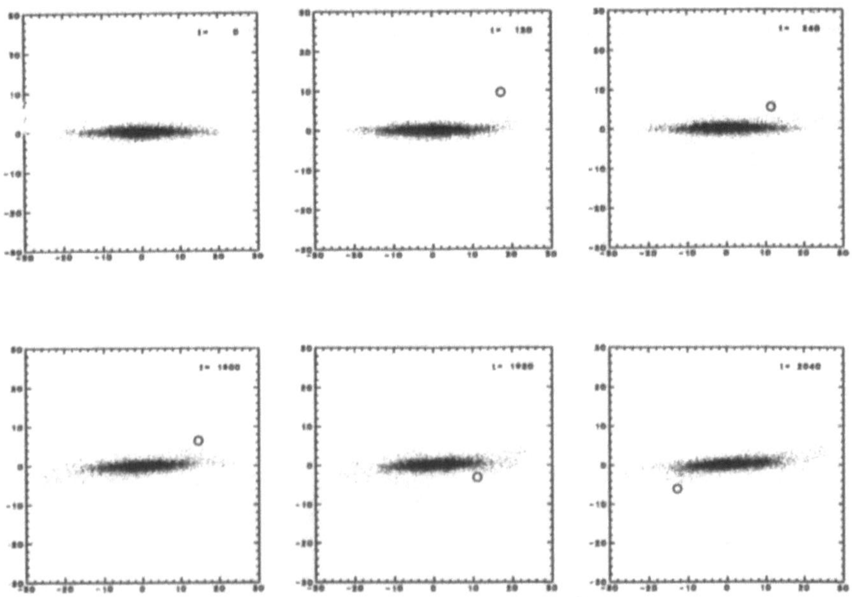

Figure 4. Final stages of the evolution shown in figure 3. The satellite is in the disk plane at a radius $\sim 1kpc$ at $t \sim 6 \times 10^9$ years.

A similar increase was found by White (1983) for three dimensional systems.

Figure 3 and Figure 4 show two time sequences for the evolution of a satellite on a direct orbit with an inclination of 30° relative to the disk plane. The satellite mass was again $0.04 M_D$. Note the tilt (warp) induced in the disk during the initial stages of the sinking eventually phase mixes out at times $\sim 6 \times 10^9$ *yrs,* and the final disk is considerably thicker in projection than the initial disk. Figure 5 shows a comparison of the initial and final mean z thickness as a function of radius. Even near the center of the final disk, the scale height has increased by a factor $\sim$ 50% and the disk "flares" strongly at radii $\sim 20 kpc$ by a factor $\sim$ 3.

Although further simulations are necessary to quantify the effects of self-gravity and to study more completely the thickening of the disk, it is already somewhat alarming to see the amount of damage that a normal mass satellite galaxy can produce. If several such satellites were to sink into the disk, the resultant z structure may not resemble that of normal disk galaxies. This would then constrain the amount of material accreted to be less than or of the order that estimated by Tremaine(1980). Clearly more work in this area will be very fruitful and the fragile nature of disks may well constrain many theories of galaxy formation.

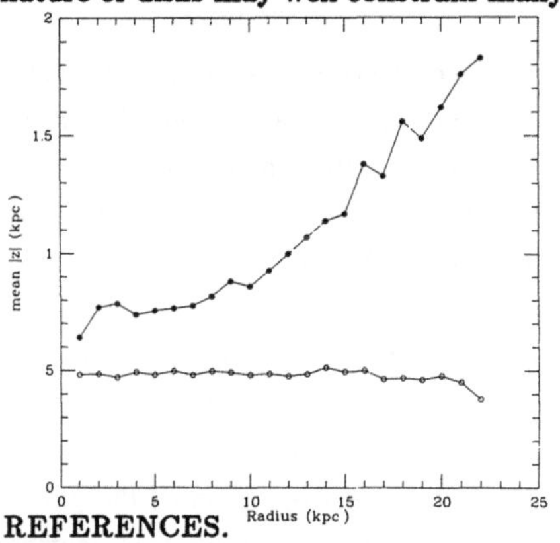

Figure 5. The mean half-thickness of the disk initially and at the end of the simulation shown in figures 3 and 4.

REFERENCES.

Bahcall,J.N. and Soneria,R.M., 1980, *Ap.J. Supp*, **44**, 73.
Bothun,G.D. and Sullivan,W.T., 1977, *P.A.S.P.*, **89**, 5.
Einasto,J., Saar,E. and Kassik,A., 1974, *Nature*, **252**, 111.
Gaskell,C.M., 1985, *Nature*, **315**, 386.
Goldreich,P. and Tremaine,S., 1982, *Ann. Rev. Astr.& Ap.*, **20**, 249.
Holmberg,E., 1969, *Arkiv. Astr.*, **5**, 386.
Keel,W.C., Kennicutt,R.C.Jr., Hummel,E. and van der Hulst,J.M., 1985, *A.J.*, **90**, 708.
Lynden-Bell,D. and Kalnajs,A., 1972, *M.N.R.A.S*, **157**, 1.
Quinn,P.J. and Goodman,J., 1986, *Ap.J.*, in press.
Toomre,A., 1980, in *The structure and evolution of normal galaxies*, eds. S.M. Fall and D. Lynden-Bell, Cambridge University Press, pg. 111.
Tremaine,S., 1980, in *The structure and evolution of normal galaxies*, eds. S.M. Fall and D. Lynden-Bell, Cambridge University Press, pg. 67.
White,S.D.M., 1983, *Ap.J.*, **274**, 53.

Self-Regulating Star Formation and Disk Structure

MICHAEL A. DOPITA

Abstract.

Star formation processes determine the disk structure of galaxies. Stars heavier than about one solar mass determine the chemical evolution of the system and are produced at the rate which maintains by their momentum input, the phase stucture, pressure and vertical velocity dispersion of the gas. Low mass stars are produced quiescently within molecular clouds,and their associated T-Tauri winds maintain the support of molecular clouds and regulate the star formation rate. Inefficient cooling suppresses this mode of star formation at low metallicity.

Applied to the solar neighbourhood, such a model can account for age/metallicity relationships, the increase the O/Fe ratio at low metallicity, the paucity of metal-poor G and K dwarf stars, the missing mass in the disk and, possibly, the existence of a metal poor thick disk. For other galaxies, it accounts for constant w-velocity dispersion of the gas, the relationship between gas content and specific rates of star formation, the surface brightness / metallicity relationship and for the shallow radial gradients in both star formation rates and HI content.

1. Introduction

There can be no successful model for the evolution of disk galaxies without a correct description of the star formation processes. The essential ingredients of a star formation theory are the initial mass function (IMF), the star formation efficiency, and the rate of star formation [1]. Most models of galactic evolution [2,3] have tended to assume a constant IMF and reduce the star formation problem to a simple "prescription" of the rate in terms of the local HI gas density [4] , or of HI surface density [5,6]. However, the molecular component of gas may exceed the atomic contribution by an appreciable factor [7,8,9], and a better correlation is claimed between the total gas surface density and star formation rates [10,11]. This has been used in more recent models of galactic evolution [12,13].

The rôle of the IMF has been receiving increasing attention in recent papers. In our own Galaxy, star formation may well have a bimodal character, with high mass stars being preferentially formed in the vicinity of the spiral arms but low-mass stars being formed throughout the disk [12]. If the CO-emitting molecular clouds map star-formation regions, then their distribution in the Galaxy suggests that a bimodal hypothesis may be appropriate [14]. The CO clouds are clearly divided into two populations which reflect their kinetic temperatures. The warm molecular clouds are clustered, are associated with HII regions and form a spiral arm population. The cold core clouds are distributed throughout the disk. The apparent segregation of the high-mass and low-mass modes of star formation becomes even more pronounced in starburst regions. Here, it appears that only the high mass stars are being formed and that the low mass cutoff in the IMF is of order 3 solar masses [15,16,17,18].

In an inspiring recent paper, Larson [19] has demonstrated that, in a model of the solar neighbourhood with a decreasing global rate of star formation, the IMF must be double peaked. In such a model, the high mass mode is more important over the lifetime of the disk, and stellar remnants can account for the unseen mass in the solar neighbourhood. The division between the "high" and "low" mass sections of the IMF occurs at about one solar mass. Although, in our own Galaxy, the two modes of star formation appear to be spatially distinct, with the high-mass stars preferentially formed near spiral arms, it is not necessary, or even desirable, to associate this with a density wave trigger. Elmegreen [20] has shown that galaxies of the same Hubble types with and without a density wave have effectively identical star formation rates. He argues that the rôle of the density wave is one of spatial ordering rather than one of triggering.

In an earlier paper [21], a model for a high-mass mode of star formation was presented which relied on the idea that high-mass stars control the phase properties and the velocity dispersion in the vertical, axial or w plane, and that this produces a self-regulating feedback to the star formation process. A rather different self-regulating process has been offered for individual molecular clouds by Franco and Cox [22], who developed on ideas by Norman and Silk [23]. In this, the momentum input by winds from low mass young stars supplies the turbulent support for the cloud.In this paper, we identify these two feedback processes with the high-mass and low-mass modes of star formation respectively.

2. The High-Mass Mode of Star Formation.

The energetic proccesses (winds, ionising radiation and supernova explosions) associated with the young, massive stars exercise the fundamental control of the phase properties and pressure of the interstellar medium (ISM). This is the basis of the multi-phase models [24-28]. Cloud-cloud collisions reduce the momentum, and therefore, the velocity dispersion of the gas in the vertical direction (w-velocity dispersion). Thus, energetic processes associated with the high-mass mode of star formation must, in the steady state, feed as much momentum into the gas of the ISM as is being lost in cloud-cloud collisions. If $d\sigma_*/dt$ is the surface rate of star formation, and σ_g is the surface density of gas, then the assumption that the star formation rate is directly related to the dissipation rate can be written,

$$1/\sigma_g\,(d\sigma_*/dt) = \beta/\tau_{cc} \qquad (2.1)$$

where τ_{cc} is the cloud - cloud collision timescale and β is a constant of proportionality composed of both a "spontaneous" term and a "stimulated" term which accounts for the fact that a burst of star formation may induce a local overpressure leading to cloud crushing and induced star formation in its vicinity. These processes represent the physical basis for the model of stochastic self-propagating star formation [29-31]. Here we assume that the coefficient of stimulated star formation is linearly related to the spontaneous term, so that β is not too sensitive to the galaxian environment. The size of the component of stimulated star formation is ultimately controlled by breakout of the gas layer into the halo which drains away the local overpressure.

Since the modulus of the sum of the momentum vectors of the individual gas clouds is maintained at a constant value in the steady state, the rate of injection of momentum to the gas by radiation, mass loss and the supernova explosions of the young stellar population must just match the loss of momentum in cloud-cloud collisions. Hence:

$$\gamma v_{ej}\,(d\sigma_*/dt) = \sigma_g v_g/\tau_{cc} \tag{2.2}$$

where v_g is the vertical w-velocity dispersion of the gaseous layer, v_{ej} is a characteristic velocity of ejection of matter, and γ is a coupling constant. To the extent that the IMF and the energy yield from the high mass stellar population do not depend on metallicity, γ and v_{ej} will be independent of galaxian environment. Equations (2.1) and (2.2) together imply an immediate observational consequence, that $v_g = \beta\gamma v_{ej}$. That is to say that the vertical velocity dispersion of the gas **in all galaxies**, and **at all times** will be the same, and independent of the radial coordinate. This is an observed property of all disk galaxies, covering a wide variety of morphological type, which have so far been observed [32,33]. The observed v_g is of order 7-10 km.s^{-1} and shows little variation between the arm and interarm regions. However, it is seen to increase locally in regions of active star formation.

If we assume that the clouds are in equilibrium with the external pressure to calculate the filling factor and cloud-cloud collision timescale, and if we further assume that the gas and stars are in vertical gravitational equilibrium and that the pressure in the ISM is proportional to the star formation rate, we can solve explicitly for the star formation rate:

$$d\sigma_*/dt = 2^{-3/5}\beta^{1/5}.(f_{init}\,P_{init}/\,\alpha\gamma\,v_{ej})^{2/5}.\,(\pi G)^{1/2}.\,z_o^{-1/10}.\sigma_t^{1/10}\,\sigma_g \tag{2.3}$$

As might have been expected on purely phemenological grounds, the star formation rate depends primarily on the surface density of gas. However, there is also a fairly weak dependence on total surface density, and on the stellar scale height. Since this evolves with time due to stellar diffusion, the z_o term introduces an additional temporal dependence.

3. The Low-Mass Mode of Star Formation

The lifetime of molecular clouds is at least an order of magnitude longer than their free-fall timescales [34,35]. Typical turbulent velocities are highly supersonic,and therefore an energy source is required to give the required support. Winds from young stellar objects might provide this energetic input [22,23,36], which may also serve to regulate the rate of low-mass star formation within molecular clouds [22]. Essentially, the structure of a such a cloud at any instant can be regarded as a set of interlocking shells of compressed gas, orbiting each other under their mutual gravitational attraction. Amongst these, just a sufficient number are in a state of collapse under their self-gravity to provide enough new stars for the turbulent support. Thus, in the absence of any external perturbation, the cloud is gradually converted into stars at a nearly constant rate. At any instant, the volume of the cloud, V_c , should be filled with interacting momentum- conserving shells each of volume V_{int}; $V_c = N_s V_{int}$. If the interaction timescale is t_{int}, and the cooling timescale of the gas after interaction is t_{cool} then the rate of star formation per unit volume, S_V, will be given by:

$$S_v = \epsilon \, [V_{int} (t_{int} + t_{cool})]^{-1} \tag{3.1}$$

where ϵ is an efficency factor. The major coolant is CO [37]. Assuming that the cloud is wholly molecular, the CO abundance, Z(CO), cannot exceed the product of the C and O abundances. Thus the cooling rate declines rapidly toward low abundances and thus the effect of the cooling term in (3.1) is to suppress low-mass star formation at low metallicity. In the steady-state, the interacting shells must be marginally unstable to their self-gravity. We can therefore apply the Elmegreen and Elmegreen [38] criterion;

$$\sigma_s \approx 0.91 \, (P_{int} / \pi G)^{1/2} \quad \text{gm.cm}^{-2} \tag{3.2}$$

where σ_s is the surface density in the shock-compressed shell. Equating the pressure, P_{int}, to the ram pressure of the interacting shells, and solving the equation of motion of the momentum - conserving shells yields:

$$S_v = \eta \, \rho_c^{(3\xi-1)/2} / [1 + \Theta . \, Z(CO)^{-1} . \, \rho_c^{-(\xi+1/2)}] \tag{3.3}$$

where ρ_c is the cloud density, and η and Θ are constants. Once again, the star formation rate is determined primarily by the available gas. The value of ξ depends on the equation of motion of the shells. It is 1/4 for momentum conserving shells in the "snowplough" phase, 1/2 for T-Tauri winds that are still being driven at the time of interaction, and about 3/8 for rotationally driven winds. All of these values imply a very weak dependence of star formation rate on the cloud density. For ξ =1/4,1/3,3/8 and 1/2, $(3\xi-1)/2$ = -1/8, 0, 1/16 and 1/4, respectively. Thus the star formation rate depends very little on the cloud density.

A rough value for the second term in the denominator of (3.3) is about 0.2 in the solar neighbourhood at present . This is estimated from the current ratio of cooling to free-fall timescales for a fragment of the cloud forming into stars.

4. Structural Evolution of the Disk

In order to use the results of the previous two sections in a model of galactic evolution, we require to know how the gaseous and stellar disks evolve with age. Essentially, this resolves itself into two separate problems, stellar diffusion and collapse / infall.

Consider the first of these. A classic problem in stellar dynamics is that the velocity dispersion and scale height of stars appears to increase with age [39]. One possibility is that the observed velocity dispersions reflect the velocity dispersion at birth. However, it appears that the vertical velocity dispersion of the gaseous thin disk is constant for all galaxies, and it is difficult to sustain the argument that velocity dispersions have remained unchanged with time. The one possible exception to this is the "thick disk" of our own Galaxy, which is distinct both in its metallicity, dynamics and scale height. The alternative explanation for a scale height / velocity dispersion / age relationship is that the velocity dispersion is increased with time by stellar orbital diffusion caused by gravitational scattering by giant molecular clouds [39-43]. According to this theory, stars are born with some initial velocity dispersion V_0, which we can take to be the same as the dispersion in the gaseous thin disk. At time t, they will have aquired a velocity dispersion $V_*(t)$ given by:

$$V_*(t) = V_0 [1 + t / \tau]^{1/3} \qquad (4.1)$$

If the scattering clouds and the stars do not remain in the same layer, the exponent will be somewhat different. Empirical values lie in the range 0.25-0.5. The scattering timescale τ depends on the characteristic mass of the molecular clouds. Locally, $\tau \approx 5\times10^7$ years, which implies a cloud mass of a few by 10^5 solar masses, in good agreement with the most recent observations [14]. Similar results apply to the planetary nebula system of the LMC [33].

Van der Kruit and Searle [44-46] found that the stellar scale height , z_0 , is constant with radius in disk galaxies. Since the disk mass density and therefore the restoring force decreases rapidly with radius, this result can only be true if the axial velocity dispersion decreases with radius, V_z = const. exp [$-r / 2R_0$], where R_0 is the disk scale length. If we take equation (4.1) with an exponent of 1/2, this implies that, for $t/\tau >> 1$; t / τ = const. exp [$-r / R_0$]. However, τ scales inversely as the surface density of scattering clouds, i.e. as exp [r / R_0] at the epoch of formation of the majority of the disk stars. Hence the lack of any variation of z_0 with r is consistent with uniform disk age and the operation of stellar diffusion; see also [47].

The existence of a kinematically and chemically distinct "thick disk" [48-51],and papers presented at this conference, suggests that star formation occurred in a thicker layer early in the history of the galaxy. The free-fall timescale is $\tau_{ff} = (z_i / \pi G \sigma)^{1/2}$, where z_i is the scale height and σ is the surface density. The solar neighbourhood values are $\sigma = 67\ M_0\ pc^{-2}$ and z_i =1.5kpc, giving $\tau_{ff} = 4.0\times10^7$ yr. Thus the final collapse to the thin disk would be very rapid unless some form of support was generated. A possible source of such support would be the onset of star formation in the developing thin disk. This would generate a very energetic galactic fountain, similar to starburst regions today (see papers in this conference). The thick disk stars would be formed in the high velocity HI clouds associated with this fountain with the metallicity which is characteristic of this epoch.

5. The Galactic Evolution Model

The initial condition of our disk models is one of a purely gaseous thick disk of zero metal content having a surface density of matter which follows an exponential distribution [44-46].The collapse timescale to the thin disk configuration is taken as a free parameter.

The Scalo [52] IMF has been adopted as a universal law and has been fitted to a power law. For the most massive stars, with masses in excess of 12-15 solar masses, the prompt recycling approximation will be valid. These are the main source of oxygen enrichment [53], but in view of the theoretical uncertainties we take the effective oxygen yield, Y_O, as a free variable. A small fraction, about 0.09, of these stars remain as dark stellar remnants (neutron stars and black holes). For less massive stars we must take into account the delayed yield due to the finite main sequence lifetimes [54]. These are the major source of iron enrichment through carbon deflagration supernovae ([55] and references therein), they also produce carbon by the dredge up and ejection of CN processed material in planetary nebulae [56].

a.The Solar Neighbourhood

The region of the Galaxy around the sun is the only region in which the details of the structural and chemical evolution can be studied, as well as the evolution in stellar content. We can therefore put more restraints on a model of galactic evolution by comparing its predictions with the observations of the solar neighbourhood.

Surface Density, Gas Content ,Gas Depletion Timescale and Metallicity. The gravitational acceleration K_Z, perpendicular to the plane defines the surface mass density in the solar neighbourhood. This yields a value of 67 $M_O pc^{-2}$ about 40% of which is in an unseen form, with a scale height not exceeding 700 pc [57,58]. The local gas content has been estimated in the range from 4.1 to 7.8 $M_O pc^{-2}$ with a mean of about 6 $M_O pc^{-2}$[12,13]. Thus the gas fraction ranges from 0.055-0.104. The local star formation rates are variously computed at between 3 and 8 M_O $pc^{-2} Gyr^{-1}$ [12,59]. Thus gas passes through a stellar generation in a timescale of order 1-3 Gyr in the solar neighbourhood.

These figures, together with the current carbon and oxygen abundances in the solar neighbourhood, put tight constraints on the absolute efficiency of both the high and low mass modes of star formation and on the yields of both carbon and oxygen. This fixes the free variables of the theory, with the exception of the initial collapse timescale. Figure 1 shows a possible model for the solar neighbourhood.

The Age-Metallicity and Metallicity-Metallicity Relationships. There are conflicting estimates for the age-metallicity relation in the solar neighbourhood [60,61]. Fortunately Nissen, Edvardsson and Gustafsson [62] have undertaken an extensive program of high resolution spectrophotometry to determine accurate abundances of many elements in the F and early G type stars. This work shows that the iron abundance has probably varied over a factor of ten over the age of the disk, about 15Gyr. This work also amply confirms the work of Clegg, Lambert and Tomkin [63] showing systematic variation in the ratio of [O/Fe] with [Fe/H], fitting a linear relation; [O/Fe] = -0.5[Fe/H]. By the time that "halo" abundances are reached, at [Fe/H]< -1.0, this relationship appears to flatten and reach an asymptotic value of about [Fe/H] ≈ 0.7 [64].

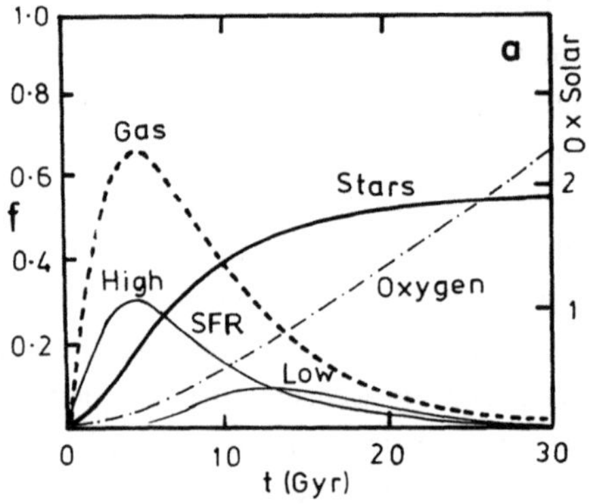

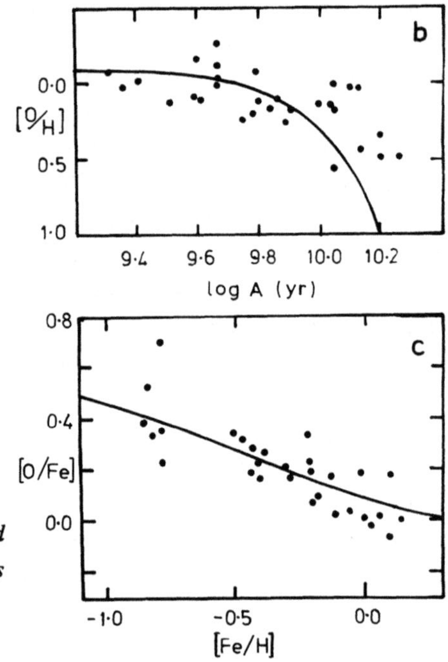

Fig 1a: Evolution of gas content, metallicity and star formation rate in a model of the solar neighbourhood.
b,c: Age/metallicity and metallicity/ metallicity relationships in the solar neighbourhood. Solid lines from the model in(a). Points observations by Nissen et al [62].

Stars producing carbon appear to have the same mass range as those producing iron, since no variation in the [C/Fe] ratio with [Fe/H] is detected, at least for [Fe/H] > -2.0 [64]. This result implies that both Fe and C are produced in the relatively long-lived stars at the lower end of the intermediate mass range, and that the [O/Fe]; [Fe/H] relationship is a consequence of the delayed Fe enrichment vs.the prompt enrichment of O. The theory gives an excellent description of both the age metallicityand O,Fe,H relationships (figure 1b,c), since it is determined only by the lifetime of the carbon and iron producing stars with respect to the gas depletion timescale.

The Metallicity Distribution in the Lower Main Sequence. The problem of the metallicity distribution of stars in the lower main sequence is a key observational constraint on models of galactic evolution. These stars appear to have formed preferentially at a particular epoch, when the metallicity was about 0.4 that of today [4,65,66].

This has a natural explanation in terms of the model, since the low-mass star-formation rate goes through a maximum at a metallicity determined by the product of the carbon and oxygen abundances on the one hand, and by the surface density of gas on the other (see figure 1). Agreement between theory and observations is satisfactory within the limitations of the data.

b.Other Disk Galaxies

Radial Gradients in Star Formation Rates. Kennicutt [67] showed that the H-Alpha flux is effectively a measure of the surface area of O-B stars, and can be used to measure the absolute rate of star formation in galaxies, provided that the shape of the upper IMF is invariant. We have recently undertaken a study of the radial dependence of star formation in normal disk galaxies, using narrowband on-band and off-band imaging photometry at H-Alpha. In all cases yet observed, the radial gradient of star formation is less steep than that of the stars already formed. Such a result can be understood in the context of any which predicts star formation rate to be dependent on a power of the surface density of gas with an index somewhat greater than one. If the power was less than or equal to one, the radial gradient of star formation would scale as the stellar disk. On the other hand, if the index was much greater than one, the gas surface density and the star formation rate would rapidly become flat with radius. In the case that star formation rates depend to any degree on the surface density of matter, then in the absence of radial flows [68], the gas in the central regions of galaxies would rapidly be eaten out, resulting in outwardly increasing gradients of star formation.

Our model has an effective power law index of 1.11, and a radially decreasing gradient in gas content and star formation rate is retained until the gas fraction is less than about 0.01.

Absolute Rates of Star Formation. Dopita [21] used the UV and radio data [69] to show that, for a very wide range of Hubble types, there is a good correlation between the gas content and specific rate of star formation. Theory shows that a certain amount of scatter is induced by the size of individual galaxies in scalelengths, and by the galaxian age. Large galaxies have higher gas content as a result of the contribution of the gas-rich outer portions of the disk, and, naturally, old systems contain less gas. The absolute rate of star formation implied by the local star-formation efficiency in the solar neighbourhood appears to be generally applicable to all galaxian environments.

Vertical Velocity Dispersion of the Gas. As discussed above, the vertical velocity dispersions of the HI at all radii in all disk galaxies and at all times is predicted to be constant. No observational counter-examples are known by this author.

Surface-Brightness / Metallicity Relationship. To the extent to which the elemental yields are constant and to which galactic disks are coeval, there should be a direct relationship between the metallicity and the surface brightness of the disk. Such a relationship has been discovered [70,71]. Dopita and Evans (in press) have quantified this relationship using the re-calibration of the abundance scale for extragalactic HII regions, with results in good agreement with the theory.

7. Conclusions

The model of star-formation proposed in this paper offers a physical model for bimodal star formation which appears to be capable of explaining, in a general way, many of the physical and chemical characteristics of disk galaxies and has been successfully applied to a wide variety of observational material.

References

1. J.Silk: In IAU Symposium #115, M.Peimbert and J.Jugaku eds.: Star Forming Regions, Tokyo, Japan 11-15 Nov. Reidel Dordrecht (1986)
2. J. Audouze and B.M. Tinsley: Ann. Rev. Ast. Ap. 14, 43 (1977)
3. J.P. Vader and T. de Jong: Ast. Ap. 100, 124 (1981)
4. M. Schmidt: Ap. J. 129,243 (1959)
5. N.Sanduleak: Ast. J. 74,47 (1969)
6. K. Hamajima and M. Tosa: Pub. Astr. Soc. Japan 27,561 (1975)
7 J.S. Young and N.Z. Scoville: Ap. J. 258, 467 (1982)
8. N.Z. Scoville and J.S. Young: Ap. J. 265,148 (1983)
9. D.B. Sanders, P.M. Solomon and N.Z. Scoville: Ap. J. 276,182 (1984)
10. R.F Talbot: Ap. J., 235, 821 (1980)
11. K.DeGioia-Eastwood G.L.Grasdalen, S.E.Strom and K.M.Strom: Ap.J.278,564 (1984)
12. R. Güsten and P.G. Mezger: Vistas in Astronomy 26, 159 (1982)
13. N.C. Rana and D.A. Wilkinson: *M.N.R.A.S.* 218, 497 (1986)
14. D.B.Sanders, N.Z. Scoville, and P.M. Solomon: Ap. J. 289, 373 (1985)
15. G.H. Reike, R.M. Catri, J.H. Black, W.F. Kailey, C.W. Mc Alary, M.J. Lebofsky and R.Elston: Ap. J. 290, 116 (1980)
16. G.H.Rieke, M.J. Lebofsky, R.I. Thomson, F.J. Low and A.T. Tokunaga: Ap.J. 238,24 (1985)
17. K. Olaffsson, N. Bergrall and A. Ekman: Ast. Ap. 137, 327 (1984)
18. R. Augarde and J. Lequeux: Ast. Ap. (in press) (1986)
19. R.B. Larson: M.N.R.A.S. 218, 409 (1986)
20. B. J. Elmegreen: In IAU Symposium #115, M. Peimbert and J. Jugaku eds.: Star Forming Regions, Tokyo, Japan 11-15 Nov. Reidel Dordrecht (1986)
21. M.A. Dopita: Ap.J. (Lett.) 295, L5 (1985)
22. J. Franco and D.P. Cox: Ap.J. 273, 243 (1983)
23. C. Norman and J. Silk: Ap.J. 238, 158 (1980)
24. G.B. Field, D.W. Goldsmith and H.J. Habing: Ap.J. (Lett.) 155, L149 (1969)
25. D.P. Cox and B.W. Smith: Ap.J. (Lett.) 189, L105 (1974)
26. C.F. McKee and J.P. Ostriker: Ap.J. 218, 148 (1977)
27. D.P. Cox: Ap. J. 234, 863 (1979)
28. D.P. Cox: Ap.J. 245, 534 (1980)
29. H. Gerola and P.E. Seiden: Ap.J. 223,129 (1978)
30. P.E. Seiden and H. Gerola: Ap. J. 233, 56 (1979)
31. J.V. Feitzinger, A.E. Glassgold, H. Gerola and P.E. Seiden: Ast. Ap. 98,371 (1981)
32. P.C. van der Kruit and G.S. Shostak: Astr. Ap. 134, 258 (1984)
33. M.A. Dopita, S. Meatheringham, H.C. Ford. and B.L. Webster: Ap. J. (in press) (1986)
34. J. Kwan: Ap.J. 229, 567 (1979)
35. L. Blitz and F.H. Shu: Ap. J. 238, 148 (1980)

36. J. Franco: Ap. J. 264, 508 (1983)
37. D. Hollenbach and C.F. McKee: Ap. J. Suppl. 41,555 (1979)
38. B.G. Elmegreen and D.M. Elmegreen: Ap.J. 220, 1051 (1978)
39. R. Wielen: Astr. Ap. 60, 263 (1977)
40. L. Spitzer and M. Schwartzschild: Ap. J. 114,106 (1951)
41. ____________________: Ap. J. 118, 106 (1953)
42. C.G. Lacey: M.N.R.A.S., 208, 687 (1984)
43. J.V. Villumsen: Ap. J. 290, 75 (1985)
44. P.C. van der Kruit and L. Searle: Astr. Ap. 95, 105 (1981)
45. ____________________: Astr. Ap. 95, 116 (1981)
46. ____________________: Astr. Ap. 110, 61 (1982)
47. C.G. Lacey and S.M. Fall: M.N.R.A.S. 204, 791 (1983)
48. G. Gilmore and I.N. Reid: M.N.R.A.S. 202, 1025 (1983)
49. G. Gilmore and R.F.G. Wyse: Astr. J. 90, 2016 (1985)
50. J. Norris M.S. Bessell and A.J. Pickles: Ap. J. Suppl. 58, 463 (1985)
51. J. Norris: Ap.J.Suppl. (in press) (1986)
52. J. Scalo: Fund. Cosmic Phys. 11, 1 (1986)
53. Woosley, S.E. and Weaver, T.A.in eds C.A.Barnes, D.D. Clayton, and R.D. Schramm Essays in Nuclear Astrophysics Cambridge University Press, p377 (1982)
54. I. Iben Jr. and A. Renzini: Ann. Rev. Ast. Ap. 21, 271 (1983)
55. F-K. Thielemann, K. Nomoto and K. Yokoi: Astr. Ap. 158, 17 (1986)
56. A. Renzini and M. Voli: Astr. Ap. 94, 175 (1981)
57. J. N. Bahcall: Ap.J. 276, 169 (1984)
58.__________: Ap.J. 287, 926 (1984)
59. L.F. Smith, P. Biermann and P.G. Mezger: Astr. Ap. 66, 65 (1978)
60. B.A. Twarog: Ap. J. 242, 242 (1980)
61. R.G. Carlberg, P.C. Dawson, T. Hsu and D.A. Vandenberg: Ap.J. 294, 674
62. P.E. Nissen, B. Edvardsson and B. Gustafsson In eds I.J. Danziger, F. Matteuci and K. Kjär 21st ESO Workshop "Production and Distribution of C,N,O Elements", ESO Garching, p 151 (1985)
63. R.E.S. Clegg D.L. Lambert and J. Tomkin: Ap.J. 250, 262 (1981)
64. C. Sneden: in eds I.J. Danziger, F. Matteuci and K. Kjär 21st ESO Workshop "Production and Distribution of C,N,O Elements", ESO Garching, p 1 (1985)
65. H.E. Bond: Ap. J. Suppl. Ser. 22, 117 (1970)
66. B.E.J. Pagel and B.E. Patchett: M.N.R.A.S. 172, 13 (1975)
67. R.C. Kennicut: Ap.J. 272, 54 (1983)
68. C.G. Lacey and S.M. Fall: Ap. J. 290, 154 (1985)
69. J. Donas and J.M. Deharveng: Ast. Ap. 140, 325
70. M.L. Mc.Call: Ph.D. Dissertation, U. of Texas, Publ. in Astronomy, No. 20 (1982)
71. B.M.H.R. Wevers: Ph. D. Dissertation, Groeningen (1983)

Halo Response to Galaxy Formation

JOSHUA E. BARNES

"Flat" rotation curves appear to require correlations between the parameters of the disk and halo of typical galaxies. These correlations can arise naturally when a disk forms by slow accretion within a pre-existing dark halo. As the disk grows, the inner part of the halo responds to the changing gravitational field. The resulting correlation gives a factor of ~ 2 relief from the "fine tuning" problem posed by flat rotation curves. Models with large initial halo core radii ($r_c^0 \gtrsim 8\alpha^{-1}$) give approximately flat rotation curves over the widest range of disk mass. Such large r_c^0 imply that most of the halo orbits contributing to the mass within the optical radius of the galaxy have large radial excursion, extending to much greater r. Consequently, approximations based on circular orbits fail to correctly describe the halo response, and the disk has little effect on the *shape* of the halo.

1. Introduction

The rotation curves of most disk galaxies are either flat or slowly rising out to the last point measured (Bosma 1981, Burstein & Ruben 1985). This has long been interpreted as evidence for extended halos of dark matter around disk galaxies. Dark matter seems required to explain the flatness of rotation curves, barring unusual disk M/L profiles or non-Newtonian gravitation, but it may not naturally explain why rotation curves are so flat. Bahcall & Casertano (1985) have argued that the parameters of the dark and luminous components must be correlated to account for the limited range of rotation curve shapes observed. For example, they find ratios $M_h(r_H)/M_d(r_H)$ of halo to disk mass within the optical radius r_H between 0.84 and 1.5 for a sample of 8 late-type spirals spanning a range of ~ 100 in mass. The two components "conspire" to always produce flat or rising rotation curves (Athanassoula, these proceedings).

What mechanism generates this correlation? The experiments described here were designed to study the hypothesis that the *gravitational* interaction of the disk and halo provides these two components with the means to conspire. Specifically, they model the formation of galaxies along the lines suggested by White & Rees (1978): a luminous galaxy results when a baryonic fraction $f_b \sim 0.1$ collects and forms stars at the center of a pre-existing dark halo. In such a picture, one might imagine that as the baryonic mass began to dominate within a given radius, the dark mass would be pulled in so as to "track" the baryons. This effect is in the right direction to help explain the correlation inferred from the observations.

2. Models

These experiments simulate the dynamical response of a halo of dark matter to the field of an exponential disk. The halo is modeled almost self-consistently with an N-body code in which the gravitational field is expanded to quadrupole order (*e.g.*, Barnes & White 1984). Higher moments are not important when calculating the halo field, since the halo is nearly spherical. An additional term $\mathbf{a}_d(\mathbf{r}) = -\nabla\phi_d(\mathbf{r})$, where $\phi_d(\mathbf{r})$ is potential of an exponential disk with time-dependent mass M_d and inverse scale length α, is included in the equations of motion. To smooth discontinuities in the force field, a small amount of softening (of the usual $(r^2+\epsilon^2)$ form) is used; in all cases, $\epsilon \leq 0.2\alpha^{-1}$, producing no significant effect on halo structure.

Several simplifying assumptions are made in these N-body models. First, the halo is taken to be an isolated system in equilibrium, while real halos may continue growing by mergers and infall throughout galaxy formation. However, mergers seem likely to disrupt disks, and mass accreted by infall may have too much angular momentum to contribute to the central halo density. Thus, an isolated equilibrium halo may not be unreasonable for a model of the optical regions of disk galaxies. Second, the disk is grown very slowly, so that the halo responds adiabatically. In this regime, the final halo depends only on the final disk and not on the history of disk formation (Barnes & White 1984), which greatly simplifies the set of models to investigate. Limited experiments with sudden disk formation do not yield significantly different results. Third, the halo mass is held fixed while the disk mass is slowly increased; galaxies are modeled as open systems which accrete baryons from outside. Alternatively, Blumenthal *et al.* (1985) treated galaxies as closed systems in which the baryons settle out of an initially well-mixed system; this approach is more self-consistent but may not be any more realistic, and requires additional assumptions. However, both models should give similar results near the center (*e.g.*, within the optical radius).

3. Results

A typical simulation is shown in fig. 1. Initial conditions for $N = 4096$ particles were generated from a King (1966) model with mass $M_h \equiv 1$, potential energy $U_h \equiv 1$, and dimensionless central potential $W_c = 5$. As always, $G \equiv 1$. With this scaling, the virial radius R_V, velocity V_V and crossing time $T_V \equiv R_V/V_V$ of the initial halo are all unity; the initial core radius and truncation radii are $r_c^0 \simeq 0.204R_V$ and $r_t^0 \simeq 2.18R_V$ respectively. The N-body realization was first run without a disk for $5T_V$ to insure that the system was well relaxed. Next, the model was run for $20T_V$ in the field of a disk with scale $\alpha^{-1} = 0.025 R_V$ and time-dependent mass $M_d = 0.005(t/T_V)M_h$. Finally, the model was run for $5T_V$ with the disk mass held fixed at $M_d = 0.1$ to let the halo come into equilibrium. "Before and after" plots (fig. 1) show that the halo contracts significantly in the radial direction, but does not become appreciably flattened. The corresponding mass profiles plotted in fig. 2 quantify the halo response; the total halo mass within $r_H \simeq 4.5\alpha^{-1}$ more than doubles.

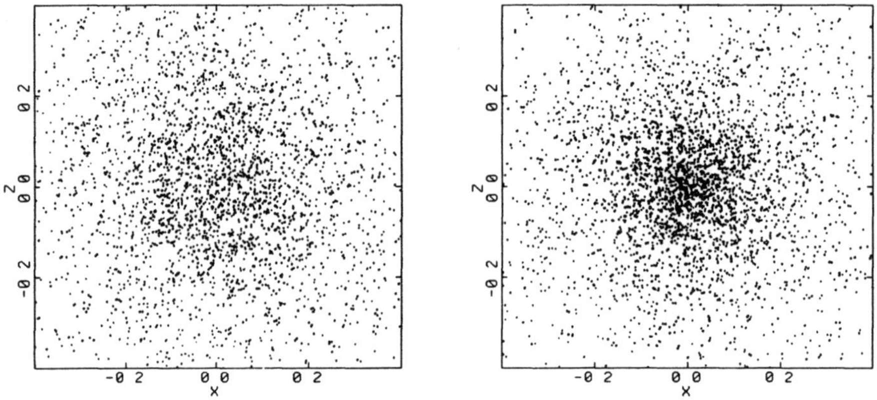

Figure 1. Galactic halo shown before (left) and after (right) the imposition of a disk with mass $M_d = 0.1M_h$ and scale $\alpha^{-1} = 0.025R_V \simeq r_c^0/8.2$. The disk is perpendicular to the **z** axis; note lack of halo flattening.

As an alternative to N-body simulations, a "circular-orbit" model was used by Ryden & Gunn (1984) and Blumenthal *et al.* (1985) to calculate final halo profiles. A similar model may be developed for comparison with the numerical experiments reported here. Start with a spherical halo with initial mass profile $M_h^0(r)$. Consider a circular test orbit with initial radius r^0. Instead of a flat disk, impose a spherical component with the mass profile of an exponential disk,

$$M_d(r) = M_d[1 - (1 + \alpha r)e^{-\alpha r}]; \tag{1}$$

let the resulting halo profile be $M_h(r)$. Adiabatic invariance implies that the final radius r of the test orbit is given implicitly by

$$\frac{r^0}{r} = \frac{M_h(r) + M_d(r)}{M_h^0(r^0)}. \tag{2}$$

Now assume that the bulk of the halo contracts in the same way that circular test orbits do; then $M_h(r) \simeq M_h^0(r^0)$, and

$$\frac{r^0}{r} \simeq 1 + \frac{M_d(r)}{M_h^0(r^0)}, \tag{3}$$

which may be solved by iteration to obtain $r^0(r)$ and hence $M_h(r)$. In the present case, this final assumption appears to be unwarranted. This is shown by the dotted line in fig. 2, which is the halo profile predicted using eq. (3). The circular-orbit calculation overestimates the halo response within r_H by as much as a factor of ~ 2 for models with very large cores ($\alpha r_c^0 \simeq 26.0$).

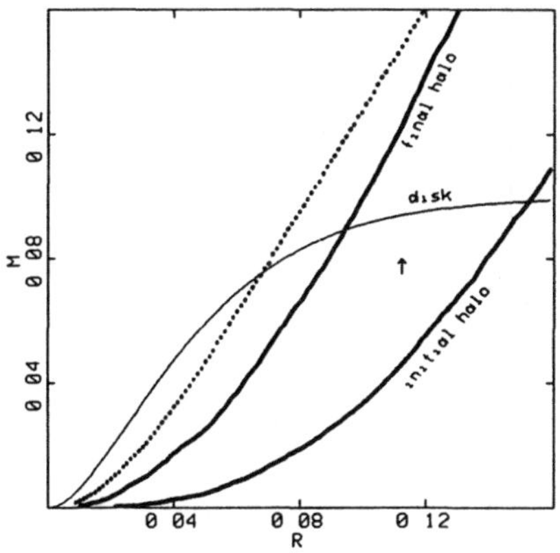

Figure 2. Cumulative mass profiles (mass M within radius R) for the model shown in fig. 1. Arrow indicates the optical radius $r_H \sim 4.5\alpha^{-1}$. Dotted line is prediction of the circular-orbit model.

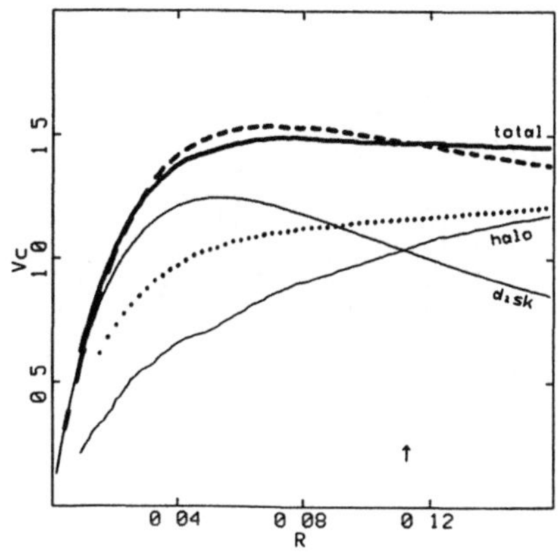

Figure 3. Rotation curves for the model shown in fig. 1, with $\alpha r_c^0 = 8.2$ (solid lines), a model with $M_d = 0.05 M_h$ (dotted line), and a model with $\alpha r_c^0 = 2.06$ (dashed). Arrow indicates optical radius.

Circular velocity profiles $v_c(r)$, measured in the disk plane, were derived from the unsoftened force field using the exact form for the disk and a spherical approximation for the halo:

$$v_c(r) = \sqrt{r|a_d(r)| + GM_h(r)/r}. \tag{4}$$

Typical profiles are shown in fig. 3; these may be directly compared with observed rotation curves (neglecting non-circular motions). The model featured in figs. 1 and 2 yields a flat rotation curve, with no feature demarcating the disk from the halo. At r_H the two components contribute equally to the circular velocity. Models with larger halo cores also yield plausible rotation curves; less massive disks (dotted) produce gently rising rotation curves. On the other hand, models with smaller halo cores typically yield falling rotation curves (dashed), regardless of M_d. Approximately flat rotation curves appear to require large halo core radii ($r_c^0 \gtrsim 4\alpha^{-1}$), as also noted by Blumenthal *et al.* (1985).

Such large halo cores have direct consequences for the orbits of the halo particles within the visible galaxy. As fig. 4 shows, most particles within r_c^0 are moving considerably faster than the local circular velocity, $v_c^0(r) = \sqrt{GM_h^0(r)/r}$. These particles can not be on circular orbits; most reach apogalacticon far beyond the optical radius. This is completely consistent with velocity isotropy within the core. For want of a better term, these orbits will be said to have large radial excursions.

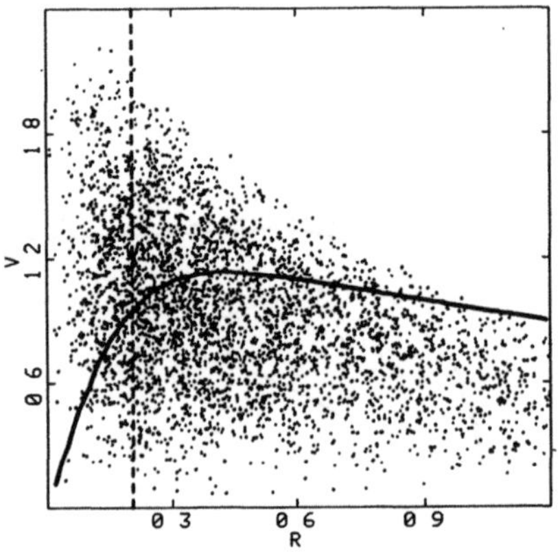

Figure 4. Radius R *vs.* speed V for particles in the initial halo, together with the local circular velocity $v_c^0(r)$. The core radius is indicated by the dashed line.

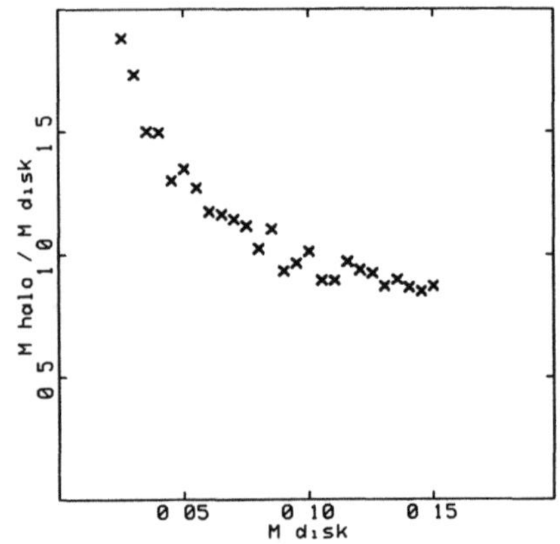

Figure 5. Halo to disk mass ratio within r_H for a one-parameter sequence of models with different M_d. All models have $\alpha r_c^0 \simeq 13.3$.

The fact that halo orbits within the core typically have large radial excursions helps explain two of the numerical results obtained above. First, the simulated halo is less responsive than predicted by the circular-orbit model. Typical halo particles within $r \leq r_H$ reach out to much greater radii at apogalacticon, and sample much more of the halo potential, than circular orbits within r do. Consequently, the *effective* mass increase when the disk forms is much less than the local mass increase within r. Second, the simulated halo is not significantly flattened by the disk potential, and no trapping of halo mass within the disk is observed. To particles on orbits reaching out far beyond r_H, the disk potential is essentially spherical, and the plane of the disk is not distinguished. In view of the physical situation, it may be better to think of the disk as immersed in an isotropic "bath" of halo particles coming in from large radii on orbits which are locally focused by the disk potential.

What do the simulations have to say about Bahcall & Casertano's (1985) generalization that the halo to disk mass ratio $M_h/M_d \simeq 1$ within the optical radius r_H? In a luminous-galaxy-forms-within-dark-halo picture, this seems to require fine tuning of the luminous mass $\sim M_d$ which manages to cool and collapse by $z = 0$. Fig. 5 shows M_h/M_d within $4.5\alpha^{-1}$ for a single simulation with time-dependent M_d and fixed $\alpha = 13.3r_c^0$, reinterpreted as a sequence of equilibrium models with different M_d (the validity of this interpretation follows from the slow disk growth). Over the range of M_d

values which yield roughly flat $v_c(r)$ profiles, the observable M_h/M_d decreases roughly as $M_d^{-0.5}$. Thus, for this sequence of models, halo response relaxes the fine-tuning requirement on M_d by a factor of ~ 2.

4. Conclusions

The dynamical response of the halo component helps to explain why rotation curves are flat, but does not provide a complete account; if the initial halo core radius is too small, it is easy to get a falling rotation curve. Additional constraints, perhaps on the core radius, baryonic fraction, or specific angular momentum of the protogalaxy, seem to be needed to explain the observed shape of rotation curves. Detailed analysis of clustering in N-body simulations (*e.g.*, Barnes & Efstathiou, submitted to *Ap. J.*) suggest that the structure and angular momentum content of protogalaxies should vary widely from one object to the next. Hierarchical clustering, regardless of the fluctuation spectrum, seems to produce only very weak correlations between various properties of typical protogalaxies. In this case, the dynamical response of the dark halo may help explain why the observed regularities appear so strong. *Halo response is a "contraction operator" on the space of initial conditions; e.g.*, the variance of $M_h(r_H)/M_d(r_H)$, the ratio of halo to disk mass within the optical radius, is found to be roughly half the variance expected if halo response is not taken into account (fig. 5). Together with constraints arising from details of galaxy formation (Fall & Efstathiou 1980, Ryden & Gunn 1984, Blumenthal *et al.* 1985), halo response may explain the conspiracy which produces flat rotation curves.

I thank J. Bahcall, G. Blumenthal, S. Casertano, S. Faber, P. Hut, K. Oh, J. Primack and S. White for helpful discussions. Some of these results were first obtained during a visit to Lick Observatory in July 1985. Support for this research came from NSF grant PHY-8440263.

REFERENCES

Bahcall, J. & Casertano, S. 1985. *Ap. J.* **293**, L7.

Barnes, J. & White, S. 1984. *M.N.R.A.S.* **211**, 753.

Bosma, A. 1981. *Astr. J.* **86**, 1791.

Blumenthal, G., Faber, S., Flores, R., & Primack, J. 1985. *Ap. J.* **301**, 27.

Burstein, D. & Ruben, V. 1985. *Ap. J.* **297**, 423.

Fall, M. & Efstathiou, G. 1980. *M.N.R.A.S.* **193**, 189.

King, I. 1966. *Astr. J.* **71**, 64.

Ryden, B. & Gunn, J. 1984. *Bull. AAS* **16**, 487.

White, S. & Rees, M. 1978. *M.N.R.A.S.* **183**, 341.

Session 5

Global Parameters of Galaxies

S.M. Faber, Chair
July 25, 1986

Cores of Early-Type Galaxies: The Nature of Dwarf Spheroidal Galaxies

JOHN KORMENDY[1]

1. Abstract and Introduction

The systematic study of galaxy cores has become possible with high-resolution CCD surface photometry. Many cores are well resolved in a photometry program with the Canada-France-Hawaii Telescope (median stellar FWHM = $0\farcs80$). A general review of core structure based on this photometry is given in [1]. Here I discuss the scale parameters of cores, i. e., the central surface brightness μ_0, the core radius r_c at which the surface brightness has fallen by a factor of two, and the central velocity dispersion σ. Core parameters are correlated: more luminous galaxies have larger core radii, fainter central surface brightnesses, and higher core mass-to-light ratios. These scaling laws provide constraints on theories of galaxy formation. The main purpose of this paper is to discuss the implications of the core parameter relations for the nature and origin of dwarf spheroidal (dSph) galaxies. There is a large discontinuity between the parameter correlations for E and dSph galaxies; the latter are more closely related to dwarf spiral and irregular (dS+I) galaxies [2]. A growing body of evidence suggests that dSph galaxies may have evolved from dS+I galaxies by ram-pressure stripping of their gas.

2. Resolution of Cores

Until recently, studies of galaxy cores were limited by inadequate resolution. This was emphasized by SCHWEIZER [3, 4], who pointed out that the tiny nuclei of M31 and M32 would not be resolved at the distance of the Virgo Cluster. In fact, with available seeing, some apparent cores in Virgo were better described by seeing-convolved $r^{1/4}$ laws than by any isothermal or KING [5] model with a physical core.

---

[1] Visiting Astronomer, Canada-France-Hawaii Telescope, operated by the National Research Council of Canada, the Centre National de la Recherche Scientifique of France, and the University of Hawaii.

Seeing smears out a "cuspy" profile like an $r^{1/4}$ law into a fake core whose apparent radius $r_{c,app}$ is determined by the effective radius r_e [3, 6]. If the apparent core is not much bigger than the radius predicted from r_e, we cannot be sure that it is resolved. It could be resolved, because seeing effects are relatively small for real cores. But the profile could be an $r^{1/4}$ law right to the center, or the galaxy could be hiding a nucleus like the one in M31. At the time of Schweizer's papers, only M31 and the ellipticals NGC 4649, NGC 4472 and M87 were securely resolved. For other galaxies, core parameters could only be derived if one *assumed* that galaxies had cores.

The key to the present photometry program is the excellent seeing at the Canada-France-Hawaii Telescope. An RCA CCD is used at the Cassegrain focus, giving a scale of 0.″215 pixel^{-1}. Figure 1 shows the histogram of seeing dispersion radii σ_*. It hardly overlaps with histograms for two published studies of cores carried out at sites known for good seeing, Mount Wilson and Lick Observatories. The median $\sigma_* = 0.''34$ is a factor of two better than in previous data. As a result, many cores are well resolved. Even SCHWEIZER's [3] best examples of previously unresolved galaxies now have measured profiles that are not consistent with seeing-convolved $r^{1/4}$ laws (e. g., NGC 4406 [1]). With $r_{c,app}/\sigma_* = 8.4$, NGC 4406 is well resolved according to SCHWEIZER's [3] criterion. Almost all bright ellipticals in Virgo and many that are several times farther away are this well resolved. LAUER [8, 9] resolved several additional ellipticals and also found nearly isothermal cores.

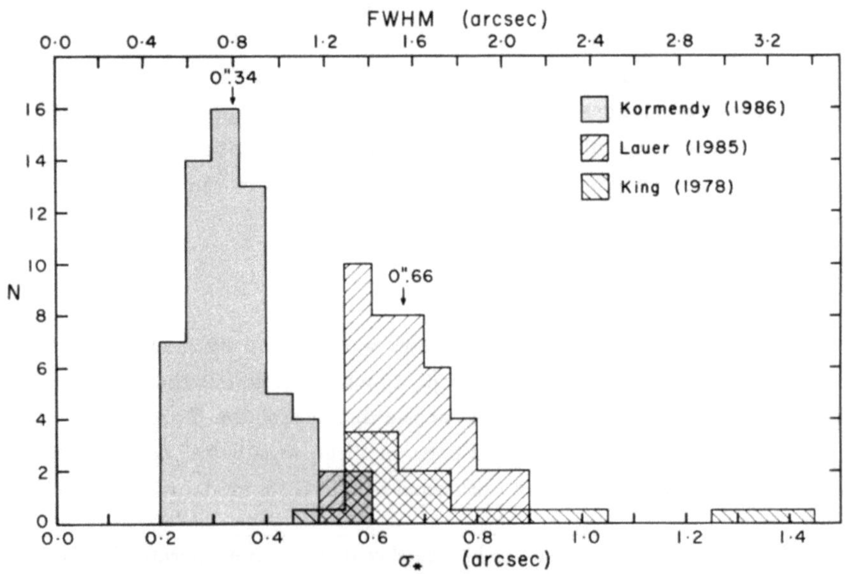

Fig. 1. – Histogram of Gaussian seeing dispersion radii σ_* for the present CFHT photometry and for two published studies of cores. The arrows indicate median values; they are equal for the KING [6] and LAUER [8] samples.

It now seems safe to assume that elliptical galaxies generally have cores. Seeing corrections based on deconvolutions will eventually be made, but here I adopt a short-cut [1, 4] calculated by convolving a King model with $\log(r_t/r_c) = 2.25$ and the point-spread function of the CFHT. It is convenient to parametrize the corrections in terms of $r_{c,app}/\sigma_*$. Galaxies with $r_{c,app}/\sigma_* \leq 3$ are poorly resolved; their seeing corrections still depend on the assumption that cores are nearly isothermal. In many cases the derived parameters are little better than limits. These galaxies are plotted with small symbols in Fig. 2. At the other extreme, galaxies with $r_{c,app}/\sigma_* > 5$ are well resolved; their profile shapes and core parameters are largely independent of assumptions. These galaxies are plotted with large symbols. Intermediate symbols are used for galaxies with moderate seeing corrections.

The following discussion is based entirely on CCD data. The use of CCDs has greatly reduced calibration problems present in nearly all photographic photometry of cores [8]. For example [1], my CFHT CCD photometry agrees well with Lauer's seeing-corrected profiles (but not with his raw data). Similarly, our seeing-corrected core parameters agree well provided that we both resolved the galaxy. This implies that Lauer's seeing corrections are accurate when they are not too large; otherwise they are underestimated. The same is probably true of my seeing corrections. In general, CCD photometry by various authors agrees to ~ 0.1 mag arcsec^{-2} over many magnitudes, allowing large improvements over photographic work.

3. Characteristic Parameter Correlations for Bulges and Elliptical Galaxies

Scaling laws satisfied by the size, density and velocity scale parameters of galaxies provide important input for theories of galaxy formation [10 – 13]. Like global parameters, core parameters correlate with each other and with total luminosity L ($M_B = -2.5 \log L + constant$). For example, more luminous ellipticals have higher central velocity dispersions, $L \propto \sigma^n$, $n \simeq 4 - 6$ [14]. Also, r_c, μ_0, and σ correlate with each other and with M_B [6, 15, 2, 9, 1]. These results are reviewed in [1]; an up-to-date version of the correlations is shown in Fig. 2.

Consider first the elliptical galaxies (*filled circles*). More luminous ellipticals have larger core radii r_c and fainter central surface brightnesses μ_{0V}. Also, r_c and μ_{0V} correlate with σ, as expected from the Faber-Jackson relation. The least-squares fits are [1]: $I_{0V} \propto r_c^{-0.82}$, $I_{0V} \propto L^{-0.86}$, $r_c \propto L^{1.09}$, and $\sigma \propto r_c^{0.18}$. The sequence of elliptical galaxies extends from M32 to cD galaxies like those in A 2029 and A 2199 (NGC 6166). M32 is very poorly resolved, but its tiny core is normal for its low L: there is no sign that tidal truncation has changed M_B by more than ~ 1 mag. Bulges of disk galaxies are almost indistinguishable from ellipticals.

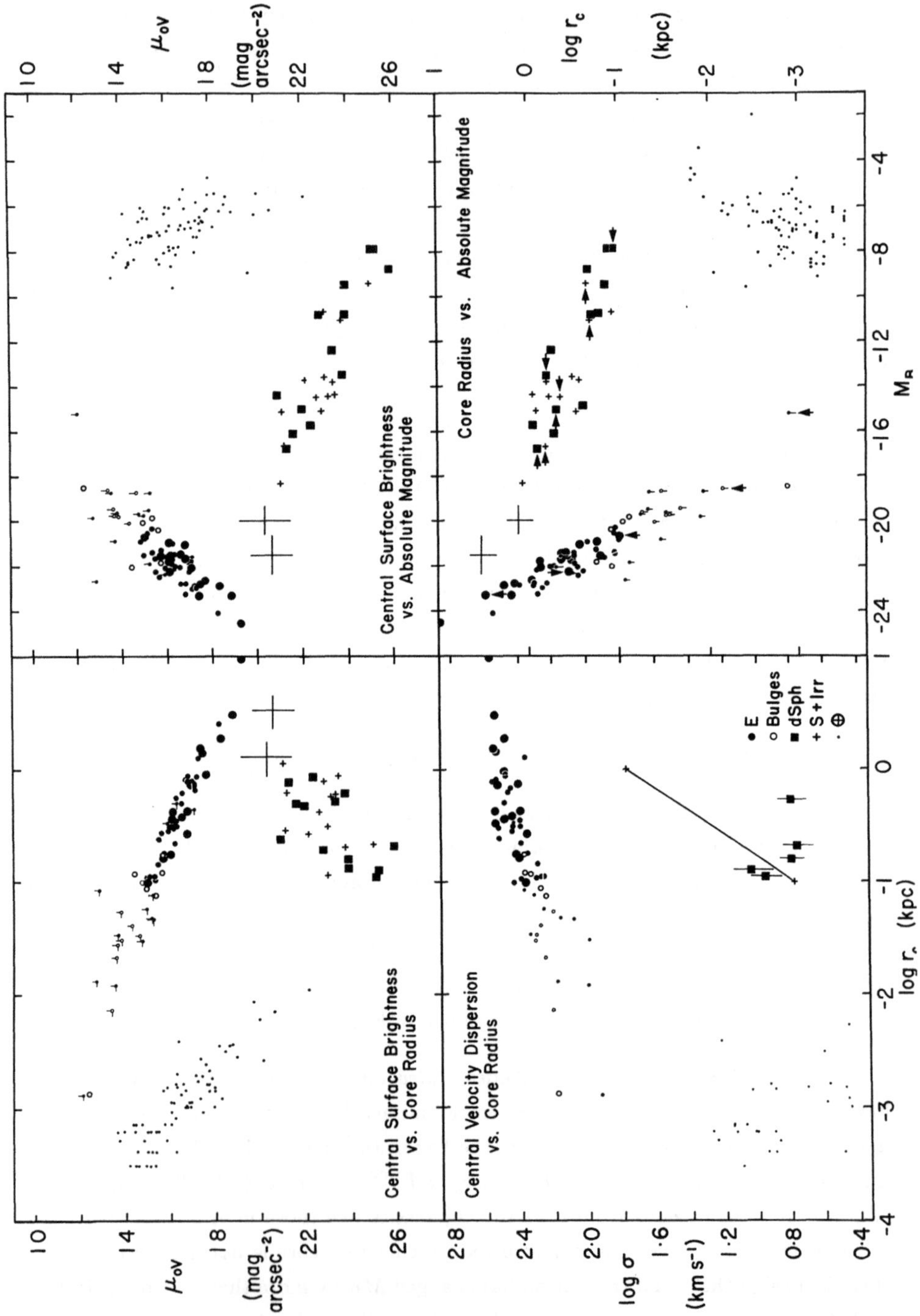
Central Surface Brightness
vs. Core Radius
Central Surface Brightness
vs. Absolute Magnitude
Central Velocity Dispersion
vs. Core Radius
Core Radius vs. Absolute Magnitude
μoV
(mag arcsec⁻²)
log σ
(km s⁻¹)
log r_c
(kpc)
log r_c (kpc)
M_B
E
Bulges
dSph
S+Irr
⊕

Fig. 2. – (Opposite) Core parameter correlations for various kinds of stellar systems [see also 2]. Distances are based on a Hubble constant of 50 km s^{-1} Mpc^{-1}. Larger points imply better resolution of the core (§ 2). Fiducial galaxies whose profiles are illustrated in Fig. 4 are labelled with arrows. The faintest 7 dSph galaxies are, in order of decreasing L, Fornax, Leo I, Sculptor, Leo II, Carina, Draco and UMi. Most of the others are in the Virgo Cluster [16]. The two large crosses show the properties of large disks; they are averages from [17] for 13 and 19 galaxies with exponential scale lengths $\alpha^{-1} > 4$ kpc and $\alpha^{-1} \leq 4$ kpc, respectively. The straight line at lower-left is the Tully-Fisher relation [18] converted to the equivalent $\sigma =$ (maximum rotation velocity)$/\sqrt{2}$.

The existence of a metallicity-luminosity relation for elliptical galaxies suggests that mass-to-light ratios M/L should increase slowly with L; TINSLEY [20] derived $M/L_B \propto L_B^{0.13}$. Figure 3 shows the observed correlation [1], where core mass-to-light ratio $M/L = 9\sigma^2/2\pi G I_0 r_c$ is calculated using King's method [21, 22]. For ellipticals, $M/L_V \propto L^{0.20\pm0.04}$, similar to Tinsley's prediction. The $M/L - L$ correlation may hold to galaxies as faint as M32. RICHSTONE [23] reports that dynamical models of M32 over a large radius range show that $M/L_V = 2.5 \pm 0.1$. This is smaller than the nuclear value, and fits the observed correlation with L. The value is interesting because it is as small as mass-to-light ratios in globular clusters. Dynamical models of globular clusters can account for all of the observed mass with stars and stellar remnants that formed with ordinary mass functions. Richstone's result shows that there is little or no dark matter in the inner parts of M32, perhaps not even the kind believed to exist in the galactic disk. The correlations of metallicity and M/L with luminosity then suggest that there is little or no exotic dark matter in the central parts of other ellipticals. We need to investigate again whether ordinary stellar populations of various metallicities can explain all of the mass in galaxy cores.

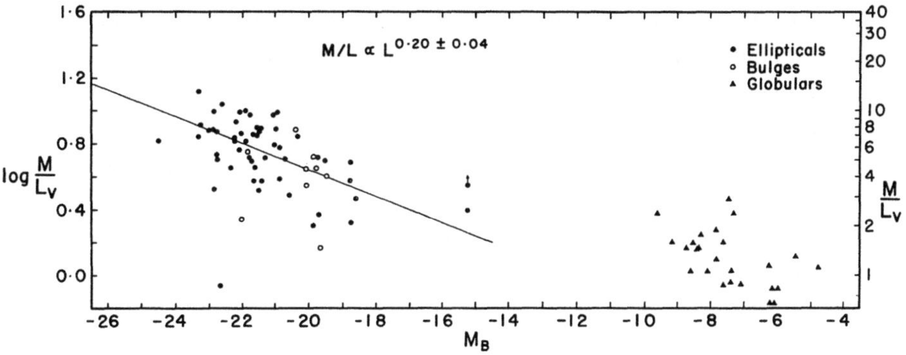

Fig. 3. – Core M/L ratios for bulges, elliptical galaxies and globular clusters [1]. The line is a least-squares fit to the ellipticals. M32 is plotted twice: the upper point is a lower limit to the core M/L [1]; the lower point is the global M/L [23]. The galaxy with the smallest mass-to-light ratio is NGC 1316.

In defining the observing program, I tried to include galaxies that were suspected to have extreme core parameters. Much of the scatter in Figs. 2 and 3 is real. Some is due to distance errors (I will eventually adopt distances based on a Virgocentric flow model). Some is due to a "second parameter" effect. Occasional large deviations suggest special events. The most deviant galaxy, NGC 1316 (Fornax A), has a nucleus that is much too small and bright for the high luminosity of the galaxy ($\log r_c \leq -1.07$, $\mu_{0V} \leq 12.9$ mag arcsec^{-2} at $M_B = -22.7$ [19]). This could have resulted if NGC 1316 swallowed a disk galaxy with a normal bulge of $M_B \sim -20$ [15]. SCHWEIZER [19] also suggested that NGC 1316 is a merger remnant, based on ripples in its light distribution, a possible tidal tail, and a gas disk whose rotation axis is different from that of the stars. If star formation were enhanced by the merger, this could explain the high central surface brightness (Fig. 2, upper left) and the small central M/L.

4.1 Families of Stellar Systems and the Origin of Dwarf Spheroidal Galaxies

Figure 2 shows that there are large discontinuities between the core parameter relations for bulges and ellipticals, dwarf spheroidal galaxies, and globular clusters. This confirms conclusions of WIRTH and GALLAGHER [24] and KORMENDY [2] that these are three very different kinds of stellar systems. The excellent CFHT seeing is critical for this conclusion: it allows sufficient resolution of small galaxies to show that M32 is not unusual for its low L. Without this resolution, the discontinuity between ellipticals and dwarf spheroidals would look much smaller. E. g., if observed in the Virgo Cluster with ordinary seeing ($\sigma_* = 0''.67$), M32 would have $\log r_{c,app} = -1.0$ and $\mu_{0V} = 16.6$ mag arcsec^{-2}, closer to the dSph sequence than to the ellipticals. We are doubly lucky to live so near M32: ellipticals this faint are rare [25, 27] and there is no hope of resolving them from the ground unless they are very nearby.

A possible criticism is that the above discussion is based on *core* parameters. Are E and dSph galaxies still different if more global parameters are considered? The answer is "yes". Core parameters *are* global parameters for dSph galaxies: since they have exponential brightness profiles [26], r_c is completely equivalent to the exponential scale length, $r_c = 0.69\,\alpha^{-1}$. In contrast, core parameters measure only a small fraction of an elliptical galaxy; this is another example of the difference between ellipticals and dwarf spheroidals. Most convincingly, Fig. 4*a* shows mean brightness profiles of galaxies chosen to span the E and dSph sequences. It is clear that E and dSph sequences defined by plausible global parameters would be very different.

The luminosity functions of E and dSph galaxies are also different. SANDAGE *et al.* [25] and BINGGELI [27] show that the luminosity function of elliptical galaxies has a peak at $M_B \simeq -17$; fainter ellipticals are increasingly rare. In contrast, dSph galaxies are rare at $M_B < -17$ but increasingly numerous at lower luminosities.

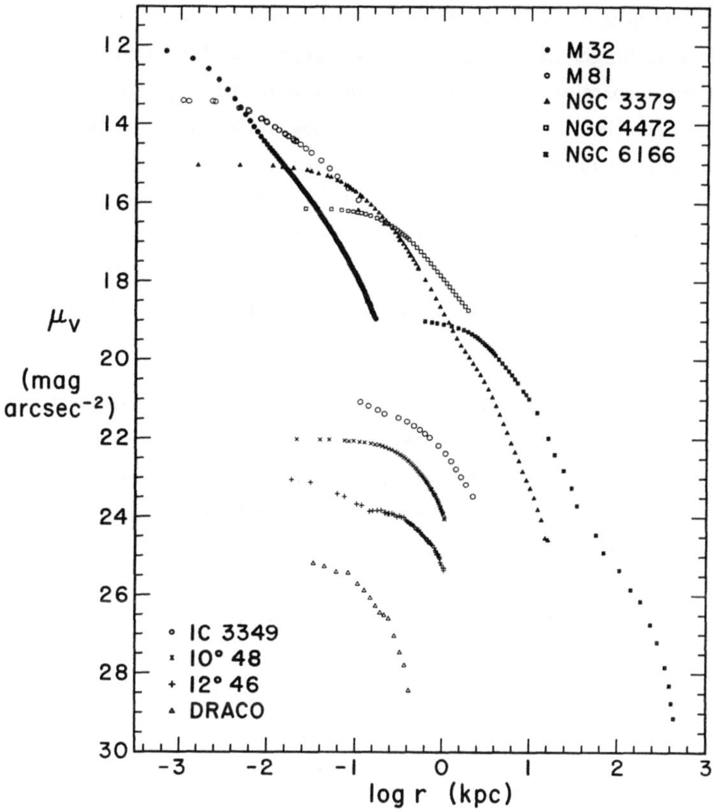

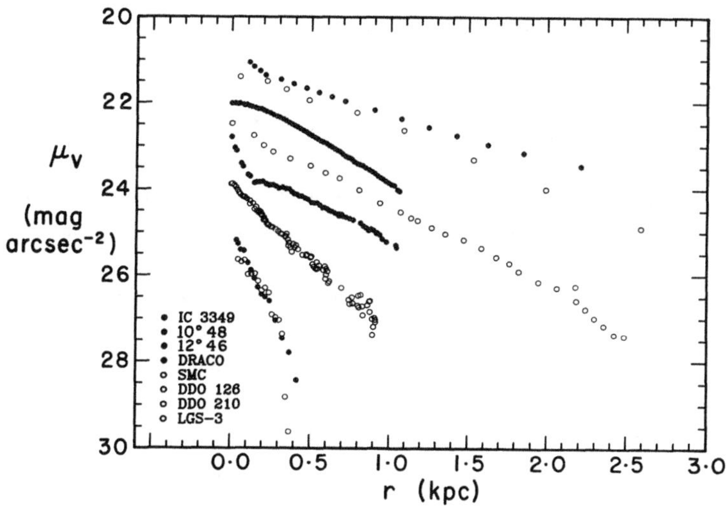

Fig. 4. – (*a*, top) Mean brightness profiles of fiducial galaxies along the E and dSph sequences. (*b*, bottom) Brightness profiles of fiducial dSph and dS+I galaxies. The galaxies are identified by arrows in Fig. 2.

The above results imply a substantial change in the goals of theories of galaxy formation. It is no longer necessary for a single process to form ordinary ellipticals, with their compact cores and deep potential wells, and dwarf spheroidals, with their remarkably low densities. The low-luminosity analog of a galaxy like M87 is still more compact (although not more tightly bound). This and the correlation of rotation with luminosity [28] may both result from increased dissipation at lower L. In fact, even the luminosity function may be related to a dissipation sequence: as L decreases, rotating ellipticals may more and more frequently form with disks, in which case we no longer call them ellipticals. Then they disappear from and contribute to a cutoff in the E-galaxy luminosity function.

Figure 2 suggests that dwarf spheroidal galaxies were formed in a different way than ellipticals. A clue to their origin is obtained by comparing them with dwarf spiral and irregular (dS+I) galaxies. A fair comparison is possible because the brightness profiles of dSph galaxies, although commonly fitted with KING [5] models, are equally well described by exponentials [26, 16, 39, 2]. So μ_{0V} and $r_c = 0.69\,\alpha^{-1}$ can be measured in the same way for dSph and dS+I galaxies. "Core" parameters for late-type dwarfs are shown in Fig. 2; new photometry not in [2] is from [29]. Figure 2 shows that galaxy disks satisfy well-defined parameter correlations of their own. The dSph galaxies fall on the faint parts of these correlations, showing that they are more closely related to dS+I galaxies than to ellipticals. The similarity of dS+I and dSph galaxies is emphasized again in Fig. 4*b*, which shows brightness profiles of typical dSph and dS+I galaxies along the sequences in Fig. 2.

4.2 Did dSph Galaxies Form From dS+I Galaxies by Ram-Pressure Stripping of Gas?

The similarity in brightness distributions and a variety of other observations has led to the suggestion that dSph galaxies formed from dS+I galaxies through ram-pressure stripping of their gas [30–36, 26, 2]. This section reviews the evidence for this hypothesis. It is possible that a different connection exists between dS+I and dSph galaxies; e. g., an internal stripping mechanism [13, 33, 50] or a process that persuades all of the gas to make stars. However, the evidence suggests that stripping makes at least some dSphs. What we need now are quantitative tests of this hypothesis.

Observations suggestive of stripping include the following:

(i) *Exponential brightness profiles* fit both dSph and dS+I galaxies.

(ii) Their *parameter correlations* are similar. These are expressed in Fig. 2 as core parameter relations, but μ_{0V}, r_c and σ are actually global parameters. The similarity in structure suggests that dSph and dS+I galaxies are related, but it does not say how. Plausible "fading vectors" after stripping have been discussed by LIN and FABER [34], by SANDAGE *et al.* [25], and by BINGGELI [36]; it is not clear that the distribution of galaxies in Fig. 2 is consistent with stripping. After stripping, a dwarf irregular will have fainter μ_{0V} and L but larger r_c, the latter because the galaxy is less tightly bound than before. Thus fading vectors are roughly parallel to the observed dSph+S+I sequence only in Fig. 2 (upper right); in the other three panels they are more nearly perpendicular to the observed sequences. It is necessary to check quantitatively that plausible fading vectors connect distributions of dS+I and dSph galaxies. This test is complicated by the fact that no unmodified progenitor population may be left in "hostile" environments like the Virgo Cluster or the vicinity of large galaxies. There are also selection effects: I omitted late-type dwarfs that are vigorously forming stars, and no galaxies like those in (iii) have been measured.

(iii) ***Huge, very low-surface-brightness dwarf galaxies*** discovered in Virgo by SANDAGE and BINGGELI [37; see also 38] have the properties expected for galaxies that lost so much of their mass that they are now almost unbound.

(iv) ***Intrinsic shapes*** of dSph and dS+I galaxies are similar [39-40], consistent with stripping. Both have shapes like those of elliptical galaxies, as expected for very low-luminosity galaxies. BINNEY and DE VAUCOULEURS [41] note the reason. The maximum rotation velocity V_{max} in dS+I galaxies correlates with L (the TULLY–FISHER [42] relation). But the internal velocity dispersion is always ~ 10 km s^{-1} as a result of local processes such as supernova heating. Therefore, as L decreases, V_{max}/σ decreases until the system is no longer flat. This is illustrated in Fig. 5, which shows that faint dS+I galaxies should have shapes like those of ellipticals even if stripping is slow enough to preserve the shape of the galaxy.

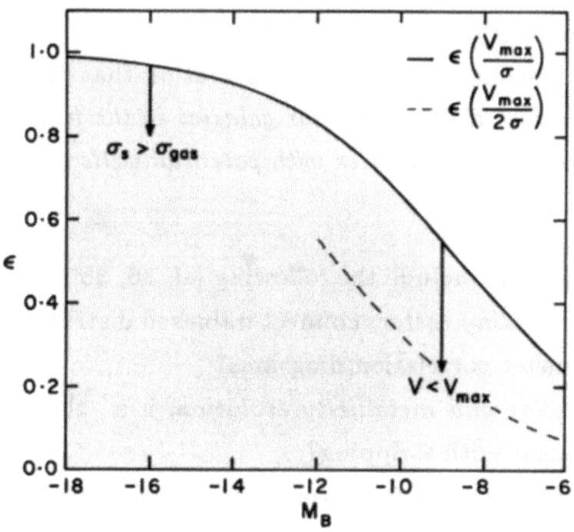

Fig. 5. – Apparent flattenings predicted for dS+I galaxies (*shaded*). The upper bound for edge-on galaxies is derived assuming that $\sigma = 10$ km s^{-1} and that V_{max} is given by the Tully-Fisher relation for Sd-Im galaxies [18]. Then the ellipticity is $\epsilon \simeq v^2/(1+v^2)$, where $v = V_{max}/\sigma$ [6]. In practice, this overestimates ϵ: at high luminosities, most of the mass is in stars with $\sigma > 10$ km s^{-1}; at low L, the rotation curve reaches V_{max} only far outside the luminous galaxy. The dashed line shows the maximum ϵ if the rotation velocity is $\sim V_{max}/2$ over the visible galaxy.

(v) The *intermediate-age stellar populations* in dSph galaxies show that stars continued to form for some time after the galaxies formed [43].

(vi) Similarly, *clumpiness in the distribution of stars in UMi* [44] is most easily understood if the galaxy was once an irregular.

(vii) The *metallicity – luminosity relation* for dSph galaxies is consistent with stripping: it is as continuous with that of S+I galaxies as with that of ellipticals [43].

(viii) The *abnormally high specific globular cluster frequency in Fornax* [45] would be less unusual if the galaxy were once brighter.

(ix) The *apparently high M/L ratios of Draco and UMi* [43] could partly result from fading after stripping. However, the dark matter distributions implied by the high dispersions are so much denser than those of dS+I galaxies that I am still worried about the observations [46].

(x) The *apparent correlation of satellite galaxy morphology with distance from parent galaxies* originally suggested the idea of stripping and is still one of the most direct pieces of evidence for it. EINASTO *et al.* [30] noted that close dwarf companions of bright spirals tend to be dSph, distant companions tend to be dI, and more luminous irregulars "survive" closer to their parent galaxies. This led them to suggest that dSph galaxies were dS+I galaxies that had been stripped by hot gaseous halos.

(xi) The *distribution of dwarf galaxies in Virgo* is suggestive of stripping. Like ellipticals, dwarf spheroidals are concentrated to the cluster center. But the distribution of dS+I galaxies is almost uniform and may even have a hole at the cluster center [25].

(xii) Similarly, the *luminosity function* of dS+I galaxies drops sharply while that of dSph galaxies rises sharply at $M_B > -17$ [25].

(xiii) Most convincing for me is the observation that *bright disk galaxies in the Virgo Cluster contain less H I than similar galaxies in the field* [47, 48]. *This suggests that stripping is important for galaxies with potential wells that are much deeper than those of dS+I galaxies.*

Unresolved questions include the following [cf. 36, 38]:

(−i) Do realistic *fading vectors* connect unbiased distributions of dS+I and dSph galaxies in the parameter correlation diagrams?

(−ii) Are the *color and metallicity evolution*, i. e., the present stellar content, quantitatively consistent with stripping?

(−iii) Are the *huge, low-surface-brightness dwarfs* really dS+I galaxies puffed up after stripping? Or are they the "tip of an iceberg" of very faint galaxies that exist even in the field where stripping cannot be important?

(−iv) *Are there too many nuclei in dSph galaxies?* The existence of some nuclei is not difficult to understand. Some dS+I galaxies have prominent nuclei (e. g., NGC 7793, [49]). Also, small bulges should be present in the brighter and earlier-type progenitor galaxies. And faint nuclei hidden in patchy dS+I galaxies may be more visible after stripping. Nevertheless, we need to understand why nuclei are present so often, especially in faint dwarf spheroidal galaxies.

In summary, a variety of observations suggest that at least some dSph galaxies originate by the stripping of gas from dS+I galaxies. However, it is not clear whether this hypothesis will survive quantitative testing. Internal gas removal processes may also be important [13, 33, 50]. What seems certain is the conclusion that dSph galaxies do not form the faint end of the sequence of elliptical galaxies; rather, they are similar in structure to dwarf spiral and irregular galaxies.

5. References

1. Kormendy, J. 1986, in *IAU Symposium 127, Structure and Dynamics of Elliptical Galaxies*, ed. T. de Zeeuw (Dordrecht: Reidel), in press.
2. Kormendy, J. 1985, *Ap. J.*, **295**, 73.
3. Schweizer, F. 1979, *Ap. J.*, **233**, 23.
4. Schweizer, F. 1981, *A. J.*, **86**, 662.
5. King, I. R. 1966, *A. J.*, **71**, 64.
6. Kormendy, J. 1982, in *Morphology and Dynamics of Galaxies*, ed. L. Martinet and M. Mayor (Sauverny: Geneva Observatory), p. 113.
7. King, I. R. 1978, *Ap. J.*, **222**, 1.
8. Lauer, T. R. 1985, *Ap. J. Suppl.*, **57**, 473.
9. Lauer, T. R. 1985, *Ap. J.*, **292**, 104.
10. Silk, J., and Norman, C. 1981, *Ap. J.*, **247**, 59.
11. Faber, S. M. 1982, in *Astrophysical Cosmology*, ed. H. A. Brück, G. V. Coyne and M. S. Longair (Vatican City: Pontifical Academy of Sciences), p. 191.
12. Silk, J. 1986, in *IAU Symp. 117, Dark Matter in the Universe*, ed. J. Kormendy and G. R. Knapp (Dordrecht: Reidel), p. 335.
13. Dekel, A., and Silk, J. 1986, *Ap. J.*, in press.
14. Faber, S. M., and Jackson, R. E. 1976, *Ap. J.*, **204**, 668.
15. Kormendy, J. 1984, *Ap. J.*, **287**, 577.
16. Binggeli, B., Sandage, A., and Tarenghi, M. 1984, *A. J.*, **89**, 64.
17. Freeman, K. C. 1970, *Ap. J.*, **160**, 811.
18. Aaronson, M., and Mould, J. 1983, *Ap. J.*, **265**, 1.
19. Schweizer, F. 1980, *Ap. J.*, **237**, 303.
20. Tinsley, B. M. 1978, *Ap. J.*, **222**, 14.
21. King, I. R., and Minkowski, R. 1972, in *IAU Symposium 44, External Galaxies and Quasi-Stellar Objects*, ed. D. S. Evans (Dordrecht: Reidel), p. 87.
22. Richstone, D. O., and Tremaine, S. 1986, *A. J.*, **92**, 72.

23. Richstone, D. O. 1986, in *IAU Symposium 127, Structure and Dynamics of Elliptical Galaxies*, ed. T. de Zeeuw (Dordrecht: Reidel), in press.
24. Wirth, A., and Gallagher, J. S. 1984, *Ap. J.*, **282**, 85.
25. Sandage, A., Binggeli, B., and Tammann, G. A. 1985, in *ESO Workshop on the Virgo Cluster*, ed. O.-G. Richter and B. Binggeli (Garching: ESO), p. 239.
26. Faber, S. M., and Lin, D. N. C. 1983, *Ap. J. (Letters)*, **266**, L17.
27. Binggeli, B. 1986, in *Nearly Normal Galaxies: From the Planck Time to the Present*, ed. S. M. Faber (New York: Springer-Verlag), in press.
28. Davies, R. L., Efstathiou, G., Fall, S. M., Illingworth, G., and Schechter, P. L. 1983, *Ap. J.*, **266**, 41.
29. Souviron, J., Kormendy, J., and Bosma, A. 1986, preprint.
30. Einasto, J., Saar, E., Kaasik, A., and Chernin, A. D. 1974, *Nature*, **252**, 111.
31. Aaronson, M., and Mould, J. 1980, *Ap. J.*, **240**, 804.
32. Frogel, J. A., Blanco, V. M., McCarthy, M. F., and Cohen, J. G. 1982, *Ap. J.*, **252**, 133.
33. Gerola, H., Carnevali, P., and Salpeter, E. E. 1983, *Ap. J. (Letters)*, **268**, L75.
34. Lin, D. N. C., and Faber, S. M. 1983, *Ap. J. (Letters)*, **266**, L21.
35. Bothun, G. D., Mould, J. R., Wirth, A., and Caldwell, N. 1985, *A. J.*, **90**, 697.
36. Binggeli, B. 1986, in *Star-Forming Dwarf Galaxies and Related Objects*, ed. D. Kunth, T. X. Thuan and J. T. T. Van (Paris: Éditions Frontières), in press.
37. Sandage, A., and Binggeli, B. 1984, *A. J.*, **89**, 919.
38. Bothun, G. 1986, in *Nearly Normal Galaxies: From the Planck Time to the Present*, ed. S. M. Faber (New York: Springer-Verlag), in press.
39. Okamura, S. 1985, in *ESO Workshop on the Virgo Cluster*, ed. O.-G. Richter and B. Binggeli (Garching: ESO), p. 201.
40. Caldwell, N. 1983, *A. J.*, **88**, 804.
41. Binney, J., and de Vaucouleurs, G. 1981, *M. N. R. A. S.*, **194**, 679.
42. Tully, R. B., and Fisher, J. R. 1977, *Astr. Ap.*, **54**, 661.
43. Aaronson, M. 1986, in *Nearly Normal Galaxies: From the Planck Time to the Present*, ed. S. M. Faber (New York: Springer-Verlag), in press.
44. Olszewski, E. W., and Aaronson, M. 1985, *A. J.*, **90**, 2221.
45. Harris, W. E., and van den Bergh, S. 1981, *A. J.*, **86**, 1627.
46. Kormendy, J. 1986, in *IAU Symposium 117, Dark Matter in the Universe*, ed. J. Kormendy and G. R. Knapp (Dordrecht: Reidel), p. 139.
47. Giovanardi, C., Helou, G., Salpeter, E. E., and Krumm, N. 1983, *Ap. J.*, **267**, 35.
48. Haynes, M. P., and Giovanelli, R. 1986, *Ap. J.*, **306**, 466.
49. de Vaucouleurs, G., and Davoust, E. 1980, *Ap. J.*, **239**, 783.
50. Vader, J. P. 1986, *Ap. J.*, **305**, 669.

Global Scaling Relations for Elliptical Galaxies and Implications for Formation

S. M. FABER, A. DRESSLER, R. L. DAVIES, D. BURSTEIN, D. LYNDEN-BELL, R. TERLEVICH, G. WEGNER

Abstract. Two recent surveys of elliptical galaxy structural properties are described. E galaxies are seen to populate a planar distribution in the global logarithmic parameter space (R_e, σ_e, I_e). Two-dimensionality implies that the virial theorem is the only tight constraint on E structure. There is an additional, weaker constraint on radius versus mass that was presumably imposed at formation. The best-fitting plane in logarithmic coordinates has the equation $R_e \sim \sigma^{1.35\pm0.07}\, I_e^{-0.84\pm0.03}$, which implies $(M/L)_e \sim L^{0.24\pm0.04} I_e^{0.00\pm0.06}$. The planar relation can be used to determine distances to E galaxies to an accuracy of $\pm23\%$. An analogous, parallel plane exists for core properties. Core and global M/Ls agree well, implying that ellipticals are mainly baryon dominated within R_e and that M/Ls are stellar. The effects of other variables such as ellipticity, aspect angle, and rotation on the basic planar relation seem to be small.

I. Introduction

Elliptical galaxies show strong regularities in structural parameters that are of interest both as clues to galaxy formation and as possible distance indicators. Early investigations seemed to show that the manifold of E galaxy structural parameters is primarily one-dimensional. For example, FISH [1] determined that potential energy varies as $M^{3/2}$, and FABER and JACKSON [2] found an L-σ^4 relation between luminosity and velocity dispersion. The data of BINGGELI et al. [3] imply that surface brightness varies as $L^{-0.8}$. KORMENDY ([4] and this volume) summarizes most of these one-dimensional relations using new surface photometry on a large sample of galaxies.

Although early studies emphasized one-parameter relations, several later investigations hinted at second-parameter effects. TERLEVICH et al. [5] claimed to see scatter in the L-σ^4 relation that correlated with intrinsic axial ratio. TONRY and DAVIS [6] confirmed the scatter but failed to find a correlation with any physical parameter including axial ratio. EFSTATHIOU and FALL [7] likewise found no correlation with axial ratio but suggested M/L variations instead. DE VAUCOULEURS and OLSON [8] claimed a correlation with global surface brightness and color. LAUER [9] found a strong correlation with surface brightness for core properties and emphasized that the distribution of both core and global structural parameters was strongly two-dimensional. As this brief summary points up, the situation was rather confused.

The present review describes new results by DJORGOVSKI and DAVIS [10,11 (DD)] and by a group of seven of us studying elliptical galaxies and the Hubble flow (BURSTEIN, DAVIES, DRESSLER, FABER, LYNDEN-BELL, TERLEVICH, and WEGNER [12].) These two studies seem now to agree with each other and Lauer's above in showing the manifold of ellipticals to be strongly two-dimensional and essentially planar.

Before describing the new results in detail, let us digress briefly to ask how many dimensions are expected *a priori* in the manifold of E galaxies. For simplicity, let us assume that all ellipticals follow the same surface-brightness law, e.g., the DE VAUCOULEURS [13] law, and obey the same fall-off in velocity dispersion away from the nucleus. For such a homologous family, three variables contain most of the structural information: the effective radius, R_e; a globally averaged velocity dispersion, σ_e; and an effective surface brightness, I_e. (For I_e, we have used the average value within R_e. The surface brightness *at* R_e, often denoted I_0, could equally well be used.) Other variables such as ellipticity, rate of isophote twist, rotation, and boxiness of isophotes may also contain information, but their importance would need to be determined empirically. Since elliptical scaling laws are basically power laws, it is convenient to use logarithms of the above variables, and reference to "(R_e, σ_e, I_e)-space," for example, denotes a coordinate system of logarithmic variables.

Consider the related space (R_e, σ_e, η_e), where mass surface density, η_e, replaces surface brightness. The virial theorem guarantees that any homologous family of self-gravitating galaxies in dynamical equilibrium will lie in a plane in this space, described by the equation

$$\eta_e \sim \sigma_e^2 R_e^{-1}. \tag{1}$$

This plane will map into a surface in the space of observables (R_e, σ_e, I_e) if, and only if, $(M/L)_e$ is a unique function of position in the plane. If $(M/L)_e$ is furthermore a power-law function of the variables, the second surface will also be planar. To summarize, if ellipticals are found to be a two-dimensional family, the virial theorem is the only strong constraint, and global (M/L) must be a well-behaved function of any two structural coordinates.

If ellipticals are instead a one-dimensional family, this would imply the existence of yet another tight constraint in addition to the virial theorem. An example would be a process that uniquely relates structural scale length to total mass. Such a constraint would have to come from formation and would be of interest because of the possible information about formation it might contain. For this reason, finding that ellipticals are basically one-dimensional rather than two would be considerably more interesting from the standpoint of cosmology.

In addition, one should also check for scatter about the base trends due to galaxy ellipticity, viewing aspect, rotational velocity, etc.

II. Two New Data Sets

The first new data set, due to DJORGOVSKI [14], consists of CCD photometry of roughly one-hundred bright, nearby ellipticals in the field and in clusters. Structural radii, surface brightnesses, and total magnitudes have been derived, plus ellipticities, ellipticity trends, isophote twists, boxiness, and other higher-order parameters. These have been supplemented by published nuclear velocity dispersions.

The second data set [15] has been assembled by our group and contains roughly four hundred galaxies, about two-thirds of which are in groups with two or more members. Existing photoelectric photometry has been homogenized and put on a consistent system, and 2000 new B, V observations have been obtained [16]. DE VAUCOULEURS effective radii and surface brightnesses have been derived from the aperture data (cf. 13), and completely new nuclear velocity dispersions and Mg_2 indices [17] have been measured.

DD base distances on Hubble velocities corrected for a standard Virgo infall model, which they optimize at the same time they fit the basic scaling relations. We find distances using the structural scaling law itself, analogous to the Tully-Fisher method for spirals. The plots below are based on our data only for group galaxies, with distances based on group medians.

III. The Two-Dimensional Manifold of Elliptical Galaxies

The basic result from both data sets is that elliptical galaxies appear to constitute a markedly two-dimensional, nearly planar family [10,11,18]. There may be a slight warp to the plane [cf. also DD], but it is very small compared to the scatter of the residuals about it. This result is illustrated in Fig. 1a,b,c, which plots our group galaxy sample projected onto coordinate axes in (R_e, σ_e, I_e)-space. Figure 1d shows the distribution of points rotated about the σ-axis to show the plane exactly edge-on. The combination (I_e, R_e) shows the plane nearly edge-on, while (σ_e, I_e) shows it nearly face-on. The latter pair thus indicates a galaxy's rough location in the manifold independent of distance. Figure 1 agrees well with corresponding figures by DD.

From remarks above, the existence of a plane as distinct from a line implies that no additional tight constraint beyond the virial theorem was operative in shaping E galaxy structure. (Additional constraints act to limit the area of the plane that is filled, but these are weaker [see below].) In addition, $(M/L)_e$ must be a well behaved power-law in the structural variables. To find the best-fitting plane in (R_e, σ_e, I_e)-space, DD minimize perpendicular residuals about the plane while simultaneously fitting the Virgo infall velocity. We utilize an alternative approach based on nine rich clusters. The distribution of points for each cluster is slid in each coordinate R_e, σ_e, and I_e until the means in each variable superimpose. A mean plane is then fit to the combined distribution. This method obviates the need to derive distances to individual objects and is unbiased as long as the surface is not significantly curved, as seems to be the case. Three least-square planes are fit in each coordinate separately and the results averaged to find a mean plane. From this plane, a power-law expression for $(M/L)_e$ can be derived.

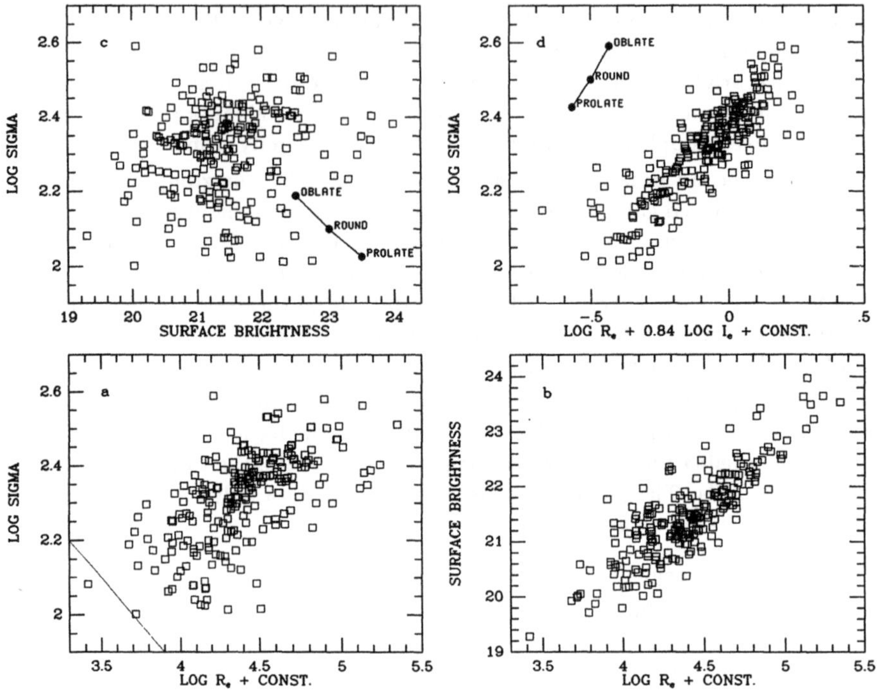

Figure 1: Structural parameters for our group galaxies. a) Velocity dispersion vs. effective radius. Diagonal line shows constant mass locus. b) Surface brightness vs. effective radius. Shows plane nearly edge-on. c) Velocity dispersion versus surface brightness. Shows plane nearly face-on. Dots show aspect effect on E3.6 oblate and prolate galaxies as seen round and at maximum elongation. d) Velocity dispersion vs. log R_e + 0.84 log I_e. Shows plane exactly edge-on. Dots as in panel c.

The two methods and data sets agree well, as shown in the following table:

Our Group:	R_e	$\sim$	$\sigma_e^{1.35\pm0.07}\ I_e^{-0.84\pm0.03}$
	$(M/L)_e$	$\sim$	$L^{0.24\pm0.04}\ I_e^{\ 0.00\pm0.06}$
DD:	R_e	$\sim$	$\sigma_e^{1.39\pm0.14}\ I_e^{-0.90\pm0.09}$
	$(M/L)_e$	$\sim$	$L^{0.22\pm0.06}\ I_e^{+0.08\pm0.09}$
Lauer (cores):	R_e	$\sim$	$\sigma^{1.48\pm0.21}\ I_c^{-0.84\pm0.09}$
	$(M/L)_e$	$\sim$	$L_c^{0.18\pm0.12}\ I_c^{-0.05\pm0.18}$

Within present accuracy, it seems fair to use the rough rule-of-thumb: $(M/L)_e \sim L^{1/4}$, or $M \sim L^{5/4}$. The variation of $(M/L)_e$ vs. L in our data is shown explicitly in Fig. 2. A real dependence seems to be present. In view of the vastly improved quality of these data, the new results ought to supersede previous, conflicting opinions on whether M/L really varies with L [2,19,20]. The one remaining uncertainty might be the use of core velocity dispersions instead of globally averaged values, a problem that will always remain at some level because of unknown velocity dispersion anisotropy.

IV. Residual Scatter Around the Plane

We find scatter in log R_e about the plane of ± 0.10 dex, corresponding to a distance error of $\pm 23\%$. DD find a comparable value. This scatter is only about half of that in the L-σ^4 relation, but it is still significant. Based on known measurement errors in our data of about 0.07 dex, an excess scatter of 0.07 dex appears to remain unexplained.

Scatter can come from various causes:

- Bad distances due to wrong group assignments
- Galaxies with anomalous structure such as S0's, tidally truncated companions, cD's, etc.
- Incorrect σ's due to use of nuclear rather than global values
- Ellipticity, aspect, and rotation effects
- M/L variations due to 1) recent star formation, 2) stellar mergers that scramble dynamical parameters but not stellar M/L, 3) intrinsic variation in stellar M/L, and 4) different relative amounts of dark and baryonic matter within R_e.

To reduce these effects, we have pruned the sample to remove all doubtful group members, plus all likely S0's, known cD's, and compacts. However, we are still left with several galaxies whose deviations are larger than plausible errors [18].

Aspect effects can be estimated quantitatively from the tensor virial theorem [21]. The effect of viewing an oblate or prolate E3.6 spheroidal galaxy as it appears round or maximally flattened is shown by the arrows in Figs. 1c,d. By chance, it appears that aspect effects move galaxies almost parallel to the plane. This is good in the sense that the distance estimate is insensitive to aspect effects, but bad because there is no way to use the residuals to distinguish pole-on galaxies from edge-on ones, or round ones from flattened ones. A plot of distance from the plane versus ellipticity confirms the lack of any correlation. Thus, scatter about the plane is probably not caused primarily by shape or aspect effects.

By elimination, it appears that the major source of scatter must be real variations in $(M/L)_e$, with perhaps some extra errors contributed by incorrect, nuclear dispersions. If M/L variations are due to differences in stellar M/L, one might hope to see some sign of this in spectral anomalies. We have not had any luck so far in finding significant correlations with Mg_2 residuals, but other features remain to be tested.

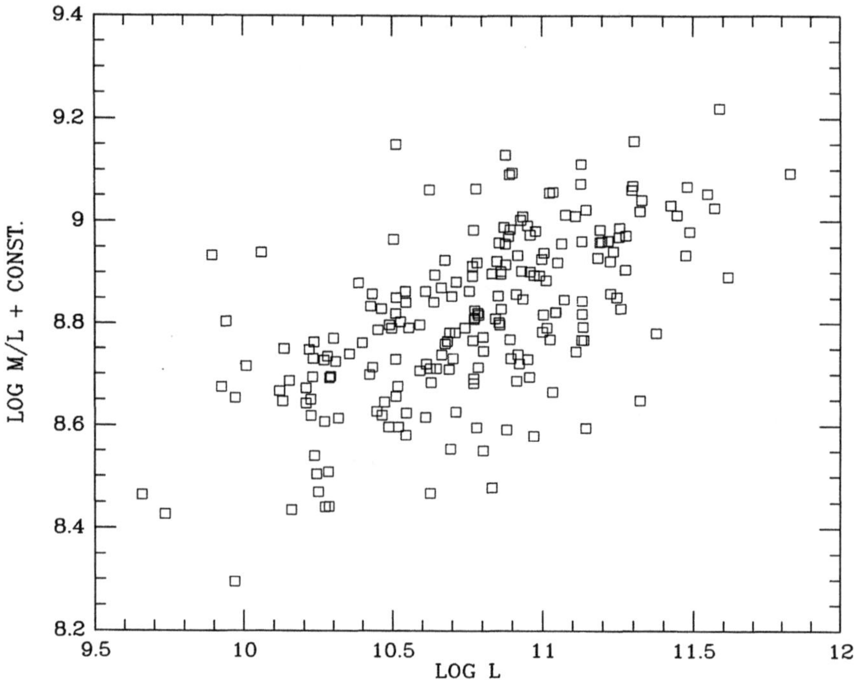

Figure 2: Log M/L_B vs. blue luminosity in solar units.

V. Core vs. Global Scaling Relations

LAUER [9] discovered a two-dimensional, planar scaling relation for the core parameters of elliptical galaxies. We ask here whether this plane is the same as that found for the global properties. We have fitted a mean plane to Lauer's core data using distances determined from global parameters. The resulting exponents are given for comparison in Table 1, where it is seen that the core and global planes are parallel within the errors. To compare zeropoints, we have used BINNEY's [22] expression for the global $(M/L)_e$ of a de Vaucouleurs spheroid based on the nuclear velocity dispersion:

$$(M/L)_e = 0.201 \frac{\sigma_{nuc}^2}{G I_0 R_e} \tag{2}$$

and the corresponding KING and MINKOWSKI [23] formula for the mass-to-light ratio of a galaxy core:

$$(M/L)_e = \frac{9}{2\pi} \frac{\sigma_{nuc}^2}{G I_c R_c} \tag{3}$$

Converting I_0 to our parameter I_e, we find that global and core M/L's agree to 0.02 dex, which is remarkably good considering the uncertainties [24].

This good agreement suggests an important conclusion. Since M/L in the core itself must certainly be dominated by stellar matter, we conclude that there is little room left for much extra non-baryonic dark matter outside the core to R_e. This result is consistent with what we already know from other sources about baryonic and dark matter in spheroidal components. Rotation velocities in the immediate outskirts of spheroid-dominated galaxies ([25]; see also the review by Fall, this volume) yield low M/Ls that are consistent with stellar dominance. Rotation curve mass models (Athanassoula, this volume; Kent, unpublished remarks at the Workshop) furthermore seem to show the steeply rising signature of a bulge in galaxies whose light has a significant bulge contribution. Both of these observations imply that the bulk of the matter within R_e must be stellar. However, we should remember that uncertainties in Eqs. 2 and 3, particularly the former, will always mean a 20-30% margin of error in zeropoint comparisons like that given here. What is really needed is a good rotation curve for at least one elliptical galaxy, but that goal, as ever, seems to remain elusive.

VI. Correlations with Other Parameters

These have been investigated extensively by DD. The following variables have been checked to see if they correlate with a galaxy's location in the plane: ellipticity, ellipticity gradient, isophotal twist rate, and rotation velocity. So far all results are negative. No one has yet tested to see whether any of these quantities correlate with distance from the plane.

Our group has briefly investigated the role of Mg_2. We find that Mg_2 correlates best with velocity dispersion but with noticeable extra scatter. So far the residuals do not correlate strongly with any other quantity including distance from the plane, but this question needs further study.

VII. Implications for Elliptical Galaxy Formation

The fact that ellipticals appear to exhibit a plane rather than a one-parameter family means that any additional constraints on radius vs. mass imposed at formation must have been fairly loose. One way of quantifying this is to note the range in radius covered at a fixed value of the mass. If the central regions are baryon-dominated, then loci of fixed mass should follow the slope shown in Fig. 1a between radius and velocity dispersion. From the figure, the range in radius at fixed mass is seen to be roughly 2.5, corresponding to a factor of 1.6 in σ, or 15 in mass density. This range is much larger than the plane thickness, but it is also much less than the *total* range of radii spanned by ellipticals, which from the same figure is closer to 30. The radius was thus not exactly constrained at formation, but its range of variation was significantly limited.

In the baryonic dissipation picture for galaxy formation, looseness in the radius could come either from differences in baryonic collapse factor or from differences in initial dark-matter radius. The former mechanism would introduce scatter between the baryonic velocity dispersion and that of the dark-matter halo. Figure 1a implies that this scatter would amount to a factor of 1.6 and would also correlate with position in the diagram. The same mechanism would also yield rotation curves with different slopes in the transition region between spheroid and halo. Both effects might show up in H I rotation curves, globular-cluster orbital motions, or X-ray observations. Neither effect is expected if galaxies start out with different initial dark-matter radii.

A major conclusion of the new studies is that stellar M/L is well correlated with total light and total mass with a scatter of only ±0.07 dex, or 16%. This implies a degree of homogeneity in the star formation process that seems hard to explain. One might plausibly expect star formation to depend on local variables such as velocity dispersion or density rather than on global variables such as total light or total mass. How indeed did local star formation "find out" about total mass? Or is the parametrization of M/L with mass found here merely a shorthand rule that conceals the role of other, more physically meaningful variables? Presently this important question remains a mystery.

Much of this review is based on a survey of elliptical galaxies by seven collaborators which was aided by many colleagues and institutions. Here we mention financial support from NSF grant AST 82-11551, a NATO Travel Grant, an ASU Grant-in-Aid, a UCSC Faculty Research Grant, Dartmouth College, the Carnegie Institution of Washington, and the Institute of Astronomy, Cambridge University.

References

1. R. A. Fish: Ap. J. *139*, 284 (1964)
2. S. M. Faber, R. E. Jackson: Ap. J. *204*, 668 (1976)
3. B. Binggeli, A. Sandage, M. Tarenghi: A. J. *89*, 64 (1984).
4. J. Kormendy: Ap. J. *295*, 73 (1985).
5. R. Terlevich, R. L. Davies, S. M. Faber, D. Burstein: Mon. Not. R.A.S. *196*, 381 (1981).
6. J. Tonry, M. Davis: Ap. J. *246*, 666 (1981)
7. G. Efstathiou, S. M. Fall: Mon. Not. R.A.S. *206*, 453 (1984)
8. G. de Vaucouleurs, D. Olson: Ap. J. *256*, 346 (1982)
9. T. Lauer: Ap. J. *292*, 104 (1985)
10. S. Djorgovski: in, T. de Zeeuw, ed.: *Structure and Dynamics of Elliptical Galaxies,* IAU Symposium 127. Dordrecht: Reidel 1987
11. S. Djorgovski, M. Davis: Preprint 1986
12. D. Burstein, R. Davies, A. Dressler, S. M. Faber, D. Lynden-Bell, R. Terlevich, G. Wegner: in, *Large Scale Structure of the Universe,* Korn 1986
13. G. de Vaucouleurs, A. de Vaucouleurs, H. R. Corwin: *Second Reference Catalog of Bright Galaxies,* University of Texas Press, Austin 1976

14. G. Djorgovski: Ph.D. Thesis, University of California, Berkeley, 1985

15. S. M. Faber, G. Wegner, D. Burstein, R. L. Davies, A. Dressler, D. Lynden-Bell, R. Terlevich: in preparation 1987

16. D. Burstein, R. L. Davies, A. Dressler, S. M. Faber, D. Lynden-Bell, R. Terlevich, G. Wegner: Preprint 1986

17. R. L. Davies, D. Burstein, A. Dressler, S. M. Faber, D. Lynden-Bell, R. Terlevich, G. Wegner: Ap. J., submitted

18. A. Dressler, D. Lynden-Bell, D. Burstein, R. L. Davies, S. M. Faber, R. Terlevich, G. Wegner: Ap. J., in press (1986)

19. P. L. Schechter and J. E. Gunn: Ap. J. *229*, 471 (1979)

20. R. Michaud: A. Ap. *121*, 313 (1983)

21. J. Binney: Mon. Not. R.A.S. *183*, 501 (1978)

22. J. Binney: Ann. Rev. Astron. Ap. *20*, 399 (1982)

23. I. R. King, R. Minkowski: in, D. S. Evans, ed.: *External Galaxies and Quasi-Stellar Objects*, I.A.U. Symposium 44, p. 87, Dordrecht: Reidel (1972)

24. D. Richstone, S. Tremaine: Preprint 1986

25. E. Bajaja, G. van der Burg, S. M. Faber, J. S. Gallagher, G. R. Knapp, W. W. Shane: A. Ap. *141*, 309 (1984)

Musings Concerning the Possible Significance of Surface Brightness Variations in Disk Galaxies

GREG BOTHUN

I. Introduction

The Theme of this conference is "Nearly Normal Galaxies" but the study of this topic is necessarily biased and influenced by selection criteria. In particular, one of the more trivial, yet perhaps fundamental observations, is that disk galaxies do not have the same surface brightness but instead exhibit a rather large range (unlike the case for luminous ellipticals). The origin of these surface brightness (SB) variations is unclear but their existence is important in a number of contexts, some of which I will briefly discuss. For instance, our understanding of the evolution (e.g. history of gas-to-star conversion) of disk galaxies is based primarily on studies of high SB actively star forming systems. To the extent that these are "normal" galaxies (e.g. they define the Hubble Sequence) is determined only by our definition. It is important to understand that large numbers of gas-rich low SB galaxies exist and that these need to be observed, in detail, before we can form a more complete picture of star formation in galaxies.

Thus, since the concept of normalcy is usually defined through inter-comparisons of what is believed to be a representative sample, it is not at all clear that we know what we are talking about since it is unlikely that a representative sample of real galaxies has yet been measured. In this contribution I intend to follow the informal nature of this conference by giving a rather open-ended discussion about LSB galaxies. In so doing, my own biases should become quite clear.

II. The SB Distribution in Galaxy Catalogs

What is known already about the SB distribution of disk galaxies comes mainly from studies of existing galaxy catalogs. Disney and Phillips (1985) have studied the distribution of SB for 500 disk galaxies catalogued in the RC2 (de Vaucouleurs et al 1976) and found 1) they exhibit a large range (5 mag) in SB and 2) there is no apparent correlation between SB and either color or luminosity. A more representative catalog is the diameter-limited UGC (Nilson 1973). Figure 1 shows the SB distribution of all UGC galaxies with measured redshifts. Once again, a large dispersion is seen. Moreover, Bothun et al. (1985a) show that many of the LSB galaxies in the UGC actually have rotational velocities > 300 km/s. In combination with the study of Disney and Phillips, there is now good evidence that LSB galaxies are not necessarily low mass systems, although it seems clear that most low luminosity galaxies are also low surface brightness (e.g. the DDO catalog - van den Bergh 1960, Bingelli et al. 1985).

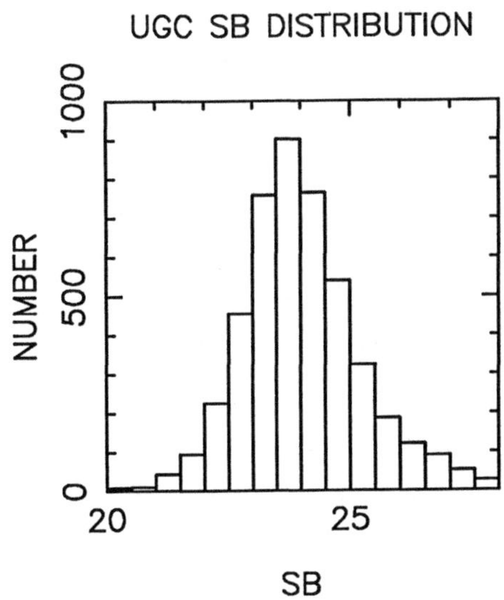

Fig. 1

Given that there does exist a large range in disk galaxy SB, then it is only natural that existing catalogs are deficient in LSB objects because they are relatively difficult to discover. For instance, the Zwicky catalog, while containing many intrinsically low luminosity galaxies, is magnitude-limited (by eye estimates) and therefore does not adequately represent the population of LSB galaxies whose apparent magnitudes are difficult (impossible) to estimate by eye (especially using Zwicky's method). Even though the Zwicky catalog doesn't contain diameter information (so SB is impossible to measure), casual inspection of 4-m plates of clusters of galaxies generally reveals that most members which are in that catalog exhibit a strong contrast with respect to the sky background. Those UGC galaxies that are in the cluster and not in the Zwicky catalog are generally quite diffuse in appearance. Moreover, the UGC must be incomplete (at some unknown level) at the LSB end due to 1) variations in the quality of the original Sky Survey and 2) human limitations stemming from such a massive undertaking. Thus, the need for new surveys is crucial (see Section IV).

III. The Proper Measure of SB

Before determining the role that SB plays in parameter correlations of disk galaxies it is necessary to develop a reliable way to measure it. To date, studies of disk galaxy SB distributions have relied upon either a mean surface brightness within some isophote or the mean surface brightness with the effective aperture. For disk galaxies, both of these measures can be strongly influenced by the B/D ratio of the galaxy. Thus, some of the spread in SB may simply be reflecting differences in B/D although Disney and Phillips (1985) show that at any morphological type, there is a large range in SB. To circumvent these difficulties, I advocate using the extrapolated central surface brightness (corrected to face-on - see below) of the exponential disk (e.g. Freeman's (1970) B(0)) as the measure of SB. This parameter is easily derived from CCD imaging data and most disk galaxies can be fit, at least crudely, by an exponential.

At present, the known distribution of B(0) (i.e. B(0) = 21.65 ± 0.3 - Freeman 1970, Boroson 1981) is strongly biased by optical selection effects. That is, HSB, face-on spirals were preferentially chosen because they yielded good S/N on photographic plates. More recent studies (Romanishin et al. 1983, Cornell et al. 1986) clearly show that late type spirals often have B(0) as low as 23.0. An example of two such galaxies with LSB disks that span a large range in physical size is shown in Figure 2. Additional data indicate that massive, late-type spirals can be found down to B(0) ~ 24.0 but below that level dwarf galaxies predominate.

For the most part, LSB spirals tend to be H I rich. Thus, while Kormendy's (1976) criticism of Freeman's result (i.e. bulges were not properly removed) is still valid, the dominant influence in the apparent constancy of B(0) is simply optical selection effects. When samples of disk galaxies are H I selected, for instance, the full range of B(0) becomes more apparent. A good example of this is provided by Pegasus I cluster where the observed mean B(0) of the H I selected spirals is ~ 22.5. Most of these spirals have inclinations in the range 45 - 70 degrees and thus correcting

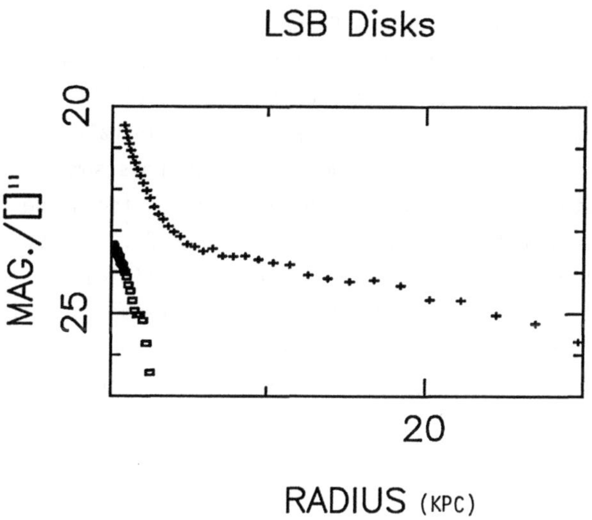

Fig. 2

the SB to face-on would produce a yet lower mean B(0). However, it is not at all clear what the exact form of this correction is since dust plays a highly uncertain role. Empirically, there is a very poor relationship between the observed value of B(0) and the inclination (see Kent 1985, Cornell et al. 1986). Even in the I-band, where the effects of dust obscuration are less (and hence the geometric path length correction becomes more important), no decent correlation appears (Bothun and Mould, 1987). Does this all mean that disk galaxies really are fairly optically thick (Burstein and Lebofsky 1986)?

IV. Searches for New LSB Objects

1) The new Palomar Sky Survey (PSS-II) is being done on much better emulsions than the old survey. Although still in its infancy, Schombert and Bothun (1987) are beginning to find and record LSB galaxies that were missed by the

UGC. Typically, 5-7 new LSB galaxies have been found per plate. Still more can be found with diameters less than 1 arc minute. Although calibration data is still minimal, it appears that the PSS-II can detect galaxies with B(0) as low as 24.0 (the night sky brightness at Palomar remains the fundamental problem).

2) Using the technique of photographic amplification (i.e. how Malin and Carter found shells), Impey, Bothun and Malin (1987) have found hundreds more LSB galaxies in fields located in the Virgo and Fornax clusters. In some of the richest fields in Virgo the density of objects (with diameters greater than 30 arcseconds) approaches 30 per square degree (but many could be background members of the Coma supercluster?). CCD data of the Malin galaxies is being acquired at Las Campanas (a dark site) and Figure 3 shows some representative surface brightness profiles. In general, Malin's technique can detect objects with central surface brightnesses as low as 27.0 mag/sq. arcsec (or ~ 2% of a dark blue sky). This survey is in its preliminary stages but plans are to look in low density regions of the universe as well, not just rich superclusters (for more details consult Impey, Bothun and Malin). The eventual goals of this survey are to address some of the questions raised in Section VI.

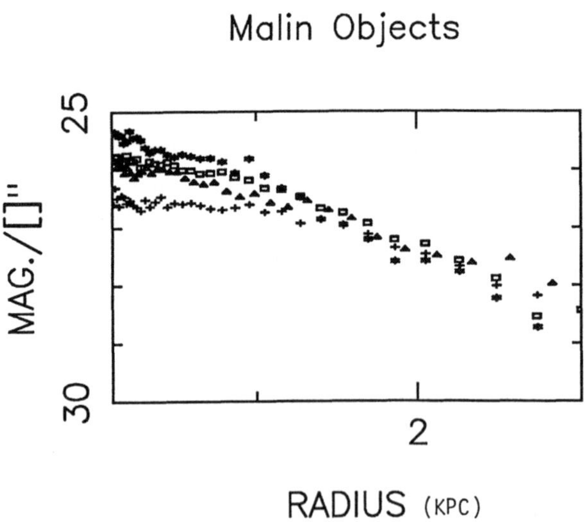

Fig. 3

There are of course, other ways to make similar surveys. The basic problem from the ground is that the terrestrial night sky is redder than most extragalactic objects (due primarily to OH emission) and thus surveys must be conducted in the blue (where the CCD response is lowest). Significantly greater sensitivity would be obtained by imaging at I (see Wright 1985) from the top of the earth's atmosphere. Similarly, blind H I imaging surveys (technically difficult and time consuming at the moment) sometimes find H I rich objects without any apparent optical counterpart (see Dickey and Salpeter 1985). However, we already know from the statistics of OFF beam measurements from Arecibo that large numbers of H I rich LSB systems with line widths in excess of 50 km/s cannot exist else some, by chance, would fall into the OFF beam and be detected as a negative signal. What is likely to have been missed, to date, are those objects with H I widths < 25 km/s, where the signal would then appear in only 1-2 channels and might be mistaken for interference. Of course, it is always possible that these newly discovered galaxies with B(0) > 25.0 don't have any gas left (which is why they are extremely LSB?).

V. Implications of Disk Galaxy SB Variations

The presence of these SB variations in disk galaxies affects a number of on-going observational investigations. It is particularly relevant in the Tully-Fisher (TF) relation where Bothun and Mould (1987) have shown that, at fixed line width, there is a large range in SB which inevitably leads to scatter in the magnitude-line width plane. This point is illustrated further in Figure 4 which depicts Gunn I-band SB profiles of a sample of disk galaxies that have the same corrected line width. Clearly, there is a large range in profile morphology and this is crucial when it comes to accurate calibration since SB differences between optically selected calibrating galaxies and H I selected cluster galaxies could produce a zero-point offset. Some, but not all, of these potential differences can be minimized by using total magnitudes where the difference in average surface brightness is not as great as in the case of some interior isophotal radius.

The existence of LSB, H I rich disk galaxies is also important in terms of star formation histories. Schommer and Bothun (1983) have called attention to a class of red, very H I rich disk galaxies. These galaxies usually have low B(0) (e.g. N3883 in Figure 2) and low values of H-alpha equivalent width. In this sense, they represent late type spirals that have very long gas depletion timescales. This strongly contrasts with the results for the optically selected sample of Kennicutt (1983) and may suggest that either 1) star formation in late type spirals is episodic or 2) some late type spirals do not make massive stars very efficiently. Of course, both could be occurring but recent data obtained by van der Hulst et al. (1986) shows that LSB H I rich spirals generally have low surface densities of H I - probably below the critical value for the formation of molecular clouds. Thus, these galaxies are destined to remain in a quiescent state until something acts to clump the H I.

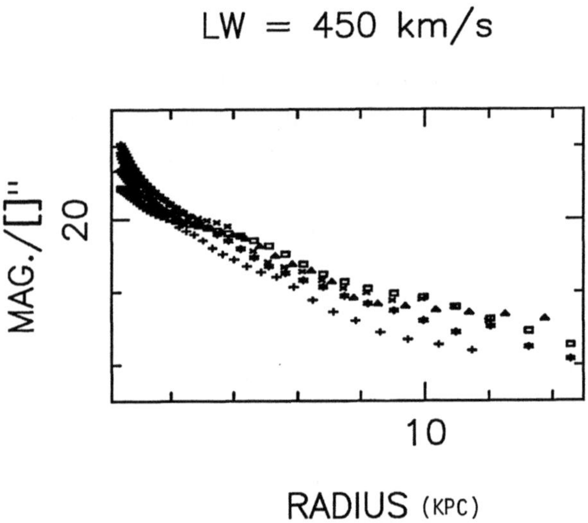

Fig. 4

Finally, if it can be established that the M/L ratio is a strong function of B(0) then stellar populations (as opposed to exotic particles) may help in understanding the mystery of dark matter in disk galaxies; that is the dark matter is likely to be in the form of remnants. Larson (1986 see also this conference) has presented a model appealing to a bimodal IMF which links the chemical evolution of galaxies to the dark matter problem. The simplicity of this model is particularly attractive and in the high mass mode of star formation, one predicts very rapid fading due to a paucity of low mass stars that will eventually feed the giant branch. Whether or not the apparently large M/L ratios observed for Draco and Ursa Minor (see Aaronson this conference) can be understood in this way remains to be seen, but the dependence of dwarf spheroidal M/L on B(0) is quite strong (see Table in Aaronson's paper this conference). In addition, for late type spirals there does seem to be a relation between B(0) and M/L where M is calculated via the H I line width. In general, Sc galaxies of normal SB have blue M/L ratios of 3-5 whereas galaxies like N3883 have M/L 10 - 20 (Schommer and Bothun 1983). This is consistent with ~ 1.5 magnitudes of disk fading.

VI. Unanswered Questions

Given that many LSB galaxies exist, but have been neglected (for the most part) one naturally wonders if they shed light on any of the following outstanding questions regarding the nature and evolution of disk galaxies:

1) What is the space density of galaxies as a function of B(0)? Does the ratio of LSB to HSB disks depend upon local galaxy density?

2) What is the bivariate optical and H I luminosity function and does H I content (mass) depend on optical surface brightness?

3) Is the disk galaxy luminosity function dependent upon B(0)? Is B(0) a function of mass?

4) Can the color distribution as a function of B(0) tell us anything upon the possible nature of episodic star formation in disk galaxies?

5) Do M/L ratios depend upon surface brightness in a manner that can be reconciled with fading stellar populations? At what level of B(0) does a galaxy cease to have a light distribution that can be adequately fit by an exponential fall-off.

6) Given question 1 - Do LSB galaxies trace out a different pattern of large scale structure than HSB galaxies (i.e. do they fill the voids?).

7) Are low surface brightness galaxies a manifestation of a) low surface mass density, b) faded stellar populations, c) heavy extinction, or d) anomalous mass functions?

At best only partial answers can be given to the above questions and many of them have been discussed either directly or indirectly in this conference. More complete answers will require more systematic surveys for LSB galaxies but my preliminary (i.e. things that I know but cannot prove) answers to these questions are:

1) The space density is unknown at present but the ratio of LSB-to-HSB galaxies does seem to increase with decreasing local galaxy density. In particular, the Coma cluster appears to be devoid of LSB objects whereas the Pisces Supercluster is littered with them (as is Virgo).

2) Unknown (wait n years) but most H I data sets (e.g. Bothun et al. 1985b, Haynes and Giovanelli (1984) show some increase in H I content with decreasing B(0) for late type disk galaxies.

3) Luminosity function dependence is hard to estimate. It does appear that the Bingelli et al (1984) effect of decreasing surface brightness with decreasing absolute magnitude falls apart when B(0) > 23.5. The extremely LSB galaxies discovered to date in Virgo and Fornax generally have -15 < M < -12. Most low mass galaxies are LSB but LSB analogs (down to B(0)) ~ 24) can be found at virtually any mass (line width).

4) I hope so. Presently, the colors of very LSB systems are relatively blue and similar to metal-poor galactic globulars (Impey, Bothun and Malin 1987).

5) In view of the Larson (1986) model this may be possible. More data (difficult to obtain) will be required in order to make a proper test. From data in Caldwell and Bothun (1986) it would seem that fainter than B(0) = 25, luminosity profiles are generally flat over much of the radius.

6) Both the surveys of Bothun et al. (1986) and Oemler et al (1986 - see this conference) suggest that LSB galaxies do not fill the voids. If biased galaxy formation (Dekel and Silk 1986) has occurred then it would appear that the voids must be filled with mass that is below SB = 27.0 (i.e. it really is invisible).

7) Only alternative c can be ruled out with any confidence since LSB galaxies in general, do not have excessively large ratios of far IR flux to blue flux (based on preliminary IRAS data - Persson 1986 private communication). Distinguishing between the other alternatives is difficult although it is curious that these LSB objects are generally blue. Possibly this means that they are relatively young and unevolved. If LSB systems truly are galaxies with lower surface mass densities than other galaxies then they serve as laboratories to study the character of star formation in a low density environment. Perhaps such systems are then only capable of supporting

one generation of star formation before the gas is lost. In that case, they are destined to fade (especially if this initial epoch was dominated by a high-mass mode) into the background of the night sky. In that sense, they serve as forerunners to what most galaxies will look like after the next Hubble time has elapsed.

I would like to thank the organizer, Sandra Faber for inviting me to this conference. I further thank my principal collaborators, Nelson Caldwell, Chris Impey, Jeremy Mould and James Schombert for helping to explore the mysteries of LSB galaxies. As always, S. Strom has been an indirect source of inspiration. Lastly, I wish to thank D. Malin for making all of this possible.

References

1. M. Aaronson: (1986), This Conference
2. B. Bingelli, A. Sandage, M. Tarenghi: Astron. J. 89, 1 (1984)
3. B. Bingelli, A. Sandage, G. Tammann: Astron. J. 90, 1681 (1985)
4. T. Boroson: Ap. J. Suppl. 46, 177 (1981)
5. G. Bothun, T. Beers, J. Mould, J. Huchra: A.J. 90, 2487 (1985a)
6. G. Bothun, R. Schommer, M. Aaronson, J. Mould, J. Huchra, W. Sullivan: Ap. J. Suppl. 57, 423 (1985b)
7. G. Bothun, T. Beers, J. Mould, J. Huchra: Ap. J., in press (1986)
8. G. Bothun, J. Mould: Ap. J., in press (1987)
9. D. Burstein, M. Lebofsky: Ap. J. 301, 683 (1986)
10. N. Caldwell, G. Bothun: In preparation (1986)
11. M. Cornell, M. Aaronson, G. Bothun, J. Mould: preprint (1986)
12. A. Dekel, J. Silk: Astrophys. J. 303, 39 (1986)
13. G. De Vaucouleurs, A. De Vaucouleurs, H. Corwin: (1976) - RC2
14. J. Dickey, E. Salpeter: Ap. J. 292, 426 (1985)
15. M. Disney, S. Phillips: M.N.R.A.S. 216, 53 (1985)
16. K. Freeman: Ap. J. 160, 811 (1970)
17. M. Haynes, R. Giovanelli: A.J. 89, 758 (1984)
18. C. Impey, G. Bothun, D. Malin: In preparation (1987)
19. R. Kennicutt: Ap. J. 272, 54 (1983)
20. S. Kent: Ap. J. Suppl. 59, 115 (1985)
21. J. Kormendy: Ph.D. Thesis, Caltech (1976)
22. R. Larson: M.N.R.A.S. 218, 409 (1986)
23. P. Nilson: UGC (1973)
24. G. Oemler: (1986), This Conference

25. W. Romanishin, S. Strom, K. Strom: Ap. J. Suppl. 53, 105 (1982)
26. J. Schombert, G. Bothun: In prearation (1987)
27. R. Schommer, G. Bothun: A.J. 88, 577 (1983)
28. S. van den Bergh: The DDO catalog (1960)
29. T. van der Hulst, E. Skillman, R. Kennicutt, G. Bothun: Preprint (1986)
30. E. Wright: P.A.S.P. 97, 451 (1985)

The Luminosity Function: Dependence on Hubble Type and Environment

BRUNO BINGGELI

1. Introduction

Since the appearance ten years ago of SCHECHTER's [1] well-known analytic expression for the galaxian luminosity function (=LF), much emphasis has been put on the existence of a universal LF (Fig. 1). Galaxies in the field (= outside of clusters) and galaxies in rich clusters (with little cluster-to-cluster variation) seem to obey a Schechter-type LF with nearly the same characteristic form parameters, M^*, the magnitude at the break (the "knee"), and α, the faint-end slope of the LF [1-6]. Different galaxy types are usually lumped together because there have been hints of only modest type dependence of the LF at the bright end [5, 7]. And so working on the galaxian LF has largely become a "search for two numbers" (to paraphrase Sandage [8]) - the "universal", or "standard" M^* and α (neglecting the normalization parameter, the universal mean density).

But there is a different approach to the galaxian LF which retains the rich morphological variety of galaxies from the beginning. This approach is more difficult because galaxies of faint absolute magnitudes have to be included in reasonable numbers and with good morphological resolution. Not surprisingly, HOLMBERG's early work [9] found very little follow-up [10, partially 5]. For Virgo cluster galaxies the full details of the LF dependence on type have recently been explored by SANDAGE et al. [11]. There is a wealth of type specifity, as will be seen below.

On the other hand, DRESSLER's general morphology-density relation [12, 13], which shows that the type mixture varies strongly and systematically with local density over a range of 10^6 in density, makes us suspect that the type specific LFs also vary with local density. Hence the most general approach should treat the LF as a function of Hubble type and environment (=local density). To outline these functional dependences is the purpose of this paper. It is not intended here to discredit the "universal" LF altogether, which is likely to remain essential for general applications in extragalactic astronomy and cosmology. But more can be learned about the galaxies by focussing on the type and environment specific LF.

The LF type dependence for Virgo cluster galaxies is summarized in Section 2 (Fig. 2). The dependence on environment is less well-known. This problem is tackled in Section 3 by first varying the LFs within the Virgo cluster as a function of local density, and then by comparing the Virgo LFs with those of extreme environments : the core of the Coma cluster (high density), and nearby groups (low density). Surprisingly little variation of

Fig. 1 Schechter's analytic form and the "universal" approach to the galaxian LF. This paper tries to recover some of the details.

the LF shape is found (Figs. 3 and 4), and the hypothesis is put forward, in Section 4, that the generalized LF can be separated into a dependence on type (which determines the LF shape) and a dependence on environment (which determines the LF normalization), at least as a first order approximation. If true, this is an important constraint on models of galaxy formation and evolution. Finally, as an addendum, LFs for spheroids (E's and bulges) and for disks are shown in Section 5 (Fig. 5). A distinction between galaxy components rather than galaxy types may be useful for the theorist.

2. LF Dependence on Hubble Type (Virgo Cluster)

For obvious reasons the Virgo cluster of galaxies is the ideal place for a detailed LF study. For less obvious reasons this task has been accomplished only recently by SANDAGE et al. [11], based on the Las Campanas Photographic survey of the Virgo cluster [14]. This survey uncovered ~ 1300 cluster members down to the limiting magnitude of $B_T = 20$, or $M_{B_T} \sim -12$ (assuming throughout a distance of 22 Mpc for the Virgo cluster) with a limit of completeness of $B_T \sim 18$, or $M_{B_T} \sim -14$. Owing to the large plate scale employed (10$''$.8/mm), all galaxies could be classified with great morphological resolution. The resulting type specific LFs for Virgo galaxies are given in [11], and the reader is referred to this paper for all details. Here we give a brief summary. The main features - shown in Fig. 2 - are as follows.

i) All three basic giant types - spirals, S0's, E's - have a maximum in their LF. Spirals and S0's follow a similar Gauss-type distribution, whereby S0's are brighter than spirals in the mean by less than 1 mag. E's have a very flat LF at the bright end. The brightest galaxies are E's. All LFs essentially cut off at $M_{B_T} \sim -16$.

ii) Dividing spirals further into Sa through Sm reveals an important trend. Early-type spirals are brighter than late-type spirals. This does not apply to Sa - Sc, however, since Sb's are the brightest spirals on average. But clearly from Sb or Sc to later we deal with a luminosity sequence that continues right down into the dwarf irregular (Im) domain. The faint end for Im's, and for spirals+Im's (=star forming galaxies) is not clear; recall the completeness limit at $M_{B_T} \sim -14$; but a maximum is reached well above this limit in both cases.

iii) There is a class of faint ($M_{B_T} \sim -17$) galaxies with characteristics of both S0 and dE which was called dS0 [15]. It is not yet clear how these galaxies are physically related to the classical S0's, nor to dE's.

iv) Dwarf elliptical galaxies (dE) make up ~ 70% of the total cluster population. They set in at $M_{B_T} \sim -18$ and grow exponentially in number with decreasing brightness down to the completeness limit and probably beyond. There is substantial overlap in total magnitude with the classical, or compact E's, against which dE's stand out by their diffuse appearance [15]. E's and dE's clearly form two distinct sequences of galaxy forms, and they may have nothing more in common than their smooth elliptical image [16, 17]. The sharp cut-off at $M_{B_T} \sim -16$ of faint compact E's (for which M32 in the Local Group may stand as a prototype) is probably real. The same was observed in the Fornax cluster [18].

In general we note that only dwarf ellipticals follow an exponential LF - the LFs of all other types have a maximum and reach zero at a sufficiently

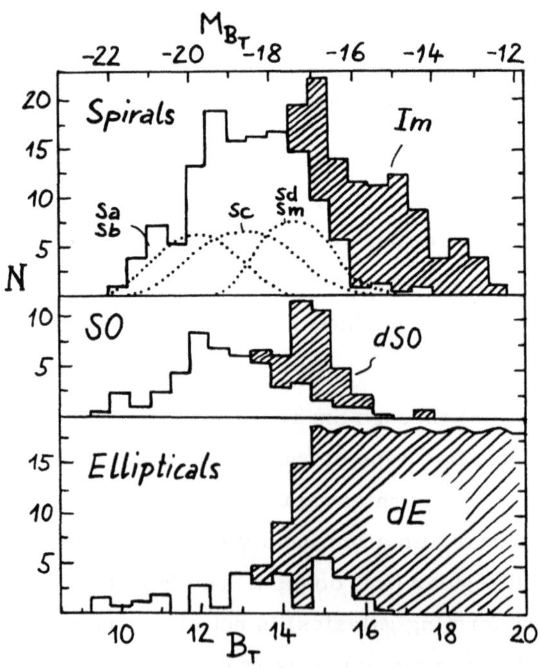

Fig. 2 The type specific LFs for Virgo cluster galaxies. Shown are the three main types : spirals, S0's, and E's, and (shaded) their dwarf counterparts, or dwarf extensions : Im's, dS0's, and dE's. The exponential growth of the dE's had to be broken. Spiral subtypes are indicated as dotted Gaussians. Apparent magnitudes (bottom scale) and absolute magnitudes (top scale) are related by $(m-M)^{o}_{Virgo} = 31.7$.

faint magnitude (neglecting "Im/dE" galaxies [15, 11]). At the bright end the type differences are rather small (they become larger as one goes fainter). This is of course the ground on which the universal Schechter LF could be erected because most LF investigations do not go very much beyond M^*. In this light it is surprising that the total (all-type) LF for Virgo cluster galaxies is well represented by a Schechter function with nearly "standard" M^* and α down to the faintest limit [11]!

3. LF Dependence on Environment

The Virgo cluster is one particular galaxy environment, and there is no a priori reason why the LFs of spirals, ellipticals etc. should be the same in a different environment. The relative frequency of these types is known to be a strong function of environment, i.e. local density [12, 13]. So why not generalize the morphology-density relation to a LF-density relation ? To take a first step in this direction (since this is still no man's land) we have compared the Virgo LFs with LFs for the Coma cluster core (a very high-density place), and for nearby groups (low-density spots). But the Virgo cluster (which is of medium mean density) is a very inhomogeneous place itself, spanning a factor of several hundred in local spatial density with distinct morphological segregation [19]. Hence it is worthwhile to look first for

LF variations within the Virgo cluster.

To test for internal consistency we have determined the Virgo LFs separately for a number of density bins, where the local (projected) densities were calculated after DRESSLER's [12] recipe. The details of this analysis will be presented elsewhere, but the results can be summarized as follows.

The LFs of spirals, irregulars, nucleated dE's (dE, N), and non-nucleated dE's do not vary appreciably with local density. By this we mean that the LF shape is constant. The absolute numbers are of course a strong function of local density [19]. For E's and S0's there is a slight but significant tendency for the most luminous galaxies to be found in high density regions.

Little variation of the LF shape was thus indicated. This had to be tested by going to more extreme environments.

High density : Coma cluster

The core of the Coma cluster is probably the densest and evolutionary most pressing galaxy environment we know of. We would expect this to be reflected in the Coma cluster LFs. THOMPSON and GREGORY [20] give separate Coma LFs for E+dE, S0, and S+Irr galaxies, based on a complete sample of ~ 400 Coma core galaxies with M_{B_T} brighter than ~ -17.5 (H_o = 55). These LFs are easy to compare with the corresponding Virgo LFs within the same magnitude limit. The result is shown in Fig. 3.

Notice how remarkably similar the LFs of Coma and Virgo cluster galaxies are for all three Hubble types ! At the same time recall how much these clusters differ in their galaxy content. The type mix of Coma is 44% E, 49% S0, and 7% S+Irr, while for Virgo it is 12% E, 26% S0, and 62% S+Irr. Yet the LF shapes are nearly constant ! Only the relative abundances of the

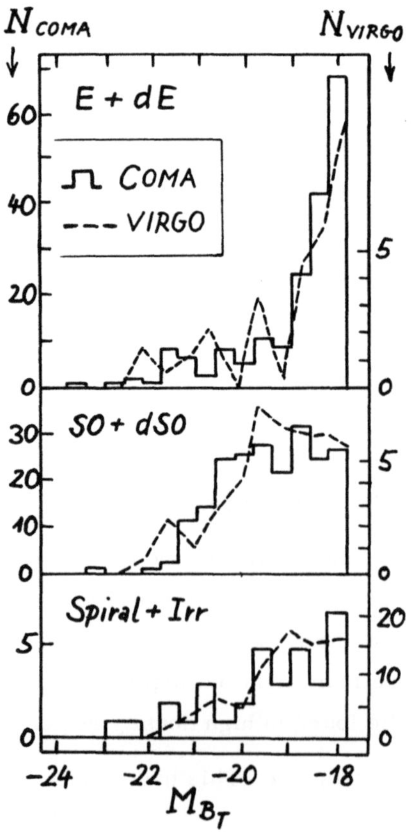

Fig. 3 Coma cluster and Virgo cluster LFs for elliptical, S0, and spiral+irr galaxies. Data are from [20] for Coma, and from [11] for Virgo (cf. also Fig. 2). A relative distance modulus of $\Delta(m - M) = 4$ has been adopted. The LFs are scaled to each other to give the same numbers in total. Absolute numbers are given on the left-hand ordinate for Coma, and on right-hand ordinate for Virgo.

types, or the LF normalizations, seem to vary with the environment (density).

The hypothesis of constant LF shape for the separate types has already been adopted by THOMPSON and GREGORY [20] to explain the scattering of Schechter LF parameters among clusters of galaxies [3]. The idea is that most of the cluster-to-cluster variation of the total LF is not due to evolution but simply reflects the different type mixture of the clusters, with each type obeying a specific LF of constant shape. The type mix itself, which is governed by density, is probably not a result of evolution but of formation (=very early evolution) [12, 21]. Of course, there remain systematic distortions in certain (mostly cD) cluster LFs which have to be assigned to late evolution [21]. In this context we have also to point to a slight (evolutionary?) difference between the Virgo and Coma LFs : the brightest Coma cluster galaxies are clearly brighter than the brightest Virgo cluster galaxies (see Fig. 3).

Low density : nearby groups in the field

At the opposite extreme of the density spectrum we find the general "field", defined as the galaxy field outside of rich clusters, which comprises groups of galaxies (like the Local Group), and "isolated" galaxies (that cannot easily be assigned to a group). Many studies of the "field" LF have been carried out, but mostly with the purpose to establish a standard LF with all types mixed together [6]. And those few studies which do address type differences [7, 5, 10] cannot easily be compared with the Virgo cluster LFs. We have therefore used an updated version of KRAAN - KORTEWEG's [22] "10 Mpc Catalog" to derive new local "field" LFs with high morphological resolution [23]. The catalog is a list of all known galaxies with $v < 500$ km/s (corrected for Virgo-centric infall), totalling 176 galaxies at present, of which 121 are members of well-known nearby groups (LG, M81, M101, etc.), and 55 are "isolated" systems. Here we use only the group data because the groups have been searched for faint members to an estimated completeness limit of $M_{B_T} \sim -15$.

In Fig. 4 we show the group LFs for three spiral subtypes, and for irr's. (There are only a handful of E and S0 galaxies in the sample, as expected from low-density regions.) Absolute magnitudes are based on the mean group velocities (not for LG galaxies) with H_o=55 and a local Virgo infall velocity of 220 km/s. Superposed on the group data are the LFs of Virgo spirals (Sa - Sm), and Virgo spirals + irregulars (Sa - Im), both of which are normalized to give the same number of spirals as the groups for $M_{B_T} < -15$.

The first thing to note is the overall similarity of the spiral LFs of Virgo and the nearby groups. Secondly, we see again a clear progression in mean luminosity along the spiral+irr sequence, just as in Virgo (Fig. 2). But there is also an important difference! The relative frequences of spiral subtypes in the groups are : 13% Sa-Sbc, 27% Sc, and 60% Scd-Sm. In the Virgo cluster the corresponding numbers are 32%, 36%, and 32% (cf. Fig. 2). Interpreting this as a density effect, this means that the environment not only governs the relative frequency of spirals (and the other main Hubble types) - but also the type mix within the spiral class! (Such a fine morphological segregation has also been reported by GIOVANELLI et al. [24].) This effect should of course distort the shape of the spiral LF as a function of environment; but with the present data we do not see this. The density effect is clearly there, however, if we add the irregulars (which are the low-luminosity extension of spirals) : The groups have a considerably larger irr population than the denser Virgo cluster (see Fig. 4)

In summary, comparing late-type galaxies in nearby groups (low density)

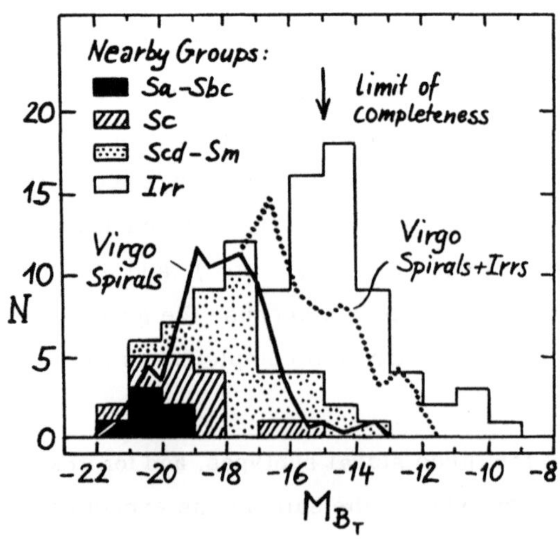

Fig. 4 LFs for late-type galaxies in nearby groups (D<Mpc), including the Local Group. The sample is divided into four morphological classes (three spiral subtypes + irregulars). For comparison the spiral and spiral + irr LFs of Virgo cluster galaxies are added, normalized to give the number of group spirals within the completeness limit.

and in the Virgo cluster we see again no clear variation of the LF shapes, if the morphological binning is sufficiently fine.

4. Discussion

The whole range of environmental density for galaxies spans a factor of a million, and the samples we have looked at (Virgo, Coma, nearby groups) are only a few "windows" in this great continuum. But what we have seen - by comparing the type specific LFs in these few different environments - is clearly pointing to a simple universal relationship that has already been found by TAMMANN in his own LF analysis [25]: While the relative frequency of a given Hubble type is a strong function of environment (density), the shape of its LF is nearly constant (as far as we can tell). This means that the general LF can be written as a product of a type-dependent part (the LF shape) and a density-dependent part (the LF normalization for a given type). More formally:

$$\Phi(M,\rho)\,dM = \sum_T \varphi^*_T(\rho)\cdot\varphi_T(M)\,dM = \Phi^* \sum_T f_T(\rho)\cdot\varphi_T(M)\,dM\,, \qquad (1)$$

where $\Phi(M,\rho)\,dM$ stands for the number of galaxies of all types per unit volume in an environment of local density ρ with absolute magnitudes between M and M+dM (at present ρ means a symbol, not a continuous parameter) ; $\varphi_T(M)$ specifies the LF shape of type T, and $\varphi^*_T(\rho)$ is the LF normalization of T as a function of environment (local density ρ). The total LF is a sum over all types. Normalizing the total LF (by Φ^*), we can introduce a type fraction as a function of local density, $f_T(\rho)$, which is the well-known morphology-density relation [12, 13].

It is important to note that all normalization factors and the type fractions, $f_T(\rho)$, depend on a magnitude limit which has to be specified (ideally $M_{lim}=\infty$). Hence in general the morphology-density relation is a function of the adopted magnitude cut - as a result of the type dependence of the LF (a fact stressed also by DE SOUZA et al. [26]).

We have not yet defined the clear meaning of the type parameter T. - Should we merely distinguish between the main Hubble types - E (+ dE), S0 (+ dS0), and spiral + irr ? If so, we know that the constancy of the LF shape will break down at some point because there is a morphology-density relation even for the spiral subtypes (cf. Sec. 3). On the other hand, if T stands for the finest classification bins, it may be easier to defend the concept of constant LF shape. In that case, however, early-type and late-type galaxies would not be treated on an equal footing; star-forming galaxies have a much richer morphology and thus can be classified with much greater detail than the dead galaxies. And - pushed to extremes - there is no point in breaking up the total LF into a thousand delta functions.

Consequently, to what extent the LF shape is independent of environment is a function of morphological resolution. It can be debated what T resolution is best suited for the LF discussion. For the moment we suggest a resolution into the following classes : E, dE, S0, dS0, spiral, and irregular, - and we hypothesize that for these types the LF shape is independent of environment - as a first order approximation.

There have been indications of what might be called secondary effects : Galaxies of higher luminosity may slightly prefer denser environments. But this is still very uncertain because DE SOUZA et al. find just the opposite trend for "field" S0's [26, 27]. In any case, the hypothesis of constant LF shape has of course to be tested further by studying the type specific LFs of many more clusters of galaxies, following THOMPSON and GREGORY [20], and SANDAGE et al. [11] ; and by exploring the LF-density relation (as an extension of the morphology-density relation) for "field" galaxies.

The simple relationship we advocate, if true, leads of course to a partial rehabilitation of the "universal approach" to the LF. The difference is that not the total LF but the type specific LFs are universal in shape. This new universality (as far as it goes) is likely to tell us a lot about galaxy formation, while the deviations from it may tell us something about galaxy evolution. But the theorist should also feel challenged to explain why the specific LFs are as they are, and not only why they are the same everywhere. For instance - How can the LFs of so widely different systems as E, S0, and spiral galaxies be so similar at the bright end ?

5. Addendum : LFs for Spheroids and Disks (Virgo Cluster)

Galaxies are made up of two principal components : spheroids and disks. E galaxies are purely spheroidal, galaxies later than Sd are pure disks, with a systematic mixing ratio of the two components (bulge-to-disk ratio) between. In a sense the distinction between components seems more fundamental than between Hubble types, and so it may be useful to construct separate LFs for spheroids (E's + bulges of S0's and spirals), and for disk components (of S0's and spirals). This has been done in a simple, statistical way for the sample of Virgo cluster galaxies used in Sec. 2. Galaxies of a given type were split into components by applying a mean bulge-to-disk ratio for that type. The disks were statistically corrected for internal absorption. The resulting LFs for spheroids and disks, having summed over the contributions from all types, are shown in Fig. 5. For completeness we have added the LF of "diffuse dwarfs" (=dE + dS0 + Im/dE). These dwarfs can be regarded as a third species because they have spheroidal and disk characteristics, and they are quite distinct by their low surface brightness. Also shown is the total LF, comprising spheroids, disks, and the diffuse dwarfs, which is of course identical to the total LF of all Virgo cluster galaxies lumped together.

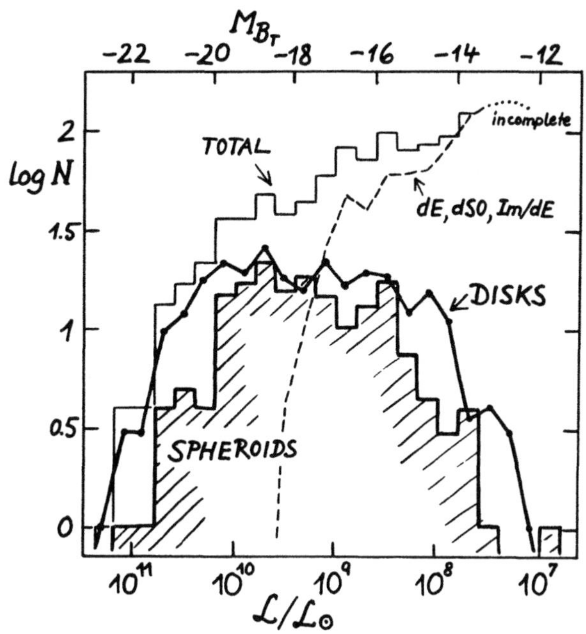

Fig. 5 Logarithmic LFs for spheroids (E's and bulges), and for disk components (of S0's and spirals), based on a complete sample of Virgo cluster galaxies. Also shown are the "diffuse dwarfs" (dE + dS0 + Im/dE), and the grand total.

Certainly the most striking feature in Fig. 5 is that both spheroids and disks have a broad Gaussian LF with a peak around $M_{B_T} = -18$ (or a few 10^9L The distribution of disks is somewhat broader than that of spheroids, but overall the two LFs are surprisingly similar. This is one more simple fact for us to ponder about.

On the other hand, the LF of diffuse dwarfs is essentially exponential with a cut-off at $M_{B_T} \sim -18$. And as far as the total LF..... compare with Fig. 1 !

I am very much indebted to Prof. G. A. Tammann for putting me on the right track regarding the simple hypothesis of constant LF shape. Dr. R. Kraan-Korteweg has generously given of her time to derive updated local LFs for the purpose of comparison with Virgo LFs. Finally, it is a pleasure to thank the Swiss National Science Foundation for financial support.

References

1. P. Schechter : Astrophys. J. 203, 297 (1976)
2. J. E. Felten : Astron. J. 82, 861 (1977)
3. A. Dressler : Astrophys. J. 223, 765 (1978)
4. R. P. Kirshner, A. Oemler, and P. L. Schechter : Astron. J. 84, 951 (1979)
5. G. A. Tammann, A. Yahil, andA. Sandage : Astrophys. J. 234, 775 (1979)
6. J. E. Felten : Comm. Astrophys. 11, 53 (1985)
7. C. G. Christensen : Astron. J. 80, 282 (1974)
8. A. Sandage : Phys. Today 23, No. 2, p. 34 (1970)
9. E. Holmberg : Ark. f. Astron. 5, No. 20, p. 305 (1969)
10. G. A. Tammann and R. Kraan : in The Large Scale Structure of the Univers M. S. Longair and J. Einasto (eds.), Dordrecht : Reidel, p. 71 (1978)
11. A. Sandage, B. Binggeli, and G. A. Tammann : Astron. J. 90, 1759 (1985)
12. A. Dressler : Astrophys. J. 236, 351 (1980)
13. M. Postman and M. J. Geller : Astrophys. J. 281, 95 (1984)
14. B. Binggeli, A. Sandage, and G. A. Tammann : Astron. J. 90, 1681 (1985)
15. A. Sandage and B. Binggeli : Astron. J. 89, 919 (1984)
16. J. Kormendy : Astrophys. J. 295, 73 (1985)
17. J. Kormendy : this volume
18. A. Wirth and J. S. Gallagher : Astrophys. J. 282, 85 (1984)
19. G. A. Tammann, B. Binggeli, and A. Sandage : to appear in Astron. J. (1987)
20. L. A. Thompson and S. A. Gregory : Astrophys. J. 242, 1 (1980)
21. A. Dressler : Ann. Rev. Astron. Astrophys. 22, 185 (1984)
22. R. C. Krann-Korteweg : Astron. Nachr. 300, 181 (1979)
23. R. C. Kraan-Korteweg and B. Binggeli : in preparation
24. R. Giovanelli, M. P. Haynes, and G. L. Chincarini : Astrophys. J. 300, 77(198
25. G. A. Tammann : in Star Forming Dwarf Galaxies and Related Objects, D. Kunth, T. X. Thuan, and J. T. T. Van(eds.), Paris: Ed. Frontières, p. 52(19
26. R. E. deSouza, G. Chincarini, and G. Vettolani : preprint (1986)
27. R. E. deSouza, G. Vettolani, and G.Chincarini : Astron. Astrophys. 143, 143 (1986)

Core Properties of Elliptical Galaxies

TOD R. LAUER

With the advent of high resolution surface photometry of elliptical galaxies [1, 2] there are many interesting problems that we can now investigate – in this brief paper I will confine my discussion to the basic core structure parameters such as core radius r_C and central surface brightness, I_0, and their relationship to global properties of elliptical galaxies. Most of the discussion presented here is based on the analysis of slightly sub-arcsecond resolution CCD photometry of 42 galaxies presented in LAUER [3]. The main result presented therein is that while core properties depend in part on total galaxy luminosity, L, they exhibit well over an order of magnitude variation in central luminosity density ($\rho_C \equiv I_0/2r_C$) at any luminosity; in fact, central density acts as a "second parameter" in addition to luminosity for the description of the core properties of any galaxy. This result is sketched out below.

LAUER [3] and KORMENDY [4] find similar core parameter relationships as a function of luminosity. Brighter galaxies tend to have bigger and dimmer cores – specifically I get $r_C \propto L^{1.2}$ and $I_0 \propto L^{-1.0}$. There is significant large scatter in both relationships, however, but with tightly correlated residuals; at any luminosity excessively large cores are dim in the center and excessively compact cores are brighter. This is shown in Fig. 1,which plots residuals in both relationships against each other. Dashed lines mark out lines of constant central density excesses or deficits with respect to the nominal central density expected at any luminosity from the individual $r_c - L$ and $I_0 - L$ relationships. As can be seen, the scatter in core parameters at any luminosity can be attributed to even larger scatter in the core central densities. Luminosity independent variation in core structure is shown directly in Fig. 2,which displays core profiles for five galaxies of the same luminosity (and at nearly the same distances). The cores of

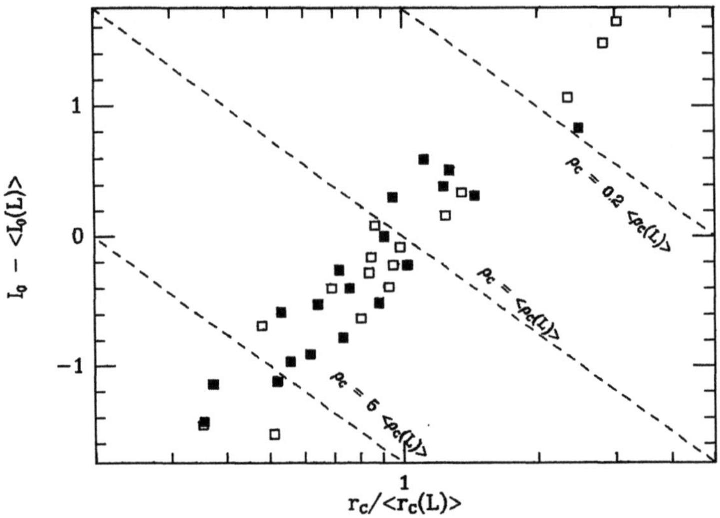

Figure 1: Residuals of the mean $r_C - L$ relationship are plotted against residuals of the mean $I_0 - L$ relationship. Solid symbols are well-resolved cores; open symbols are unresolved cores and will move towards the lower left corner of the diagram upon better resolution. Dashed lines mark out constant values of ρ_c excesses or deficits from the average ρ_C at any luminosity.

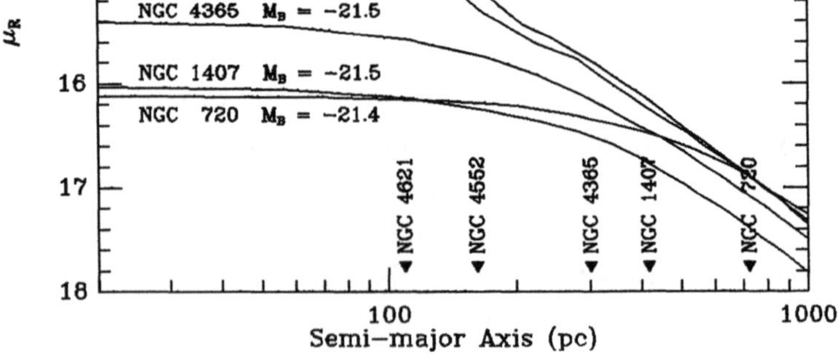

Figure 2: Core surface brightness profiles for five galaxies with $M_B \approx -21.5$ and at similar distances. Core radii of the galaxies are indicated on the horizontal axis.

NGC 4621 and NGC 4552 are unresolved and are thus even more compact than shown here.

The picture presented here I believe suggests dissipational formation of the cores. The cores of giant ellipticals contain only $10^8 - 10^9 M_\odot$, a small fraction of their total masses. Further, since the cores sit at the bottom of their galaxies' potential wells,they are sensitive to the central accretion of even small amounts of matter which would otherwise have no effect on the global structure of their galaxies. In this context, it is interesting to note that there appears to be no relationship between the ratio R_E/r_C and galaxy luminosity. This argues against the formation of elliptical galaxies by any sort of dissipationless hierarchy of merging, since simulations of such processes predict R_E/r_C to increase as a strong function of L. A naive picture of merging *with* dissipation might be expected to produce even more compact cores for a given final effective radius; however, variations in the amount of "stuff" allowed to sink to the center of the final galaxy produced by the merger might obscure any basic dependence of R_E/r_C on L. Central mass-to-light ratios also appear to be weakly dependent on ρ_C (although this needs to be verified) as $M/L_B \propto \rho_C^{-0.2}$ Again this relationship is suggestive of dissipational formation of galaxy cores – the central stellar population depends on a local property of the core perhaps implying that the stars in the core formed there.

I would now like to turn to the problem of how core structure depends on total luminosity. So far I have emphasized the ρ_C dependence of the core parameters; it is also possible to isolate the pure luminosity dependence of the core parameters. A principal components analysis gives $L \propto r_C^{2.4} I_0^{1.7}$. As argued by DRESSLER *et al.* [5] central velocity dispersion σ correlates better than L with global structural properties of the galaxies; this also appears to be true with the core parameters and works best with the central Mg_2 index tossed in. Fig. 3 shows the results of a fit involving central velocity dispersion, σ, r_C, I_0, and Mg_2 on the assumption that the later three variables determine σ. The resulting fit using just the well resolved galaxies gives:

$$\sigma \propto r_C^{0.4} I_0^{0.3} 10^{1.4 Mg_2}. \quad (1)$$

This fit is good; dispersions predicted from the core parameters of the well resolved and nearly resolved galaxies fall very close to their true values. Note that r_C is the only distance dependent parameter in (1); the goodness of the fit suggests that cores might be used as a metric distance indicator if (1) can be confirmed with a larger sample of galaxies with well resolved cores.

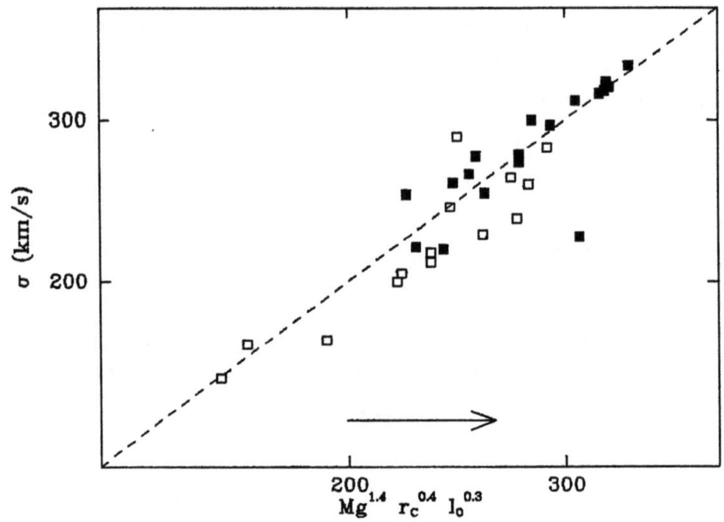

Figure 3: Central velocity dispersion plotted against that calculated from a fit to r_C, I_0, and Mg_2 for the sample. Symbols are as in fig. 1. The arrow shows the change in calculated σ for a factor-of-two change in r_C.

In summary, two parameters L and ρ_C determine the observed values of r_C and I_0. The scatter in ρ_C at any galaxy luminosity may be evidence for dissipational formation of the cores. Despite this scatter, however, the luminosity dependent component of core structure can be extracted and may be useful as a metric distance indicator.

1. T. R. Lauer: Ap. J. Suppl. 57, 473 (1985)
2. J. Kormendy: Ap. J. (Letters) 292, L9 (1985)
3. T. R. Lauer: Ap. J. 292, 104 (1985)
4. J. Kormendy: Ap. J. 295, 73 (1985)
5. A. Dressler, D. Lynden-Bell, D. Burstein, R. L. Davies, S. M. Faber, R. J. Terlevich, G. Wegner: preprint (1986)

Session 6

Galaxies in Relation to Larger Structures

A. Dekel, Chair
July 28, 1986

Voids and Galaxies in Voids

AUGUSTUS OEMLER, JR.

1. Introduction

In the past few years, the properties of voids have become central to the study of the large-scale distribution of matter in the universe. The reasons for this are partly observational- very large voids have been discovered- and partly due to their potential usefulness as measures of the clustering of galaxies. Such a use is important, because the most popular statistic describing the distribution of galaxies, the galaxy autocorrelation function, has proved remarkably uninformative. Distributions of galaxies which have very different appearances may nevertheless have indistinguishable autocorrelation functions; and very different scenarios for the production of structure in the universe produce autocorrelation functions which vary by much less than the uncertainties in the observations [1]. The statistics of voids are a very different measure of the clustering than the two-point correlation function, as WHITE [2] has shown, and they have the potential of much greater sensitivity to different clustering models.

The relevance of voids to the astrophysics of galaxies, a topic closer to the subject of this conference, is a consequence of their use as a measure of clustering. The existence of at least some voids appears to be inconsistent with all current clustering models, a result which some have suggested is due to the failure of galaxies to trace the distribution of mass on large scales. If that is correct, it implies that galaxy formation has a sensitivity to environment which will have significant astrophysical implications.

2. Evidence for Voids

Before discussing the available information on voids, I should make a semantic point. "Voids and galaxies in voids" must seem an odd title, since voids are, by definition, where galaxies are *not*. However, it must be remembered that what are called voids are only regions is which the density of galaxies is too low to be detected in the available surveys; and in most cosmological models a large region of low density is as significant as one totally devoid of matter. Since the more accurate phrase "region of very low galaxy density" is much too clumsy for continual use, I shall call such regions voids.

Among the first to point out the existence of regions devoid of galaxies were EINASTO and his collaborators [e.g. 3], who insisted on the cell-like structure of the universe at a time when most astronomers (or at least most western astronomers) resisted the idea. Clear evidence for the existence of voids with diameters between $10h^{-1}$ and $30h^{-1}$ Mpc has emerged from redshift surveys in the direction of nearby superclusters [4,5,6]. The recent two-dimensional redshift survey by LAPPERANT, GELLER, and HUCHRA [7] suggests that roughly spherical voids of comparable scale fill a large fraction of space.

The largest known void remains that in Bootes [8]. It was discovered from a gap in the distribution of redshifts of galaxies located in 3 small fields about 30° apart. We have now surveyed its structure by a redshift survey in 283 small fields spread over the region between the three original fields [9]. Our data are insufficient to delineate the void's complete structure, but we do know that the largest empty sphere which can be placed in this region is centered at $\alpha = 14^h\ 50$, $\delta = 46^o$, $v = 15500$ km s^{-1}, and has a volume, assuming $H_o = 50$ km s^{-1} Mpc^{-1}, of 10^6 Mpc^3. If our survey galaxies had been uniformly distributed, this sphere would have contained 31 sample galaxies.

Curiously, this empty sphere does not extend as far as any of the three fields, on the basis of whose redshift distributions the void was originally discovered. It is possible, but very unlikely, that that redshift gap is unrelated to the void which exists between the three fields. More probably, the empty spherical region is surrounded by a much larger region of very low density. This possibility is supported by the recent survey of a region to the southeast of the Bootes void, conducted by POSTMAN, HUCHRA, and GELLER [10], which showed a velocity structure identical to that in Bootes, and by the distribution of clusters studied by BAHCALL AND SONEIRA [11], which have a similar velocity structure over an even larger angular scale.

Evaluating the significance of this structure is complicated by the problem of doing *a posteriori* statistics on a sample of one, but an attempt to fairly calculate the statistics suggests that the probability of such an object being found in a survey such as ours is of the order of 10^{-3} in any of the currently popular clustering models, including hierarchical, neutrino pancake, and cold dark matter models. Also, the Bootes void is not the only one of such scale known; we have discovered a very similar void in a preliminary survey in the constellation of Phoenix [12].

3. Do Galaxies Trace the Mass?

One might conclude, from the existence of such improbable voids, that a new theory of clustering is needed, but there are other possibilities. One which has become very popular among those who construct clustering models is *biased galaxy formation.* Since theorists calculate the distribution of matter, and observers study the distribution of galaxies, theory and observations can be reconciled if galaxies do not trace the mass. Such a picture is doubly attractive because it also removes the embarrassment that measurements of the cosmic density parameter uniformly produce results much less than the value of unity preferred by most theorists.

The idea of biased galaxy formation is clearly a fudge, but not an arbitrary one, as there are both theoretical and observational reasons why it might be plausible. KAISER [13] has demonstrated that, *if* galaxies formed from random density fluctuations, and *if* these fluctuations were superimposed on uncorrelated larger scale fluctuations of small amplitude, and *if* galaxies are rare events, which, for example, only form from 3 σ fluctuations, then galaxies will be more clustered than the underlying matter distribution.

Observationally, it has been known since the time of Hubble and Zwicky that elliptical galaxies are more clustered than are spirals. Both GELLER and DAVIS [14], and, more recently, SADLER and SHARP [15] have shown that the covariance function of elliptical galaxies is of higher amplitude than that of spirals. Equivalently, DRESSLER and others [16,17] have demonstrated that the morphological mix of galaxies is a function of local density, ellipticals being more prevalent at higher densities. These observations suggest that the sequence from ellipticals through spirals to irregulars may be a sequence from high-density, rare events to low-density, less rare events. If this were so, ellipticals would be the most biased, and irregulars the least biased galactic tracers of the cosmic mass distribution.

Some apparent confirmation of this idea comes from a recent study of the relative distribution of high and low surface brightness galaxies [18]. DAVIS and DJORGOVSKI calculated the angular correlation functions of high and low surface brightness objects in the UGC catalog [19]. They approximated the results by power laws and, using the distributions of measured redshifts of galaxies in the two classes to determine their distances, inverted the angular correlation functions to obtain the spatial functions. The result was a correlation function of high surface brightness galaxies which was steeper, and higher in amplitude than that of low surface brightness galaxies.

Davis and Djorgovski interpret this result as support for the idea of biased galaxy formation, but this analysis has several problems. It has been pointed out [20] that the two redshift distributions which Davis and Djorgovski used to invert the angular correlation functions were incorrect, and the large differences between them were due only to the incompleteness of the data. This error has artificially depressed the spatial correlation function of the low surface brightness galaxies. In fact, the true redshift distributions of the two classes are so similar that one can directly compare the angular covariance functions. Such a comparison shows that a power law is a bad fit to the data. While the high surface brightness galaxy correlation function is higher and steeper on small scales, the two correlation functions are identical on scales larger than about 1°, implying that high and low surface brightness galaxies have the same distribution on scales larger than a few megaparsecs. This conclusion is confirmed by the work of BOTHUN, BEERS, MOULD, and HUCHRA [20], who have obtained new velocities for a large sample of low surface brightness UGC galaxies. A direct comparison of the spatial distributions of the two classes of objects show them to be identical on large scales. In particular, both low and high surface brightness galaxies delineate superclusters and voids equally well.

This result does not, however, close the door on the idea of biased galaxy formation. As Bothun, Beers, Mould, and Huchra point out, most of the low surface brightness UGC galaxies are not, contrary to expectation, dwarfs: their HI line widths show that many are quite massive galaxies. It is possible that only the dwarfs are unbiased and fill the voids, and one particular theory, by DEKEL and SILK [21], suggests that this might be so. To test this possibility, EDER, SCHOMBERT, DEKEL, and OEMLER [22] have studied the distribution of dwarfs around a nearby void centered at $\alpha = 1^h$, $\delta = +15^o$, $v = 3500$ km s^{-1}. This void, with a diameter of about $12h^{-1}$ Mpc, was discovered in the CFA survey, and is seen clearly in the survey of Giovanelli and Haynes, discussed elsewhere in this volume.

We have selected dwarf irregular galaxies from 15 Palomar Schmidt plates taken in the vicinity of the void and have measured redshifts of 99 objects with diameters greater than 20 arc sec, using the Arecibo 1000 foot radio telescope. These galaxies are true dwarfs, with a median hydrogen line width of 100 km s^{-1}, and absolute visual magnitudes of -17. The results are presented in Fig. 1. We divide the 15 fields into on and off void groups, and compare the velocity distributions of the dwarfs with those of bright galaxies, using data made available to us in advance of publication by Drs. Giovanelli and Haynes. The velocity distributions of the two types of galaxies are virtually identical. Both equally well delineate the void. The one dwarf which falls in the redshift gap of the void is located near the void's edge, only $1h^{-1}$ megaparsec from the nearest bright galaxies.

Thus, neither normal bright galaxies, nor low surface brightness galaxies nor dwarfs inhabit voids. There is, however, one type of galaxy which apparently does. Beginning with the work of BALZANO and WEEDMAN [23], several studies have been made of the distribution of strong emission line galaxies in the vicinity of the Bootes void [24,25,26]. Seven are now known, most found from objective prism searches. Because it is very difficult to calibrate the depth of surveys of this type, the space distribution of these galaxies is not well known. Nevertheless,it is clear that their space density is lower inside the Bootes void than it is outside, but also that they are overabundant, relative to normal galaxies, inside. Can these be the missing objects, products of low amplitude density fluctuations, which fill the voids? Undoubtedly not: these are very uncommon objects, comprising on average only about 5 percent of the general galaxy population. They are probably not even a fundamental galaxy type, but rather the accidental products of mergers or gas infall. Their overabundance in the void is probably due to the enhanced formation of starburst galaxies in low density regions rather than to biased galaxy formation.

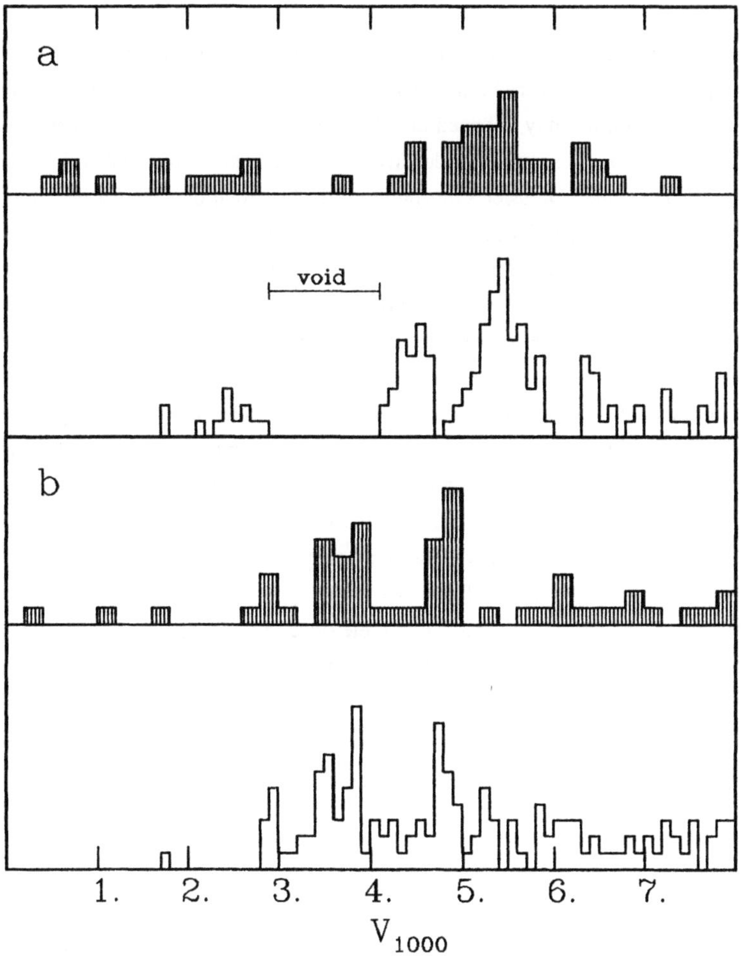

Figure 1. Velocity distribution of dwarf (filled histogram), and giant (open histogram) galaxies in 15 fields in the area of a nearby void. a- on-void fields; b- off-void fields.

4. Discussion

The example of the emission line galaxies within Bootes does illustrate an important point: variations in the mix of galaxy types with environment could be due to any of several processes. One is biased galaxy formation, in which the space distribution of one type may follow that of the underlying mass, while the distribution of others differs. Another possibility is that the distribution of the entire galaxy population may be the same as that of the mass, but the relative proportions of different types varies, due to the effect of environment at the time of formation. It is also possible that galaxy populations were originally homogeneous, and all inhomogeneities have arisen because of the effect of environment on the evolution of galaxies.

These alternatives have profoundly different implications for the interpretation of the distribution of galaxies, but are very hard to distinguish observationally. Fortunately, for the cosmological problem, we may not need to. Whatever the process, there is no evidence that it operates anywhere but in small, high density regions. The study of POSTMAN and GELLER [17], which extended Dressler's morphology-density relation to lower densities, showed that there is *no* dependence of galaxy populations on density in regions whose density is so low that the crossing time is comparable to the Hubble time, a result amplified by the studies of the distribution of low surface brightness and dwarf galaxies described above [19,20,22]. Therefore, all types of galaxies equally well delineate large scale structure.

If, then, biased galaxy formation is to explain the large scale distribution of galaxies, the restrictions imposed by the observations are quite severe. A mechanism must be found which is capable of suppressing the formation of galaxies in some environments, and which works *equally* well on galaxies of all luminosities and all morphological types. Given the large range in the physical properties of galaxies, that will be very hard to do. Unfortunately, our understanding of the processes of galaxy formation is sufficiently poor that we cannot demonstrate that it is impossible. And, a complete observational disproof of biased galaxy formation is a hopeless task: it is easy to reformulate the missing mass in more and more unobservable forms as observations progress. However, given the present lack of observational evidence *for* biased galaxy formation, it seems reasonable to demand that some be provided before we are asked to take the idea seriously.

References

1. Oemler, A., Schechter, P.L., Shectman, S.A., and Kirshner, R.P., in preparation.
2. White, S.D.M. 1979, *Mon. Not. Roy. Astron. Soc.*, **186**, 145.
3. Joeveer, M., and Einasto, J. 1978, In *The Large Scale Structure of the Universe*, IAU Symposium No. 79, ed. M.S. Longair and J. Eiansto, (Dordrecht:Reidel), p. 241.

4. Gregory, S.A., and Thompson, L.A. 1979, *Astrophys. J.* , **222**, 784.
5. Tarenghi, M., Chincarini, G., Rood, H.J., and Thompson, L.A. 1980, *Astrophys. J.*, **235**, 724.
6. Gregory, S.A., Thompson, L.A., and Tifft, W. 1981, *Astrophys. J.*, **243**, 411.
7. Lapparent, V. de, Geller, M.J., and Huchra, J.B. 1986, *Astrophys. J. (Lett)*, **302**, L1.
8. Kirshner, R.P., Oemler, A., Schechter, P.L., and Shectman, S.A., *Astrophys. J. (Lett.)*, **248**, L57.
9. Kirshner, R.P., Oemler, A., Schechter, P.L., and Shectman, S.A., *Astrophys. J.* in press.
10. Postman, M., Geller, M., and Huchra, J.B. 1987, *Astron. J.* in press.
11. Bahcall, N.A., and Soneira, R.M. 1982, *Astrophys. J. (Lett.)*, **258**, L17.
12. Kirshner, R.P., Oemler, A., Schechter, P.L., and Shectman, S.A., in preparation.
13. Kaiser, N. 1984, *Astrophys. J. (Lett)*, **284**, L9.
14. Davis, M.J., and Geller, M. 1976, *Astrophys. J.*, **208**, 13.
15. Sadler, E.M., and Sharp, N.A. 1984, *Astrophys. J.*, **287**, 80.
16. Dressler, A. 1980, *Astrophys. J.*, **236**, 351.
17. Postman, M., and Geller, M.J. 1984, *Astrophys. J.*, **281**, 95.
18. Davis, M.J., and Djorgovski, S. 1985, *Astrophys. J.*, **299**, 15.
19. Nilsen, P. 1973, *Uppsala Astron. Observ. Annals*, **6**.
20. Bothun, G.D., Beers, T.C., Mould, J.R., and Huchra, J.B. 1986, preprint.
21. Dekel, A., and Silk, J. 1986, *Astrophys. J.*, **303**, 39.
22. Eder, J., Schombert, J.M., Dekel, A., and Oemler, A. 1986, in preparation.
23. Balzano, V.A., and Weedman, D.W. 1982, *Astrophys. J. (Lett)*, **255**, 4.
24. Sanduleak, N., and Pesch, P. 1982, *Astrophys. J. (Lett)*, **258**, L11.
25. Tifft, W.G., Kirshner, R.P., Gregory, S.A., and Moody, J.W. 1986, preprint.
26. Moody, J.W. 1986, PhD thesis, University of Michigan.

The Large Scale Distribution of Galaxy Types

MARTHA P. HAYNES

1. Introduction

Theories of galaxy formation and evolution must explain the observed differences not only in the morphological appearance of galaxies but also in the degree of clustering of the elliptical, lenticular and spiral classes. Although of fundamental importance, the epoch of the enforcement of morphological segregation is uncertain. DRESSLER [1] has recently reviewed the evidence concerning models for various schemes that lead to morphological segregation, mostly in high density environments. It is still much debated whether the observed differentiation of morphologies is inbred during the era of galaxy formation or shortly thereafter, or whether it is a continuing process, likely resulting from the alteration of morphologies. Evidence supporting both hypotheses exists; much of the argument revolves around the possibility of converting spirals into lenticulars in clusters. While there seem to be fundamental differences between S0's and spirals, spirals that pass through the cores of X-ray clusters do lose large fractions of their interstellar HI [2]. Such gas sweeping, however, only occurs in those rare high density environments containing hot intracluster gas that make up only a small fraction of the volume of the universe.

The segregation of galaxy morphologies with varying local density has been long recognized, but its quantitative description has been made only recently. The monotonic variation in population fraction with increasing local density was presented in the study of 55 clusters undertaken by DRESSLER [3]. While some 80% of field galaxies are spiral, that percentage drops to only 15% in evolved, spiral-poor clusters like Coma.

The variation in population fraction with local density extends in a uniform manner from the high-density regimes of the rich clusters studied by Dressler to those of typical nearby groups of galaxies. Nearby loose groups are almost always dominated by spirals; in fact, excluding the Virgo region, the only two groups in the list of DEVAUCOULEURS [4] that are dominated by early type galaxies - the systems around NGC 3607 and NGC 5846 - both possess a higher density core with a central X-ray galaxy, surrounded by a more diffuse cloud of spirals.

The availability of redshifts for large numbers of galaxies makes possible the derivation of space densities and the identification of dynamical aggregates of galaxies, so that a better quantification of the morphology-density relation is possible. In their examination of galaxy groups, POSTMAN and GELLER [5] have shown that the apparently universal morphology density relation quantified by DRESSLER [3] extends over six orders of magnitude in space density. Only at very low densities, where the dynamical timescale is comparable to or greater than a Hubble time, does the population fraction not reflect variations in local density.

The most commonly-used statistic of clustering is the two-point correlation function $\xi(r)$ and its angular form $w(\theta)$. Comparison of the amplitude and slope of $w(\theta)$ for various subsets of galaxies provides a quantitative measure of the differences in clustering scale. For studies of the distribution of galaxy morphologies, the largest homogeneous compilation of galaxies which includes morphological classification is the Uppsala General Catalog (UGC) [6]. Using the UGC, DAVIS and GELLER [7] have compared the two-point angular correlation function of galaxies brighter than $m = +14.5$ for different morphologies and found that clustering is indeed a function of morphology: elliptical-elliptical clustering is characterized by a power law with a slope steeper than that appropriate for spiral-spiral clustering, while the case of lenticulars is intermediate. For that sample, the number of galaxies found within 1 Mpc of a random elliptical is about twice that found within the same distance of a random spiral.

As established by Dressler, the variation in population fraction with galaxy space density is monotonic but slow. In the highest density environments, the possible morphology-altering mechanisms are many and include both galaxy-galaxy, galaxy-intracluster gas and galaxy-cluster interactions. The scale over which segregation occurs and the degree of continuity of the variation throughout all regimes of density will constrain models predicting the relative formation times of galaxies and the large-scale structure in which they are embedded.

2. Morphological Segregation in the Perseus Supercluster

The Pisces-Perseus supercluster is one of the most prominent enhancements visible in the two-dimensional distribution of galaxies seen in the local universe, as traced by the distribution of galaxies contained in the Catalog of Galaxies and Clusters of Galaxies (CGCG) [8]. The vast extent of the supercluster was suggested by EINASTO ET AL. [9] and verified by GREGORY ET AL. [10]. Today, Pisces-Perseus is most suitable as the subject of detailed study because the apparent magnitude corresponding to the characteristic luminosity function absolute magnitude M^* is +14.9, so that its membership is well represented both in the UGC and the CGCG and because, at its characteristic distance of $50h^{-1}$ Mpc, redshifts are easily measured by both optical and HI line methods.

To date, redshifts are available for some 4000 galaxies in the region surrounding the supercluster. The majority of those redshifts have been obtained via 21 cm HI line observations undertaken with the 305 m telescope at the Arecibo Observatory[1] and the 91 m telescope of the National Radio Astronomy Observatory[2] by GIOVANELLI, HAYNES and co-workers. The overall structure of the supercluster is summarized by HAYNES and GIOVANELLI [11]. It consists of a prominent ridge, easily visible in the surface density maps, which connects the rich clusters A262, A347 and A426 as well as Pisces and Pegasus, and numerous other connective structures of lower contrast. Figure 1 shows the enhancement in the number of galaxies observed in different redshift intervals, compared to that expected for a homogeneously-distributed population, in the region of the prominent supercluster ridge. The dominance of the redshift regime

[1] The Arecibo Observatory is part of the National Astronomy and Ionosphere Center which is operated by Cornell University under contract with the National Science Foundation.

[2] The National Radio Astronomy Observatory is operated by Associated Universities, Inc. under contract with the National Science Foundation.

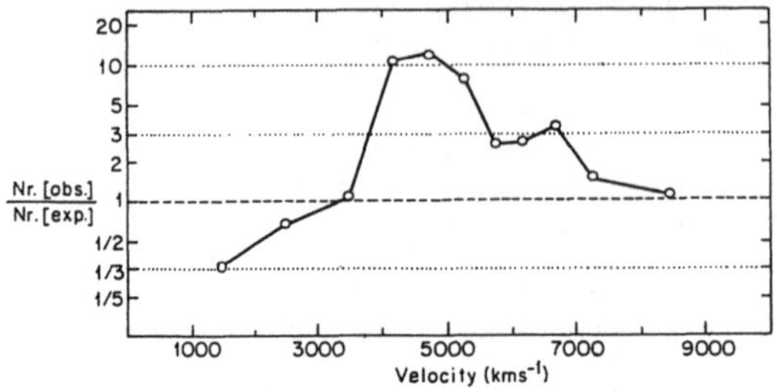

Figure 1: Enhancement in the redshift distribution in the region of the Pisces-Perseus ridge.

around 5000 kms^{-1} is further magnified by the presence in the foreground of a significant void. The regimes encompassed by structures within the supercluster extend over a wide range of space density from cluster cores to the far periphery where the typical interaction timescale is much longer than a Hubble time. Pisces-Perseus then serves as an excellent laboratory for the study of environmental effects and morphological segregation [12].

The degree of morphological segregation in the supercluster can be seen even in the sky distribution of UGC galaxies of differing morphology shown in Figure 2. Galaxies of progressively earlier morphology are segregated to higher density regions; the supercluster ridge is most evident in the distribution of elliptical and SO galaxies. Furthermore, the comparison among the spiral classes themselves suggests a monotonic differentiation of the formation loci with type.

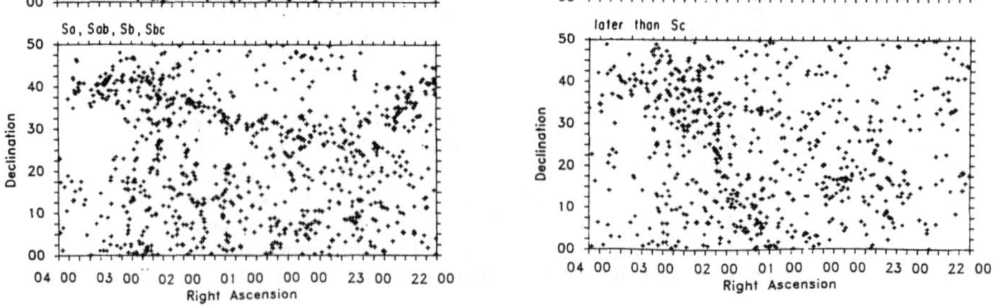

Figure 2: Sky distribution of UGC galaxies of different morphological classes in the Pisces-Perseus region from [12].

GIOVANELLI, HAYNES and CHINCARINI [12] have analyzed the dependence on morphological type of the two-point angular correlation function $w(\theta)$ for galaxies in the Pisces-Perseus supercluster. The parameters A, the amplitude, and β, the exponent of the power law, of the angular correlation function estimates derived by those authors are summarized in Table 1. In addition to the global variation of clustering characteristics among ellipticals, S0's and spirals noted by DAVIS and GELLER [7], the analysis of Pisces-Perseus galaxies shows the the spirals classes themselves show a smooth variation: spirals of type Sa and Sab cluster on smaller angular scales than do later spirals. The correlation analysis quantifies the visual impression seen by merely plotting the sky distribution of varying morphologies.

Table 1
Parameters of Angular Correlation Function Estimates [12]

type	Number	A	β
E	227	2.60	-1.06
S0,S0a	423	1.75	-0.84
Sa,Sab	312	0.29	-0.81
Sb,Sbc	566	0.50	-0.63
Sc	678	0.62	-0.47
later than Sc	681	0.58	-0.30

The variation of the population fraction with projected surface density, after subtraction of the foreground component, for the Pisces-Perseus region is shown in Figure 3. In agreement with the well-established trend seen in rich clusters by DRESSLER [3] and extended to groups by POSTMAN and GELLER [5], the population fractions of elliptical and S0 galaxies in Pisces-Perseus grow with increasing density, while the corresponding spiral fraction decreases. In this region, a significant degree of morphological segregation exists over the wide range of volume densities characteristic of the supercluster. Indeed, the continuous variation in population fraction is traceable over scales which well exceed those of individual groups or even clusters of galaxies, and into density regimes much poorer than those that characterize such aggregates. For most morphologies, a gradient in the population fraction is evident at nearly all density regimes. Moreover, the morphological segregation is not only limited to the three principal categories - elliptical, gas-poor systems (S0's), and gas-rich systems (spirals and irregulars) - but is also seen to vary within the range of spiral types as well. The smooth change in the observed population fraction among the spiral types implies that the conditions that lead to the currently observed morphological distribution arise to a large extent from the matter density at the time of formation, or at least, at an early phase of galaxy evolution. Diffusion of galaxies from high to low density regions is not viable within a Hubble time.

3. Luminosity versus Mass

The study of the distribution of dwarf galaxies may prove critical to the understanding of morphological segregation and the possibility of biased galaxy formation. SHARP, JONES and JONES [13] have examined the distribution of DDO galaxies relative to galaxies in the CGCG. They found that dwarfs tend to avoid clusters, are preferentially found in the vicinity of bright galaxies and are intrinsically different from ordinary galaxies of the same apparent magnitude.

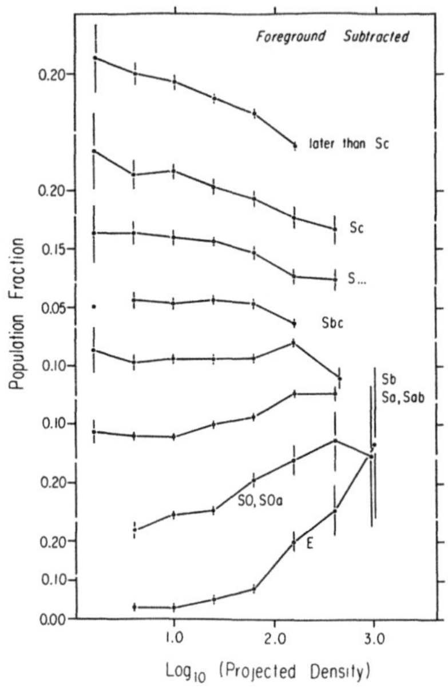

Figure 3: Variation in the population fraction with galaxy density after correction for the foreground component in Pisces-Perseus, from [12].

If the distribution of luminosity in the universe is a good tracer of the mass, then the correlation function $\xi(r)$ should be the same for both giant and dwarf galaxies. Although environmental effects will result in the enhanced clustering of ellipticals with respect to spirals, different clustering properties can also be examined in terms of the galaxies' surface brightness properties, after allowing for the obviously different morphological dependence on clustering.

Two recent reports have addressed the clustering of galaxies as a function of surface brightness. DAVIS and DJORGOVSKI [14] have shown that the angular correlations are weaker and shallower for the low-surface brightness (LSB) objects. They find that the spatial clustering length scale of LSB galaxies is about a factor of two smaller than that of higher surface brightness objects. Making use of redshift data for a selection of LSB galaxies, BOTHUN et al. [15] confirm the general conclusions of DAVIS and DJORGOVSKI but argue that there is <u>no</u> evidence that redshift voids are filled by LSB objects.

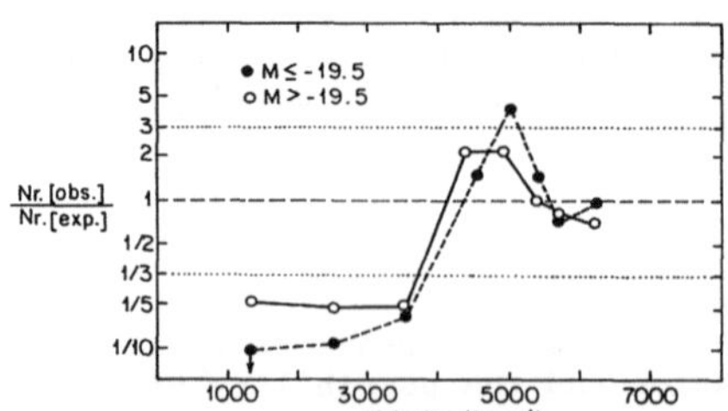

Figure 4: Distribution relative to random of galaxies in the region of the Pisces-Perseus void. Filled circles indicate higher luminosity galaxies; open circles, lower luminosity ones.

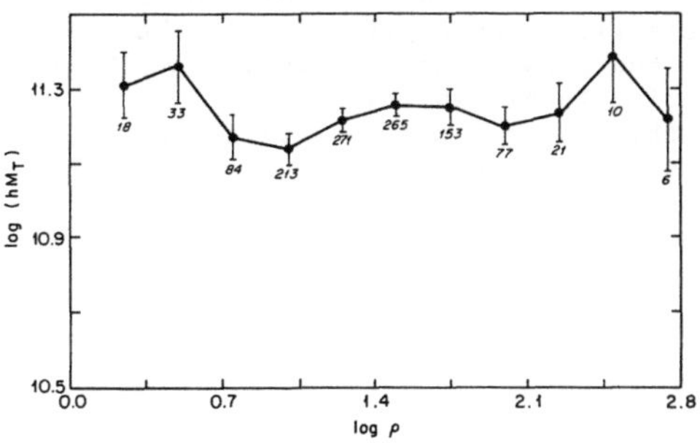

Figure 5: Distribution of total mass hM_T for inclined spirals as a function of projected surface density in the Pisces-Perseus supercluster.

In the region of Pisces-Perseus, the lower luminosity objects are in fact more widely distributed than the higher luminosity ones, but they do not fill in the prominent void [11]. Similar to Figure 1, Figure 4 shows the ratio of the number of observed objects per redshift interval to that expected for a random distribution, but for the void in the Pisces-Perseus region. The filled circles show the high luminosity objects, those with M brighter than -19.5, while the fainter galaxies are indicated by open circles. The lower luminosity galaxies show a more diffuse distribution. On-going studies by several groups, including the work reported by OEMLER in this volume, will investigate the distribution of low luminosity galaxies in specific voids and to even fainter absolute magnitudes.

Further evidence that the bright galaxies do not trace the distribution of mass is suggested by the variation in intrinsic galaxy properties as a function of local density seen in Pisces-Perseus. Figure 5 shows the lack of variation of the total máss, as derived from the 21 cm line profile width, for systems with inclinations greater than 30° as a function of projected density. In contrast, Figure 6 shows clear evidence for luminosity segregation, with an increase in the mean luminosity of galaxies in higher density environments.

The consequence of the trends evident in Figures 5 and 6 is that, for spirals at least, the mean mass-to-light ratios M/L, measured in different volumes of the universe, may not be the same; higher values of M/L will be observed in less dense regions because of variations in L, not in M. Making use of a much richer body of data, this result confirms the suggestion made previously by GIOVANELLI [16] that a universal M/L may not apply for spirals. Note that the observed trend is opposite that to be

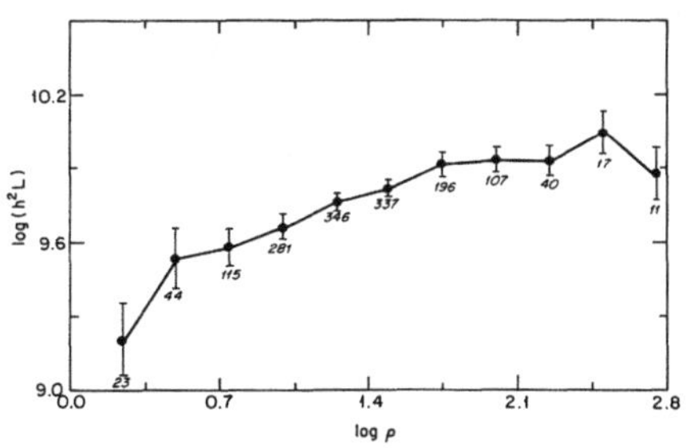

Figure 6: Distribution of luminosity h^2L for spirals as a function of projected surface density in the Pisces-Perseus supercluster.

expected from morphological segregation alone, since the M/L for Sa's is known to be higher than that for Sc's [17]. The application of the Tully-Fisher relation must consider the possibility of varying calibration of the mean M/L in regimes of differing local density.

As pointed out by BINGGELI in this volume, the possibility of luminosity segregation for earlier type galaxies is less clear. But the variation in the luminosity function introduced by morphological segregation should be considered in the comparison of samples derived from volumes characterized by varying densities.

4. Summary

The segregation of galaxy morphologies can be traced across all density regimes and shows distinctions among all Hubble types. To a large extent, the observed morphological segregation must reflect the environmental conditions at early epochs. Yet, it is also clear that on-going interactions that secularly affect the evolution of galaxies also play a role in reinforcing the segregation characteristics, at least in the highest density environments. Currently-popular gas removal mechanisms are generally not efficient over the range of densities which show segregation. At the same time, both the topology of the universe and its variation with Hubble type seems to reflect differing environmental conditions at early times. The large-scale variations in clustering characteristics must be reproduced in models that describe the pregalactic era or shortly thereafter and may play an important role in constraining such scenarios.

All of this work has been undertaken in collaboration with Riccardo Giovanelli whose contribution has often been the greater.

References

1. A.Dressler: Ann.Rev.Astr.Ap. 22, 185 (1984)
2. R.Giovanelli, M.P.Haynes: Ap.J. 292, 404 (1985)
3. A.Dressler: Ap.J. 236, 351 (1980)
4. G.deVaucouleurs: Galaxies and the Universe, A.Sandage, M.Sandage and J.Kristian eds. (U. of Chicago Press, Chicago), p.457 (1975)
5. M.Postman, M.J.Geller: Ap.J. 281, 95 (1984)
6. P.Nilson: Uppsala General Catalog, Uppsala Astr. Obs. Ann. 6 (1973)
7. M.Davis, M.J.Geller: Ap.J. 208, 13 (1976)
8. F.Zwicky, E.Herzog, M.Karpowicz, C.T.Kowal, P.Wild: Catalog of Galaxies and Cluster of Galaxies (Calif. Inst. Tech., Pasadena) (1961-68)
9. J.Einasto, M.Joeveer, E.Saar: M.N.R.A.S. 193, 353 (1981)
10. S.A.Gregory, L.A.Thompson, W.G.Tifft: Ap.J. 243, 411 (1981)
11. M.P.Haynes, R.Giovanelli: Ap.J.(Lett.) 306, L55 (1986)
12. R.Giovanelli, M.P.Haynes, G.L.Chincarini: Ap.J. 300, 77 (1986)
13. N.A.Sharp, B.J.T.Jones, J.E.Jones: M.N.R.A.S. 185, 457 (1978)
14. M.Davis, S.Djorgovski: Ap.J. 299, 15 (1985)
15. G.D.Bothun, T.C.Beers, J.R.Mould, J.R.Huchra: A.J.(to appear) (1986)
16. R.Giovanelli: The Comparative HI Content of Normal Galaxies, M.P. Haynes and R. Giovanelli eds. (N.R.A.O., Green Bank), p.105 (1982)
17. V.C.Rubin, D.Burstein, W.K.Ford, N.Thonnard: Ap.J. 289, 81 (1985)

Coherent Orientation Effects of Galaxies and Clusters

S. DJORGOVSKI

Properties and distribution of galaxies on large scales appear to be an exercise in randomness. Our favorite quantifiers of these things, the correlation functions, are a little more than noise spectrum descriptors. It is thus not surprising that we are suspicious about any signs or reports of systematic coherence in that noisy world. Galaxy alignments in particular are prone to evoke skepticism. This is in the best spirit of a healthy scientific conservativism, but with a smattering of hypocrisy as well. The order out of chaos is really what we are hoping for, after all. To be more specific, any systematic behavior of "inborn" galaxian properties (that is, not acquired by subsequent evolution, e.g., by the ram-pressure stripping and its consequences) that depend on the large-scale environment would be a strong evidence in favor of the top-down formation scenarios; at most, galaxies and their parent large-scale structure could form at the same time, but the bottom-up scenarios would be in great difficulty to explain any such phenomena. Therein lies the deeper reason why most reports of large-scale coherence were greeted with doubts or controversy: due to a misfortunate historical accident, the theoretical cosmogony in english-speaking countries was stuck upon the hierarchical, bottom-up formation scenarios, and there is perhaps some inertia of belief, which mere data cannot shake too easily. To be fair, the burden of the proof is with the observers.

There is now an ever increasing evidence for a coherence in the distribution and motions of galaxies at large scales (cf. GIOVANELLI ET AL. [1], DE LAPPARENT, GELLER & HUCHRA [2], or DRESSLER ET AL. [3], to quote only a few). We know that the galaxy morphology is a strong function of environment (DRESSLER [4], DAVIS & GELLER [5], and others), which is not necessarly or completely explainable by the gas stripping; there is also a connection between the galaxy surface brightness and clustering (DAVIS & DJORGOVSKI [6,7]). I will assume an attitude of a cautious optimism: the evidence for coherent orientation effects (alignments) of galaxies and clusters is now at least as good (or bad) and as abundant as the evidence pertaining to the two-point correlation functions at scales larger than $\sim$ 10 Mpc; that is, it can be better.

The importance and the physical meaning of alignment effects are based on the fact that the geometric axes of galaxies (and clusters?) reflect their internal dynamical axes: spin vectors in the case of disks, and the principal axes of velocity dispersion tensor in the case of ellipticals. We know that the shapes of most ellipticals are determined by their velocity anisotropy, which is frozen in the stellar orbits once the dissipation ceases. Any anisotropy of the protocluster material would be thus inherited by the simultaneously or subsequently forming galaxies. If the formation of disks occurs in a collapsing flattened structure, colliding shock waves will tend to produce spin vectors aligned with the protocluster plane. Alternatively, primordial vorticity or torquing in an expanding flattened protocluster may produce spin vectors which are perpendicular to the protocluster plane [8,9,10,11]. All these processes may go on simultaneously. In almost all environments, the post-formation-epoch tidal effects would not be sufficient to perturb significantly these dynamical axes. Thus, the orientation of galaxian axes may be a dynamical fossil, probing directly the epoch of galaxy and cluster formation: possible anisotropies of the protocluster material, primordial torquing, or turbulence. It is a subject which deserves a more careful attention. An adequate and full theory does not yet exist, and the N-body simulations so far are almost incapable of modeling such processes, or making any testable predictions of coherent orientation effects. (See, however, the interesting paper by BOND, elsewhere in this volume.)

The subject has a rich history, which begins with what we would now call statistics of spin vectors in the Local Supercluster (LSC), but it went under the name of distribution of apparent poles of spiral nebulæ. ABBE [12], KNOX SHAW [13], MEYERMANN [14], and GREGORY [15] correctly concluded that the apparent flattening of the spiral nebulæis reflective of thin, rotating disks seen at some aspect angles; thus, one can plot two possible pairs of poles on the celestial sphere. These authors more or less have found that such poles tend to cluster around a particular plane, which is close to the plane of the Milky Way (and almost perpendicular to the plane of the LSC, but they did not know that). To this day, we have renewed attacks on this problem, which will be briefly described below.

The subject of galaxy alignments was kept stirring by the efforts of a gentleman from Surrey by the name of BROWN [16,17,18]. He used Shapley's data for galaxies in the Horologium cluster, where he found a striking preference for major axes position angles (PA's) to pile up around 30°. His findings were challenged in an abstract by REAVES [19], who never published his data; it may be worthwhile to repeat this study with more modern material. Then WYATT & BROWN [20] found a mild evidence for prefered orientations around PA $\simeq$ 130° in Cetus, and a possible, very weak preference for PA $\simeq$ 30° in Pisces. Brown's efforts in this field were crowned with two major studies [17,18] of PA distributions of $\sim$ 9000 galaxies in several clusters. He did some blind tests, and claimed a measurement accuracy of $\sim$ 2°. His PA histograms occasionally show some striking deviations from the isotropy, but alas, he did something unforgivable: he smoothed his histograms in a semi-systematic way, and did not publish the original measurements! This smoothing made the apparent anisotropies look more significant than what they really were. HAWLEY & PEEBLES [21] later on took on this error, and pounced hard – perhaps too hard. Even if most of his work is now called into doubt, Brown remains as a pioneer in this field.

The pre-history of this field ends with a work of ROOD & SASTRY [22] on Abell 2199. They found that there is a very mild excess of early-type galaxies with PA $\sim$ 40°, including the central cD, with the PA = 42°, which is close to the PA of the cluster itself. The modern history of the subject begins with the work of HAWLEY & PEEBLES [21], which introduced the Fourier method of analysis of PA histograms, used by almost all subsequent investigations. That work is a very model of a carefully controlled experiment, and is a required reading for all aspiring workers in this field. Hawley & Peebles investigated several areas and clusters, and did not find anything very striking, but still concluded that their "results differ from a random isotropic distribution of orientation angles at a level close to but above the statistical uncertainty". They also found that the galaxies in Coma tend to point towards the center of the cluster, which was later also claimed by THOMPSON [23]. However, Hawley & Peebles chose in the end not to claim explicit detection of any significant alignment effects.

Then THOMPSON [23] published his study of galaxy orientations in eight Abell clusters. The only cluster in which he found a significant preference in PA distribution was Abell 2199. He repeated his measurements with an independent plate material, and found the same thing. He also found a tendency for galaxies in Coma to point towards the center of the cluster.

At that time, the pancake theory was gaining in popularity, and searches for the alignments were deliberately conceived as probes of cosmogonical genetics. In this picture, it is reasonable to expect the strongest effects in the most anisotropic clusters. Thus, ADAMS, STROM & STROM [24] completed a study of galaxy alignments in seven Rood-Sastry L-type Abell clusters. In all clusters which they investigated, they found weak tendencies for galaxy PA's to concentrate along the cluster PA, and *perpendicular* to it. The only cluster in which the alignments were very prominent was again Abell 2197, in full agreement with the Thompson's result. However, when the data for all clusters are combined (Fig. 1), the effects become clear and strong: the ellipticals tend to align with the clusters' major axes, and the disks tend to align or be perpendicular. The effects were stronger for the brighter galaxies, where the background contamination is less important, which further supports the reality of the effects. It should be emphasized that theirs were machine measurements of well resolved galaxies, in several clusters with different PA on the sky, using a well-working scanner on a good plate material, and thoroughly debugged software. Their study remains to this day some of the best evidence for alignments of galaxies and clusters.

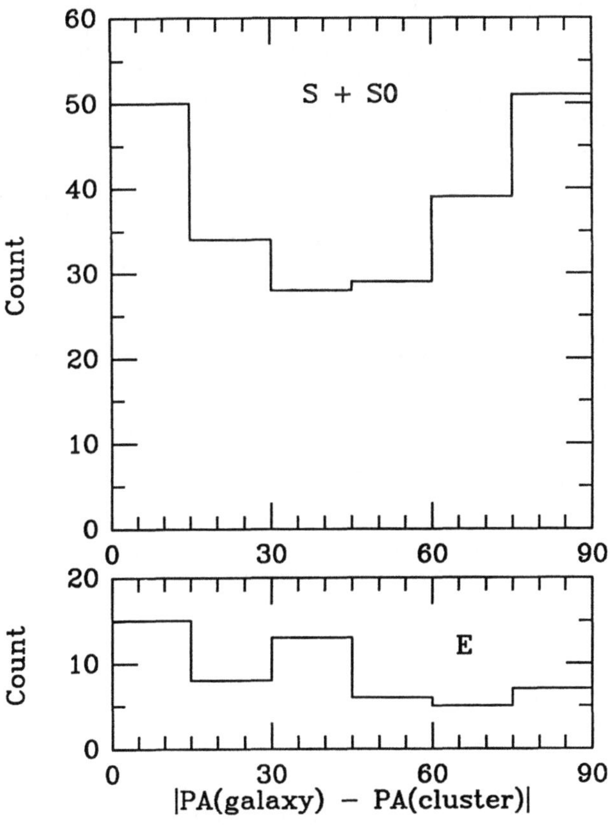

Figure 1. PA difference histograms for galaxies in all seven clusters studied by Adams *et al.* The ellipticals show a weak tendency to align (be parallel) with their clusters, and the disks tend to be either parallel or perpendicular, but avoid the intermediate angles. This may be suggesting that there are two different mechanisms by which disk galaxies acquire their angular momenta.

MACGILLIVRAY & DODD [25,26,27,28] investigated several groups and clusters, and found weak tendencies for galaxies to be aligned with, or perpendicular to, their radius vectors to the cluster or group center; this is reminiscent of the Hawley & Peebles's and Thompson's results for the Coma cluster.

Coma deserves special attention. ROOD & BAUM [29] measured by eye PA's for 135 E, S0 and SB0 galaxies; there was a very weak preference for PA's of ellipticals to pile up near 90°, but nothing really significant. GODWIN, METCALFE & PEACH [30] published machine measurements of 6724 galaxies in the Coma region, 4217 of which had $\epsilon > 0$, and thus measurable PA's. I have analysed their data [31], and found very prominent alignment effects. Galaxies tended to align in the E–W direction, with a high degree of significance. Moreover, the size of the sample allowed for investigations of secondary effects: brighter galaxies tended to align better than the fainter ones, those with the larger distances from the cluster center more than those in the cluster core, redder galaxies more than the bluer ones, and galaxies in the N and S quadrants (above and below the cluster plane) more than those in the E and W quadrants (in the cluster plane). Finally, there was a marked tendency for galaxies to point towards their nearest neighbor (in projection).

Unfortunately, there was an unreported defect in the data: the PDS scanner used by Godwin, Metcalfe & Peach suffered from asynchronicity in the raster scan: the left and the right scan rows were slightly offset from each other, which introduced a slight stretching in the E–W direction. This would introduce the PA modulation, as observed, and would affect the fainter galaxies more than the bright ones. With some contriving, it may even account for the positional dependence within the cluster. But, there is no easy way to understand the color dependence, or the neighbor-pointing effect. The ellipticity histogram, and the ellipticity – magnitude scatter plots suggest that the stretching could not have been very strong. Thus, it may be reasonable to say that even though all of my results are suspect, and some are partly or largely wrong, there are residual effects which cannot be explained away as being due to the instrumental artifacts. There still may be some alignment in Coma, and that needs to be checked again with some better data.

Perseus-Pisces Supercluster, and the Perseus cluster proper should be even more promissing grounds for the search for large-scale coherence. STROM & STROM [32] found a distinct preference for the Perseus cluster ellipticals to align with the main cluster chain, at PA $\sim 70° - 80°$. This was reconfirmed by DJORGOVSKI & VALDES [33], who analysed the Strom & Strom plate scans in an independent way. GREGORY, THOMPSON & TIFFT [34] found alignment and perpendicularity at the larger scales in the Perseus-Pisces Supercluster; their work is now being checked in a CCD survey by LAUBSCHER [35].

Orientation of galaxies in the LSC has been a subject of many modern investigations. REINHARDT & ROBERTS [36] detected positive effects in Virgo and Ursa Major. JAANISTE & SAAR [37] have found out that the spin axes of LSC spirals tend to lie in the LSC plane, but MACGILLIVRAY ET AL. [38] found a marginally significant tendency for the galaxies to be parallel to the LSC plane, with some secondary dependence on the supergalactic latitude. KAPRANDIS & SULLIVAN [39] found no significant effects. Then FLIN & GODLOWSKI [40] again found that the spin axes tend to lie in the LSC plane, and MACGILLIVRAY & DODD [27] found that there are systematic effects both in alignment and winding direction of LSC spirals, which are strongest for the intermediate types.

An interesting study of mutual orientations of spin vectors in spiral pairs was done by HELOU [41]. He found that the spin vectors tend to be *anti*parallel, and the effect was stronger for the pairs with the lower M/L. This would support the primordial torquing picture, and would argue against a primordial turbulence as a source of angular momentum in spirals.

Probably the largest data sets on galaxy PA's are the UGC and EU catalogs [42,43]. These are eye measurements, and thus prone to systematics (this is particularly apparent in the UGC).DEKEL [44] did a very thorough study of possible alignment effects in these catalogs, which he also subdivided in many subsamples (see also the paper by MANDZHOS ET AL. [45]). Dekel computed angle differences betwen the galaxy PA's and PA's of eigenvectors of galaxy density in concentric shells around them. In some subsamples (ellipticals) there were detected alignment effects, but mostly the results were negative. Dekel did a very thorough job, but I must wonder whether the design of his experiment was an optimal one: moments of inertia of *shells* must be too noisy, and perhaps the data were used too indiscriminately. It may be worth redoing this study by first preselecting only those galaxies whose PA measurements are deemed reliable, and then comparing the galaxy PA's with the PA's of density distribution in solid ellipsoids or spheres around them, or with the moments of inertia of a fixed number of their nearest neighbors; furthermore, only the galaxies in environments with well-determined and stable PA's should be considered. Failing to prune the samples in this way may swamp any possible alignment effects with too much unnecessary noise.

Finally, I should address the so-called Binggeli effects. The first Binggeli effect, that cD galaxies tend to align with their parent clusters, was first investigated by CARTER & METCALFE [46], and then by BINGGELI [47] (Fig. 2). It is not clear how much of this effect is primordial, and how much of it is produced by the dynamical evolution – the answer depends on whether cD galaxies were created with their clusters, or made later on by the cannibalism. The second Binggeli effect, that the neighboring clusters tend to point toward each other is still not too well documented, and may be a subject of an interpretational controversy. The works by STRUBLE & PEEBLES [48] and ARGYRES ET AL. [49] provide some new data and analysis.

To conclude, the evidence for alignments is neither definitive nor very dramatic, but there are far too many positive indications for the subject to be dismissed easily. Moreover, some systematics are begining to emerge: the most persuasive evidence

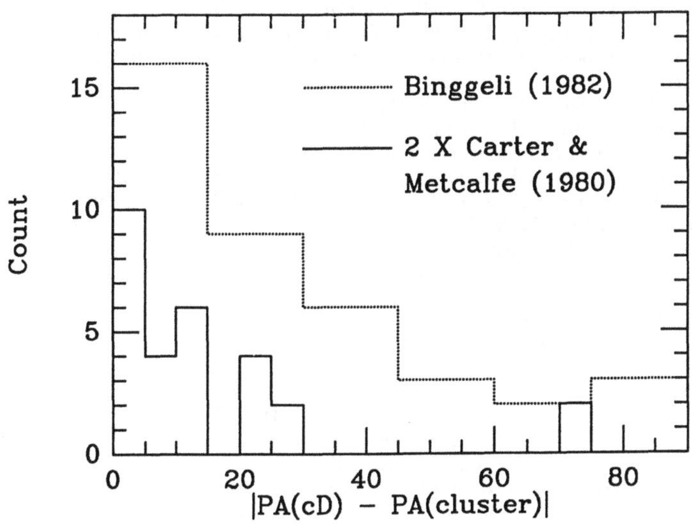

Figure 2. PA difference histograms for cD galaxies and their clusters show a strong evidence for alignments, *viz.* the Carter–Metcalfe–Binggeli effect. The Carter & Metcalfe counts were multiplied by two, for presentation clarity.

for alignments is found in highly flattened large-scale structures (Perseus, L-type Abell clusters), which is exactly what a physical intuition would suggest, if these effects indeed probe the genetics of galaxy and clusters formation. In more spherical structures, the prefered orientation of galaxian axes seem to be related to their radius vectors toward the cluster center, but the overall evidence for such effects is weaker. Second, there is a clear dependence upon the morphological type: ellipticals always tend to align with the principal axis of the cluster (or the radius vector towards the cluster center), and the disks tend to be either parallel or perpendicular. The Carter–Metcalfe–Binggeli effect for cD's may be a special case of this general trend, but may also reflect other phenomena, e.g., anisotropy induced by the post-cluster-formation merging. All this is not easily explainable by selection effects or instrumental artifacts.

What next? We need more of good quality, carefully controlled, *machine* measurements of hundreds or thousands of galaxies in different large-scale structures. Large numbers of galaxies are needed in order to populate the histogram bins adequately. The alignment effects are both easy to miss (due to a poor experimental design or sample selection), and easy to hide (e.g., by using too narrow PA histogram bins). The most promising places to look should be the highly anisotropic clusters, where the evidence is best so far: Perseus, A2196, other L-type clusters. The studies should also be done in conjuction with the redshift surveys, and the structures discovered there (e.g., shells and bubbles of DE LAPPARENT, GELLER & HUCHRA [2]). Some definite theoretical predictions and commitments would be most welcome, too. We have a potentially powerful cosmogonical test on our hands, and we should try to use it well.

References:

1. Giovanelli, R., Haynes, M., Myers, S., and Roth, J. 1986, *Astron. J.* **91**, 250.
2. de Lapparent, V., Geller, M., and Huchra, J. 1986, *Astrophys. J. Lett.* **302**, L1.
3. Dressler, A., Faber, S., Burstein, D., Davies, R., Lynden-Bell, D., Terlevich, R., and Wagner, M. 1987, *Astrophys. J. Lett.* (submitted).
4. Dressler, A. 1980, *Astrophys. J.* **236**, 351.
5. Davis, M., and Geller, M. 1976, *Astrophys. J.* **208**, 13.
6. Davis, M., and Djorgovski, S. 1985, *Astrophys. J.* **299**, 15.
7. Djorgovski, S., and Davis, M. 1986, in *Distances to Galaxies and Deviations from the Universal Expansion*, B. Madore and B. Tully (eds.), p. 135. Dordrecht: D. Reidel.
8. Ozernoi, L. 1972, *Sov. Astron.* **15**, 923.
9. Doroshkevich, A. 1973, *Astrophys. Lett.* **14**, 11.
10. Shandarin, S. 1974, *Sov. Astron.* **18**, 392.
11. Doroshkevich, A., and Shandarin, S. 1974, *Sov. Astron.* **18**, 24.
12. Abbe, C. 1875, *American J. of Science and Arts*, **9**(109), 3rd ser., p. 42.
13. Knox Shaw, H. 1908, *M.N.R.A.S.* **69**, 72.
14. Meyermann, B. 1923, *Astron. Nachr.* **219**, 133.
15. Gregory, C. 1924, *M.N.R.A.S.* **84**, 456.
16. Brown, F. G. 1939, *M.N.R.A.S.* **99**, 534.
17. Brown, F. G. 1964, *M.N.R.A.S.* **127**, 517.
18. Brown, F. G. 1968, *M.N.R.A.S.* **138**, 527.
19. Reaves, G. 1958, *Publ. Astron. Soc. Pacific* **70**, 461.
20. Wyatt, S., and Brown, F. 1955, *Astron. J.* **60**, 415.
21. Hawley, D., and Peebles, J. 1975, *Astron. J.* **80**, 477.
22. Rood, H., and Sastry, G. 1972, *Astron. J.* **77**, 451.
23. Thompson, L. 1976, *Astrophys. J.* **209**, 22.
24. Adams, M., Strom, K., and Strom, S. 1980, *Astrophys. J.* **238**, 445.
25. MacGillivray, H., and Dodd, R. 1979a, *M.N.R.A.S.* **186**, 69.
26. MacGillivray, H., and Dodd, R. 1979b, *M.N.R.A.S.* **186**, 743.
27. MacGillivray, H., and Dodd, R. 1985a, *Astron. Astrophys.* **145**, 269.

28. MacGillivray, H., and Dodd, R. 1985b, in the proceedings of the ESO workshop *The Virgo Cluster of Galaxies*, O.-G. Richter and B. Binggeli (eds.), p. 217. München: European Southern Observatory.
29. Rood, H., and Baum, W. 1967, *Astron. J.* **72**, 398.
30. Godwin, J., Metcalfe, N., and Peach, J. 1983, *M.N.R.A.S.* **202**, 113.
31. Djorgovski, S. 1983, *Astrophys. J. Lett.* **274**, L7.
32. Strom, S., and Strom, K. 1978, *Astron. J.* **83**, 732.
33. Djorgovski, S., and Valdes, F. 1987, in preparation.
34. Gregory, S., Thompson, L., and Tifft, W. 1981, *Astrophys. J.* **243**, 411.
35. Laubscher, B., 1987(?), Ph. D. Thesis, University of New Mexico, Albuquerque.
36. Reinhardt, M., and Roberts, M. 1972, *Astrophys. Lett.* **12**, 201.
37. Jaaniste, J., and Saar, E. 1977, Tartu Observatory preprint A-2.
38. MacGillivray, H., Dodd, R., McNally, B., and Corwin, H., 1982, *M.N.R.A.S.* **198**, 605.
39. Kaprandis, S., and Sullivan, W. 1983, *Astron. Astrophys.* **118**, 33.
40. Flin, P., and Godlowski, W. 1984, preprint (publ. in *Acta Cosmologica*?).
41. Helou, G. 1984, *Astrophys. J.* **284**, 471.
42. Nilson, P. 1973, *Uppsala General Catalogue of Galaxies*, Uppsala: University of Uppsala.
43. Lauberts, A. 1982, *The ESO/Uppsala Survey of the ESO (B) Atlas*, Münich: European Southern Observatory.
44. Dekel, A. 1985, *Astrophys. J.* **298**, 461.
45. Mandzhos, A., Tel'nyuk-Adamchuk, V., and Gregul', A. 1985, *Sov. Astron. Lett.* **11**, 206.
46. Carter, D., and Metcalfe, N. 1980, *M.N.R.A.S.* **191**, 325.
47. Binggeli, B. 1982, *Astron. Astrophys.* **107**, 338.
48. Struble, M., and Peebles, J. 1985, *Astron. J.* **90**, 582.
49. Argyres, P., Groth, E., Peebles, J., and Struble, M. 1986, *Astron. J.* **91**, 471.

Galaxy Formation and Large Scale Structure

SIMON D. M. WHITE

This article deals with galaxy formation within current models for the evolution of structure. I review these models and their relationship to models which were popular 10 years ago. I then concentrate on the cold dark matter model. I give an overview of its present status and of how and where galaxies are expected to form in it. Finally I discuss its predictions for large scale structure.

Almost all our information about structure on large scales comes from the galaxy distribution. Observations of the microwave background give us upper limits on the structure present at early epochs, X-ray data allow us to infer the mass distribution in a few dense, massive clumps, and quasar absorption line measurements give tantalizing clues to the residual neutral gas distribution at moderate redshifts. Nevertheless, our current picture is based overwhelmingly on studies of the morphology, statistics, kinematics and dynamics of the distribution of galaxies. Thus theories of cosmogony must specify how and where galaxies form, not only because of the great intrinsic interest of these processes, but also because they determine how what we see is related to the overall distribution of matter. The most fundamental properties of galaxies to be explained are their characteristic sizes, velocity dispersions and luminosities. In addition, an explanation is needed for the broad morphological division into ellipsoidal and disk systems with their differing angular momenta. An important clue to this latter question is the relation between the morphology of galaxies and the dynamical state of their environment. Most current theories of cosmogony are too incomplete to predict any of these properties reliably.

In the 1970's cosmogony was a relatively simple subject. There were two major competing theories, usually known as the adiabatic and isothermal models. Both were based on analysis of the gravitational evolution of a random phase distribution of small amplitude fluctuations. Both assumed a universe containing ordinary baryonic matter only. The adiabatic theory adopted the plausible *a priori* assumptions that only the fastest growing adiabatic fluctuation mode would emerge from the early universe, and that the amplitude of fluctuations would not diverge rapidly on either large or small scales. These assumptions lead to a model in which damping processes wipe out small scale fluctuations [1]. The first nonlinear structures to form are then shock fronts, or "pancakes", much larger than individual galaxies [2]. Galaxies are supposed to form from gas that cooled and fragmented behind the shock front; unfortunately simple estimates of the fragment masses suggest that they would be much smaller than typical galaxies [3].

The competing isothermal model supposed that some unspecified process suppressed the adiabatic mode. Fluctuations at late times are then dominated by the more slowly growing isothermal mode which is not strongly damped. As a result fluctuations on all scales survive. There was no strong motivation for any specific fluctuation spectrum so a power law with adjustable index was usually adopted. Nonlinear evolution proceeds via hierarchical clustering; clumps of matter gradually aggregate into larger and larger systems [4,5]. Within this model, galaxy masses are most naturally explained as a reflection of the cooling times of large gas clouds [6,7,8]. The standard model (hereafter the WR scheme) invokes galaxy formation by the cooling of gas within an evolving hierarchy of dark matter halos [9]. This model predicts the luminosity function of galaxies with moderate success when normalised to produce groups and clusters of galaxies like those observed [9]. It also accounts for the angular momenta of disk galaxies [10,11], and fits galaxy clustering and galaxy properties quite well for a fluctuation spectrum, $\delta_k^2 \propto k^n$, with $-2 < n < -1$ [12-15]. However, it is unclear whether galaxies will survive the merging of their halos so that galaxy clusters, rather than single merged galaxies, can be produced [9]. Some merging seems desirable to explain the origin of elliptical galaxies [16,10].

Since 1980 this subject has been greatly enriched by a number of ideas from high energy physics. The inflationary model provided a physical model for the origin of structure and gave a strong impetus to old *a priori* arguments in favor of a universe with $\Omega \simeq 1$ [17]. If these latter arguments are accepted, the standard theory of cosmological nucleosynthesis [18] can be retained only by assuming that at most 20% of the energy density of the present universe is contributed by ordinary baryonic matter. Here again particle physics comes to the rescue by providing many weakly interacting particles which might possibly contribute the other 80%. These developments led immediately to new versions of the old cosmogonies. A neutrino dominated universe is attractive because neutrinos are the only dark matter candidate known to exist, and because of a claim that the electron neutrino has a mass in the interesting range [19]. Structure in such a universe forms by the collapse and fragmentation of large pancakes [20,21]. Alternatively dark matter candidates which had lower velocities than neutrinos at early times may form structure via hierarchical clustering [22-24]. Such particles are collectively known as cold dark matter (CDM).

In addition to these extensions of earlier theories, models have been considered in which the universe is dominated by a cosmological constant [25], or by relativistic decay products of unstable particles [26,27], in which perturbations are induced by cosmic strings rather than by quantum fluctuations [28], and in which structure results from recent hydrodynamic [29,30] or radiative [31,32] processes rather than from the gravitational amplification of small fluctuations. Most of these newer theories are still in a preliminary stage of development. They are motivated partly by the desire to show that the older models do not exhaust the possibilities and partly by perceived flaws in those models. Predictions tend to be qualitative, and are good, at best, to order of magnitude. They should therefore be treated with indulgence, but also with some suspicion. The same is not true for the neutrino and CDM models. Both are fully specified if the simplest predictions of inflation and cosmological nucleosynthesis are adopted; the only undetermined parameter is the overall amplitude of fluctuations. In each the growth of structure can be followed in full generality and with fair accuracy by N-body methods. Although galaxy formation involves radiative and hydrodynamic processes which cannot be treated in detail, quantitative predictions can be made in both models, and can be compared directly with observation. The predictions of such well specified models should be brought fully into the observational plane and compared directly with real data.

The qualitative properties of the neutrino model appeared to agree with the apparent cell-like distribution of galaxies [20]. However quantitative study showed that sufficient gas to make galaxies is unlikely to cool and fragment in pancakes [33,34], and that the natural scale of structure is too large to be acceptable [35,36]. The latter difficulty is often described by saying that the neutrino model requires galaxy formation to be unreasonably recent. This statement is misleading and is based on comparing the observed galaxy distribution to the theoretical neutrino distribution. In fact the model predicts the galaxy distribution to have a very large characteristic scale no matter when galaxies form. Regions entirely devoid of galaxies and of typical diameter at least 4000 km/s are expected to fill most of space. Such structure conflicts with that seen in galaxy surveys [37].

The CDM model has proven considerably more successful. BLUMENTHAL *et al.* [38] used scaling laws to show that the WR scheme works well in this model. CDM not only provides the dark matter; it also predicts the fluctuation spectrum needed to produce structures with the characteristics of halos, groups and clusters of galaxies. Cooling arguments lead to galaxy luminosities, binding energies and angular momenta, and to scaling relations between these quantities, in good agreement with observation. A fully general treatment of structure formation in the CDM model requires numerical simulation, and Marc Davis, George Efstathiou, Carlos Frenk and I have undertaken an extensive N-body investigation over the last few years. The rest of this article, together with Frenk's contribution to this volume [39], describes the present status of this project. I begin with a general overview of the most attractive CDM model. This has $\Omega = 1$, $\Omega_{baryon} = 0.15$, $H_o = 50$ km/s/Mpc, and constant curvature, adiabatic initial fluctuations; it thus agrees with all our *a priori* theoretical prejudices. However, agreement with observation requires that galaxies should be more strongly clustered than the underlying mass. As detailed by Frenk, a bias of this kind is a direct consequence of extending the WR scheme to a CDM universe. Below I show that if a convenient analytic description of the bias is adopted, the resulting model is in good agreement with most of the observed morphological and statistical properties of the large scale galaxy distribution.

In a CDM universe, constant curvature, adiabatic initial fluctuations lead to a post-recombination power spectrum which is not a power law. Rather its slope varies from $\delta_k^2 \propto k^{-3}$ for masses smaller than those of galaxies to $\delta_k^2 \propto k$ for masses greater than those of superclusters [22,24]. Scaling arguments suggest that structure forms hierarchically in such a universe, but that early evolution (up to galactic scales) is very rapid; objects with a wide range of masses form almost simultaneously [38,40]. The validity of the simple hierarchical model is in doubt in such a situation because of "crosstalk" between different spatial frequencies. A substantial contribution to the rms overdensity in (say) 10^8 $M_\odot$ regions comes from density fluctuations associated with the 10^{11} $M_\odot$ regions in which they are embedded. As a result the properties with which small clumps form are a strong function of the large-scale environment in which they find themselves. In principle one might hope to investigate these effects numerically. However, difficulties arise because of the very wide range of scales involved. To study the formation of 10^8 $M_\odot$ clumps during the collapse of a 10^{12} $M_\odot$ object requires accurate force calculation and orbit integration for a system of at least 10^5 particles.

These problems are less severe on larger scales because of the steepening of the power spectrum. Nevertheless, significant crosstalk can still occur; thus the formation of galaxy halos is affected by their location in "protocluster" or "protovoid" regions. In the WR scheme the distribution of bright galaxies reflects the distribution of the centres of those halos which were massive and dense enough, and which survived further merging long enough, for cooling to produce a large stellar system. It is far from obvious how the statistics of such sites should be related to those of the overall mass distribution. In general some kind of bias is to be expected. In protocluster regions, dense, massive halos form more readily than in protovoids, but they are also more rapidly disrupted. In a CDM universe such effects lead to a substantial overrepresentation of bright galaxies in high density regions. The density contrast of galaxy clustering is thus enhanced relative to that of the dark matter, and the mass per galaxy in clusters is lower than that of the universe as a whole [39].

A convenient quasianalytic model for this bias is obtained by convolving the initial linear density field with a smoothing kernel corresponding to the halo mass of a bright galaxy; the sites of galaxy formation are then identified with high peaks of the resulting field. This idea stemmed from the use of a similar model to study the distribution of Abell clusters [41]. Its purely analytic properties have been considered at length by BARDEEN *et al.* [42;BBKS] and it has become the standard model for "biased galaxy formation". Its purpose is to provide a simple and plausible means for locating galaxies within the overall dark matter distribution. It can be fully justified only by comparison with a more detailed treatment of galaxy formation. In reality galaxy formation is unlikely to depend only on local properties of the linear density field, a large galaxy probably forms from a high fluctuation over a larger region of space than a small galaxy, and a sharp threshold is implausible. Nevertheless, numerical models show the location of high peaks to correspond surprisingly well to the centres of massive halos [39]. Thus the high peak model does seem to represent the result of applying the WR scheme to a CDM universe. Notice that such biasing does *not* require failed galaxies or inhibited galaxy formation.

The high peak model for galaxy formation was applied by DAVIS *et al.* [43; DEFW] to our simulations of a flat CDM universe. We found that plausible values for its two parameters lead to a galaxy distribution with two and three point correlations and pairwise relative velocities in agreement with observation. Relative velocities of close pairs turn out to be somewhat too large, a discrepancy which also shows up in a comparison of groups in such models with the CfA survey [44]. However, this problem does not appear serious and may reflect any of a number of simplifications inherent in the modeling. Studies of the small scale properties of this same model showed it to produce dark matter halos with the abundance, structure and binding energy inferred for the halos of bright galaxies; in addition they suggested the possible physical origin for the bias discussed above [39,45]. DEFW were unable to study large scale structure in this model because our simulations were too small, and because our technique for specifying the galaxy distribution required individual galaxy halos to be resolved. We have recently used the analytic machinery of BBKS to circumvent the latter requirement, and as a result we have been able to model much larger regions of space [46;WFDE]. As I now discuss, the flat CDM model comes remarkably close to fitting all available data on the large scale galaxy distribution.

It has become popular to use large scale structure to discriminate between theories of galaxy formation. Apart from the obvious temptation to focus on the biggest objects in the universe, this is motivated by the idea that nonlinear effects may be least important on such scales so that the initial conditions may be most apparent there. In addition, the pancake and hierarchical clustering models suggest qualitatively different morphologies which are evident in any comparison of simulations of the two theories. Unfortunately the available data sets are far from ideal for studying such questions. Most surveys appear to contain structure comparable to the size of the volume studied, but rather poorly sampled by the galaxies or galaxy clusters which delineate it. Since the form of this structure is irregular, obtaining a realistic estimate of its significance can be difficult. Theoretical models are best compared with such surveys by generating artificial data with similar selection criteria.

The local number density of peaks of a smoothed Gaussian random field can be related to the linear overdensity of the surrounding region [42]. Within the high peak model the relative number of galaxies associated with each simulation particle can thus be derived from the linear initial conditions. WFDE take the galaxy luminosity function to have the standard Schechter form independent of position, and adjust its overall normalisation to give the correct luminosity density for the universe [47]. The strength of the bias is determined by the threshold of the high peak model which, following DEFW, we took as 2.5 σ. Magnitude or volume limited "galaxy" catalogs can then be constructed for any simulation and compared directly with the corresponding real data sets. WFDE applied these techniques to an ensemble of 32768-particle models of 280 Mpc cubic regions, and to one 216000-particle model of a 360 Mpc cube (all scales quoted assume H_o = 50 km/s/Mpc). We made artificial "Abell cluster" catalogs of rich clusters. We made "Zwicky catalogs", "de Lapperent slices" and "deep pencil beam surveys" by constructing magnitude-limited catalogs over appropriate "sky" areas. Finally we made "KOSS void surveys" by sampling in the pattern adopted by KIRSHNER *et al.* [48,49;KOSS].

The Abell catalog provides some of the most reliable data on large scale structure [50]. The abundance, luminosity and velocity dispersion distribution of rich clusters is relatively well determined, and the catalog has been the prime data base for statistical studies of superclusters. For a standard luminosity function, Abell's definitions of cluster richness correspond to a luminosity within 3 Mpc exceeding 42 L_* for $R \geq 1$ and exceeding 86 L_* for $R \geq 2$. For comparison the Coma cluster has about 120 L_* within this radius. Figure 1 plots the fraction of rich clusters in our flat CDM

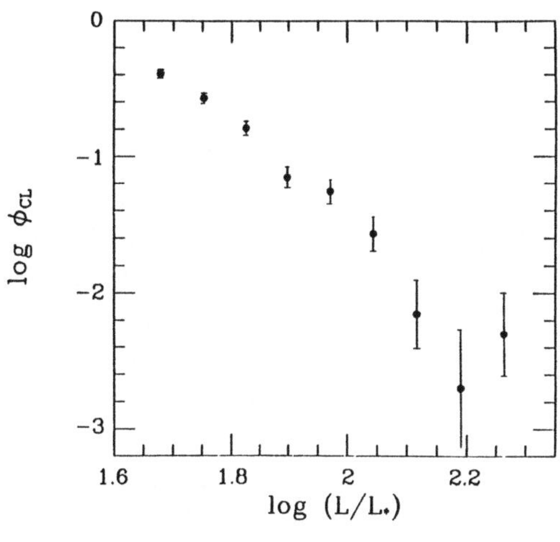

Figure 1. Luminosity function of rich clusters in a flat CDM universe.

models in a series of logarithmic luminosity intervals. The total number density of clusters brighter than 42 L_* is 8.0 $\times 10^{-7}$ Mpc^{-3}, which should be compared with

7.5×10^{-7} Mpc^{-3} for Abell clusters with $R \geq 1$ [51]. As a result of the bias, the mean M/L of the clusters is 20% of that of the CDM models as a whole, in good agreement with the fact that the M/L of observed clusters is about 20% of that required to close the universe. Within 27 Mpc of a cluster centre the mean M/L is 65% of the global value; applying this factor to obtain the mass overdensity of the Local Supercluster from its observed luminosity overdensity results in a value $\Omega \approx 1$ being inferred for an infall velocity of about 300 km/s. As shown in Fig. 2 the rich clusters in our models are more clustered than the galaxies but are less clustered than the available sample of 100 Abell clusters [51,52]. Monte Carlo studies suggest that this discrepancy is significant at about the 90% level and thus is likely to be real.

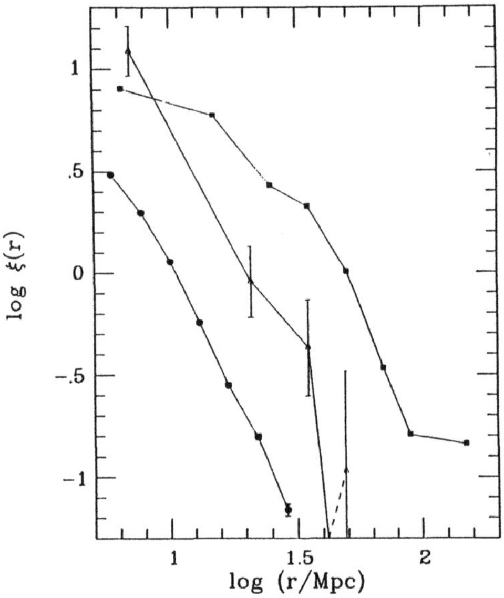

Figure 2. Correlation functions for galaxies and rich clusters. The filled circles give the galaxy correlation function in a flat CDM universe and are consistent with the observed function. The triangles with (Poisson) error bars are for clusters with $R \geq 1$ in the same model. The filled squares give the results of Bahcall and Soneira [51] for a sample of 100 Abell clusters.

For a standard luminosity function and mean luminosity density [47] the galaxy density at the median distance in a complete magnitude limited survey is 9×10^{-4} Mpc^{-3}. In volume limited catalogs made at this galaxy density, WFDE find the largest empty sphere in a 280 Mpc cubic region to have a diameter between 56 and 61 Mpc. However, in their survey of the Bootes void KOSS only sampled galaxies along a series of parallel "skewers". The largest empty sphere they were able to place within their galaxy sample had a diameter of 124 Mpc and was expected to contain 31 galaxies. When our volume limited catalogs are sampled in the same way and at the same density, all 280 Mpc cubes contain an apparent "void" exceeding 100 Mpc in diameter, and three of them contain "voids" larger than 112 Mpc. This shows not only that the CDM model comes close to reproducing the observations, but also that the size of empty regions in a galaxy catalog depends critically on the sampling strategy. A further example of this is given by simulated catalogs of small areas of sky to 21 apparent magnitude. The number of galaxies in such deep pencil beam surveys fluctuates by more than a factor of two, and the morphology of their redshift distribution is also highly variable. Some fields show a relatively smooth distribution while others contain large clumps separated by empty intervals up to 20,000 km/s across. This behavior is very similar to that in as yet unpublished deep redshift surveys (Koo, Kron and Szalay; Ellis *et al.* ; private communications).

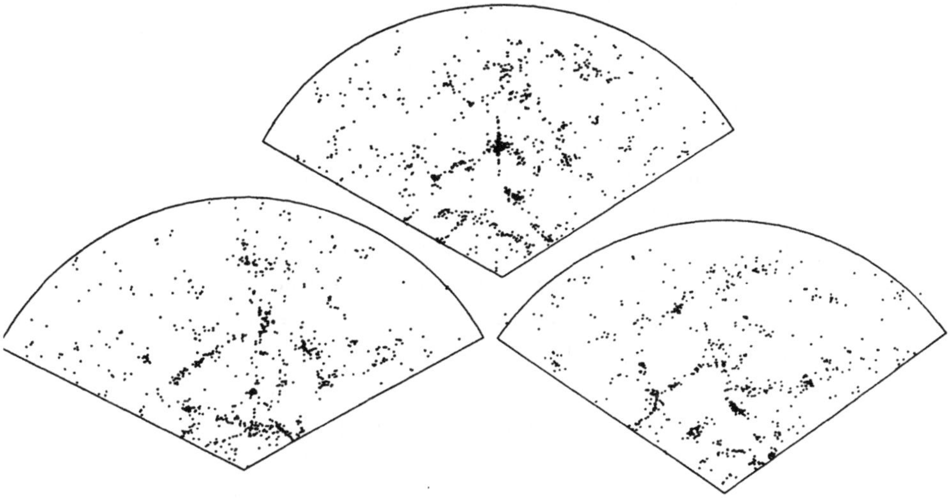

Figure 3. Wedge surveys of a flat CDM universe with the geometry and selection criteria of de Lapparent *et al.* [53]. These are equal area plots bounded at 15000 km/s.

Considerable interest has been generated recently by the morphology of the galaxy distribution in surveys by DE LAPPARENT *et al.* [53] and by HAYNES and GIOVANELLI [54]. De Lapparent *et al.* studied the redshift distribution of galaxies in a strip of sky passing through the Coma cluster and commented on the bubble-like distribution of galaxies in this particular slice of the universe; many objects appear to lie on the boundaries of a few large, near circular, low density regions. The Perseus-Pisces sample of Haynes and Giovanelli contains more than twice as many galaxies. As presented in their paper it does not appear to have as much large-scale coherence as the Coma strip, except that the Perseus supercluster is extremely prominent. Figure 3 shows three artificial redshift surveys of a flat CDM universe. The central one has exactly the geometry of the survey of de Lapparent *et al.* ; its orientation was chosen so that a rich cluster (unfortunately much less rich than Coma) is at the centre at 7000 km/s. The two other surveys show the "galaxy" distribution in 6° wide strips immediately above and below that containing the cluster. Filamentary structures can be seen in each of these surveys. The most prominent can be traced from one survey to the next, showing that they are sections through sheets. The large filament in the foreground is more than 6000 km/s long and so is comparable to the observed Perseus-Pisces feature; the largest low density regions are about 5000 km/s across, are quite sharply defined, and do not differ greatly from those in the de Lapparent *et al.* survey.

The extent to which pictures like Figure 3 agree with the real data appears to be a matter of personal judgement, and to depend sensitively both on how the data are displayed and on individual prejudice. However, the fact that differences are not large shows that it is not yet time to conclude that the observations require nonrandom phase initial conditions or explosive evolution. Similarly the fact that the flat CDM model is quite a good fit to the dynamics of small groups, produces the observed M/L of rich clusters, and is consistent with the observed infall to Virgo, shows that present data do not require the universe to be open. It is remarkable how well the model agrees with observation considering that its motivation comes from *a priori* theoretical considerations. If our physical explanation for the bias is fully substantiated, only one adjustable parameter will remain in the model—the initial fluctuation amplitude. At present its most serious disagreement with observation may be that it predicts weaker superclustering and smaller large scale streaming motions than are inferred from current data sets. As Dick Bond discusses in this volume, these discrepancies can be rectified by a minor phenomenological modification of the initial conditions, but there is no obvious physical justification for any such modification. At present the discrepancies are not well enough established to be fatal to the model, and are probably outweighed by its successes. For no other model is quantitative testing currently possible at a comparable level.

This research was supported in part by NSF Presidential Young Investigator award #AST-8352062.

References:

1. J.I. Silk: Astrophys.J. 151, 459 (1968)
2. Ya.B. Zel'dovich: Astron.Astrophys. 5, 84 (1970)
3. R.A. Sunyaev, Ya.B. Zel'dovich: Astron.Astrophys. 20, 189 (1972)
4. P.J.E. Peebles, R.H. Dicke: Astrophys.J. 154, 891 (1968)
5. P.J.E. Peebles: Astrophys.J.Lett. 189, L51 (1974)
6. J. Binney: Astrophys.J. 215, 483 (1977)
7. M.J. Rees, J.P. Ostriker: Mon.Not.R.astr.Soc. 179, 541 (1977)
8. J.I. Silk: Astrophys.J. 211, 638 (1977)
9. S.D.M. White, M.J. Rees: Mon.Not.R.astr.Soc. 183, 341 (1978)
10. S.M. Fall: Nature 281, 200 (1979)
11. S.M. Fall, G. Efstathiou: Mon.Not.R.astr.Soc. 193, 189 (1980)
12. J.R. Gott, E.L. Turner: Astrophys.J. 216, 357 (1977)
13. S.M. Fall: Rev.Mod.Phys. 51, 21 (1979)
14. S.M. Faber: in *Astrophysical Cosmology*, H.A. Brück, G.V. Coyne, and M.S. Longair, eds., p. 219, Pont.Acad.Sci. 1982
15. J.E. Gunn: in *Astrophysical Cosmology*, H.A. Brück, H.A., G.V. Coyne, and M.S. Longair, eds., p.219, Pont.Acad.Sci. 1982
16. A. Toomre: in *Evolution of Galaxies and Stellar Populations*, B.M. Tinsley, and R.B. Larson, eds., p.410, Yale Univ. Obs. 1977

17. A. Guth: in *The Very Early Universe* G.W. Gibbons, S. Hawking, S.T.C. Siklos, eds., p.171, Cambridge Univ. Press 1983
18. J. Yang, M.S. Turner, G. Steigman, D.N. Schramm, K.A. Olive: Astrophys.J. 281, 493 (1984)
19. V.A. Lyubimov, E.G. Novikov, V.Z. Nozik, E.F. Tretyakov, V.S. Kozik: Phys. Lett. *B94*, 266 (1980)
20. A.G. Doroshkevich, M.Yu. Khlopov, R.A. Sunyaev, A.S. Szalay, Ya.B. Zel'dovich: Ann.N.Y.Acad.Sci. 375, 32 (1980)
21. J.R. Bond, G. Efstathiou, J.I. Silk: Phys.Rev.Lett. 45, 1980 (1980)
22. P.J.E. Peebles: Astrophys.J. 258, 415 (1982)
23. G.R. Blumenthal, J.R. Primack: in *Fourth Workshop on Grand Unification*, H.A. Weldon, P. Langacker, P.J. Steinhardt, eds., p.256, Birkhauser 1983
24. J.R. Bond, G. Efstathiou: Astrophys.J.Lett. 285, L45 (1984)
25. P.J.E. Peebles: Astrophys.J. 284, 439 (1984)
26. A.G. Doroshkevich, M.Yu. Khlopov: Soviet Astr.Lett. 9, 171 (1983)
27. M.S. Turner, G. Steigman, L.M. Kraus: Phys.Rev.Lett. 52, 2090 (1984)
28. A. Vilenkin: in *Inner Space/Outer Space* E.W. Kolb, M.S. Turner, D. Lindley, K.A. Olive, D. Seckel, eds., p.269, U. Chic. Press 1986
29. J.P. Ostriker, L.L. Cowie: Astrophys.J.Lett. 243, L127 (1981)
30. S. Ikeuchi: Pub.astr.Soc.Japan 33, 211 (1981)
31. C.J. Hogan, N. Kaiser: Astrophys.J. 274, 7 (1983)
32. C.J. Hogan, S.D.M. White: Nature 321, 575 (1986)
33. P.R. Shapiro, C. Struck-Marcell, A.L. Melott: Astrophys.J. 275, 413 (1983)
34. J.R. Bond, J. Centrella, A.S. Szalay, J.R. Wilson: Mon.Not.R.astr.Soc. 210, 515 (1984)
35. S.D.M. White, C.S. Frenk, M. Davis: Astrophys.J.Lett. 271, L1 (1983)
36. S.D.M. White, M. Davis, C.S. Frenk: Mon.Not.R.astr.Soc. 209, 27P (1984)
37. S.D.M. White: in *Inner Space/Outer Space*, E.W. Kolb, M.S. Turner, D. Lindley, K.A. Olive, D. Seckel, eds., p.228, U. Chic. Press 1986
38. G.R. Blumenthal, S.M. Faber, J.R. Primack, M.J. Rees: Nature 311, 517 (1984)
39. C.S. Frenk: this volume.
40. P.J.E. Peebles: Astrophys.J. 277, 470 (1984)
41. N. Kaiser: Astrophys.J.Lett. 284, L9.
42. J.M. Bardeen, J.R. Bond, N. Kaiser, A.S. Szalay: Astrophys.J. 304, 15 (1986)
43. M. Davis, G. Efstathiou, C.S. Frenk, S.D.M. White: Astrophys.J. 292, 371 (1985)
44. R. Nolthenius, S.D.M. White: Mon.Not.R.astr.Soc., in press (1987)
45. C.S. Frenk, S.D.M. White, G. Efstathiou, M. Davis: Nature 317, 595 (1985)
46. S.D.M. White, C.S. Frenk, M. Davis, G. Efstathiou: Astrophys.J., in press (1987)
47. J.E. Felten: Comm.Astrophys.Sp.Sci. 11, 53 (1985)
48. R.P. Kirshner, A. Oemler, P.L. Schechter, S.A. Shectman: Astrophys.J.Lett. 248, L57 (1981)
49. R.P. Kirshner, A. Oemler, P.L. Schechter, S.A. Shectman in *Early Evolution of the Universe and its Present Structure*, G.O. Abell, G. Chincarini, eds., p.197 Reidel (1983)
50. G.O. Abell: Astrophys.J.Supp. 3, 211 (1958)
51. N. Bahcall, R. Soneira: Astrophys.J. 270, 20 (1983)
52. A.A. Klypin, A.A. Kopylov: Soviet Astr.Lett. 9, 41 (1983)
53. V. De Lapparent, M.J. Geller, J. Huchra: Astrophys.J.Lett. 302, L1 (1986)
54. M.P. Haynes, R. Giovanelli: Astrophys.J.Lett. 306, L55 (1986)

Scenarios of Biased Galaxy Formation

AVISHAI DEKEL

1. Why 'Biased Galaxy Formation'?

The naive assumption that the galaxies trace the underlying mass distribution makes life easy: it enables us to estimate the total mass density in the universe and to interpret the large-scale structure based on the observed distribution of galaxies. But when assuming that the mass lies where we see light, aren't we playing the role of the drunk who is searching for his lost keys under the street light because it is easier to look there? I would argue that based on the observed correlation of the type of a galaxy and its environment [1], it would be astonishing if galaxy formation itself was not affected by environmental effects segregating the galaxies from the underlying mass. I will first summarize the motivation behind this suspicion and then, together with Martin Rees in the following talk, discuss bias mechanisms in the various cosmogonies, arguing that a segregation of one sort or another is a natural outcome of almost every cosmogony (although the bias is not always of the desired sort). I will conclude by trying to point at key observational tests.

On the observational side, the dynamically inferred mass in galactic halos and in clusters of galaxies indicates $\Omega \sim 0.2$ (if M/L is similar everywhere), which, although corresponding to dark matter (DM) which overwhelms the luminous mass by a factor of ~ 20, is still short of 'closing' the universe by a factor ~ 5. If the matter were all baryonic, $\Omega \simeq 0.1$ were compatible with the constraints imposed by standard big bang nucleosynthesis [2] if $h \leq 0.65$ (but note that these constraints could be challenged by non-conventional nucleosynthesis scenarios; see Hogan, this volume).

Motivated by particle theories, however, we are willing to consider non-baryonic forms of DM, which can amount to $\Omega = 1$ and thus solve the old 'flatness' problem. With the Friedmann equation for an open universe and a zero cosmological constant,

$$1 - \Omega = (RH)^{-2}, \qquad (1)$$

it would be a remarkable coincidence if the ratio between the curvature radius R and the horizon radius H^{-1}, which decreases in time approaching unity asymptotically, is just now $\sim 2-3$; this would correspond to a very special time where Ω is no longer very close to unity, but is not yet much smaller. The Inflation scenario, being so appealing in resolving certain other cosmological problems, such as the causality problem and the smoothness problem, gives a natural explanation for the present flatness of the universe: the exponential growth has stretched any small part of an initial chaotic hypersurface so that it became essentially flat over our present horizon scale ($RH \gg 1$), yielding $\Omega = 1$ to a very high precision. For inflation to yield $\Omega \sim 0.2$, the inflation factor had to be 'just' 10^{25}, and the pre-inflation radius of curvature 'just' comparable to the Hubble radius then, making the present radius of curvature 'just' of order the Hubble radius. This demands a *fine-tuning* which one would prefer to avoid.

Can the universe be flat and just misleading us to believe that $\Omega \sim 0.2$? If $\Omega = 1$, the rising trend of M/L observed in galaxies and clusters must continue out to scales of $\sim 10^{-1}$ Mpc. The mass estimates on such scales are based either on modeling the infall of the Local Group into the Local Supercluster (LSC) [3,4], or on applying a cosmic virial theorem (CVT) to pairs of galaxies or to cluster-galaxy pairs [5,6]. In a linear, spherical model for the LSC

$$\Omega \simeq \delta^{-1.7}(3v/Hr)^{1.7}, \tag{2}$$

where δ is the density enhancement within the Local Group virgocentric radius r, and v is the infall velocity at r. The CVT provides a dependence

$$\Omega \propto \xi(r)^{-1}(v/r)^{2}, \tag{3}$$

where $\xi(r)$ is the two-point correlation function and v is the mean pair-velocity. The results apparently suggest again $\Omega = 0.1 - 0.3$. But, the crucial point is that they are obtained using the quantities corresponding to *galaxies*: their number overdensity δ_g or the galaxy-galaxy correlation function $\xi_g(r)$. If, indeed, the galaxies cluster more than the underlying matter, such that $\delta_g = f\delta$ and accordingly $\xi_g(r) = f^2\xi(r)$, the real value of Ω, as obtained by (2) or (3), is larger by a factor $f^{1.7} - f^2$. The data would be compatible with $\Omega = 1$ if the degree of bias corresponds to $f \simeq 2 - 3$.

The emerging picture based on recent data (see Oemler, this volume) is that *'voids'* $\sim 50\,\mathrm{h}^{-1}$ Mpc in diameter are actually quite common in the galaxy distribution. The typical number density of (bright) galaxies in these 'voids' seems to be less than 10% of the mean (e.g. one expects 32 galaxies in the Bootes 'void' and finds none). Based on spherical 'void' models [7], such an underdensity in the mass distribution would correspond at recombination to $|\delta| \geq 10^{-2}$ if $\Omega \simeq 1$, and $\geq 5 \times 10^{-2}$ if $\Omega \simeq 0.1$. This would be incompatible with the isotropy of the microwave background on angles $10' - 1^{\circ}$ in any of the cosmogonic scenarios, unless post-recombination reionization has washed out the fluctuations. The large-scale N-body simulations demonstrate the difficulty: even in 'pancake' simulations [8-10] such large regions are never found with an underdensity less than 25% of the mean; they cannot be substantially evacuated dynamically by the present epoch, which is defined by matching the correlation functions of the simulated mass and the observed galaxies. The situation is worse if $\Omega < 1$, where the 'voids' are evacuated even less efficiently.

What is the real mass density in the 'voids'? Consider a 'toy' universe which consists of 'top-hat' superclusters ('sc') and uniformly underdense 'voids'. The equation

$$(\delta_g/\delta)_{void} = (\delta_g/\delta)_{sc} = f \tag{4}$$

results automatically from the definition of the δ's. Assuming that the LSC and the Bootes 'void' are typical, we can adopt the corresponding observed values for the *galaxies*, $\delta_g \simeq 2.5$ and -0.9 respectively. If $\Omega = 1$, the real mass overdensity in the LSC must be $\delta \simeq 0.85$ ($f \sim 3$), so using (4) we get for the *mass* in the 'voids' $\delta \simeq -0.32$. (The fractional volume in the 'voids' is then $\sim 73\%$ and the corresponding mass is $\sim 50\%$.) The obtained mass densities in superclusters and in 'voids' are both compatible with $|\delta| \simeq 9 \times 10^{-4}$ at recombination ($z \simeq 10^3$). If most of the mass is non-baryonic, this corresponds to $\delta T/T \simeq 3.5 \times 10^{-5}(\Omega h^2)^{-1}$, which is compatible with the observed isotropy constraints if Ωh^2 is not much smaller than unity. An open universe with $\Omega \simeq 0.2$ would be in trouble here: it would require no bias, $f \simeq 1$, and therefore, by (4), a real deep mass underdensity of $\sim 10\%$ in the 'voids'. The corresponding $\delta T/T \sim 5 \times 10^{-3}$ would be hard to reconcile with observations without substantial reionization.

From the view point of galaxy formation, its relative 'efficiency' in an environment 'x' can be quantified by $\epsilon = (n_x/\rho_x)/(n/\rho)$, where n corresponds to galaxy number density and ρ to matter density. In 'voids', using the values obtained above, the efficiency is only $\epsilon \simeq 0.15$ while in superclusters $\epsilon \simeq 1.9$.

Finally, studies of the cosmogonies in which the universe is either dominated by 'cold' DM (CDM) or by massive neutrinos have led to the conclusion that (if $\Omega \sim 1$) neither can reproduce the observed large-scale distribution of galaxies unless galaxy formation is biased. In both cases the matter two-point correlation function steepens in time, and, if galaxies trace mass, the stage of the simulation to be regarded as the present epoch can be determined by matching its logarithmic slope to the observed $\gamma = 1.8$ of galaxies. The *CDM* correlation length at this time turns out to be only $r_0 \simeq 1\ (\Omega h^2)^{-1}$ [11] – too small if compared directly with the $r_0 \simeq 5\ \mathrm{h}^{-1}$ Mpc observed for galaxies, unless $\Omega h \leq 0.2$. Thus, the specific scenario of CDM with $\Omega = 1$ would require an enhanced clustering of the galaxies relative to the DM such that $\xi_g(r) = (5 - 20)\ \xi(r)$ for $0.5 \leq h \leq 1$, which is consistent with the values of f deduced above from more general considerations. In the case of ~ 30 eV *neutrinos*, at the time when $\gamma = 1.8$ the large scale structure is still young: collapse to pancakes must have occurred after $z = 2$. This poses a timing difficulty for any 'dissipative pancake' scenario which assumes that galaxies are 'daughters' of pancakes, if one considers the evidence suggesting that galaxies had formed at $z > 3$ (e.g. the existence of high-redshift quasars and galaxies, the old ages of globular clusters, evolutionary models for galaxies, and the need to 'hide' newly formed galaxies). Also, the neutrino correlation length, by the time when $\gamma \simeq 1.8$, has already grown to be $\simeq 8\ (\Omega h^2)^{-1}$ [8-10] – too large in comparison with r_0 unless $\Omega h > 1$. If the galaxies form only in the collapsed regions as one would naively assume in this scenario, these constraints become even tighter, so the bias required here is of an opposite sense: the galaxies should somehow be *less* clustered than the neutrinos.

In conclusion, a large-scale segregation between galaxies and the dominant mass seems *inevitable.* If $\Omega = 1$, as favored by theory, most of the (non-baryonic) mass must hide in the 'voids' to escape detection in clusters and superclusters. Big regions of low densities as observed in the 'voids' cannot form by the present, especially if $\Omega < 1$. Finally, neither of the two popular cosmogonic scenarios can match the observed large-scale structure unless galaxy formation is biased.

2. Bias Scenarios

The bias mechanism could be of one of the following general types: (i) the 'voids' may be filled uniformly with DM different from the clustered DM such that the bias is 'built-in', or (ii) there is only one important kind of DM but the baryonic component is segregated from the non-baryonic DM even on scales $\sim 30\ \mathrm{h}^{-1}$ Mpc (e.g. from large-scale 'isothermal' fluctuations, from gas dissipation relative to a collisionless DM, or from energetic winds pushing the gas over large distances), or (iii), less extravagant in energy, the large-scale baryon distribution does trace the DM on scales $>1h^{-1}$ Mpc but the efficiency with which baryons turn into luminous galaxies is sensitive to other environmental effects. The bias could be determined in each protogalaxy *autonomously*, for example by its background density, or it may be a result of *feedback* from other galaxies. This feedback influence may propagate by gas transport to limited distances, or by radiation or fast particles to larger distances. The result might be *destructive*, suppressing galaxy formation *locally* (causing 'under-clustering') or *far away* (causing 'over-clustering'), but it could also be *constructive*, enhancing galaxy formation in the neighborhood of other galaxies (e.g. explosions).

2.1. A Uniform Component

The universe may be dynamically dominated by 'ultrahot' weakly interacting particles which do not cluster because they have velocities $> 10^3$ km s^{-1}. For example, the 'ultrahot' DM may be non-baryonic and the clumped DM all baryonic. But if the 'ultrahot' particles are relics of an early epoch, their mass would have always been dynamically dominant over the baryons, and would have inhibited gravitational clustering altogether [12]. This would also yield an unacceptably fast expansion timescale during nucleosynthesis. A way around this difficulty involves supposing that these particles arise from nonradiative *decay of heavy particles* with lifetimes only slightly shorter than the age of the universe (see Flores, this volume). Assuming that these decay products (presumably neutrinos) are substantially lighter than their unstable parents, they would be very 'hot'. Galaxies (halos) and clusters formed during the era of matter domination by the unstable particles, but they then expanded or even became unbound as a result of the decay. An elaboration of this idea suggests that the universe finally becomes dominated by a stable primordial CDM species [13] which helps explain the survival of structure on both small and large scales, but it requires certain *ad-hoc* fine-tuning among various DM components. The following astrophysical considerations constrain the allowable parameters in this scheme, and they may already eliminate it all together. Upper limits of $1+z_D \leq 0.5$ on the decay epoch are obtained from the isotropy of the microwave background [14] and the requirement that galaxies and the cores of rich clusters remain bound after the decay [15]. The observed gravitational lenses, if due to typical galaxies, require $1+z_D \leq 3$ [16]. A lower limit of $1+z_D \geq 5$ can be obtained from the dynamics of the Local Supercluster [15,17] but this limit is very model dependent. On the other hand, assuming a quite general scenario for galaxy formation, we [18] find that galactic rotation curves would not have remained flat if the universe were dominated by relativistic decay products.

The universe could be flat ($k = 0$) with $\Omega < 1$ if a non-zero *cosmological constant* contributed to the curvature such that $\Omega + \Lambda/(3H_0^2) = 1$. In some respects this idea resembles the alternative discussed above, but contrariwise, the Λ-term is unimportant at early epochs and so it would not have had such a serious inhibiting effect on galaxy formation. It is found in N-body simulations [11] that the large-scale structure in a flat CDM scenario with $\Lambda \neq 0$ is quite successful in reproducing the two and three point galaxy correlation functions and their peculiar velocities (with no further bias in the galaxy formation). It is also compatible with the isotropy of the microwave background [19]. However, for the Λ contribution today to be comparable to the ordinary matter, the required fine-tuning is as *ad-hoc* as the one we intended to avoid adopting $\Omega = 1$ [20].

2.2. Bias In Hierarchical Clustering (e.g. Cold Dark Matter)

An enhanced clustering of galaxies over the background matter can arise in a 'bottom-up' scenario if galaxies formed only from exceptionally *high peaks* of the density distribution smoothed on galactic scales; peaks with an overdensity δ above a threshold $\nu\sigma$, where $\sigma^2 \equiv \langle\delta^2\rangle$. If the local distribution function of δ has a steeply decreasing tail, like a *Gaussian*, and the power spectrum is not a white-noise, high peaks occur with enhanced probability in the crests rather than the troughs of a large-scale fluctuation mode, so they display enhanced clustering [21]. In a Gaussian process, in the region where $\xi(r) \ll 1$, the enhanced correlation function of high ν peaks is approximated by [22,23]

$$\xi_{peaks}(r) \simeq exp[(\nu^2/\sigma^2)\ \xi(r)] - 1, \tag{5}$$

which becomes $\xi_{peaks}(r) \simeq (\nu/\sigma)^2\xi(r)$ where $\xi_{peaks} \ll 1$. The crucial question is what astrophysical mechanism prevents lower-amplitude peaks from also turning into galaxies, thereby neutralizing the effect. One has to come up with a mechanism that would produce a fairly sharp *cutoff* in the efficiency of (bright) galaxy formation at $\nu \sim 2.5$; the number density of such peaks in the case of CDM being comparable to that of bright galaxies [24].

N-body simulations (White, this volume) suggest that the dissipationless dark halos which are the 'parents' of bright galaxies – those with velocities > 200 km s^{-1} - are themselves more clustered than the overall mass distribution. This is essentially because the linear growth rate of galactic-scale perturbations is significantly affected by whether they are embedded in a peak or a trough of larger fluctuations. The growth rate is boosted up or suppressed if the background mimics an $\Omega > 1$ or an $\Omega < 1$ model respectively. Unfortunately, the resultant bias would show up in the distribution of bright galaxies more than in the distribution of galaxies of lower luminosities – an effect which is not supported by observations.

The dissipative gas contraction to the centers of the dark halos and the subsequent star formation would have an important role in the final bias. As a simple example, the high-ν peaks would collapse earlier, and have higher density at turnaround, than more typical fluctuations on a given mass scale. This could, in principle, in itself account for the biasing if star formation were highly sensitive to (for instance) Compton cooling on the microwave background [25] – an effect that depends on time like $t^{-8/3}$. But can this effect produce a cutoff sharp enough at ν?

The bias may result from processes intrinsic to the protogalaxies, which depend only on the local background density. For example, Silk and I [26] have argued that in a bottom-up scenario the 'normal' bright galaxies *must* originate from high density peaks ($2\sigma - 3\sigma$) in the initial fluctuation field, while typical ($\sim 1\sigma$) peaks either cannot make a luminous galaxy at all because the gas is too hot and too dilute to cool in time, or, if their virial velocity is less than ~ 100 km s^{-1}, they make diffuse dwarf galaxies by losing a substantial fraction of their mass in supernova-driven winds out of the first burst of star formation. This would lead to a *selective bias*, in which the *bright* galaxies are biased towards the clusters and superclusters, while the *dwarf* galaxies do trace the mass, and should provide an observational clue for the real distribution of the DM. The evidence for such a segregation between the high and very low surface-brightness galaxies is still inconclusive (Haynes, this volume), and in particular there are searches going on for dwarfs in relatively nearby 'voids' of bright galaxies (Oemler, this volume).

There are several ways whereby the first galaxies ($> \nu$) could have influenced their environment so as to modify the formation of later galaxies ($< \nu$). Various physical processes have been suggested [24-28], although some of them would not seem to do the job very convincingly. In order to unbind a protogalaxy one has to heat the intergalactic gas to temperatures above $\sim$ 100 eV, corresponding to the potential well of a typical galaxy. Unfortunately, photoionization by available sources (like quasars) is capable of heating the gas only to a few eV, the binding energy of hydrogen. Furthermore, in order to be relevant, any feedback influence must propagate sufficiently fast over large distances, from proto-clusters to proto-voids, and maintain a continuous suppression of galaxy formation for a long time. It would be hard to expect that any mechanical heat source, such as explosive winds, would be capable of doing the job.

UV radiation is capable of carrying the influence and perhaps affecting the IMF in protogalaxies after the redshift $z \sim 3$ corresponding to the (apparent) peak of activity of quasars[27]. The first generation of protogalaxies might have fragmented efficiently via H_2 cooling into a 'normal' stellar population. The radiation fed into the intergalactic medium then photodissociated the H_2 molecules, leaving $L\alpha$ emission as the dominant cooling mechanism, which made the fragmentation less efficient and led to a preferential formation of massive stars. The latter would be highly disruptive via supernova-driven winds, eliminating bright galaxies and leaving behind from the second generation only diffuse 'failed galaxies'. However, an anti-bias might arise instead, if the fragmentation via H_2 were so efficient that it led to a population dominated by unseen 'Jupiters', while the later inefficient fragmentation ended up with a 'normal' visible population. Also, a similar suppression of H_2 may result from shock heating in the vicinity of luminous objects, so perhaps more likely is a local negative feedback effect, which would produce 'under-clustering'.

Alternatively, 'cosmic-ray' particles from first generation galaxies may raise the Jeans mass by heating the gas (if $< 0.1c$) or raising its pressure (if relativistic), provided that they can diffuse appropriately [25]. A constant pressure gradient may produce a constant drift of the baryons which, if larger than the escape velocity from the DM potential wells, would be sufficient to prevent further galaxy formation. These mechanisms and others are discussed more comprehensively by Rees (this volume).

Back to a more general question, one may ask whether the possibility that the baryons trace the mass and the bias rely on the sensitivity of galaxy formation to the background density is consistent with what we know about the LSC. Assume that when galaxies formed, at some z_i, their mean number density contrast within the proto-supercluster was enhanced by a factor g over the matter contrast, $\delta_g^i = g\ \delta^i$. Consequently, the linear growth of δ_g would be driven by the δ of the dominant mass, leading to a present bias of [28]

$$\delta_g = f\ \delta \simeq \frac{(g + z_i)}{(1 + z_i)}\ \delta. \tag{6}$$

For f to be $\simeq 2.5$ as required by $\Omega = 1$, assuming $z_i \geq 3$, the original bias would have to be $g \geq 7$. This may look somewhat large, but one can show that it is actually consistent with the bias expected for high Gaussian peaks in an $\Omega \sim 1$ universe dominated by CDM. The bias of high peaks in a Gaussian density field is a function of the *rms* mass density contrast at z_i on comoving *galactic* scales, σ^i, and the height of the threshold, ν. Given that for galaxies today $\sigma_g \simeq 1$ at 8 h^{-1} Mpc, one can write $\sigma^i = f^{-1}(1 + z_i)^{-1}s$, where $s \equiv \sigma^i/\sigma^i(8\,\mathrm{h}^{-1}\ \mathrm{Mpc})$, a parameter describing the steepness of the linear fluctuation spectrum. The amplification at z_i is given by [23]

$$1 + \delta_g^i = exp(\nu\delta^i/\sigma^i) = exp(\nu\delta_g/s). \tag{7}$$

The initial g can be predicted from (7) using $\delta^i = \delta_g f^{-1}(1+z_i)^{-1}$, to be compared with the value deduced from (6). The epoch of galaxy formation is determined by the requirement $\nu\sigma^i \simeq 1$, i.e. $(1+z_i) \simeq s\nu f^{-1}$. Equating g in (6) and (7) we finally get

$$f^{-1} \simeq 1 \ - \ \delta_g^{-1}[exp(\nu\delta_g/s) - 1] \ + \ (\nu s)^{-1}. \tag{8}$$

Note that the present bias f is insensitive to z_i (the last term), but is quite sensitive to the shape of the spectrum through s. If we adopt for galaxies at formation a mass of 10^{11} $\mathrm{h}^{-1}M_\odot$, the standard CDM spectrum, for $\Omega = 1$ and $h = 0.5$, gives $s \simeq 6$. With $\delta_g = 2.5$ and $\nu = 2.5$ we get in (8) $f \simeq 2.6$; a value in agreement with the degree of bias required in §1. This pleasant consistency would not have occurred if the DM were not 'cold'; a steeper spectrum would give a weaker bias f (which may in turn be consistent with some $\Omega < 1$). For example, a white noise power spectrum ($n = 0$) would give only $f \simeq 1.08$. A slightly flatter power spectrum would give $f \ll 1$, and if actually $s < 5$ the bias deduced from (6) is too low to match the amplification provided by (7). Turning the argument around, if we start by assuming CDM, this consistency would require $\Omega = 1$. This is because on one hand a lower value of Ω would be associated with a smaller s in the CDM spectrum so f would be much larger in (8), while on the other hand a lower Ω needs less 'missing' mass to be reconciled with the observed galaxy distribution (§1), i.e. a smaller f. (The above estimate is not valid in the case of pancakes, where asphericity and non-linearity affect the dynamics of the LSC, and where the high local density in the pancakes, resulting from the one-dimensional collapse, may be responsible for the required bias.)

2.3. Bias in Pancake Scenarios (e.g. Neutrinos)

A bias is generated automatically in any 'top-down' scenario where the perturbations below a critical length of a few tens of megaparsecs have all damped out, as in the neutrino scenario or in the case of adiabatic perturbations in a baryonic universe. First, there are motions from 'proto-voids' to 'proto-pancakes' associated with the large coherence length; collapse into flat pancakes accompanied by streaming toward their lines of intersections ('filaments'), and toward the knots where rich clusters form. The gas then contracts dissipatively into the high density regions, within which the conditions become ripe for cooling and galaxy formation; galaxies are thus limited to very specific regions.

However, if the efficiency of galaxy formation is similar in all the collapsed regions, this natural bias makes the timing-scaling difficulties described above more severe, namely, how can 'pancakes' collapse soon enough to form galaxies at $z > 3$ without producing large-scale clustering of excessive magnitudes. Galaxy formation must be *suppressed* in the high density regions (or, less likely, enhanced in the low density regions). In particular, galaxies might have formed preferentially in the sheet-like pancakes and not in the denser filaments and clusters. This might be a result of a higher cooling efficiency behind shocks in a planar geometry, or due to higher galaxy formation efficiency at later times when most of the pancake galaxies form. N-body simulations in which the formation of a galaxy at a given position and time is determined taking into account suppressing feedback effects from nearby quasars [29] demonstrate that the required anti-bias could be easily obtained with a reasonable choice of values for the physical parameters such as the quasar output energy and lifetime and the cooling rate of the heated gas.

While an anti-bias like this can eliminate the timing-scaling problem, one has to worry about reconciling it with the indications for a positive bias summarized in §1. The solution may involve an anti-bias on scales of clusters and a bias on scales of superclusters and 'voids'.

What may help alternatively is the non-dissipative pancake scenario [30,31] which would arise from a *hybrid scenario*: If galaxies form independently of pancakes, from another component of density fluctuations, the timing constraint becomes irrelevant: galaxies could have formed at $z > 3$ and large-scale pancakes at $z \sim 1$. Galaxies would not be limited to pancakes but rather be present everywhere, subject to the biasing mechanisms that are relevant in general 'bottom-up' scenarios. Such hybrids could consist of two types of DM, baryonic and/or non-baryonic, or from two types of initial fluctuations, adiabatic and isothermal. The hybrid scenarios can be successful where the single-DM models fail, e.g., in reproducing simultaneously the observed structure on galactic scales and on supergalactic scales, and in smearing the anisotropies in the microwave background.

2.4. Bias in the Explosion Scenario

Biasing also results if galaxies *enhance* the formation of other galaxies close to them. In the 'explosions' scenario [33,34] such effects dominate the cosmogonic process. Here, nuclear energy from first generation objects helps gravity in forming further galaxies, hence enhancing their clustering. The exploding galaxies produce spherical blast waves that push the gas out of their interiors. The shells expand, cool and fragment into a new generation of galaxies, the last generation forming at $z \simeq 7$, after which the shells cannot cool efficiently anymore. The galaxies hence form in spherical shells while cool gas still fills the inter-shell volume, and the non-dissipative DM is still distributed everywhere. This introduces an automatic bias in the galaxy positions similar to that in pancake scenarios. The life time of this bias is limited though: in addition to the accretion of DM onto galaxies, the evacuation of the baryons from the shell interiors creates under-dense regions which then tend to develop gravitationally into 'voids' empty of DM. Using N-body simulations with $\Omega = 1$ and $\Omega_{baryons} = 0.1$, we find [35] that a bias of $f \geq 2.5$ persists until a universal expansion by ~ 8 after the formation of the last generation of galaxies – compatible with the favored epoch for galaxy formation in this scenario.

It is possible, although not certain, that explosive shells and their mutual interaction are responsible for the observed large-scale structure [35,36], but even if they are not, it is likely that some positive feedback effects are important at the vicinity of galaxies on scales of groups and clusters.

2.5. Non-Random Phases (Cosmic Strings)

The evidence for *very large scale* structure in the universe may call for a formation (and bias) mechanism more complicated than the type of biasing process that could plausibly account for the galaxy correlations on scales $< 10\ h^{-1}$ Mpc. The crucial observations in addition to the presence of big 'voids' are the enhanced superclustering of rich Abell clusters over the galaxies (and the matter) [37] characterized by a correlation function which is nonlinear out to $\simeq 25\ h^{-1}$ Mpc and positive at least out to $100\ h^{-1}$ Mpc, and the ~ 600 km s^{-1} drift velocity of our 'greater' local neighborhood ($> 100\,h^{-1}$ Mpc in diameter) relative to the microwave background [38]. The superclustering is hard to reproduce in any of the standard scenarios that assume initial fluctuations with a scale-invariant spectrum and random-phases [39]. In particular, the CDM scenario ($\Omega = 1$ and Zeldovich spectrum) predicts that the matter correlation function should go negative beyond $\simeq 20\,h^{-1}$ Mpc, which, if the clusters are biased as in eq. (5), implies that the cluster correlation function should also become negative there, in conflict with the observations. The predicted large -scale drift velocity in this model is ~ 150 km s^{-1} – way off the observed value.

One way to explain such very large scale structure and still be consistent with the microwave background isotropy is an open universe, where the linear spectra of fluctuations have more relative power on large scales; if $\Omega \sim 0.2$, baryons, or neutrinos, give rise to a positive feature in the spectrum at $\sim 100\,h^{-1}$ Mpc while CDM can still be responsible for the formation of galaxies [40]. But if $\Omega \sim 0.2$, no net bias is required by the dynamics of the LSC or the cosmic virial theorem, perhaps indicating that we overestimated the apparent emptiness of the big 'voids'.

If the density fluctuations began as quantum fluctuations of a free scalar field during the era of inflation they are indeed expected to be Gaussian [41], but it is also possible that the fluctuations arose from a different mechanism, in which they would not in general be Gaussian, and have non-random phases. A specific model that incorporates this feature is the scenario in which the density fluctuations were induced by *cosmic strings* (Albrecht, this volume; Turok, this volume). The strings are curvature singularities with a random-walk topology generated in a phase-transition in the early universe [associated with the gauge group U(1)], which turn into closed smoother 'parent' loops on entering the horizon and then chop themselves into (possibly) stable 'daughter' loops, all in a scale-free self-similar fashion. Based on pioneering, low resolution, string simulations [42], it was argued that the final loop-loop correlation function has a general shape close to that of galaxies or clusters [43], $\xi(r) \propto r^{-2}$, as expected from 'beads' along locally-linear 'strings'. The string-induced fluctuations in the DM (by accretion onto the loops), which would have a scale-invariant Harrison-Zeldovich spectrum, are the seeds for the formation of galaxies and clusters, which are therefore expected to be aligned in space along the same 'parent' linear structures. Studying a 'toy' string model that incorporates DM gravity, we [44] found the *phase-correlations* to have a very pronounced effect on the correlation functions of the galaxies and the clusters relative to the matter. It provides a natural galaxy biasing mechanism, as well as an explanation for the excess of cluster clustering on very large scales.

Unfortunately, I do not think that the clustering of loops is well understood yet. First, it turns out that the correlations on scales larger than the horizon are negligible because the strings display a self-avoiding random walk. Second, it is not obvious that the notion of 'beads along strings' is at all relevant; newer simulations (Albrecht, this volume) show no such effect. Also, high loop velocities tend to smear out their correlations on scales slightly smaller than the original parent loops ($\sim$ the horizon). It is therefore crucial to study this fragmentation process and the associated velocities in more detail before a serious attempt is made to understand the formation of large scale structure in the cosmic strings scenario. We are currently set up to run high resolution string simulations for this purpose.

3. Conclusions

A few key observations may be helpful in distinguishing between the possible bias mechanisms. In relation with the 'voids' one would like to answer questions like:

(i) How big are the 'voids' and how empty are they? We need to quantify the data using a meaningful statistic, to confirm (or disprove) our suspicion that no theory can account for them without bias.

(ii) Are voids empty of galaxies of all types, or only those types that are most conspicuous? Any evidence that galaxies of different types display unequal degrees of large-scale clustering would be relevant here, and most interesting would be the spatial distribution of very low surface brightness dwarfs.

(iii) How much gas is there in the voids? Absorption systems along the lines of sight to quasars passing through 'voids' may be detectable. If the gas is at $10^5 K$, too hot for 21-cm and too cold for x-ray, perhaps some features characteristic of neutral He may reveal its presence.

The relationship between 'parent' halos and 'daughter' galaxies could also have interesting implications on the bias scheme [45]:

(iv) Are there any 'barren' galactic-mass dark halos with no luminous galaxy within them? Such objects are expected either in small halos of a shallow potential well or in halos too big to let the gas cool in a Hubble time. (They may, perhaps, be candidates for invisible gravitational lenses.)

(v) Are there any 'orphan' galaxies which lack dark halos? Such galaxies may form from regions where the baryons had been compressed to ~ 10 times the DM density, indicating a certain type of bias mechanism

I have tried to argue here that the idea of biasing is not just an *ad-hoc* idea introduced by theorists to save the attractive $\Omega = 1$ model when confronted with apparently conflicting evidence. A bias is essential in order to understand even the gross features of the large-scale structure, in particular the big 'voids' and the superclustering of clusters, and to reconcile any of the cosmogonic scenarios under current discussion with the observed universe. Therefore, the search for an appropriate physical bias mechanism, and especially for confirming observational evidence, are of great importance. I believe that what might have looked at first as a frustrating idea for astronomers, who can only observe the luminous 'tips' of the 'icebergs', should become an exciting area of observational search for the relevant evidence. It would be circumstantial, though, therefore requiring non-trivial interpretation.

On the theoretical side, finding the 'correct' bias mechanism is intimately related to the 'correct' cosmogonic scenario or the 'correct' nature of the DM. Although some of the proposed bias mechanisms may seem *ad-hoc*, others are very plausible. In some cases the bias improves the consistency of the cosmogonic scenario and the observations, and in others it introduces new problems. The moral is, in any case, that the default assumption to be made when a theory is compared to observations is not necessarily that galaxies trace the mass. Instead, if possible, a physical 'bias' scheme should be applied ([46] for more details).

References

1. A. Dressler: Ap. J. 236, 351 (1980).
2. J. Yang, M.S. Turner, G. Steigman, D.N. Schramm and K.A. Olive: Ap. J. 281, 493 (1984).
3. M. Davis, J. Tonry, J. Huchra and D.W. Latham: Ap. J. Lett. 238, L113 (1980)
4. A. Yahil, A. Sandage and G. Tammann: Ap. J. 242, 448 (1980).
5. M. Davis and P.J.E. Peebles: Ap. J. 267, 465 (1983).

6. A.J. Bean, G. Efstathiou, R.S. Ellis, B.A. Peterson and T. Shanks: M.N.R.A.S. 205, 605 (1983).
7. G.L Hoffman, E.E. Salpeter and I. Wasserman: Ap. J. 263, 485 (1982).
8. S.D.M. White, C.S. Frenk and M. Davis: Ap. J. Lett. 274, L1 (1983).
9. J. Centrella and A. Melott: Nature 305, 196 (1983).
10. A. Dekel and S.J. Aarseth: Ap. J. 283, 1 (1984).
11. M. Davis, G. Efstathiou, C.S. Frenk and S.D.M. White: Ap. J. 292, 371 (1985).
12. Y. Hoffman and S. Bludman: Phys. Rev. Lett. 52, 2087 (1984).
13. K. Olive, D. Seckel, and E. Vishniac: Astrophys J. 1985, in press.
14. J. Silk and N. Vittorio: Berkeley preprint (1985).
15. G. Efstathiou: M.N.R.A.S. 213, 29 (1985).
16. A. Dekel and T. Piran: Ap. J. (Lett.), submitted (1986).
17. Y. Hoffman: M.N.R.A.S., in press (1986).
18. R. Flores, G.R. Blumenthal, A. Dekel and J.R. Primack: Nature, in press (1986).
19. N. Vittorio and J. Silk: Berkeley preprint (1985).
20. P.J.E. Peebles: Ap. J. 284, 439 (1984).
21. N. Kaiser: Ap. J. (Lett.) 284, L9 (1984).
22. D.H. Politzer and and M.B. Wise: Ap. J. Lett. 285, L1 (1984).
23. Jensen, L.G. and Szalay, A.S.: Fermi Lab. preprint (1986).
24. Papers by J. Bardeen, N. Kaiser, and S.D.M White: in E.W. Kolb and M.S. Turner eds., Inner Space/Outer Space University of Chicago Press 1985.
25. M.J. Rees: M.N.R.A.S. 213, 75P (1985).
26. A. Dekel and J. Silk: Ap. J., 303, April 1.
27. J. Silk: Ap. J. 297, 1 (1985).
28. P.J.E. Peebles: Nature, in press (1986).
29. E. Braun, A. Dekel and P. Shapiro: in preparation (1986).
30. A. Dekel: Ap. J. 264, 373 (1983).
31. A. Dekel: in G. Domokos and S. Koveski-Domokos eds., Proceedings of the Eighth Johns Hopkins Workshop on Current Problems in Particle Theory, p. 191. World Publishing Co.: Singapore 1984.
32. A. Dekel, G.R. Blumenthal and J.R. Primack: in preparation (1986)
33. J.P. Ostriker and L. Cowie: Ap. J.(Lett.) 243, L127 (1981).
34. S. Ikeuchi: Pub. Astron. Soc. Japan 33, 211 (1981).
35. S. Saarinen, A. Dekel and B. Carr: Nature, in press (1986).
36. D. Weinberg, A. Dekel and J.P. Ostriker: in preparation (1986).
37. N. Bahcall and R. Soneira: Ap. J. 270, 20 (1983).
38. D. Burstein *et al.* : in B.F. Madore ed., Galaxy Distances and Deviations from Universal Expansion, Dordrecht: Reidel 1986.
39. J. Barnes, A. Dekel, G. Efstathiou and C. Frenk: Ap. J., 295, 368 (1985).
40. A. Dekel, G.R. Blumenthal and J.R. Primack: in preparation (1986).
41. J. Bardeen, J.R. Bond, N. Kaiser and A. Szalay: Ap. J., in press (1986).
42. A. Albrecht and N. Turok: Phys. Rev. Lett. 54, 1868 (1985).
43. N. Turok: ITP preprint (1985).
44. J.R. Primack, G.R. Blumenthal and A. Dekel: in B. F. Madore, ed. Galaxy Distances and Deviations from Universal Expansion, Dordrecht: Reidel 1986.
45. M.J. Rees: in J. Kormendy and G. Knapp eds., Dark Matter in the Universe, Dordrecht: Reidel 1986.
46. M.J. Rees and A. Dekel: Nature, submitted (1986).

Biasing and Suppression of Galaxy Formation

MARTIN J. REES

1. Classification Scheme

It is all too easy to think of reasons why bright galaxies should be poor tracers of the overall cosmic mass distribution. The following "morphological" classification indicates the range of possibilities and perhaps helps to categorise various biasing mechanisms that merit more discussion:

Do baryons trace mass?

- yes ↓
- no:
  - large-scale explosions
  - primordial n_b/n_γ fluctuations

Are initial galactic-scale perturbations gaussian?

- yes ↓
- no:
  - string-induced perturbations
  - "secondary" fluctuations (eg in pancake model)

Are feedback processes from the first galaxies important?

- yes ↓
- no:
  - "autonomous" biasing: the high-amplitude peaks convert into luminous galaxies more efficiently than lower-amplitude peaks of the same mass

Non-local negative feedback (rather than local positive)?

- yes ↓
- no:
  - "epidemic" models of galaxy formation

Are baryons prevented from accumulating in galactic-mass potential wells?

- yes:
  - Heating of gas to $\gtrsim 10^6$ °K
  - Cosmic ray pressure
  - Trapped Lyman α pressure
  - Pushing at $\gtrsim 100$ km s^{-1} relative to dark matter
- no:
  - Star-formation efficiency (or IMF) in later-forming galaxies modified by (e.g) UV background, destruction of H_2, etc.

2. Biasing in the 'Cold Dark Matter' Cosmogony

A cosmogonic scheme that has been explored in considerable detail, and discussed by several speakers at this meeting, is one in which the dark matter takes some "cold" non-baryonic form and the initial curvature fluctuations are scale-independent (the Harrison-Zeldovich spectrum). If $\Omega = 1$, this cosmogony cannot be reconciled with observations without some form of biasing. Davis et al. [1] carried out detailed N-body simulations of such models by hypothesising in an "ad hoc" fashion that galaxies form only from regions which are $\gtrsim 2.5\sigma$ peaks of the initial mass distribution smoothed over a galactic scale; Frenk and White have reported follow-up studies at this meeting. They obtain an impressively good fit with the covariance function, the M/L ratio in clusters, and the Virgo infall. This is encouraging and gives us the motivation to seek a genuine physical justification for the biasing.

The CDM cosmogony will not be able to predict the luminosity function of galaxies until we develop a physical understanding of several distinct phases of galaxy formation (see Blumenthal et al. [2] for fuller details).

(i) Baryons are presumed to condense in viriralised "halos" of dark matter in the approximate mass range $10^8 - 10^{12} M_\odot$. For larger masses, dissipative cooling may be inefficient; below $\sim 10^8 M_\odot$ the potential wells may be too shallow to capture primordial gas (especially if this gas has been ionized and heated to $\sim 10^4\ {}^oK$). The mass distribution of isolated virialised systems can in principle be learnt from N-body simulations. In practice, progress is impeded by the large dynamic range involved; galactic-mass halos build up hierarchically, via a complicated series of mergers, between $z \cong 5$ and the present epoch (see Frenk et al. [3]).

(ii) Even if the dissipationless clustering of the dark matter were accurately known, the fate of the baryonic component — how much gas falls into each potential well, and how much is retained - involves complex gas dynamics.

(iii) The resultant luminosity and brightness profile of the galaxy depends on the initial mass function (IMF) of the stars that form from the gas and on the efficiency of star formation.

The biasing could in principle involve any (or all) of these three aspects of galaxy formation. The N-body simulations suggest that even stage (i) may lead to substantial biasing: Frenk (these proceedings) reports that the dark halos associated with bright galaxies – those with circular velocities $v_c > 200$ km s^{-1} – are themselves more clustered than the overall mass distribution, essentially because the binding energy of galactic mass perturbations is significantly affected by whether they are in a peak or a trough of the fluctuation spectrum at larger wavelengths (i.e. if they are, in effect, in a background whose overall expansion mimics an $\Omega > 1$ or an $\Omega < 1$ universe).

This effect alone might account for all the requisite biasing especially if lower-luminosity galaxies are more widely distributed. Any of the other mechanisms involving autonomous biasing or feedback could also, however, be operative (modifying stages (ii) and/or (iii) of the formation process), and several have been considered.

Autonomous biasing

The high-amplitude peaks would collapse earlier and have higher density at turnaround than more typical fluctuations on a given mass scale. This could, in principle, itself aid capture and retention of baryons. For instance, collapse and fragmentation of gas within protogalaxies is highly sensitive to the ratio of the cooling time ($\underset{\sim}{\propto} \rho^{-1}$ for bremsstrahlung and recombination cooling) to the collapse time ($\propto \rho^{-\frac{1}{2}}$): it occurs more readily when this ratio is small. Moreover, if the high-ν perturbations collapsed at $z \gtrsim 10$, Compton scattering of the microwave background would guarantee efficient cooling; this process, whose efficiency is proportional to $(1+z)^4$, would be unimportant for lower-amplitude perturbations collapsing at later times.

Protogalaxies that formed from high-amplitude fluctuations have higher velocity dispersions (and escape velocities) than later-forming systems of similar mass. This makes it easier for them to retain gas that might otherwise be expelled by, for instance, supernova-driven winds.

Dekel and Silk [4] suggest that 'normal' bright galaxies originate from high density peaks (2σ - 3σ) in the initial fluctuation field; typical ($\sim 1\sigma$) peaks either fail to make a luminous galaxy at all because the gas is too hot and too dilute to cool; or, if their virial velocity is less than ~ 100 km s^{-1}, they become diffuse dwarf galaxies after losing a substantial fraction of their gas in supernova-driven winds. This would lead to a selective bias, big galaxies being concentrated in clusters and superclusters, whereas dwarf galaxies trace the mass.

Feedback mechanisms

The formation of galaxies from lower-amplitude perturbations (including dwarf galaxies) may be further impeded owing to some kind of negative feedback induced by the first galaxies. If these galaxies were able to heat the intergalactic medium to > 10^6 °K, the Jeans mass would then rise high enough to prevent further protogalaxies from freely collapsing. The medium could readily be photoionized by UV from the first objects, but this process does not in itself raise the temperature much above 10^4 °K. Although the energy required to heat the gas to $\gtrsim 10^6$ °K is modest (a few hundred ev per baryon) there are difficulties in understanding how the requisite energy could be injected in a sufficiently uniform way: blast waves (triggered by galaxies) with post-shock temperatures $\sim 10^6$ °K do not propagate fast enough to yield homogeneous heating throughout a void; and if the blast waves did travel fast enough to cross a proto-void region, they would generate temperatures $\gtrsim 10^8$ °K and thus require much more energy. It is possible that the first galaxies could supply $\sim$ 100 ev per baryon in some form which diffused rapidly but was nonetheless sufficiently well coupled to the intergalactic gas to exert an effective pressure and thereby raise the Jeans mass. Cosmic rays, whose streaming speed depends on intergalactic magnetic fields, or UV photons longward of the Lyman limit (which propagate freely, even through a neutral IGM, until the cosmological redshift brings them near the Lyman α resonance) are among the possibilities that have been discussed.[5]

The required amount of trapped energy is modest: if in photons or relativistic particles, where individual energies go as (1+z), the energy density needed to inhibit collapse of a (baryonic) mass $10^{11} M_{11b}$ solar masses is $\sim 3 \times 10^{-5} M_{b11}^{1/3}$ ev cm^{-3} This is below the background radiation limits in any waveband and less than the likely cosmic ray density in intergalactic space. The fraction of the total intergalactic energy that is effective in contributing to pressure gradients may, however, be small — for Lyman α, for instance, it is only $\sim 10^{-3}$.

Capture of baryons could be inhibited if the intergalactic gas acquired a streaming velocity $\gtrsim 100$ km s^{-1} relative to the dark matter (even if the gas stayed cold). This could happen if the first galaxies or quasars created large-scale pressure gradients (via radiation or cosmic ray pressure). The required streaming velocities, comparable to the escape velocities from the halo potentials, are of course much less than would be needed to evacuate voids completely of baryons; the energy requirements are therefore not excessive.

Even if feedback efforts cannot inhibit the collapse of galactic-mass gas clouds, they can perhaps influence the internal evolution of protogalaxies, especially the stellar IMF. UV radiation is capable of affecting the IMF in protogalaxies after the redshift $z \sim 3$ corresponding to the (apparent) peak of activity of quasars, and UV could build up earlier if other sources were important. Silk [6] suggested that first-generation protogalaxies fragmented efficiently via H_2 cooling into a 'normal' stellar population. The radiation fed into the intergalactic medium would photodissociate the H_2 molecules and prevent further formation during protogalactic collapse, leaving $L\alpha$ emission as the dominant cooling mechanism. Fragmentation would then be less efficient, leading to a preferential formation of massive stars. The latter would inject energy into the gas, and expel most of it via supernova-driven winds before it had time to form stars, eliminating bright galaxies and leaving a second generation of diffuse 'failed galaxies'. However, an <u>anti</u>-bias might arise instead if the fragmentation via H_2 were so efficient that it led to a population dominated by unseen 'Jupiters', while the later inefficient fragmentation ended up with a 'normal' visible population.

Biasing also results if galaxies enhance the formation of other galaxies close to them. In the 'explosion ' scenario [7,8] such effects dominate the cosmogonic process. Even if explosions are irrelevant to large-scale structure, it would be surprising if positive feedback effects ("epidemic" galaxy formation) did not, even in the CDM cosmogony, play some role on the scale of individual groups and clusters.

Pregalactic influences

Although galaxy formation in the CDM cosmogony is a relatively recent cosmic phenomenon [2,3] occurring at $z \lesssim 4$, bound systems of subgalactic scale would appear somewhat earlier. The first such systems would have condensed at $z \cong 10 - 20$ from high-amplitude peaks in the density distribution on scales $\sim 10^6 M_\odot$. Even a fraction 10^{-4} of the universe, condensed into such objects, could release enough energy to reionize all the matter, thereby raising the Jeans mass to $> 10^8 M_\odot$ and quenching the formation of further objects of $10^6 M_\odot$. The only $10^6 M_\odot$ objects that ever formed could therefore have arisen from > 3.5σ peaks, and consequently display enhanced clustering. In particular, they will be clustered on galactic scales (10^{11} - $10^{12} M_{o}$) in such a way that there are more than twice as many in a 2.5σ galactic-scale peak as in a 1σ peak on the same scale. If this led to an environmental difference betwen "high" and "typical" galactic-scale perturbations (by, for instance, producing a significant heavy element abundance, or triggering some runaway process, in the former) this would provide another physical reason for Davis et al.'s prescription that only the high peaks evolve into galaxies — high and low amplitude protogalaxies would have developed internal differences even before turnaround [9].

[It is perhaps worth noting, parenthetically, that even though the first galaxies to form in the CDM cosmogony are small and larger dark halos build up hierarchically, this does not necessarily mean that the dwarf galaxies observed now need necessarily be especially old. The present-day dwarfs would be atypical "stragglers". Most of the low-mass systems which formed early would have experienced mergers and lost their identity; those that survive would tend to have formed from low amplitude perturbations or in regions of large-scale underdensity.]

3. Some Limitations of Proposed Mechanisms

Time-lags and feedback effects

A non-trivial constraint on any mechanism involving negative feedback is perhaps worth mentioning. Obviously the influence of the first bound systems must propagate fast in order to be effective. The time-lag between the turnaround of bound proto-galaxies in an incipient cluster ($(\delta\rho/\rho)_c > 0$) and the turnaround of an equivalent fraction in an incipient void (with $(\delta\rho/\rho)_v < 0$) is $\Delta t \cong \langle |(\delta\rho/\rho)|_c + |(\delta\rho/\rho)_v| \rangle\, t_{exp}$, when t_{exp} is the cosmological timescale; unless the influence pervades the void within a time Δt it plainly cannot pre-empt the formation of galaxies there. However, it would seem even easier for a galaxy to quench formation of new neighbours - whatever the actual feedback mechanism was, it would surely be strengthened by propinquity - and the energy released within an incipient cluster might then be inadequate to exert any feedback on larger scales. This difficulty would not arise if there were a delay t_{delay} between the onset of protogalactic collapse (defined as the latest stage at which negative feedback could stop the protogalaxy from evolving into a luminous galaxy) and the stage when it could first exert negative feedback. The first galaxies would then be unable to quench the formation of others (even close neighbours) unless they lagged the first ones by longer than t_{delay}. Remote negative feedback would still occur if the influence of the first galaxies reached the void in a shorter time than $\Delta t - t_{delay}$. The physical interpretation of t_{delay} depends on the actual feedback mechanism. If, for instance, this involved quasar-like activity in the first galaxies, t_{delay} would be the time taken for a young galaxy to develop violent activity in its nucleus.

Large-scale structure

Biasing mechanisms based on favouring high-ν peaks can enhance the amplitude of correlation functions but cannot reverse their sign. In particular, they cannot, in the CDM model, yield positive correlations on scales exceeding 18 h_{100}^{-2} Mpc, since the mass correlation function is then negative, because the spectrum on large scales falls off more steeply than "white noise" (unless, as Bond has suggested in his contribution, there is an unexpectedly large amplitude on scales 10 - 100 Mpc). However, there are some feedback processes that could in principle generate apparent large scale positive correlations in the distribution of galaxies. An example is the following.

We know that some galaxies at $z \cong 3$ develop powerful quasar-type activity in their nuclei — indeed, the UV or kinetic energy output from these objects is widely believed to have a dominant influence on the thermal history of the intergalactic medium. The active lifetime of individual quasars is unknown; it could be as long as $\sim 2 \times 10^9$ yrs, the timescale for cosmological evolution at $z \cong 3$. A quasar emitting $> 10^{46}$ erg s^{-1} for this timespan would release enough energy to ionize all material within a sphere of radius 15 Mpc (corresponding to 60 Mpc when expanded to the present epoch), and could supply $\sim$ 100 ev per baryon (sufficient to raise the Jeans mass to galactic scale) out to almost as large a radius. For different assumptions about the quasar luminosities L and lifetimes t_Q, the range of influence of course scales as $(Lt_Q)^{1/3}$. The efficiency of galaxy formation would then be modulated on scales corresponding to the range of influence of each quasar. If the destruction of H_2 somehow modulated galaxy formation or the IMF, then the range of influence of a single quasar (whose output of 10 - 13 ev photons would bring this about) could be larger still. Moreover, there are good reasons to believe that the output from a quasar is beamed rather than isotropic: each object could imprint correlations (e.g. large scale linear domains, or fan-like sheets if the beam precesses) on still larger scales than for the isotropic case. This is just an example (and models involving cosmic strings can provide others) to illustrate that the large scale distribution of galaxies could be modulated by non- gravitational effects: it would then reveal essentially nothing about the overall mass distribution on those scales, and no non-Hubble velocities need be associated with the apparent inhomogeneities.

REFERENCES

1. M. Davis , G. Efstathiou, C.S. Frenk and S.D.M. White: Astrophys.J. 292, 371 (1985)
2. G.R. Blumenthal, S.M. Faber, J.R. Primack and M.J. Rees: Nature 311, 517 (1984)
3. C.S. Frenk, S.D.M. White, G. Efstathiou and M. Davis: Nature 317 595 (1985)
4. A. Dekel and J.I. Silk: Astrophys.J. 303, 39 (1986)
5. M.J. Rees: MNRAS 213, 75P (1985)
6. J.I. Silk: Astrophys.J . 297, 1 (1985)
7. J.P. Ostriker and L. Cowie: Astrophys.J(Lett) 243, L127 (1981)
8. S. Ikeuchi: Proc.Astron.Soc.Japan 33, 211 (1981)
9. H.M.P. Couchman and M.J. Rees: MNRAS 211, 53 (1986)

Session 7

Distant Galaxies

R. Kron, Chair
July 29, 1986

Stellar Populations in Distant Galaxies

GUSTAVO BRUZUAL A.

Abstract
In this contribution I review the problem of the detection of changes in the population content of distant galaxies with respect to the population content of nearby galaxies. The basic assumptions made by most authors working in the field are examined and criticized. I evaluate the comparisons of theoretical models with observations of distant and nearby galaxies that lead to different lines of evidence, both in favor and against the detection of evolution of the stellar populations dominating the light in distant and nearby galaxies.

1. Introduction

The term distant galaxy is used in this paper to denote a galaxy that is observed at a cosmological redshift (z) which is high enough for significant changes in its stellar population content to occur during the light travel time from the galaxy to us. The value of z at which evolutionary effects become significant depends on the cosmological model. Significant changes in a galaxy stellar population are expected to occur over a period of at least 5 to 6 Gyr. In Friedmann cosmologies with zero cosmological constant this travel time corresponds roughly to $z = 0.4 - 0.5$ ($h = 0.5$), or $z = 1.4 - 1.6$ ($h = 1.0$).

At present it is not possible to derive the population content of distant galaxies from direct observations. The stellar populations present in a distant galaxy must be inferred from the indirect information available to us. The most frequently observed property for a large number of distant galaxies is their magnitude through one or several broad band filters, using either photographic or photon counting detectors. The spectral energy distribution (s.e.d.) is available only for a small number of distant galaxies. The stellar population present in a galaxy is then derived from these data using either population or evolutionary synthesis techniques.

Since the changes in the population content of galaxies translate into changes in their s.e.d.'s, the concept of evolution of stellar populations is closely related to the concept of spectral evolution. In the remainder of this paper both terms will be used with the same meaning.

2. Basic Assumptions

Our current understanding of the evolution of the population content of galaxies rests on several basic assumptions, not all of them necessarily correct. These assumptions are listed below and discussed in the following section.

a) We understand the population content of nearby galaxies.

b) We know how to select a uniform sample of galaxies at all observable redshifts. The selected galaxy sample is such that the only difference from galaxy to galaxy is age (related to z by cosmology).

c) We understand stellar evolution to such extent that the population content of a model galaxy can be computed for any epoch, provided that we know the distribution of stellar populations in this galaxy at an earlier or later age.

d) To a first approximation, chemical evolution, dynamical evolution, the interactions of a galaxy with its environment, etc., affect neither the evolution nor the distribution of stellar populations in galaxies up to a z of about 1. Hence, these effects can be neglected.

3. Discussion of the Basic Assumptions

In this section I indicate some of the weak points of the basic assumptions listed in the previous section.

3a. Population Content of Nearby Galaxies

We still do not know the stellar population responsible of emitting the UV flux (shortward of 2000 A) in elliptical galaxies and in central bulges of spirals. This question is still open even for the closest galaxies: M32 and the central bulge of M31. A large number of elliptical galaxies have been observed with IUE [1] and a wide dispersion in the UV flux level is observed in otherwise identical E-galaxies (e.g. giant ellipticals in Virgo).

The possible stellar sources for this flux have been reviewed by WU et al. [2]. The most important sources are:

- Horizontal branch stars (intermediate to blue),
- Post asymptotic giant branch stars (nuclei of planetary nebulae),
- Upper main sequence of a recent generation of stars.

O'CONNELL [3,4] and PICKLES [5,6] using population synthesis (see below) derive the age associated with the main sequence turnoff in M32 and in E galaxies in the Fornax cluster. They favor young turnoff ages: 6 - 10 Gyr. In their interpretation, this age represents the last major epoch of star formation that took place in these galaxies. The stars at the main sequence turnoff are then the source of the UV flux.

BRUZUAL [7] has pointed out that the UV flux in elliptical galaxies can be explained if star formation has been going on at small rates (that vary from galaxy to galaxy) up to the present epoch (12 - 16 Gyr). The continuous star formation hypothesis used in these models implies that the corresponding main sequence turnoff is even younger than the one derived in [3-6].

Work in progress by BERTOLA et al. [8] seems to rule out the assumption [3-7] that recently formed stars are the source of the UV flux. BERTOLA et al. [8] correlate the magnesium line-strength index Mg2 with the UV flux level [9], and find that galaxies that show clear signs of ongoing star formation do not follow the same relationship as the large majority of the data points. The observed trend indicates that for normal ellipticals the UV flux increases with the metallicity of the galaxy [1]. The meaning of this result is still unclear.

Recent studies by FROGEL [10 and references therein], WHITFORD [11] and RIEKE [12] have clearly shown that in the central bulge of our galaxy there exists a distinctive population of metal rich, very cool M giants. We do not know how common to other galaxies these stars are. RIEKE [12] concludes that they are present in the central bulge of M31. The s.e.d.'s of these stars are not available (only broad band colors have been measured), and hence they cannot be included in the stellar libraries used for population synthesis with the same degree of accuracy as other stellar types. The evolutionary stage of these stars, needed to treat them correctly in the evolutionary synthesis approach, is not known either.

The population content of spiral and irregular galaxies is an even more complicated subject. There are large variations from galaxy to galaxy and from region to region in a given galaxy. By studying the simpler populations present in elliptical galaxies one hopes to make progress in understanding stellar populations in general.

In summary, there are still many unanswered questions pertaining to the population content of nearby elliptical galaxies. As long as these problems are not solved, it is difficult to establish the correctness of the models built to explain the population content of both nearby and distant galaxies.

3b. Sample Selection

Very different criteria have been used by different authors to select samples of galaxies in which to look for evidence of population or spectral evolution in galaxies. Some of these samples are:

- Field galaxies [13-16],
- Cluster galaxies [17-21],
- Powerful 3CR sources [22-23],
- Weak radio galaxies [25-27],
- Reddest galaxies at a given z [28].

To what extent these criteria select galaxies that are similar in all respects but seen at different cosmological times has not been established in all cases. In the case of active galaxies (3CR and weak radio sources) it is clear that physical processes different from passive stellar evolution are taking place in these systems. We may expect that these processes affect the evolution of the stellar population (star formation in random bursts, variation in the IMF, etc.) as compared to the simple approach followed in theoretical models (see below).

In order to derive the evidence for evolution or lack of evolution of the sample been studied, one has to know which are the local counterparts of the galaxies in the sample, i.e., one has to compare distant galaxies with some local standard and then establish evolution with respect to this standard. All the selection criteria that enter into the sample should be carefully evaluated before any conclusion in one sense or another can be established.

Different authors, working with different samples, have come to opposite conclusions concerning the detection of spectral evolution of galaxies. It should be clearly understood that any such conclusion refers only to the sample been considered and to the volume of space been sampled, and not to galaxies in general.

Despite the diversity of sampling criteria listed above, most authors use the same set of theoretical models [7] to study evolution of galaxies in their samples. Probably this is due to the fact that Bruzual's models [7] were explicitly built to make comparisons with observations very easy. This is not the case for most of the models available (see below).

3c. Stellar Evolution

There is no doubt that the physical processes taking place in stars during the main stages of their evolution (Main Sequence, Red Giant Branch) are well understood.

There are some evolutionary stages, such as the position of a star in the Horizontal Branch, that depend critically on less well understood physical processes (e.g. mass loss) which makes it more difficult to build accurate evolutionary tracks through these stages. The inclusion of these stages in the synthesis models is then not as accurate as the Main Sequence evolution.

Some evolutionary stages are often neglected in population synthesis (e.g., Asymptotic Giant Branch or Post-Asymptotic Giant Branch phases) on grounds that they are very short lived. However, the important parameter is the amount of nuclear fuel used during the phase [29]. WYSE [30] has made a simplified attempt to include the AGB stars in population synthesis models (see below).

For some evolutionary stages empirical data based on stellar populations in the solar neighborhood (Tinsley and Gunn Red Giant Luminosity Function [31]) and in the galactic center (Frogel and Whitford M Giants [10-12]) are used. We do not know if these stellar groups are equally present in all galaxies.

The same comment applies to the calibration of stellar evolution codes by comparison to galactic or globular clusters, to stars in the solar neighborhood, and to clusters in the Magellanic Clouds. Given the differences found in the properties of these stellar systems from galaxy to galaxy, the evolutionary tracks and isochrones derived by adjusting physical parameters to fit these specific systems, may not be general enough to represent all possible systems in the universe. The globular clusters in the local group galaxies show a wide dispersion in properties, such as metallicity and age distributions [32-33].

3d. Neglecting Important Effects

The dynamical evolution of a galaxy, as well as its interactions with the environment will influence, and most likely determine, the history of star formation and the subsequent evolution of the stellar population in a galaxy. The chemical evolution of a galaxy must also have important effects on its observable properties. These effects are neglected in present models just to allow us to make progress in the prediction of the properties of distant galaxies.

It should be kept in mind that most of the existing galaxy evolution models should represent the evolution of an average population of galaxies, which have not been highly affected by interactions with the environment, nor chemical evolution, at least from $z = 1$ to $z = 0$. In particular, these galaxies are characterized in the models by an Initial Mass Function that is assumed to be both universal and constant in time.

4. Theoretical Models and Results

Despite the difficulties mentioned in the previous section, considerable effort has gone into the problem of understanding the stellar populations and interpreting the photometric properties of both nearby and distant galaxies. There are three independent lines of approach to this problem used by different authors. In this section I summarize the main results obtained by these authors.

4a. Cluster Template

RENZINI and BUZZONI [29] advocate that at present most of the effort should go into understanding the population content of stellar clusters in the Galaxy, and especially, in the Magellanic Clouds. This procedure permits the correct calibration of theoretical evolutionary tracks and isochrones over different ranges of metallicity and age. Then, the calibrated isochrones fitted to these clusters can be used as a template for population synthesis in other galaxies.

This procedure insures that at least all the stellar populations present in the MC clusters will be included in the synthesis of distant galaxy populations. In particular, the AGB, PAGB, SGB, and HB stars [29] will be included. Since the population in the clusters has been dated, the evolution in time of the synthetic population and its observational properties can be followed.

The drawback with this method is that the template will contain only those stellar populations found in the systems that were studied. The MC's may not contain clusters, and most likely do not, that are representative of all the stellar populations present in giant elliptical galaxies. Thus, it may be necessary to extend the cluster template to include metal richer populations not found in the clouds or in the Galaxy.

RENZINI and BUZZONI's project [29] is still in progress and it has not been applied to distant galaxies.

4b. Population Synthesis

Most of the recent work in this field has been done by O'CONNELL [3] and PICKLES [5]. They find that the age associated with the main sequence turnoff groups in M32 [3] and Fornax cluster ellipticals [5] is quite young, between 6 and 10 Gyr. In their interpretation, this number represents the last major epoch of star formation, and this result does not imply that galaxies formed such a short time ago.

However, if this major event of star formation is common to all elliptical galaxies, it should show as rapid luminosity, color, and spectral evolution in the range of z from 0.5 to 1. This rapid evolution has not been detected [17-28,34]. The stellar red giant branch in genuine elliptical galaxies up to z of 1 seems to be dominated by low-mass passively-evolving stars, consistent with a constant giant luminosity function [31]. In the color-z plane the scattering in the colors is larger than expected for this kind of giant luminosity function. This may indicate that some stellar types are not accounted for properly in this luminosity function.

If the main sequence turnoff is only 6 - 10 Gyr, then at 6 Gyr lookback time, these galaxies should have the colors of a current epoch spiral galaxy that is actively forming stars. This fact is not inconsistent with the data, provided that galaxies evolve at different rates and/or that galaxy formation is not coeval. The reddest galaxies up to z of 0.5 - 1 do not show as blue colors as would be expected from this major star forming event that dominates the main sequence turnoff. Up to z of 1 there are galaxies as red as nearby ellipticals [17-28].

It should be pointed out that the turnoff age is not a direct result of the population synthesis models. Once a given mixture of stellar types is obtained from the model, its age may be inferred from the distribution of stellar types in the HR diagram (by fitting theoretical isochrones). In this step several external assumptions about the (mean) metallicity of the stellar mixture and the adequacy of the isochrones must be made. Different sets of isochrones (independently calibrated) may result in a different turnoff age for the same stellar mixture.

PICKLES [5] finds evidence for a significant contribution from upper main sequence stars in N1404, N1399, and N1379, which he attributes to star formation going on in these systems. He cannot explain the UV flux in these galaxies as due to PAGB stars. PICKLES confirms the well known trend of decreasing mean metallicity with decreasing galaxy luminosity.

Population synthesis is a powerful technique that gives us clues about the stellar populations present in galaxies. Population synthesis models do not predict the star formation history of a given galaxy, and thus are not very useful for studying galaxy evolution. The technique cannot be applied yet to distant galaxies due to the lower quality of the available spectra.

4c. Evolutionary Synthesis

The evolutionary synthesis technique [7,28,36, and references therein] provides simple models in which assumptions about the SFR, IMF, and stellar evolution can be tested against the observational data for a large range of values of z. In this approach it is not easy to include in the models any stellar type for which we do not have a clear understanding of its evolutionary stage. In the latter case, the empirical population synthesis is more appropriate.

There are many recent models for the spectral evolution of galaxies built using the evolutionary synthesis approach: BRUZUAL [7,40], HAMILTON [28], WYSE [30], BARBARO and OLIVI [37], and ARIMOTO and YOSHII [38-39]. In the rest of this paper I summarize and compare the results of some of these papers.

ARIMOTO and YOSHII [38-39] have computed detailed evolutionary models that follow the chemical and luminosity evolution of elliptical galaxies. They conclude that elliptical galaxies are composite stellar systems, composed of stars with a wide variety of metallicities. In their view, elliptical galaxies form a one-parameter family of their initial mass, in which all galaxies have the same age and IMF, but different metallicities. Finally, ARIMOTO and YOSHII [38-39] conclude that the color and luminosity evolution of elliptical galaxies depends strongly on metallicity, and, hence, on the initial galaxy mass.

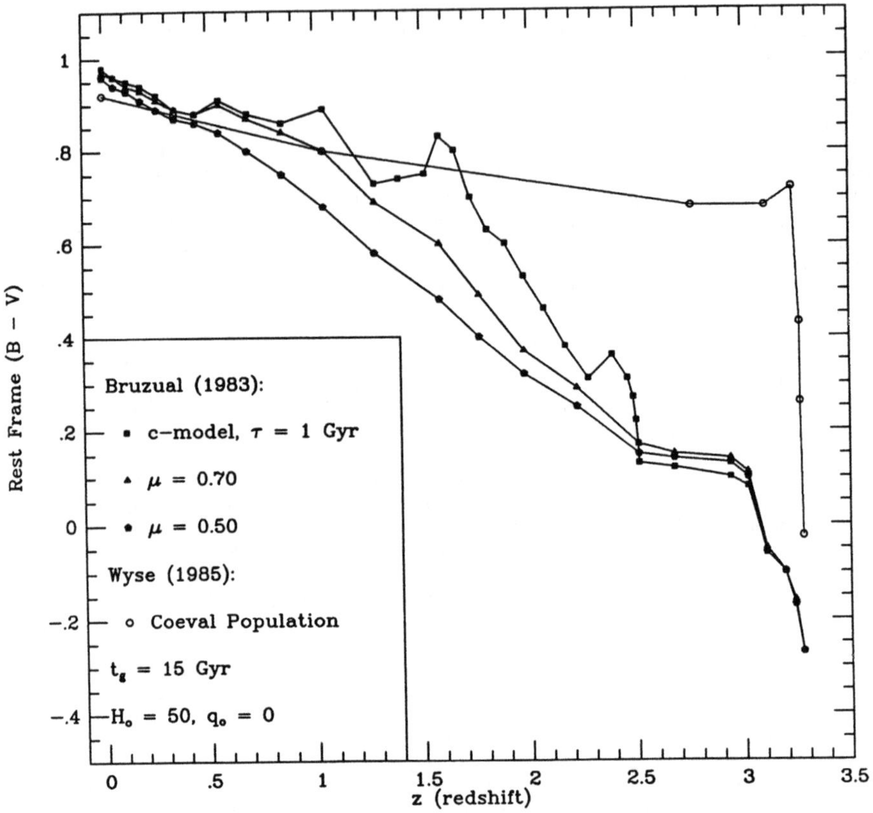

Fig. 1. Rest frame B-V color as a function of galaxy redshift for the E-galaxy models by WYSE [30] and BRUZUAL [7,40 for the Salpeter IMF].

ARIMOTO and YOSHII [38] compare the UBV magnitudes and colors predicted by their models with a small sample of nearby galaxies. In [39] they extend their computations to include all the bands in the UBVRIJKL system. Detailed comparisons with the observations, including high z galaxies, are needed in order to judge how accurate these models represent elliptical galaxies.

WYSE [30] addressed the question of how recently elliptical galaxies could have formed. She includes in her models the AGB and HB evolutionary phases, following the prescriptions in [29 and references therein]. She finds that when the AGB stars are included, elliptical galaxy models become red at earlier times than when these stars are not included. Her model reaches B-V = 0.7 at an age of 0.1 Gyr. This color is reached at age 2 - 3 Gyr in BRUZUAL's [7,40] models, which do not include the AGB phase. WYSE's [30] models allow recent massive galaxy formation to be consistent with the observed colors of present day ellipticals.

In Fig. 1 I compare WYSE [30] rest frame B-V color as a function of z with the same quantity derived by BRUZUAL [7,40]. The models used in [40] are identical to those described in [7], but the optical stellar spectra used to build the galaxy spectra are taken from GUNN and STRYKER [41]. Fig. 1 shows clearly how the rest frame B-V color increases from 0 to 0.7 when the AGB stars first appear. This fact allows for galaxy formation at redshifts of order 1. The red B-V colors are reached slowly in Bruzual's models. This

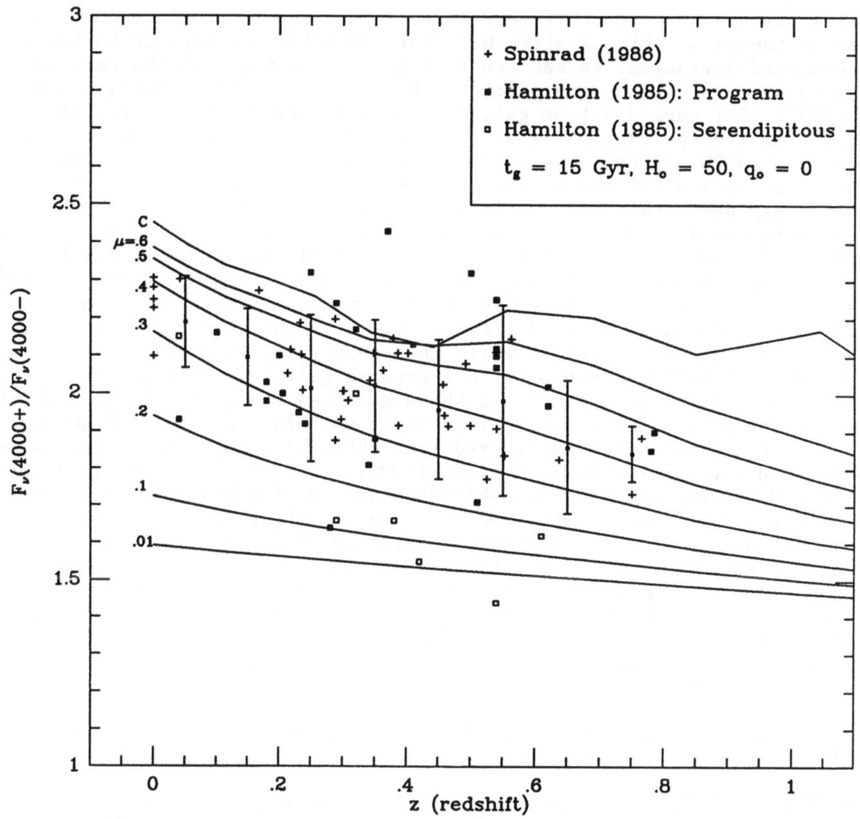

Fig. 2. Amplitude B(4000) for the galaxies measured by HAMILTON [28] and SPINRAD [22]. The expected behavior of this discontinuity according to a set of the evolutionary models of BRUZUAL [40] using the Salpeter IMF is shown for the indicated cosmology. Each model is identified by the number in the left side of each solid line. See text for details.

requires higher z's of galaxy formation. In WYSE's [30] models there still remains a systematic difference in the predicted B-V color for present age ellipticals. Her models only reach B-V = 0.92. This is most likely due to the choice of the bolometric corrections and colors for the stars used in her library. To what extent the behavior shown in Fig. 1 depends on this choice should be explored in more detail.

HAMILTON [28] studied a sample of galaxies chosen by photographic color. He chose at any redshift the galaxies with the reddest J-F (optical J) color defined by Koo [14]. He used the amplitude B(4000) of the spectral discontinuity at 4000 A [7,22] as an indicator of spectral evolution. He concludes that there is no evidence for spectral evolution in his sample up to z = 0.8, contrary to SPINRAD's [22] results.

In Fig. 2 I plot the amplitude B(4000) for the 33 galaxies measured by HAMILTON [28] and the 38 galaxies measured by SPINRAD [22]. The expected behavior of this discontinuity according to a set of the evolutionary models of BRUZUAL [40] is shown for the indicated cosmology. Except for some of Hamilton's program galaxies, both data sets seem to be well mixed and it may be meaningful to average the values of B(4000) measured at a given z. The small squares with error bars represent the average value of B(4000) for the galaxies inside a z-bin of width 0.1. The error bar corresponds to twice the standard deviation. We can see that there is a trend in the data in the same sense as the models predict. This result is less significant if HAMILTON's [29] serendipitous galaxies are not included in the sample.

HAMILTON [29] program galaxies up to $z = 0.8$ have similar spectra and values of B(4000) to nearby galaxies. This indicates that the reddest galaxies up to $z = 0.8$ are as red as nearby ellipticals. This fact by itself does not say anything about spectral evolution. Simply, chosing the reddest galaxies at any z insures us in advance that the galaxies will have similar spectra. In this sense, the lack of spectral evolution found by HAMILTON [29] is due to his selection criterium. Galaxies selected without color criterium (serendipitous) are found in different evolutionary stages, and show values of B(4000) characteristic of different stellar population mixtures. These galaxies will not necessarily evolve into nearby ellipticals.

However, HAMILTON's result for his program galaxies still poses the question of how the populations in these galaxies manage to evolve without changing the galaxy s.e.d. There is a tight correlation between B(4000) and B-V. This correlation can be derived from BRUZUAL's [40] models. For galaxy models with $0.9 < B-V < 0.98$, I find $B(4000) = 3.25(B-V) - 0.76$. This implies that the 3 program galaxies in HAMILTON's sample with $B(4000) > 2.3$, have rest frame B-V colors between 0.94 and 0.98. As shown in Fig. 1, the B-V color of model galaxies when the AGB stars are included remains at high values for most of the life of the galaxy (the exact values depend on the stellar spectra chosen). Thus, the expected signature of spectral evolution represented by the decrease of B(4000) due to the contribution to the s.e.d. of the main sequence stars, may be hidden by the evolution of other stellar groups, such as the AGB. The question of why some galaxies do seem to have low values of B(4000) still remains.

<u>BRUZUAL's</u> [7,40] evolutionary models have been compared to a large number of observations of distant galaxies. The results of some of the most recent comparisons are listed here.

WINDHORST, KOO, and SPINRAD [27] studied weak radio galaxies. In their sample they detect luminosity evolution in the optical F band. They do not detect color evolution in J-F (optical J) for the elliptical (red) radio galaxies. The blue radio galaxies have the colors of galaxies that will evolve into galaxies as nearby spirals. At a given z, galaxies in different evolutionary stages coexist. From the reddest galaxies in their sample they constrain the epoch of galaxy formation and derive ages of 14 - 15 Gyr for elliptical galaxies. This age is model dependent, as explained in 4. As with all active galaxies, there is the possibility that non-thermal radiation contributes to the s.e.d. In this case the comparison with [7,40] is weakened, as the models only include stellar sources.

LEBOFSKY and EISENHARDT [35] obtained near IR photometry of 61 elliptical galaxies up to z of 1.1. They find no color evolution in the near IR. There is luminosity evolution in their sample. Galaxies at z of 1 are brighter in the K band than nearby ones. The scatter in their data is partially due to observational errors (pointing accuracy), but part of it is intrinsic to the galaxies. This implies that there are differences between the stellar populations sampled on each galaxy. BRUZUAL's [7,40] models based on the local red giant luminosity function [31] fail to explain this scatter for any choice of the SFR and the IMF.

DJORGOVSKI and SPINRAD [23 and references therein] have studied the colors of the optical counterparts of the 3CR radio galaxies up to $z = 1.8$. SPINRAD [22] argues in favor of considering 3CR sources as normal galaxies that can be compared with nearby elliptical galaxies, despite the unknown source of the radio power. They detect substantial amounts of evolution in the 3CR galaxies. The V-R color and the V magnitude change with redshift cannot be accounted for just by the K-correction derived from nearby ellipticals. In

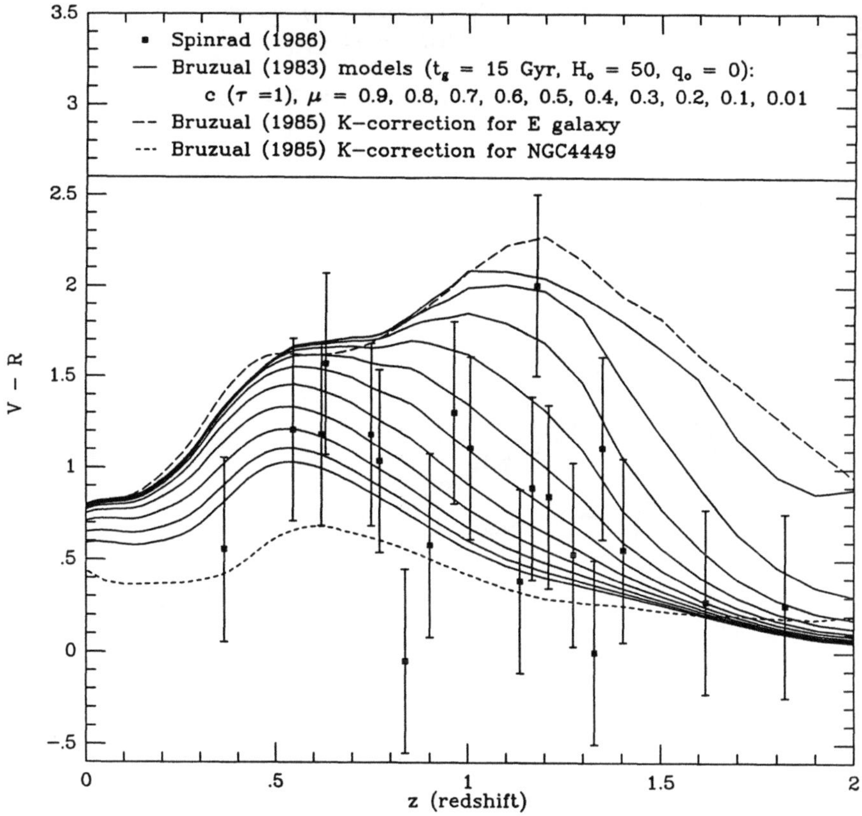

Fig. 3. V-R vs z for the 3CR sources [22]. The expected behavior of this color according to a set of models of BRUZUAL [40] is shown for the indicated cosmology. The different models used are listed in the top part of the figure (listed from reddest to bluest). All models were computed using the Salpeter IMF.

Fig. 3 I compare the V-R data [22] with evolutionary model predictions. The models and the cosmology used are indicated inside the figure. The V-R colors of most of the galaxies in this sample follow the behavior of BRUZUAL [7,40] μ= 0.01 - 0.6 models. The very low μ models are almost unevolving and remain very blue for the lifetime of the galaxy. Some galaxies are as blue as NGC4449 [36] will look at their respective z. However, it is not clear if the 3CR sources are representative of normal galaxies (dominated by stellar populations). In particular, we do not know their local counterparts (see [22 and DJORGOVSKI, this conference] for more details).

The amount of spectral and luminosity evolution implied by the V-R data is quite large. This is apparent from Fig. 4 in which I plot, in the galaxy rest frame, the s.e.d.'s corresponding to the μ = 0.50 model (Salpeter IMF) at age = 2, 3, 5, 8, and 15 Gyr. For the cosmology indicated in Fig. 3, these s.e.d.'s will be seen at approximately z = 2, 1.5, 1, 0.5, and 0, respectively.

5. Conclusions

The detection of spectral evolution of galaxies remains still uncertain. In the recent past the following facts have emerged:

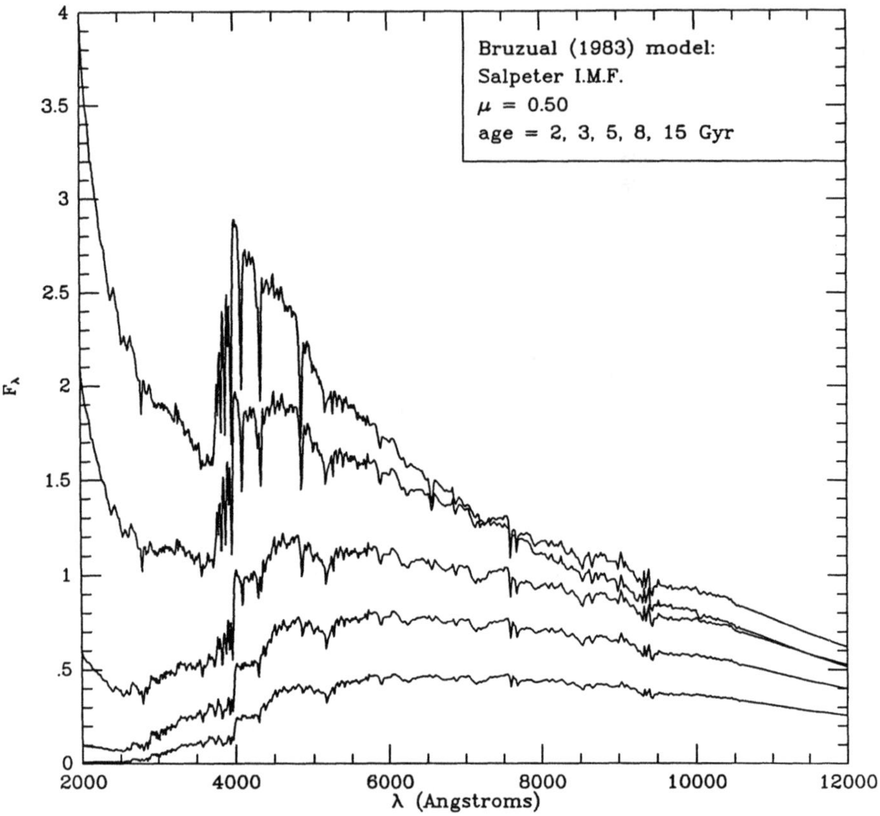

Fig. 4. S.e.d.'s corresponding to the $\mu = 0.50$ model (Salpeter IMF) at age = 2, 3, 5, 8, and 15 Gyr in the galaxy rest frame.

- Galaxies bluer than expected from the K-correction of nearby elliptical galaxies have been observed up to z of order 1. However, up to the same z's there are galaxies which show no sign (or very little sign) of evolution, and are as red as a K-corrected elliptical. Thus, galaxies of different types coexist at a given epoch, as they do at present.
- The problem of how to select a sample of truly identical galaxies in order to study their evolution has not been solved yet.
- Any claim about the detection or not of spectral evolution depends on the sample being studied and on the volume of space being sampled. It should not be generalized to all galaxies in the universe.
- Spectral and population evolution seems to be a slow process. The early stages of fast galaxy evolution occur long before the epoch that is being sampled.
- The age of galaxies deduced from synthesis models depends on the set of evolutionary tracks used to build the models. In general, a synthetic single population will be representative of the population in an elliptical galaxy at an age similar to that derived from the same set of tracks for the galactic globular clusters.
- There seems to be real scatter in the luminosity function of red giant stars responsible for the near IR colors.
- It is imperative that the existing spectral evolution models be generalized to include a mixture of stellar metallicities, more complete sets of evolu-

tionary tracks, and better stellar s.e.d.'s.
- Galaxy interaction with its environment surely plays a role in spectral evolution, but its treatment adds unlimited complications to the problem.

References
1. Bertola, F. 1986, in Spectral Evolution of Galaxies, eds. C. Chiosi and A. Renzini, (Dordrecht: Reidel), p. 363.
2. Wu, C. C., Faber, S. M., Gallagher, J. S., Peck, M., and Tinsley, B. M. 1980, Ap. J., 237, 290.
3. O'Connell, R. W. 1986, in Spectral Evolution of Galaxies, eds. C. Chiosi and A. Renzini, (Dordrecht: Reidel), p. 321.
4. O'Connell, R. W. 1983, Ap. J., 267, 80.
5. Pickles, A. 1986, in Spectral Evolution of Galaxies, eds. C. Chiosi and A. Renzini, (Dordrecht: Reidel), p. 345.
6. Pickles, A. 1985, Ap. J., 296, 340.
7. Bruzual A., G. 1983, Ap. J., 273, 105.
8. Bertola, F., Burstein, D., Buson, L. M., Faber, S., M., and Lauer, T. R. 1987, in preparation.
9. Faber, S. M. 1983, in Highlights of Astronomy, 6, 165.
10. Frogel, J. A. 1986, (this conference).
11. Whitford, A. E. 1986, in Spectral Evolution of Galaxies, eds. C. Chiosi and A. Renzini, (Dordrecht: Reidel), p. 157.
12. Rieke, M. J. L. 1986, (this conference).
13. Kron, R. G 1978, Ph.D. thesis, University of California, Berkeley.
14. Koo, D. C. 1986, Ap. J., (in press).
15. Tyson, J. A., and Jarvis, J. F. 1979, Ap. J. (Letters), 230, L153.
16. Ellis, R. S. 1982, in The Origin and Evolution of Galaxies, eds. B. J. T. Jones and J. E. Jones (Dordrecht: Reidel).
17. Oke, J. B., and Wilkinson, A. 1978, Ap. J., 220, 376.
18. Dressler, A., and Gunn, J. E. 1983, Ap. J., 270, 7.
19. Koo, D. C. 1981, Ap. J. (Letters), 280, L43.
20. Dressler, A. 1986 (this conference).
21. Schild, R. 1984, Ap. J., 286, 450.
22. Spinrad, H. 1986, P. A. S. P., 98, 269.
23. Djorgovski, S., and Spinrad, H. 1986, Ap. J., (in press).
24. Koo, D. C. 1985, A. J., 90, 418.
25. Kron, R. G., Koo, D. C., and Windhorst, R. A. 1985, A. A., 146, 38.
26. Windhorst, R. A. 1984, Ph.D. thesis, University of Leiden.
27. Windhorst, R. A., Koo, D. C., and Spinrad, H. 1986, in Galaxy Distances and Deviations from Universal Expansion, ed. B. F. Madore, (Dordrecht: Reidel).
28. Hamilton, D. 1985, Ap. J., 297, 371.
29. Renzini, A., and Buzzoni, A. 1986, in Spectral Evolution of Galaxies, eds. C. Chiosi and A. Renzini, (Dordrecht: Reidel), p. 195.
30. Wyse, R. F. G. 1985, Ap. J., 299, 593.
31. Tinsley, B. M., and Gunn, J. E. 1976, Ap. J., 203, 52.
32. Searle, L. 1984, in Structure and Evolution of the Magellanic Clouds, IAU Symposium No. 108, eds. S. van den Bergh and K. S. de Boer, (Dordrecht: Reidel), p. 13.
33. Burstein, D., Faber, S., Gaskell, C., Krumm, M. 1984, Ap. J., 287, 586.
34. Lilly, S. J., and Longair, M. S. 1984, M. N. R. A. S., 211, 833.
35. Lebofsky, M. J., and Eisenhardt, P. R. M. 1986, Ap. J., 300, 151.
36. Bruzual A., G. 1985, Rev. Mex. Astron. y Astrofis., 10, 55.
37. Barbaro, G., and Olivi, F. M. 1986, in Spectral Evolution of Galaxies, eds. C. Chiosi and A. Renzini, (Dordrecht: Reidel), p. 283.
38. Arimoto, N. and Yoshii, Y. 1986, in Spectral Evolution of Galaxies, eds. C. Chiosi and A. Renzini, (Dordrecht: Reidel), p. 309.
39. Arimoto, N. and Yoshii, Y. 1986, A. A., (in press).
40. Bruzual A., G. 1986, (in preparation).
41. Gunn, J. E., and Stryker, L. L. 1983, Ap. J. Suppl., 52, 121.

Evolution of Cluster Galaxies Since z = 1

A. DRESSLER

1. Using Cluster Populations to Investigate Galaxy Evolution

Evidence for galaxy evolution based on observations of galaxies at great lookback times continues to be incomplete and difficult to interpret. The 3C radio samples studied by DJORGOVSKY and SPINRAD [1], LILLY and LONGAIR [2], and LEBOVSKY and EISENHARDT [3], show substantial color evolution when compared with Bruzual's models, but there still remains some question as to whether these are representative of galaxies in general. We have also heard at this workshop a report by M. Yates who claims that, at least for the Lilly and Longair sample, a selection effect based on radio power accounts for most if not all of the observed "evolution." Similarly strong claims for evolution have been based on the number density and colors of very faint ($m_r > 22$) galaxies, but interpretation of these data is very model dependent, for example, redshift distributions are as yet unknown.

I use these examples to reiterate the point made by Bruzual: the principal difficulty in this area is the selection of representative objects. How do we make sure that galaxies in our high-z samples are typical of the low-z galaxies with which they are compared? Clusters of galaxies provide perhaps the best tool at the present time. Because of their high densities and predominance of early type galaxies, clusters do not offer us a sample of average galaxies in typical environments, but they do provide a volume limited sample of galaxies under similar conditions for a range of epoch $z \lesssim 1$. Though there may have been some question at first of whether distant clusters were, by selection, all extraordinarily rich or dense, more complete surveys like that done by GUNN, HOESSEL, and OKE [4], have provided us with clusters that appear, based on their space densities and luminosity functions, to be the ancestors of at least the richer present-epoch clusters like Coma or Hercules.

A Greater Fraction of Active Galaxies in High-z Clusters

Based on studies of the colors of galaxies in clusters, BUTCHER and OEMLER [5,6] claimed that, in comparison to the low-z clusters, high-z clusters contain a larger fraction of galaxies bluer than that expected for a purely old stellar population. In order to test this result, which rested on the assumption of cluster membership in samples where field contamination is necessarily large, Gunn and I began a spectroscopic study of high-z cluster galaxies in order to verify the membership of the galaxies in question and to identify the reason for their bluer colors. After five years and about 25 clear nights with the Hale 5m and TI CCD's, we have accumulated spectra with redshifts for 215 galaxies $20 < m_r < 23$, of which 135 are members of six clusters we have been studying. I will discuss these data here, which basically confirm the conclusion of Butcher and Oemler: clusters at $z \gtrsim 0.4$ have a higher frequency of "active" galaxies, i.e., those with signs of recent star formation or nuclear activity. These are to be contrasted with what I will call "passive" galaxies, those with a KIII spectrum typical of an old stellar population, rest frame $B-V \approx 0.9$, and little or no sign of star formation within the previous 5

Gyr. At the present epoch these are the spectral characteristics of E or S0 galaxies which account for about 95% of the galaxies in the inner regions of dense, concentrated clusters. By z ~ 0.5, this fraction has dropped to about 80%.

The spectra of the "active" population can be divided into those with and without emission lines. Those with emission lines include galaxies with long-term, relatively steady star formation like spirals, "starburst" galaxies with greatly enhanced star formation rates (SFRs), and high-excitation spectra typical of active nuclei (AGNs). Examples of these are shown in DRESSLER, GUNN, and SCHNEIDER [7]. The other common type, first noted by DRESSLER and GUNN [8] in the population of the 3C295 cluster, is a basically old stellar spectrum, perhaps about 0.2 mag bluer than a typical passive galaxy, with strong Balmer absorption lines and little or no emission. We called these "E+A" spectra because one could match them by adding A stars to an elliptical-type (passive) spectrum. I have from time to time also called these "post-starburst galaxies," based on our interpretation that many of these galaxies had a significant increase in the SFR which subsided about 1 Gyr before the epoch of observation.

Out of the sample of 135 galaxies in six clusters with $0.37 < z < 0.75$, 31 are emission-line galaxies (23%) and 22 have E+A spectra (16%). However, we have undersampled passive (red) galaxies by about a factor of two (by preferentially choosing bluer galaxies), which lowers these fractions to ~14% for the emission-line galaxies and ~10% for the E+A type. How do these percentages compare with the occurrence of such types in present-epoch clusters? DRESSLER, THOMPSON, and SHECTMAN [9] found a fraction of 7% for such emission-line galaxies in a sample of about 1000 galaxies in low-z clusters. The distribution of luminosity and average surface brightness for this low-z sample studied by Shectman and me, which I will make more use of below, is fortunately very similar to the characteristics of the high-z sample. However, the low-z sample comes from an area in each cluster that is about four times as large as the area covered in the distant clusters, so I estimate that the appropriate fraction is certainly less than 5%. This implies an increase of at least a factor of three in the frequency of emission-line galaxies at $z \gtrsim 0.4$ compared to present-epoch clusters.

The E+A case is more dramatic. I identified 20 candidate objects from about 1000 low signal-to-noise spectra from the Dressler-Shectman sample, and then obtained better spectra for these with the du Pont telescope at LCO. I found that only 3 or 4 were comparable to the high-z E+A galaxies; from our low-z sample, at least, the fraction is under 1%! Therefore, an order of magnitude increase is found in the frequency of E+A galaxies in high-z clusters. This is not to say that there are not examples of such galaxies in both clusters and the field at low-z, but that they are relatively rare.

It seems, then, that active galaxies make up a substantially larger, though still small, fraction of high-z clusters when compared with their low-z analogs. Although the relative contribution of emission-line vs. E+A spectra varies somewhat from cluster-to-cluster, there is no compelling evidence that this variation is fundamental. It could be due simply to an accident of timing. If, for example, many of the emission-line galaxies are starbursts, and these turn into E+A galaxies for 1 Gyr or so, the variation in frequency may only reflect whether a cluster has been observed during, immediately after, or well after, a period of activity. A particularly interesting example is the cluster Cl0016+16 (z=0.54), discovered by Kron and claimed by KOO [10] to have a negligible blue population. As such it was believed to be a counter-example of the Butcher-Oemler effect. Among our spectroscopic sample of 33 cluster members, Gunn and I have found 9 E+A spectra, the largest number of these found in any cluster. Because these "active" galaxies are not very blue, it is not surprising that the study of galaxy colors in Cl0016+16 by Koo did not uncover this population. But the spectroscopic data reveal that the population of this extremely rich cluster, far from being like that of a present-epoch cluster like Coma, is the most unusual we have found. One Gyr earlier its color distribution must have looked strikingly different.

The 4000 Angstrom Break as an Indicator of Galaxy Evolution

SPINRAD [11] pioneered use of the amplitude of the 4000 Å break as an indicator of spectral evolution and found a large evolutionary effect for a heterogeneous sample. On the other hand, HAMILTON [12], who has recently published results of a more systematic attempt to measure the amplitude of the break as a f(z), found negligible evolution. However, as BRUZUAL [13] has pointed out, Hamilton's procedure of selecting only the reddest galaxies at every epoch carries a built in selection effect if the break amplitude correlates with color (as it is shown to in Hamilton's own paper) and if, as the volume increases with z, redder (and presumably more luminous) galaxies are preferentially selected. After looking at Hamilton's data and the Heckman sample which he compiles, I suggest that his nearby sample is underrepresented in the reddest, most luminous galaxies (those with break amplitudes ~2.5), and conversely his most distant galaxies are likely unusually red. This will be difficult to verify until the data themselves (magnitudes and colors) are published.

Our cluster sample again provides a unique arena in which to investigate this issue. I measured the break amplitude, using Hamilton's definition, from spectral plots of galaxies in four of the distant clusters. Though this is not a highly accurate way of proceeding, the large excursions in these spectra due to poor subtraction of night sky lines or cosmic ray events limit the applicability of the otherwise more accurate method of actually averaging the counts. I estimate that a typical by-eye measurement of this sort is good to 20%, which should be adequate for statistical purposes. I also applied this procedure to about 200 spectra of galaxies in A1644 and A548, two low-z Abell clusters. Figure 1 shows histograms of these data. Specifically, the Abell sample includes all galaxies for which Shectman and I have spectra, over 90% of which would be called "passive." For the four other clusters, I have included only the passive spectra in Fig. 1.

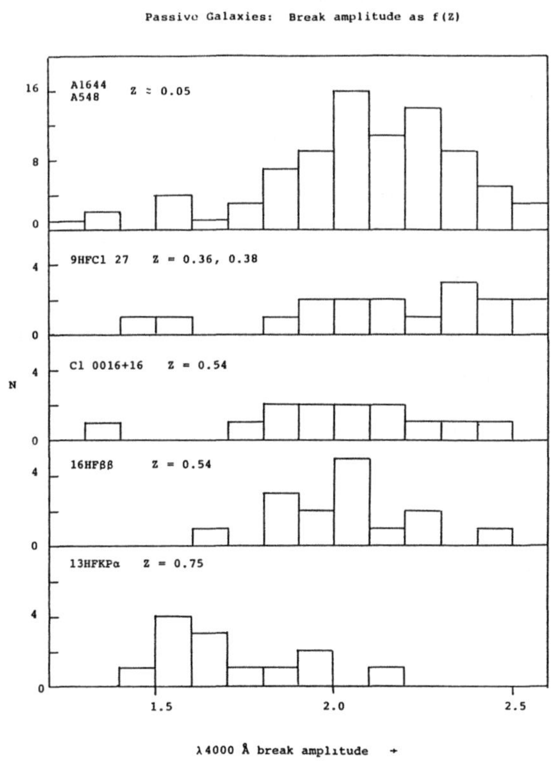

Fig. 1: Amplitude of the λ4000 Å break for galaxies in low- and high-redshift clusters. The sample in A1644 and A548 includes both active and passive galaxies, though there are very few of the former. The other histograms show the distribution for passive galaxies only. There is at most a weak trend from the low-z to intermediate-z sample, but the cluster at z = 0.75 shows a clear shift in its distribution.

Although there is little or no trend up to z ~ 0.5, in agreement with Hamilton's conclusion, there seems to be a significant effect by z = 0.75. There are two additional reasons to take the z = 0.75 data very seriously. First, though these galaxies are very faint, their spectra are surprisingly good around the 4000 Å break since the spectrum of the sky at ~7000 Å is not very bright nor full of structure. Second, we have selected, in general, the brightest galaxies in the cluster because of the difficulty in obtaining redshifts for such distant objects. This means that, relative to the other three clusters (and even more so the low-z sample), the z = 0.75 should be biased towards the spectra with the largest break amplitude. Even with this selection effect working against us, the lower break amplitude is quite apparent. It is worth mentioning that Hamilton's data, similarly biased against finding such a trend, seems to show it as well (with admittedly small number statistics). From these data it seems quite clear that by z = 0.75, spectral evolution is evident for not only the active galaxies, but the passive galaxies as well.

In Fig. 2, I include all of our cluster galaxies, both active and passive. In order to do this, I counted each active galaxy at half weight, since, as mentioned earlier, passive galaxies are underrepresented by about a factor of two. I have added the data for the three intermediate-redshift clusters together to allow a more significant statistical comparison with the larger low-z sample. With these representative ensembles of cluster galaxies that cover the same range in luminosity and surface brightness, a strong epoch dependence emerges that is obvious even by z ~ 0.5. This is, in fact, the Butcher-Oemler effect displayed with spectroscopic data only, which removes the confusion of non-cluster galaxies.

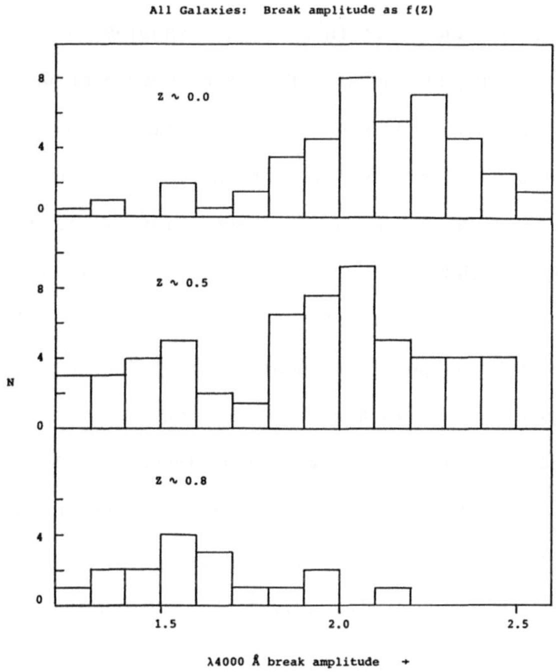

Fig. 2: Same as Fig. 1 but for all galaxies (both active and passive). Active galaxies in the intermediate- and high-z samples have been counted at half weight to compensate for the fact that passive galaxies are underrepresented as a result of the selection procedure. The three intermediate-z clusters have been combined to improve the statistics. A clear trend is now seen in that the break amplitude increases with increasing z. This can be thought of as the Butcher-Oemler effect derived from spectroscopic rather than photometric data, which, as a result, is uncontaminated by non-members.

What are the E+A Galaxies?

As we showed in our study of the E+A galaxies in the 3C295 cluster, the relatively red colors but strong Balmer absorption lines of these galaxies are indicative of a population of A stars added to an older, KIII population. The scarcity of O and B stars associated with ongoing star formation is indicated by the lack of forbidden emission from HII regions excited by such stars, and the low UV flux. These spectral differences set the E+A galaxies apart from those forming stars continuously such as normal spirals. Furthermore, we showed that the equivalent width of Hγ and Hδ in these first examples, W ~ 7-8 Å, is difficult to explain unless the star formation rate rose well above the past average. I should add now that this claim assumed: (1) a normal initial mass function and (2) that the galaxies in question were old, say ~10 Gyr, at the epoch of observation. It is clear that a much younger galaxy that had been forming stars continuously (and vigorously) for only several billion years, and then stopped, would be indistinguishable from a "post-starburst galaxy." This point is relevant to the question of whether the SFR actually increased, or was merely terminated, as might happen with a stripped spiral. Similarly, a galaxy with a high rate of ongoing star formation which included no very massive stars might be identified as a post-starburst galaxy.

These points are shown graphically in Fig. 3, which plots the strength of the Balmer absorption lines vs. B-V color for coeval stellar populations with different star formation histories. These models, constructed with synthetic spectra from A. Manduca, all start at lookback time of 15 Gyr. For a galaxy with a constant rate of star formation from t=0, as might be appropriate for a present-epoch Sc galaxy, the Balmer lines decrease from 10 Å at about 4 Gyr (B-V ~ 0.3) to about 6 Å at 14 Gyr (B-V ~ 0.6). I have indicated on the plot the trajectory followed if star formation is abruptly terminated at 4, 9, or 14 Gyrs. From these it appears that such a situation is only able to produce W ~ 6 Å Balmer lines if the typically observed color B-V ~ 0.7 is to be matched. This is to be compared with the distribution of Balmer line strengths for all 23 examples we have found, shown on the right hand side of Fig. 2), for which less than half have strengths of 6 Å or less. In order to achieve stronger Balmer lines a burst is necessary, as shown by the curved track in Fig. 3 for which a high SFR was introduced from 8-9 Gyr into a passive galaxy which formed the rest of its stars at 0-1 Gyr. Through this path, quite high Balmer line strengths can be reached for 1-2 Gyr in galaxies that remain fairly red. As shown in the diagram, little trace of this era of activity would be visible in its present-epoch descendant.

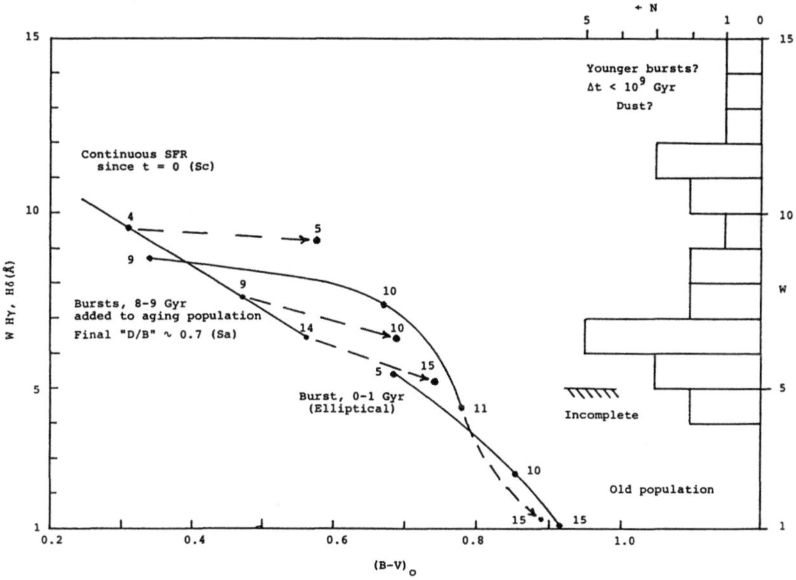

Fig. 3: Models of the rest-frame B-V color vs. the average equivalent width of Hγ and Hδ absorption for galaxies with different star formation histories. The numbers next to the points give the age of the systems over a time scale 0-15 Gyr. The histogram on the right shows the distribution of the equivalent widths for 23 E+A spectra observed in the high-z clusters. B-V = 0.7 and W ≲ 6 Å are characteristics of systems with constant SFRs since t = 0 which have been abruptly terminated 1 Gyr before the epoch of observation. Stronger Balmer absorption, as is quite common in the data, seems to require a burst of star formation (SFR rises above the past average).

I am not certain that even this mechanism can account for "yellow" galaxies with W ≳ 10 Å. Perhaps internal reddening plays a role in these cases. But it is important to note that with so many examples of strong Balmer absorption, one expects that these will decay slowly into the weaker examples, which may account for the number of W < 7 Å examples. In other words, if bursts are required to explain the strongest examples, they probably account for the weaker cases too, leaving little or no room for a model in which some E+A galaxies are produced without bursts, like suddenly terminating star formation in an Sc.

But if E+A galaxies are post-starburst galaxies, what has caused the starbursts, and why are they more prevalent in high-z clusters? Plunging headlong into the realm of pure speculation, Gunn and I [8] suggested that gas-rich galaxies falling into a dense intracluster medium might well be stimulated to form stars at a markedly increased rate. BOTHUN and DRESSLER [14] may have found low-z examples of this phenomenon in the Coma cluster, though most of these are less luminous than the high-z examples. Figure 4 presents some circumstantial evidence that such a velocity-dependent process may be responsible for both active starburst and post-starburst galaxies. Combining all the data for six clusters, it is clear that active galaxies in the sample have a much larger velocity dispersion than the passive galaxies. We also note that the active galaxies in Cl0016+16 seem to map out a doughnut shape with a diameter ~1 Mpc, as if activity is triggered when a galaxy reaches a certain radius in the cluster. This is, again, circumstantial evidence that the action begins when the highest density of the intracluster gas is encountered.

Since a cluster that formed at high-z has a deeper potential well, and thus higher velocity dispersion and denser intracluster medium, than a cluster forming today, it might be reasonable to expect that this ram-pressure induced star formation would decline significantly with time. Declining gas fractions, as galaxies use up their reservoirs, should also work in the same direction. Whether this is enough to explain the dramatic decrease in E+A frequency since $z = 0.5$ is the toughest question faced by this model.

Of course, galaxies might simply be more prone to starbursts in the past completely independent of environmental conditions, perhaps due to the putative higher gas fractions. A test of this idea that internal processes alone are responsible for the increased likelihood of a starburst seems straightforward. One simply compares the statistics for clusters to field galaxies immune to the enviromental effects. In practice, however, selection effects play a big role in determining how many E+A galaxies will be picked up since they occur in a narrow color range, and the k-corrections for galaxies with a broad redshift distribution are not known a priori. Thus it is unclear at this time what large, high-z field samples like that of Kron and Koo have to say about this matter.

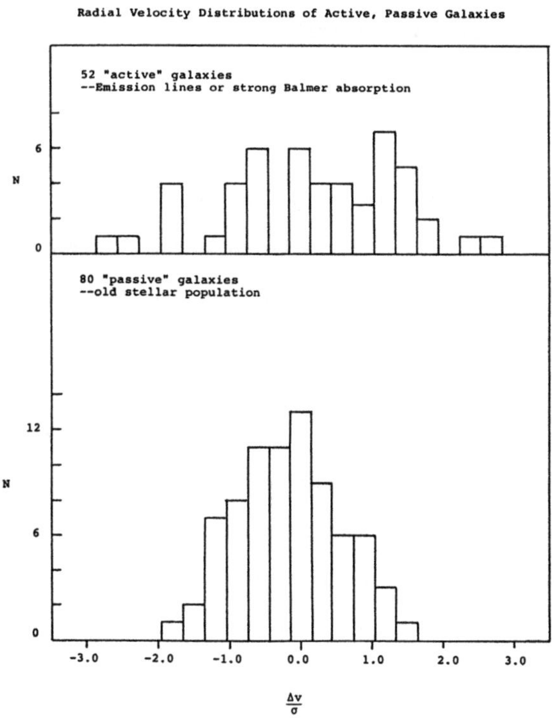

Fig. 4: Radial velocity distributions, expressed in terms of $\Delta V/\sigma$, where $\Delta V = V_{gal} - V_{clust}$ and σ is the cluster velocity dispersion. The active galaxies show a much higher dispersion, indicative of a correlation between galaxy activity (star formation or nuclear) and kinematics in the cluster. This would be the case, for example, if strong ram-pressure on the galaxy from the ICM, or high velocity encounters with other galaxies, triggered the activity.

Nor can we rule out the possibility that mergers or violent encounters are driving the activity in the cluster galaxies. Although one might think that the velocities of approach are generally too high in clusters, such interactions might be occurring in subclusters where the dispersion is a factor of two lower. The poor resolution from the ground of the present sample reveals only that a non-negligible fraction of the active galaxies have close companions. One expects that images with HST will be decisive in resolving this question. They could be crucial in testing the ram-pressure induced star formation model as well, since the pattern of star formation in such cases is unlikely to be as well-ordered as the spiral structure seen in galaxies with continuous star formation.

Summary

Cluster galaxies do evolve. There is a higher fraction of active galaxies in clusters at $z \gtrsim 0.4$, although these clusters are still dominated by passive galaxies. Judging from the amplitude of the 4000 Å break, even the passive galaxies show evidence for evolution by $z = 0.75$.

Cluster-to-cluster variations in the types of active galaxies may be significant, but could be a only an accident of the epoch of observation of a population that is in a state of flux. For example, many of the AGN and active starburst galaxies may signal an epoch of activity (like the infall of a subcluster of gas-rich galaxies), which is later seen as an increase in E+A galaxies. We have suggested that E+A galaxies are post-starburst galaxies, perhaps produced by ram-pressure induced star formation. Other models, such as galaxy interactions or environment independent bursting are also consistent with the data at this time.

Crucial questions remain. What is the fraction and makeup of active galaxies in the field population? What are the Hubble types of the active galaxies? Do internal parameters like gas fraction or the initial mass function explain the major component of the observed evolution with lookback time? These and related issues provide an unnecessary reminder that our investigation of galaxy evolution is just beginning.

References

1. S. Djorgovski, H. Spinrad: Ap. J. in press (1986)
2. S. J. Lilly, M. S. Longair: M.N.R.A.S. 211, 833 (1984)
3. M. J. Lebofsky, P. R. M. Eisenhardt: Ap. J. 300, 151 (1986)
4. J. E. Gunn, J. G. Hoessel, J. B. Oke: Ap. J. 306, 30 (1986)
5. H. Butcher, A. Oemler: Ap. J. 219, 18 (1978)
6. H. Butcher, A. Oemler: Ap. J. 285, 426 (1984)
7. A. Dressler, J. E. Gunn, D. P. Schneider: Ap. J. 294, 70 (1985)
8. A. Dressler, J. E. Gunn: Ap. J. 270, 7 (1983)
9. A. Dressler, I. B. Thompson, S. A. Shectman: Ap. J. 288, 481. (1985)
10. D. C. Koo: Ap. J. Letters 251, L75 (1981)
11. H. Spinrad: "Spectroscopy and Photometry of Faint Galaxies" in G. O. Abell and P. J. E. Peebles, eds.: IAU Symposium 92, Objects of High Redshift, pp. 39-48 Reidel, Dordrecht 1980
12. D. Hamilton: Ap. J. 297, 371 (1985)
13. G. Bruzual A.: this conference (1986)
14. G. D. Bothun, A. Dressler: Ap. J. 301, 57 (1986)

Galaxies at Very High Redshifts (z > 1)

S. DJORGOVSKI

1. Introduction

Observational cosmology is partly based upon wishful thinking. (I am not even going to hint on what is the theoretical cosmology based.) This is, perhaps, inevitable: most, if not all, cosmological tests seem to require idealized test particles or conditions, and there is much skill in coming up with a setup which is the least unrealistically oversimplified. Some classical tests, such as the m *vs.* z diagram, require the "standard candles", that is, sources of light whose intrinsic behavior is well understood, and which span a large baseline in redshift. QSO's are much too finicky, variable, and poorly understood to serve in this role; and ever since the days of Hubble, giant elliptical galaxies were a favorite choice. They seem to qualify, since they are luminous, and can be spotted at high redshifts; this is made easier by the facts that they are often found in rich clusters, and sometimes have powerful radio lobes. Moreover, here and now ($z \simeq 0$) they seem to live apparently clean and quiet lives, with no ongoing star formation, no optically thick dust, etc. On a closer look this simplicity disappears, but we cannot be too picky now...

Galaxies are made of stars, and since the stars evolve, the spectral energy distributions (SED's) of galaxies must change in time. The principal unknown is the star formation history, but the chemical differences and the shape of the IMF play important roles as well. In addition, there are possible dynamical evolution effects: merging, which by definition increases the amounts of the luminous material, and may either stimulate bursts of star formation, or extinguish any ongoing star formation in the participating galaxies. The evolutionary effects compete and mix with the subtle cosmological effects that were to be measured in the first place, and both are expected to increase with the redshift. Thus, some ten years ago, the focus of observational cosmology shifted to studies of galaxy evolution, as a necessary step before *any* galaxies can be used as cosmological probes. In a sense, cosmology moved from eschatology to Darwinism, with occasional monkeying and many trials on the way.

Understanding of galaxy evolution is probably even more interesting and important than the exact value of q_0. Much has been achieved with a wholesale approach, in which photometry is done on complete samples of thousands of field galaxies, though the redshifts are obtained only for a small fraction thereof, and typically reach only to $z \sim 0.6 - 0.8$ (cf. KRON [1], KOO [2], or BRUZUAL & KRON [3]). However, the evolutionary and cosmological effects become strong only at $z \sim 1$, and beyond, and there we are back to the special, giant galaxies. Galaxies at high redshifts are faint and hard to come by, and we have to satisfy ourselves with simple global measurements, such as their colors and magnitudes. Surely, as the technology advances, we will get more fancy. At the intermediate redshifts ($z \sim 0.5$, say), we can now investigate finer points of SED evolution; some of the best examples are from the work of DRESSLER & GUNN [4], discussed also by DRESSLER elsewhere in this volume. A good recent review of these topics was given by SPINRAD [5]. Here I will only briefly address the new results pertaining to galaxy evolution at high redshifts; I will talk more about the new phenomena which we are seeing beyond z ~ 1, and some exciting new prospects.

2. Radio Galaxies Far, Far Away and Long Time Ago

The problem of sample definition is an essential one: we need a sample of galaxies of the same kind at different redshifts, before any systematic detected changes can be attributed to evolution or cosmology. An alternative is a "cosmic conspiracy" in which galaxies of one kind are gradually replaced by the galaxies of another kind as the redshift increases; this is a logical *reductio ad absurdum* of observational cosmology, and no sample is guaranteed to be safe from this, not even the complete field samples. In other words, the selection effects must be well understood.

The 3CR sample may be a good one: it is a sample well defined through its *radio* properties, and at the low redshifts there is no known important correlation between the radio and the optical luminosities, at least for the narrow-line radio galaxies (NLRG), which will be the only ones considered here. The identification content of this sample is now almost complete (SPINRAD ET AL. [6]), and it is practically the only sample of galaxies reaching up to z ~ 1.8 . The 3CR NLRG at low and intermediate redshifts are invariably brightest cluster or group members, giant E or D galaxies. They are certainly a more heterogeneous bunch than the optically selected brightest cluster members (BCM's), but perhaps not too much so. With enough work, we can see clusters or groups around all 3C NLRG up to z ~ 0.9 or so. A tentative detection of a cluster around 3C324 (z=1.206) by SPINRAD & DJORGOVSKI [7] was recently confirmed by LE FEVRE [8] at CFHT. LEBOFSKY & EISENHARDT [9] find that there is no distinction between the powerful 3CR galaxies, intermediate-power radio galaxies, and radio-quiet brightest cluster members in their infrared properties, up to z ~ 1. Thus, our view is that 3CR galaxies at all redshifts are gE, D, or cD galaxies, or progenitors thereof, although something may be happening to them, which does not happen to all BCM's. After all, these are some of the most powerful radio sources in the Northern half of the observable Universe, and we must be wary.

Stellar populations, as they get older, become more and more similar. It is possible to have galaxies which look very similar today (gE, say), but had very different star formation histories. This may be the case here; we will never know about any possible wild past of NGC 4374 (= 3C272.1), or M87 (= 3C274). Among the low-z 3CR galaxies, Cyg-A (= 3C405) may be a good counterpart of its high-z cousins.

Assuming that the sample of 3CR NLRG is well defined, we can look at the behavior of their colors and luminosities with increasing redshift. Based on the *BVR* color and magnitude data, DJORGOVSKI ET AL. [10,11] have found very prominent evolution effects, in the sense that these galaxies were both brighter and bluer in the past. This may be interpreted as the evidence for a more vigorous star formation, which diminishes in time. Comparison with the evolution models by BRUZUAL [12] indicates that the preferred models are those in which the star formation rate (SFR) decays exponentially, with the *e*-folding times of the order of 1 – 1.5 Gy. The passive evolution models, in which the star formation is confined to the initial burst only (of 1 Gy duration, say), are too red and too faint. Somewhat similar results were achieved by LILLY & LONGAIR [13], who used near-IR data. They find less prominent evidence for an active evolution, but the redder part of the spectrum which they probe is considerably less sensitive to the presence of young stars, than the redshifted UV which we observe in *BVR*. It must be emphasized that the images of these distant galaxies are well resolved (several arcsec). Our good-seeing frames, surface brightness profiles, and color gradients do not show any indications of luminous active semistellar nuclei. Even though some of the high-ionization lines, when present, probably do originate in active nuclei, the strongest, low ionization lines are spread over the whole galaxy. Thus, we are confident that most of the detected light is starlight, and that the relative contributions from suspected active nuclei must be small or negligible.

We must be careful when comparing the data with the models, which are known to be perhaps too simple (RENZINI & BUZZONI [14], BRUZUAL, this conference). In fact, the models fit better than they have any right to do. We still do not understand what is the origin of UV light in z ~ 0 ellipticals [15,16]. We should not rush with detailed interpretations as yet, but it is probably safe to say that prominent evolutionary effects have been detected for the 3CR sample.

With the tremendous uncertainties about galaxy evolution, and the observed scatter, it is doubtful whether the Hubble diagram will ever be a viable cosmological test, at least in the observed visible light. One glimmer of hope is the near-IR: from the compiled data in the *K* band, SPINRAD & DJORGOVSKI [17] found that the evolutionary effects become small in comparison with the cosmological effects. It may be possible to introduce evolution corrections derived from the *BVR* work, where evolution dominates over the cosmology, and still attack the q_0/Λ_0 combination in the near-IR. One possible problem in all this may be existence of a so-far unknown correlation between the optical/IR luminosity of a galaxy, and the power in radio lobes. YATES ET AL. [18] suggested that there is such correlation for the highest-power radio galaxies. I do not find the evidence presented so far very persuasive, but this is an important question, and we need more data, more intermediate- and low-power radio galaxies at high redshifts. One very promising avenue is a study of the "1-Jy" sample of ALLINGTON-SMITH ET AL. [19], which reaches to the same high redshifts as the 3CR sample.

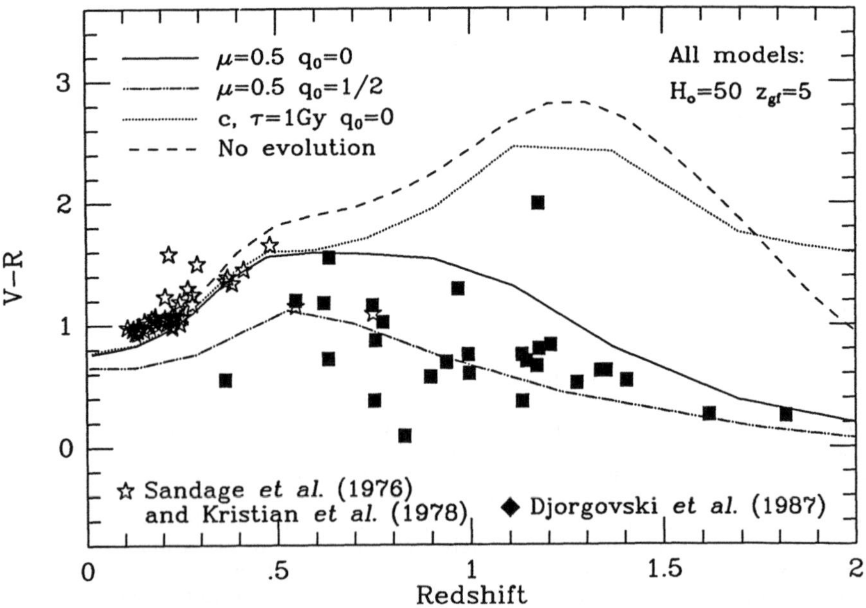

Figure 1. The observed $(V - R)$ color of 3CR radio-galaxies and some BCM's as a function of redshift. Several Bruzual models are drawn for comparison. Actively star-forming μ-models provide a better fit to the data than the passive *c*-models. The color–redshift diagrams are a safer evidence for evolution than the Hubble diagrams, since the colors are less affected by the aperture and seeing corrections, and possible luminosity selection effects, than the single band magnitudes alone.

It remains to be seen whether similar evolutionary effects persist for the low-power radio galaxies, and radio-quiet BCM's at high redshifts. For example, one possible interpretation (favored by LILLY & LONGAIR [13]) is that we are seeing sporadic bursts of star formation atop of passively evolving gE galaxies, which make the 3CR galaxies appear temporarily bluer and brighter. Our observed effects may be due to an exponentially decaying sequence of ever weaker starbursts, rather than to a smoothly changing SFR. This interpretation would require a good correlation between the radio emission from the lobes, and the star formation, but this need not be a direct causal connection:a possible mechanism would be gas-rich mergers, which would feed both the central engine, and stimulate the star bursts in the main body of the galaxy. This may be exactly what we are observing.

3. Mergers at High Redshifts: the Formation of cD Galaxies?

These distant 3CR galaxies often have elongated, or even multimodal shapes. While the multiple nuclei are a common occurrence in the low-z BCMs, most low-z radio galaxies tend to be quite round. If we position a spectrograph slit along the major axis of such galaxy, strong emission lines (mostly [O II] 3727, but sometimes also Mg II 2799 or [Ne III] 3869) become resolved, both spatially, and in velocity. One of the best examples is 3C368 (z=1.132), shown here in Fig. 2. The velocity fields

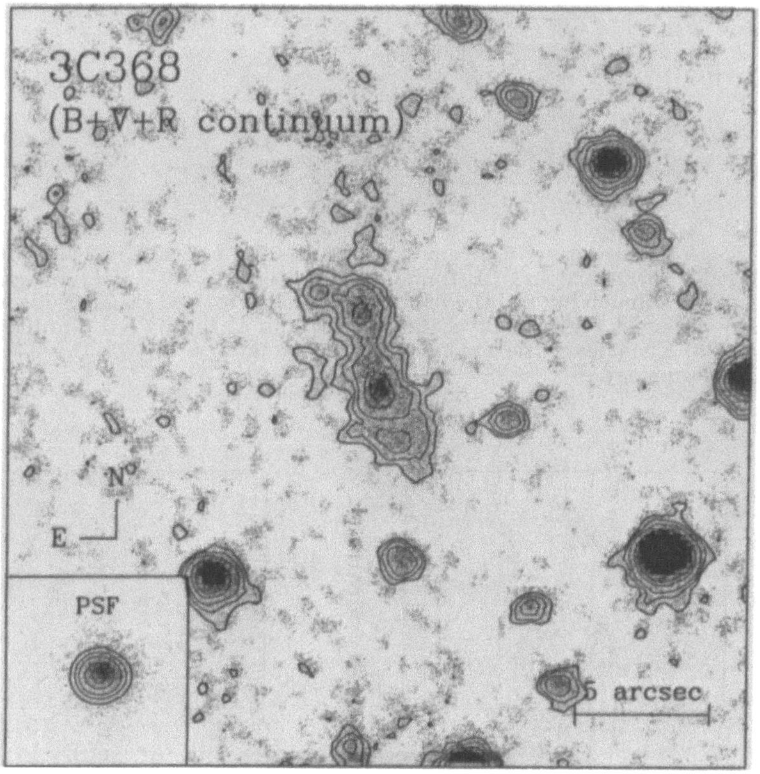

Figure 2. 3C368 (z=1.132) is an example of a multimodal galaxy with a strong [O II] emission associated with the continuum "blobs". We interpret such phenomena as most likely being due to spectacular dissipative mergers at high redshifts. This image is a stack of 3 10-minute RCA CCD exposures, obtained at CFHT in May 1986, in a half-arcsec seeing.

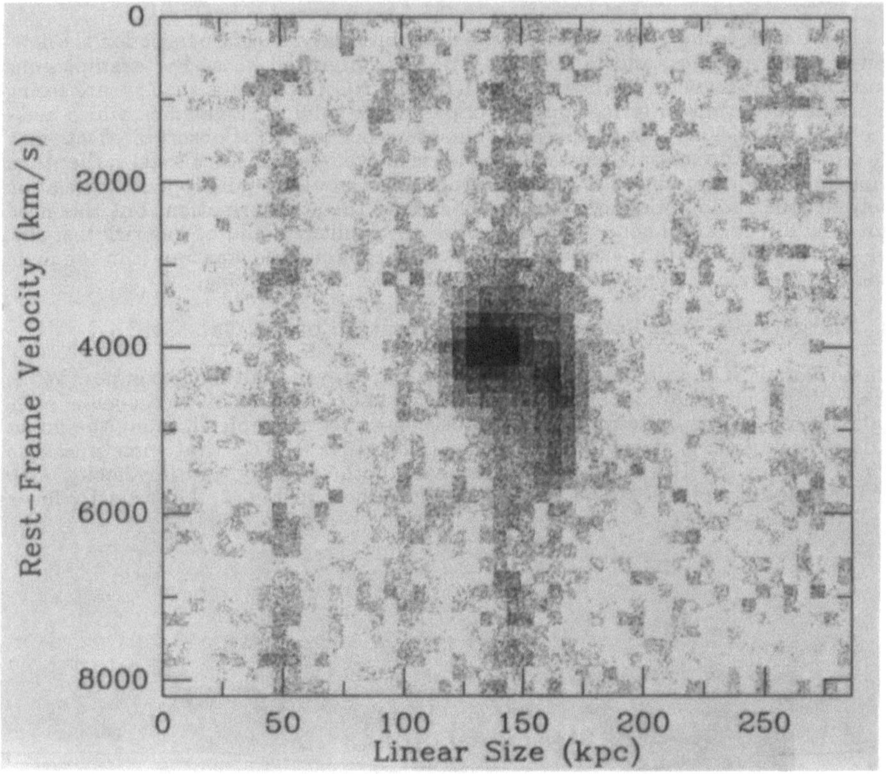

Figure 3. A "phase-space" diagram of the [O II] 3727 emission in 3C368. This is a section of a spectroscopic CCD frame, with the slit PA=10°, along the galaxy's major axis. Multimodality and the velocity shear in the ionized gas are evident. The data were obtained with the Cryocam spectrograph at the KPNO 4-m telescope, in May 1986. The linear scale was computed by assuming a H_0=75, Ω_0=0.3, Λ_0=0 Friedman cosmology.

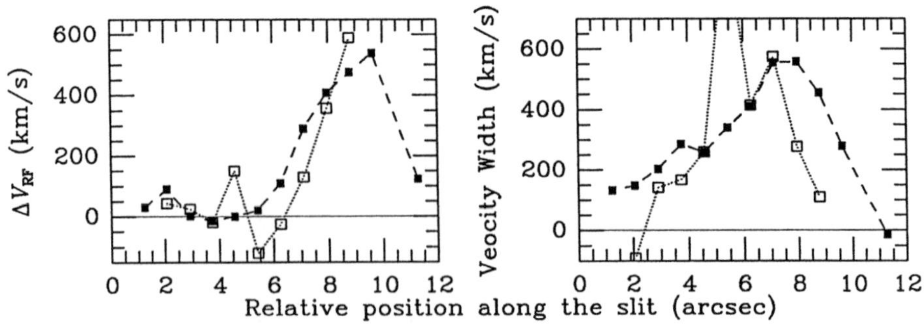

Figure 4. Velocity field (shear and turbulence) of the ionized gas in 3C368, from the spectroscopic data pictured above. Solid squares represent the [O II] 3737 emission line, and open squares the [Ne III] 3869 line.

of ionized gas (Fig. 3) span tens of kpc, and hundreds of km/s in the galaxy rest-frame, and sometimes even more (Fig. 4). The line luminosities reach a few times 10^{11} $L_\odot$ in [O II] alone! There is nothing even remotely comparable to this at the low redshifts.

One possible explanation is that we are detecting giant cooling flows in extremely distant clusters (FABIAN ET AL. [20]). However, a good correspondence between the line emission and the continuum condensations suggests that we are seeing spectacular dissipative mergers at high redshifts, in which the gas is collisionally excited. In other words, we may be witnessing the dynamical formation of cD galaxies. Clearly, such gas-rich mergers can initiate star formation over the whole giant galaxy, in agreement with our color and magnitude observations.

It is interesting that there seems to be a redshift cutoff near z $\sim$ 2 for the powerful 3CR sources. This was predicted by the models of source counts (WINDHORST [21]), and is increasingly corroborated by our observations.Perhaps the spectacular collisions which we see at z $\sim$ 1 – 1.8 are the trigger of radio activity in these objects, and the ignition of powerful radio sources seconds the formation of cD's?

4. Entering the Primeval Galaxy Land

To see the birth of galaxies is certainly one of the holy grails of modern cosmology. (I am ignoring here the objects such as I Zw 18, which are probably genuinely young, but a different kind of galaxies.) That mythical event is supposed to have happened at z > 4 (since we have not seen it unambiguously yet), $z_{gf} \simeq 5$ or 10 being the favorite guess numbers. That z_{gf} should be so high is already an important constraint for the top-down scenarios of large-scale structure formation. Direct observation of the formation of galaxies at high redshifts will certainly be one of the fundamental milestones of evolutionary cosmology; it will also tell us when the stellar population clocks of galaxies start ticking. A good recent review of primeval galaxy (PG) searches is given by KOO [22].

We already know high-redshift galaxies undergoing vigorous star formation: the 3CR sources, and Ly-α companions of z > 3 quasars. There are now four radio galaxies known with z $\simeq$ 1.8, in which we can detect Ly-α emission from the ground: 3C239, 3C256, 3C326.1, and 1141+35. Our images of 3C256 (z=1.819, Fig. 5) show that the line emission is spread over the whole galaxy, following the continuum starlight. A simple computation indicates that, at least within a factor of a few, star formation in a Bruzual μ=0.5 model, which fits the colors and magnitudes of distant 3CR galaxies, produces enough ionizing photons in order to maintain this emission. A faint blue companion galaxy to 3C239 (z=1.781) may have also been detected in Ly-α, suggesting that this is not an exclusive provenance of "weird" objects, but this result needs a confirmation.

An object of special interest is 3C326.1, at z=1.825 . SPINRAD ET AL. [23] discovered that the radio source is associated witha 10-arcsec (> 100 kpc) cloud of Ly-α gas, of very low ionization, very little continuum ($\sim 25^m$), and high velocity shear and turbulence ($\sim$ 1000 km/s in the rest frame). Ly-α emission implies over 10^{11} $M_\odot$ in the ionized hydrogen alone. There are only faint condensations in the cloud, and no apparent active nucleus. The object appears to be a "textbook example" of what a primeval galaxy should look like, and it is only for its "low" redshift of 1.825 that we are hesitant to call it that. Perhaps the galaxy formation extends to redshifts that low, and we are dealing with a late-forming galaxy?

A new method of finding extremely distant galaxies is the Ly-α interference filter imaging of QSO fields. Three galaxies were discovered so far in this way: the companion of PKS 1614+051, at z=3.218 (DJORGOVSKI ET AL. [24]), shown in Fig. 6, and two companions of MG 2016+112, at z=3.27 (SCHNEIDER ET AL. [25]). Some two dozen QSO fields were searched so far with this technique, but no

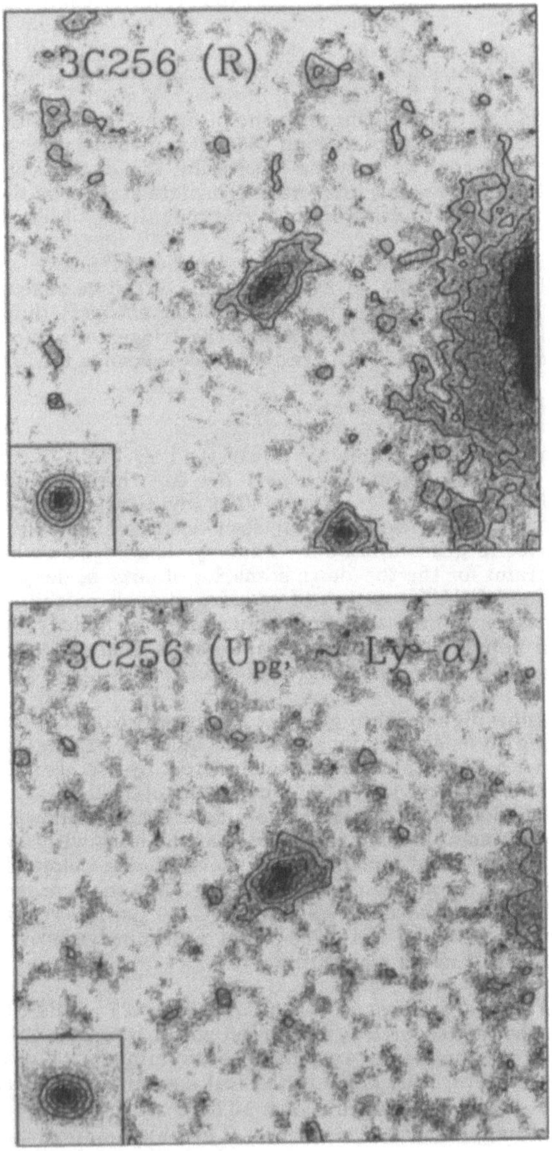

Figure 5. 3C256 (z=1.819) is one of the most distant galaxies known. Its elongated shape is well-resolved in all bands (*UBVR*), and there is no evident nucleation or bluening toward the center. The galaxy has a very strong Ly-α emission, which almost entirely dominates the *U* band, and it is spread over the whole galaxy. Computations based on a Bruzual μ=0.5 model, which fits the colors and magnitudes of such distant and evolving galaxies, indicate that the star formation alone produces enough ionizing photons to maintain this line emission. Both images were obtained with the KPNO 4-m, *R* image with a TI CCD, and *U* image from a plate taken by T. Kinman. The fields are 30 arcsec square, N to the top, E to the left.

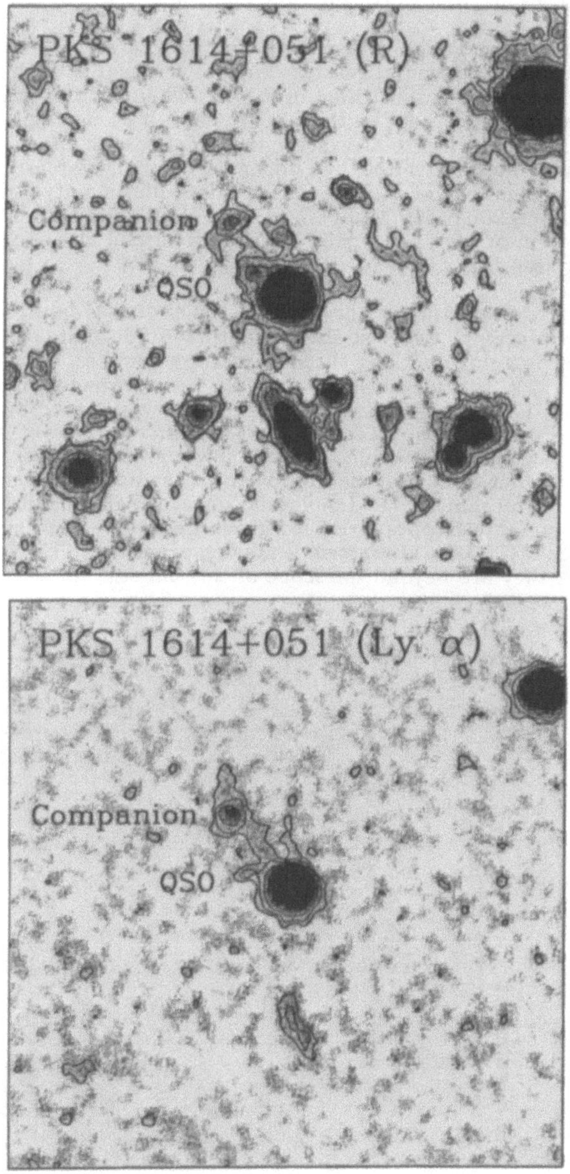

Figure 6. The galaxy companion of the quasar PKS 1614+051, at z=3.218, is one of the most distant non-QSO objects known. It has a weak continuum ($R \simeq 25$), and a very strong Ly-α. The ionization is relatively low, but the line width is several hundred km/sec in the rest frame. Both the Ly-α gas and the continuum are extended, but in a different way; there is a Ly-α-strong bridge between the object and the QSO. A fainter object due North from the QSO may be associated with a tail of Ly-α emission. The companion is radio-quiet, and has a velocity shift and the projected separation from the QSO that is characteristic of galaxy groups. We interpret it as a mildly active galaxy, probably interacting with the QSO.

new objects as bright as these three were found. Still, it is a very powerful method, and further searches are well worth the time and the effort.

Thus, we know of galaxies with $z > 3$, and QSO's at $z \simeq 4$. But when is something to be called a primeval galaxy? The point is that at high redshifts there is much Δz for little Δt, as illustrated in Fig. 7. The epoch of galaxy formation must have had some time spread at the large scales – it seems highly unlikely that the galaxies formed everywhere at exactly the same cosmic epoch. Furthermore, there was a finite time interval of that first gala burst of star formation. Let us allow 5% of the present age of galaxies for that time uncertainty (this seems to be a realistic minimum, 200 – 400 My). Given that, if galaxies form at $z_{gf} \simeq 10$, we would still call them primeval at $z \simeq 5$; and if they formed at $z_{gf} \simeq 5$ or less, then we have already seen them.

And thus, we are now on the threshold of the promised land, and may have already acquired a grail or two. We should continue to do more of the same, Ly-α interference filter imaging and long-slit spectroscopy searches, but we may have to try some other tricks. It is possible that most PG's are shrouded in an opaque ISM and/or dust, and that no Ly-α photons escape (as it was argued by HARTMANN ET AL. [26]). Perhaps most PG's are high-redshift versions of Arp 220. The new IR imaging technology, in which there is now a great progress, may open this door for us. An order-of-magnitude improvement in radio receivers may make it possible to look for megamasers and molecular lines at high redshifts. We may look through gravitational telescopes – rich foreground clusters – which would magnify distant PG's (the MG 2016+112 system is a living example). There are many other possibilities, but they are beyond the scope of this paper.

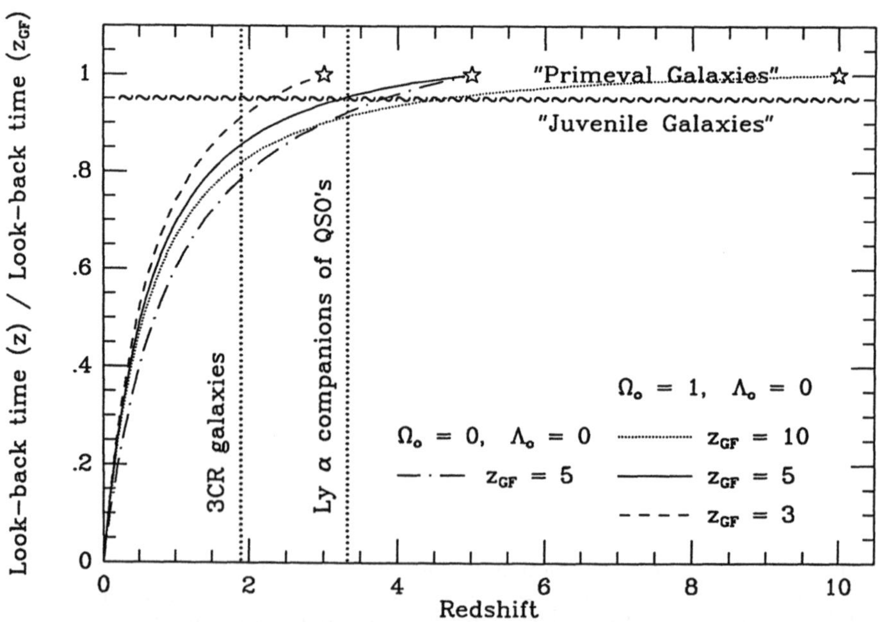

Figure 7. Relative age of galaxies, as functions of redshift in different cosmology/formation scenarios. Due to the nonlinearity of the redshift-distance relation, and the necessarily finite time interval taken by the galaxy formation, we should see the primeval galaxies at $z \sim 4 - 5$, and may have seen them already. Where does the semantics stop, and physics begin?

5. Conclusion and Acknowledgements

We are not short of problems, but it seems that we are begining to probe directly the evolution and formation of distant galaxies. But, are these far galaxies "normal"? That depends on what is "normal" at high redshifts, which we do not know. Still, I would say, probably not; all galaxies beyond $z = 1$ that we know of are strong emission line objects (otherwise we would not have known them); if *all* galaxies were doing the same thing, we would have seen many more by now. But normalcy is a cultural variable, and if we adopt the Californian meaning of the term, then perhaps these are *nearly* normal galaxies at high redshifts...

I would like to thank to my great and hard-working collaborators, whose work and ideas were freely used in this review; in particular, to Hy Spinrad, Pat McCarthy, Mark Dickinson, Michael Strauss, and many others. Thanks are also due to the staff of Kitt Peak, Lick, MMT, and CFHT observatories. Most of the work described here was completed while I was at the Berkeley Astronomy Department. Partial support from Harvard University is gratefully acknowledged.

References:

1. Kron, R. 1980, *Astrophys. J. Suppl. Ser.* **43**, 305.
2. Koo, D. 1985, *Astron. J.* **90**, 418.
3. Bruzual, G., and Kron, R. 1980, *Astrophys. J.* **241**, 25.
4. Dressler, A., and Gunn, J. 1983, *Astrophys. J.* **270**, 7.
5. Spinrad, H. 1986, *Publ. Astron. Soc. Pacific* **98**, 269.
6. Spinrad, H., Djorgovski, S., Marr, J., and Aguilar, L. 1985, *Publ. Astron. Soc. Pacific* **97**, 932.
7. Spinrad, H., and Djorgovski, S. 1983, *Astrophys. J. Lett.* **280**, L9.
8. Le Fevre, O. 1986, *IAU Circular #4233.*
9. Lebofsky, M., and Eisenhardt, P. 1986, *Astrophys. J.* **300**, 151.
10. Djorgovski, S., Spinrad, H., and Marr J. 1985, in the Proceedings of the Special IAU Colloquium *New Aspects of Galaxy Photometry*, J.-L. Nieto (ed.), *Lecture Notes in Physics* **232**, p. 193, Springer Verlag.
11. Djorgovski, S., Spinrad, H., and Dickinson, M. 1987, *Astrophys. J.* (submitted).
12. Bruzual, G. 1983, *Astrophys. J.* **273**, 105.
13. Lilly, S., and Longair, M. 1984, *M.N.R.A.S.* **211**, 833.
14. Renzini, A., and Buzzoni, A. 1986, in *The Spectral Evolution of Galaxies*, C. Chiosi and A. Renzini (eds.), Dordrecht: D. Reidel.
15. Nesci, R., and Perola, G. 1985, *Astron. Astrophys.* **145**, 296.
16. Mochkovitch, R. 1986, *Astron. Astrophys.* **157**, 311.
17. Spinrad, H., and Djorgovski, S. 1987, in the *Proceedings of the IAU Symposum #126*, G. Burbidge (ed.), Dordrecht: D. Reidel, in press.
18. Yates, M., Miller, L., and Peacock, J. 1986, *M.N.R.A.S.* in press.
19. Allington-Smith, J., Perryman, M., Longair, M., Gunn, J., and Westphal, J. 1982, *M.N.R.A.S.* **201**, 331.
20. Fabian, A., Arnaud, K., Nulsen, P., and Mushotzky, R. 1986, *Astrophys. J.* **305**, 9.
21. Windhorst, R. 1984, Ph. D. Thesis, University of Leiden.
22. Koo, D. 1986, in *The Spectral Evolution of Galaxies*, C. Chiosi and A. Renzini (eds.), Dordrecht: D. Reidel.
23. Spinrad, H., McCarthy, P., Djorgovski, S., Strauss, M., van Breugel, W., and Liebert, J. 1987, in preparation.
24. Djorgovski, S., Spinrad, H., McCarthy, P., and Strauss, M. 1985, *Astrophys. J. Lett.* **299**, L1.
25. Schneider, D., Gunn, J., Turner, E., Lawrence, C., Hewitt, J., Schmidt, M., and Burke, B. 1986, *Astron. J.* **91**, 991.
26. Hartmann, L., Huchra, J., and Geller, M. 1984, *Astrophys. J.* **287**, 487.

Dynamics of Galaxies at Large Redshift: Prospects for the Future

RICHARD G. KRON

1. Introduction

This discussion explores the feasibility of obtaining area-resolved dynamical information for galaxies to z = 1 from a combination of observations with the Space Telescope and a very large ground-based optical telescope. Information down to a disk scale length for distant galaxies can be obtained only with a heavy commitment of telescope time, both with the Space Telescope and on the ground. This can be justified because dynamical information for galaxies at large redshift will allow a fundamentally new, and arguably crucial, way to explore galaxy evolution. The observations will also allow significantly better distance indicators to be used in classical cosmological tests.

The next section will more fully develop the motivation to observe dynamics of distant galaxies. Section 3 will discuss a particular methodology. Section 4 presents examples of signal-to-noise ratio calculations. Throughout this paper $q_0 = 0.5$ and $H_0 = 50$ km sec^{-1} Mpc^{-1} are adopted.

2. Motivation

2.1 Dynamics

The circular velocity measures the interior mass, and the systematics of the proportion of dark mass in disk galaxies compared to that in spheroidal galaxies may provide some insight into the origin of different types of structures.

Secular changes are expected [1] to occur on timescales relevant to the lookback time to z = 1 or so. Since disk galaxies evolve out of phase with each other and anyway evolve at different rates, effects systematic with redshift may be much more subtle, but this is no different from trying to track the evolution of the stellar population with redshift from measurements of integrated light. (The lookback time to a galaxy reckoned in rotation periods at a given angular radius is independent of H_0, so that in this sense at least comparison with the models is easier.)

Models that are designed to make predictions suitable for comparison with observations of high-redshift galaxies seldom go beyond global spectroscopic and photometric properties. Realistic models should address the radial dependence of the stellar population and its chemical evolution, gas flows within disks, and differential star formation rates [2,3,4]. The ability to study radial gradients at large redshift will allow some direct tests of these models.

2.2 Cosmology

The other motivation for obtaining area-resolved dynamical information at high redshift is to provide refined distance indicators that can be applied to classical cosmological tests. The essential idea is that the proper distance to redshift z depends on the cosmological model, and the proper distance can be inferred by observing something that can be related to a physical scale, e.g. an angular diameter related to a metric diameter. The physical scale of the galaxy thus needs to be evaluated in some way that is independent of redshift. The following are examples of observables that do not depend on distance, yet may provide information on the relative physical scale. Later it will be argued that scales can indeed be derived that are better than standard candles so far proposed.

i) v^2. velocity measures mass, and mass traces luminosity to some degree in the central regions of giant galaxies. This of course has proven to be a highly valuable tool for the purpose of distance measurement locally, and the same idea can be applied at cosmological distances.

ii) surface brightness. Surface brightness correlates (weakly) with luminosity [5,6].

iii) line strength. There are established correlations of line strength with luminosity in ellipticals [7], S0's [8], and spiral bulges [9].

iv) energy distribution. The overall spectrum shows systematics with galaxy luminosity, as illustrated for example in the color-luminosity relation [10,11]. Some of this correlation could be driven by metallicity and would therefore not be completely independent from iii) above.

v) morphology. There are various kinds of correlations between morphology and luminosity [12]. Perhaps comparative morphology in the ultraviolet or the near infrared would show differences [cf. iv) above].

vi) environment. The physical scales of galaxies (or components within galaxies) may depend on their environment [13,14]. The mass distribution in spirals may depend on environment [15].

3. Methodology

In order to study the dynamical development of disks and to take best advantage of available information for cosmological tests, the physical scale of a galaxy needs to be derived from observables such as those just listed. A systematic way to do this was proposed by Öpik [16], who used it to obtain the distance to M 31. For a rotating disk, one simply combines the distance-dependences of the inverse-square law and Kepler's third law to arrive at

$$D = \frac{1}{4\pi G} \; \frac{1}{\alpha(\theta)\sin^2 i} \; \frac{\theta\, v^2(\theta)}{l(\theta)} \; \frac{L(\theta)}{M(\theta)} \qquad (1)$$

In this expression azimuthal symmetry is assumed and things are expressed in terms of the radial angle θ -- thus for example $l(\theta)$ is the apparent light observed within a sphere of radius θ (appropriate corrections can be made for flat systems inclined by angle i). $\alpha(\theta)$ is a geometrical index of how much the mass distribution deviates from spherical. Equation 1 is an exact relation between the distance (or equivalently a determination of the physical scale of a galaxy) and quantities that are related to both mass and light. All of the quantities on the right hand side are directly observable, except for $\alpha(\theta)$ and the mass-to-light ratio M/L. Thus the error in the distance estimate, both random and systematic, is essentially reducible to the error in M/L.

The actual value of M/L in the central regions of spiral galaxies is not too far from what is expected for a stellar population. Assuming that the majority of the mass is in the form of stars, the parameter M/L would be an observable if the stellar population could be adequately characterized. However, for normal populations most of the light comes from moderately luminous stars and most of the mass is in low-luminosity stars, and it is difficult to determine M/L directly from the integrated light. If the astrophysical processes that control star formation at greater than 1 M_0 are not closely coupled to those that regulate the formation of lower-mass stars, then such efforts would be problematical. There may be other ways to constrain M/L from finding correlations between M/L and other properties not directly connected with the light. Finding correlations of this sort is the kind of thing that extragalactic astronomers are good at doing.

Even if M/L has to be treated as a parameter that is apparently uncorrelated with the visible aspects of the stellar population, the formulation of (1) would still be superior to the usual linewidth - luminosity correlation (which is a derivative of (1)), because of the explicit angular dependence of quantities.

As mentioned above the accuracy of (1) as a distance indicator is limited by the extent to which M/L can be determined independently of distance. Figure 1 illustrates both the variation of M/L from one galaxy to another, and as a function of radius within a given galaxy. Rotation curve data were taken from [17], and photometry in the r band was taken from [18], with no correction either for absorption in the Milky Way or for absorption internal to each galaxy. M/L is the integrated mass-to-light ratio within radius θ and is expressed in solar units, where L is the luminosity in the r band and $M_0 = 4.83$ [18]. (The light growth curve $l(\theta)$ was computed from the major axis profiles and disk ellipticities, which tends to underestimate the light from the bulge.) The curves for the four galaxies illustrated are qualitatively similar: the central few points are relatively low; there is a flat region near 5 - 10 kpc; followed by a rise; and the last point in each case is high. The central one or two points correspond to angles less than 3.7 arc sec and may be subject to error.

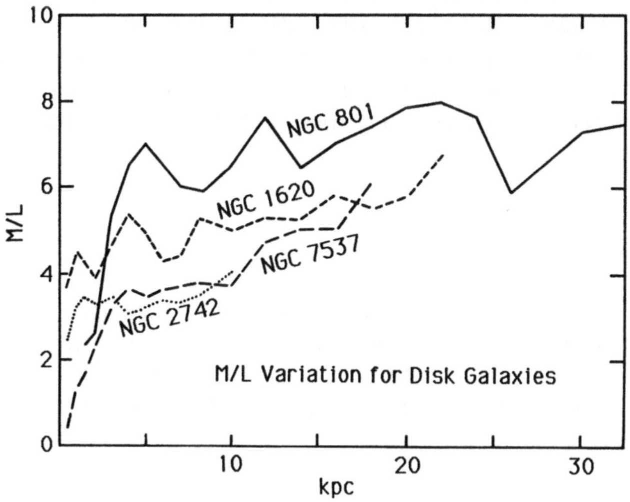

Figure 1: Integrated mass-to-light ratios for four spiral galaxies.

The four galaxies in Fig. 1 were picked because they have been considered [19] as illustrations of galaxies having the same mass distribution (NGC 7537 compared to NGC 801, and NGC 1620 compared to NGC 2742), yet having different light distributions (as judged only from the Hubble type and the bulge-to-disk ratio).

The fact that the M/L curves in Fig. 1 are flat over a range in radius suggests that some regularity in the stellar population is operating. The spread of M/L in

Fig. 1 is a factor of two, but then no corrections at all have been made (some of the spread could of course be due to distance errors). Kent's [18] study of 37 galaxies showed that the dispersion in log M/L was 0.13, after taking into account observed correlations with inclination, morphology, and mass, and after correcting for Galactic extinction. This dispersion could be reduced if additional observations were made explicitly to constrain M/L (for example, an $r-H$ color), and if the angular dependence of M/L were exploited.

Equation 1 is written explicitly for systems supported by rotation but is easily generalizable to spheroidal systems, where $v^2(\theta)$ would be some measure of the velocity dispersion. The main operational difference is that, for disk galaxies, the idea is to sample the flat part of the rotation curve, where as it turns out the surface brightness is low. Spheroidal galaxies may be easier for dynamical measurement at high redshift because the linewidth in the *integrated* light can be used, and the interpretation will be straightforward as long as velocity dispersion gradients are weak and ambiguities due to triaxiality are not of major consequence. Given measurements of velocity dispersion, surface brightness $I(\theta)$, and a characteristic angular radius, a distance can be estimated from (1) with a trivial substitution of observables, $I(\theta) = l(\theta)/\pi\theta^2$ [20,21]. The peculiar power-law dependences found in those papers of the distance estimate on each of the observables reflects either systematics of M/L with the physical scale [22] or systematics that are suppressed by ignoring changes in the velocity dispersion across the face of the galaxy.

4. Practical Application

The practical application of (1) to very distant galaxies requires the Space Telescope and a large ground-based telescope with suitable instrumentation. The following discussion assumes that the method is to be applied to large disk galaxies; this case is more challenging because it requires sub-arc sec resolution for ground-based spectroscopy.

4.1 High-resolution Imaging

Morphology obtained with the cameras of the Space Telescope would be used primarily to determine the inclination angle i and the light growth curve $l(\theta)$. It may also be that aspects of the morphology or color gradients are related to M/L, in which case the Space Telescope observations would be used also to constrain M/L at each angular position.

Expectations for Space Telescope imaging of distant galaxies are illustrated in Fig. 2, which plots the integration time required to reach the indicated signal-to-noise ratio, per pixel, *vs.* redshift. The Wide Field Camera is assumed because the red spectral region is probably more appropriate for this application;

with long exposures (2500 sec) through the F785LP filter ($\lambda_{eff} \approx 9000$ A) the pictures are likely to be limited in precision by the background sky, not the readout noise. The detection model assumes that the galaxy surface brightness in the restframe is μ_V = 22.5 mag arc sec^{-2}, i.e. similar to the value at the solar position seen perpendicularly through the Milky Way and generally characteristic of large disk galaxies one or two scale lengths out. The adopted K correction is based on spectrophotometry of the central regions of spirals [23].

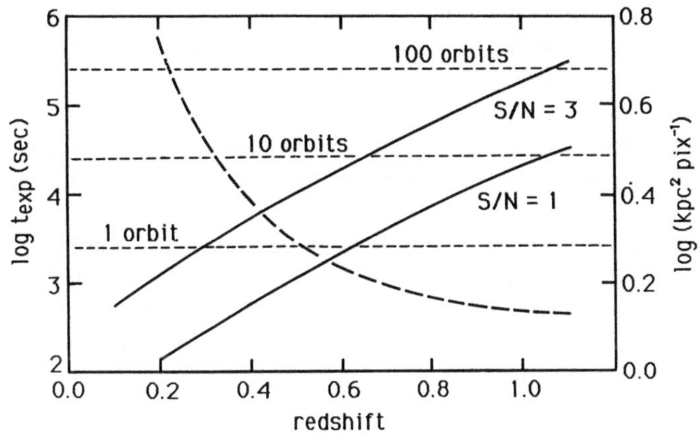

Figure 2: Space Telescope imaging of distant galaxies. The scale at the right refers to the dashed curve.

The point of Fig. 2 is that galaxy morphology (defined as reaching 22.5 mag arc sec^{-2} in the visual restframe) for $z > 0.6$ will require more than ten long exposures, thus more than ten orbits, to get S/N = 3 per pixel. If there were no systematic evolution in surface brightness with redshift, then at $z = 1$ 100 orbits would be required. Moreover structures on scales less than a couple of kpc cannot be resolved.

At the Workshop J. Gunn presented simulations that he, J. Hoessel and D. Schneider had computed for what the Wide Field Camera would see of galaxies at high redshift. The pictures showed that basic patterns -- spiral arms and such -- were still recognizable even at $z = 0.8$ in fewer than 10 orbits. This result is consistent with the above discussion because the eye is good at the kind of filtering needed to see patterns in noisy data. A good profile or color gradient measurement would certainly require higher signal-to-noise ratio, however.

4.2 Spectroscopy

One approach for distant galaxy $v^2(\theta)$ measurement would be similar to conventional slit spectroscopy of nearby galaxies, where the slit is placed on the major axis of the disk (the position angle having been determined from Space Telescope imaging). The main difference is that the interpretation is more involved because the width of the slit subtends a non-negligible fraction of the disk size and a pixel sees a mix of velocities from different parts of the disk (at $z = 1$, one arc sec is 8.5 kpc). Galaxies could be selected to have smaller inclinations, even though the projected differential velocities would be smaller. (Note that a simple linewidth measurement, as in the 21-cm approach, is in general not adequate because much of the optical light comes from regions interior to the flat part of the rotation curve, and the apparent linewidth from the integrated light would presumably have a less direct bearing on the mass.)

Another approach to obtaining velocity structures at large redshift would be Fabry-Perot imaging in some convenient emission line, assuming the redshift is already known. Depending on the circumstances, this method could be more efficient. Moreover the reduction of the data to obtain the velocity field may be more straightforward than with a slit.

If all of the noise is in the sky background, i.e. if the detector noise is ignorable, for fixed signal-to-noise ratio and velocity resolution at a particular restframe surface brightness, the required integration time scales like $(1+z)^6$. Thus to measure the equivalent region of a galaxy at $z = 1$ with a 10-meter telescope would require integrating 25 times longer than for a galaxy at $z \approx 0$ with a 4-meter telescope. In practice the relative difficulty for measuring the high-redshift galaxy is even greater because the distant galaxy is recorded with fewer pixels.

Low-noise detectors allow higher dispersions to be used and still be limited by the sky background. Higher dispersion than is normally thought appropriate for faint galaxy work can help dramatically to reduce the effective sky background for $\lambda > 7000$ A, because most of the background flux in the red is in sharp line emission. The idea is to obtain spectra at high dispersion, subtract the sky, and then bin the data to get the required signal-to-noise ratio at an acceptable velocity resolution. (The red part of the spectrum is important for studies of spectral evolution because the same restframe wavelengths need to be sampled, and in any case most of the strong features at high redshift are in the red.)

The *continuum* sky backgound at $R = 30{,}000$ would contribute 25 detected counts per pixel in a two-hour exposure with a 7.5-meter telescope, a 1 arc sec slit, and an overall efficiency of 7%. The best current charge-coupled devices can match this noise level, and lower-noise detectors should be available in the future. With these parameters, a patch with μ_r = 21 in the restframe could be measured at z = 1 with S/N = 10 in each 0.5 arc sec pixel in 25 hours, when binned to R = 1500.

Measurement of velocity fields in distant galaxies obviously requires very good angular resolution. Suppose that in one dimension the criterion is to have at least five resolution elements, and we consider a disk of diameter 20 kpc at z = 1. The disk would subtend an angle of 2.4 arc sec, requiring sampling at 0.5 arc sec. This is indeed stringent, but it is not so far out of practical experience [24,25] that the idea is completely crazy. The following are grounds for optimism that this resolution can be achieved:

-- a growing recognition that much of astronomer's seeing is locally-induced [26], leading to proper design and siting of new telescopes;

-- design of new telescopes for fast instrument and focal station changeover to take advantage of conditions of excellent seeing;

-- development of adaptive optics for wavefront correction. Affordable and otherwise practical systems may only be able to produce diffraction-limited images in the thermal infrared. However, some degree of wavefront correction may be possible as short as 2.2 μm. A sharpening of the images by a factor of two would be more than adequate.

Array detectors in the near infrared are required to take advantage of (partial) adaptive wavefront correction. Photometric measurements in the near infrared in any case are astrophysically desirable for the same reasons that *H* magnitudes are popularly used in studies of the linewidth - luminosity relation, namely the lower susceptability to problems with extinction and the likelihood that the *H* light is a relatively stable measure of the integrated luminosity of the stellar population. M/L may be essentially the same for all spirals if the luminosity is measured in the *H* band [17]. To check this requires two-dimensional photometry: it may be for instance that the spread in M/L in Fig. 1 can be understood from *r* - *H* color gradients. An example of *H*-band imaging is given in Fig. 3, which shows a mosaic of eight exposures of M 82, each 300 sec long. These pictures were obtained with a Rockwell 64 x 64 HgCdTe array with 52 μm pixels mounted on the Yerkes 0.61-meter telescope at a scale of 1.3 arc sec $pixel^{-1}$. The array is sensitive to 2.5 μm, and with a broad band it is sky-limited. Future arrays will be larger and will have lower leakage currents, allowing spectroscopic applications at low surface brightness.

Figure 3: M 82 at 1.6 μm as seen with the Yerkes near infrared array camera. The display was adjusted to show fine-scale structure. North is at the top, east to the left. The picture is about three arc min across.

5. Summary

Öpik's formulation is a good way to combine dynamical information to derive the physical scale of a galaxy, and therefore the distance, independent of redshift. The distance estimate is as good as the knowledge of M/L, and a number of redshift-independent ways to constrain M/L have been mentioned. Application of this approach to distant galaxies has great potential for both cosmological tests and for the investigation of galaxy evolution, including the dynamical development of disks. The purpose of this presentation has been to show that such measurements are feasible to z = 1, given Space Telescope imaging and long spectroscopic exposures with a very large telescope. The latter would benefit from adaptive optics. Low-noise array detectors sensitive to the near infrared will have an important place in this program.

Work with D. Koo inspired me to think about measuring velocity curves at large redshift. The near infrared imaging program at Yerkes Observatory is under the direction of D.A. Harper, and M. Hereld has been principally responsible for the development effort. I thank both for permission to reproduce Fig. 3.

References

1. Carlberg, R.G., Freedman, W.L.: Ap.J. **298**, 486 (1985)
2. Talbot, R.J., Jr., Arnett, W.D.: Ap.J. **197**, 551 (1975)
3. Lacey, C.G., Fall, S.M.: Ap.J. **290**, 154 (1985)
4. Arimoto, N., Yoshii, Y.: Astronomy & Astrophys., in press
5. van den Bergh, S.: Ap.J. **248**, L9 (1981)
6. Binggeli, B., Sandage, A., Tarenghi, M.: A.J. **89**, 64 (1984)
7. Faber, S.M.: Ap.J. **179**, 731 (1973)
8. Burstein, D.: Ap.J. **232**, 74 (1979)
9. Boroson, T.A.: Ph.D. thesis, University of Arizona (1980)
10. Tully, R.B., Mould, J.R., Aaronson, M.: Ap.J. **257**, 527 (1982)
11. Wyse, R.F.G.: M.N.R.A.S. **199**, 1P (1982)
12. Sandage, A., Binggeli, B., Tammann, G.A.: A.J. **90**, 395 (1985)
13. Strom S.E., Strom , K.M.: Ap.J. (Letters) **225**, L93 (1978)
14. Dressler, A.: Ap.J. **236**, 351 (1980)
15. Burstein, D., Rubin, V.C., Ford, W.K., Jr., Whitmore, B.C.: Ap.J. (Letters) **305**, L11 (1986)
16. Öpik, E.: Ap.J. **55**, 406 (1922)
17. Rubin, V.C., Burstein, D., Ford, W.K., Jr., Thonnard, N.: Ap.J. **289**, 81 (1985)
18. Kent, S.M.: A.J. **91**, 1301 (1986)
19. Burstein, D., Rubin, V.C.: Ap.J. **297**, 423 (1985)
20. Djorgovski, S., Davis, M.: in Galaxy Distances and Deviations from Universal Expansion, eds. B. Madore and B. Tully (Reidel, Dordrecht, in press)
21. Dressler, A., Lynden-Bell, D., Burstein, D., Davies, R.L., Faber, S.M., Terlevich, R.J., Wegner, G.: preprint.
22. Efstathiou, G., Fall, S.M.: M.N.R.A.S. **206**, 453 (1984)
23. Turnrose, B.E.: Ap.J. **210**, 33 (1976)
24. Kormendy, J.: Ap.J. (Letters) **292**, L9 (1985)
25. Thompson, L.: Ap.J. **300**, 639 (1986)
26. Woolf, N.J.: Ann. Rev. Astron. Astrophys. **20**, 367 (1982)

Session 8

Dark Matter

V. Trimble, Chair
July 30, 1986

Dark Matter in Binary Galaxies and Small Groups

VIRGINIA TRIMBLE

ABSTRACT: The main indications of dark matter in binary galaxies and small groups come from measurements of masses implying M/L ratios larger than those (typically 5-20 in solar units) found for single galaxies. Methods of measuring these masses and uncertainties inherent in them are reviewed.

1. Introduction

The existence of binary stars was proposed by MICHELL /1/ on the statistical basis that close pairs of stars in the sky were far too common to represent accidental superpositions. The conclusion was, however, not widely accepted by natural philosophers of the time until 1803-4, when Herschel succeeded in tracing out portions of several orbits, showing them to be consistent with Newtonian gravitation. Binary galaxies were similarly discovered in a statistical way /2/, and confirmation from measured orbits can be expected about the year 1,000,01987.

In the meantime, galaxies in pairs and small groups present the problem that we can know they are parts of bound systems only in a statistical sense. Any particular one for which we might want to attempt a mass determination (including, unfortunately, the Local Group) could turn out to be only a brief encounter. Even worse, the parameters used to select a sample of nominally bound systems (separations in space and velocity, R and V) are exactly the same ones that dominate calculated masses, in the form mass proportional to RV^2. Inevitably, then, the larger the range of separations admitted in your sample, the larger the masses you will find.

2. The Local Group

The mass of our own small cluster of galaxies has been estimated in three ways, using the Virial Theorem /3/, the fact that the Milky Way and Andromeda are approaching each other in an expanding universe /4-8/, and the effects of the Local Group on the velocity field of nearby galaxies /8/,

The first leads to a large, but quite uncertain, value of M/L.

The second kind of calculation was pioneered by KAHN and WOLTJER /4/, who required that the self-gravitation of the pair suffice to have turned around their initial Hubble separation velocity by the present age of the universe. The calculation is a pleasant application of classical mechanics, the resulting equations being given in convenient form by LYNDEN-BELL /5/ and by SANDAGE /8/. The chief uncertainties are the amount of time allowed for the turn-around (6-20 billion years, depending on your choice of H_o and q_o) and the present approach velocity. This is the 300 km/s heliocentric speed of Andromeda /9/ minus 80% of the local galactic rotation speed of 184-294 km/s /7,10,11/, or 65-153 km/s. The most probable range of derived masses is 2-7 X 10^{12} M_o, and M/L is about 100. Appreciable quantities of dark matter can be avoided only if the local circular speed is as improbably large as 290 km/s /6/. If it were, the inner Milky Way, with a flat rotation curve out to 20 kpc would itself contain 4 X 10^{11} M_o and have M/L = 16.

SANDAGE /8/ has applied the third method and found that the Local Group does, indeed, perturb the nearby velocity field, but only slightly, consistent with a total mass of 4 X 10^{11} M_o and M/L about 5. This is in good accord with the 2 X 10^{11} M_o for the total Milky Way reported by TREMAINE /12/, if the two component galaxies are of equal mass, but implies that the Local Group is unbound, consisting of two separate large galaxies and their satellites now approaching each other by chance. This is not horribly improbable /13/. The mass thus found for the Local Group is, inevitably, sensitive to choices of H_o, q_o, circular velocity, and the mass distribution within the Group, but is invariably less than the binding mass for the same set of assumptions.

A bound Local Group of relatively small mass is possible only if one follows ARP /14/ in attributing an appreciable fraction of the blue-shift of Andromeda to non-velocity effects.

2. Binary Galaxies

The measurement of binary galaxy masses using velocity differences and projected separations was pioneered and persistently pursued by PAGE /15,16/. He recognized most of the problems that are still with us: collecting a large enough sample that the need to average over projection angle does not produce enormous error bars; being sure that the sample includes only bound systems; and deciding the intrinsic eccentricity of the orbits. In addition, his photographically-determined velocities typically had errors comparable with the velocity differences being sought, which may be the reason that his pairs of spiral galaxies yielded much smaller values of M/L (3 ± 2) than those in later investigations. Curiously, his results for E's and SO's (90 ± 40) do not disagree.

There now exist a number of additional data sets with much more accurate optical or 21 cm velocity values /17-23/. Once differences in assumed values of H_o, orbital eccentricity, and waveband for L have been reconciled, nearly all of these are consistent with $M/L_B = 70 \pm 20(H/100)$ within radii of 100 (H/100) kpc, assuming circular orbits /24/. This drops immediately to 30 ± 10 if the orbital velocities are assumed to be distributed isotropically, and still lower for radial orbits /19/.

Worrisomely, the one discordant sample /22/ with $M/L_B = 7 \pm 1$, the same as rotation masses for the same galaxies singly, is the one that most nearly meets a strict criterion for containing only bound systems /25/. One cannot, unfortunately, resolve the issue by considering only pairs that show signs of interaction, since their members must be very close together and will be inside most of their shared dark halo, and so yield relatively low included masses, according to the conventional view of dark matter.

3. Small Groups

Early cataloguers of groups containing three to tens of comparably bright galaxies /26,27/ typically applied the Virial Theorem to their data, finding that masses near 10^{13} M_o and M/L's near 100 were necessary for binding, while at the same time they expressed severe reservations about both the permanence of the configurations and the appropriateness of the technique.

The most extensive recent data set /28/ includes 90 groups in the Center for Astrophysics redshift survey and yields a mean M/L of 170 h /29/. A number of other observational and theoretical investigations essentially concur /30-35/. Once again, a dissenting analysis of triples /36/ finds a mean M/L of only 10, and a test using redshift asymmetries /25/ suggests that many of the systems with apparently large M/L's are, in fact, unbound, as earlier proposed by MATERNE and TAMMANN /37/.

The situation for all the configurations discussed in this and the previous two sections seems to be that dark matter, exceeding that implied by rotation curves and other local measurements by about a factor of 10, is needed to guarantee their stability over a Hubble time, but that there is in every case some alternative possible interpretation of the data. Unfortunately, the only discriminants so far suggested are likely to be believed only by those who already believed one answer or the other anyway. A signature for bound, but non-interacting, systems is badly needed!

REFERENCES

1. J. Michell: Phil. Trans. Roy. Soc. 1767, 234 (1767)
2. E. Holmberg: Medd. Lund Astr. Obs. 186, 1 (1954)
3. M.L. Humason, H.D. Wahlquist: Astron. J. 60, 254 (1955)
4. F. Kahn, L. Woltjer: Astrophys. J. 130, 105 (1959)
5. D. Lynden-Bell: In. W. Shuter ed.: Kinematics, Dynamics, and Structure of the Milky Way, p. 349. Reidel, Dordrecht 1983
6. E. Herbst: Publ. Astron. Soc. Pac. 87, 827 (1975)
7. D. Lin, D. Lynden-Bell: Mon. Not. Roy. Astron. Soc. 181, 37 (1977)
8. A.R. Sandage: Astrophys. J. 307 (in press, 1986)
9. V.C. Rubin, S. D'Odorico: Astron. Astrophys. 2, 484 (1969)
10. W.H.L. Shuter: Mon. Not. Roy. Astron. Soc. 194, 851 (1981)
11. F.J. Kerr, D. Lynden-Bell: Mon. Not. Roy. Astron. Soc. (subm. 1986)
12. S.D. Tremaine: elsewhere in this volume (1987)
13. S. van den Bergh: Astron. Astrophys. 11, 154 (1971)
14. H.C. Arp: Astron. Astrophys. 156, 207 (1986)
15. T.L. Page: Astrophys. J. 116, 63 (1952)
16. T.L. Page: In A.R. Sandage et al. eds: Galaxies and the Universe, p.541 Univ. Chicago Press, 1975
17. E.L. Turner: Astrophys. J. 208, 304 (1976)
18. S.D. Peterson: Astrophys. J. 232, 20 (1979)
19. S.D. White et al.: Mon. Not. Roy. Astron. Soc. 203, 701 (1983)
20. G. Lake, R. Schommer: Astrophys. J. 279, L19 (1984)
21. L.P. Blackman, G.A. van Morsel: MNRAS 208, 91 (1984)
22. I.D. Karachentsev: Sov. Astron. AJ 29, 243 (1985)
23. L. Schweizer: Astrophys. J. Suppl. (subm. 1986)
24. J.P. Ostriker: IAU Symp. 117 (in press, 1986)
25. M.J. Valtonen, G.G. Byrd: Astrophys. J. 303, 523 (1986)
26. E.M. Burbidge, G.R. Burbidge: Astron. J. 66, 541 (1961)
27. G. de Vaucouleurs: in A.R. Sandage et al. eds: Galaxies and the Universe, p. 557 (1975)
28. M.J. Geller, J. Huchra: Astrophys. J. Suppl. 52, 61 (1983)
29. J. Huchra, M.J. Geller: Astrophys. J. 257, 423 (1982)
30. J. Heisler et al.: Astrophys. J. 298, 8 (1985)
31. J. Barnes: Mon. Not. Roy. Astron. Soc. 208, 873 (1984)
32. F.H. Briggs: Astrophys. J. 300, 613 (1986)
33. F. Hammer, L. Notale: Astron. Astrophys. 155, 420 (1986)
34. D. Carter et al. : Mon. Not. Roy. Astron. Soc. 212, 471 (1985)
35. A.E. Evrard, A. Yahil: Astrophys. J. 296, 299 & 310 (1982)
36. V.E. Karachentseva, I.D. Karachentsev: Astrofiz. 18, 1 (1982)
37. J. Materne, G. Tammann: Astron. Astrophys. 37, 383 (1974)

Dark Matter in Dwarf Galaxies

K. C. FREEMAN

1. Introduction

For the brighter spirals ($M_B < -14.5$), the rotation curves give direct and convincing evidence for the presence of extended dark halos. For ellipticals, the situation is not so clear. The X-ray data for the brighter ellipticals, and the data on shells around one elliptical strongly suggest that these systems also contain dark matter. Other dynamical data for ellipticals, such as the radial distribution of the observed line of sight velocity dispersion, are inconclusive, and will remain so until more is known about how the shape of the velocity ellipsoid changes with radius in these galaxies.

Do the faintest galaxies contain significant amounts of dark matter? If they do, then there are some interesting possible consequences:

(1) the existence of galaxies with negligible or zero luminous mass.

(2) the parameters for the dark matter distribution in these small galaxies (core radius, central density, velocity dispersion) constrain the kinds of particles that make up the dark matter.

(3) the luminous matter and the gas in the smallest dwarfs may not be selfgravitating, and this could lead to inhibition of star formation.

(4) DEKEL and SILK [1] argued that dwarf galaxies form and eject gas within dark matter halos. With this picture, they were able to reproduce the observed scaling relations of surface density, abundance, velocity dispersion and M/L ratio with absolute magnitude. Selfgravitating models, without dark matter, were unable to give all of these scaling laws at once. Direct observational evidence for dark matter in dwarfs would support their picture.

This review will concentrate mainly on the true dwarf galaxies, ie those with $M_B > -14$. First I will discuss the dwarf irregulars, and then the dwarf spheroidal systems.

2. Dwarf Irregular Galaxies

The dwarf irregular galaxies are believed to be the low mass end of the family of disk galaxies. We should therefore look first at what is known about the dark matter content of the brighter disk galaxies. The really strong results on dark matter in spirals come from a few systems for which the rotation curves can be measured out to a radius of 6 or more disk scale lengths (see FREEMAN [2]). Fortunately, this small sample covers a wide range of M_B, from about -20 to -14.5. For these galaxies, it is possible to estimate the density and length scales of their dark matter. The procedure (see for example KALNAJS [3]; CARIGNAN and FREEMAN [4]) is now well known. Accurate surface brightness distributions I(r) and rotation curves V(r) are needed. From the I(r) distribution, the expected rotation curve $V_{exp}(r)$ is calculated, assuming a uniform M/L ratio. This $V_{exp}(r)$ curve is then fitted to the observed rotation curve V(r) in the inner parts of the galaxy, where the disk is probably selfgravitating; this fitting procedure ties down the M/L ratio for the luminous disk.

In every case, the observed rotation curve is flat or rising in the outer parts of the galaxy (r > 3 disk scale lengths), while the expected rotation curve falls. The difference $(V^2 - V^2_{exp})(r)$ is then attributed to the gravitational field of the dark halo. It is necessary to assume some form for the structure of this dark halo (eg an isothermal sphere). The parameters for this halo (eg the central density ρ_o and the core radius r_c) can then be estimated by fitting the (calculated circular velocity)^2 for the halo to the difference $(V^2 - V^2_{exp})(r)$.

This procedure maximises the contribution of the disk to the observed rotational velocity, and gives in this sense a maximum disk or minimum halo solution. Table 1 shows some parameters derived in this way for five well observed disk galaxies. The rows give: (1) the absolute magnitude M_B, (2)-(4) the core radius r_c, velocity dispersion σ and central density ρ_o of the dark halo (only two of these are independent; the third is given by $4\pi G\rho_o r_c^2 = 9\sigma^2$), (5)-(6) the maximum distance R_{lim} to which the rotation curve was measured, in kpc and in units of the disk scale length h_R, (7) the ratio of the total halo to disk mass within R_{lim}, (8) the integrated M/L_B ratio within R_{lim}; this is given for comparison with the virial M/L values of the smaller dwarf irregular systems, for which no more detailed data is available, (9) reference to the source of the data.

These galaxies, with $-19.8 < M_B < -14.5$, give strong evidence for dark matter because (1) they have negligible bulges, so their structure is simple (flat disk plus dark halo), (2) their kinematics are unambiguous (pure rotation and known orientation), (3) their HI rotation data reaches out into the region where the halo dominates the potential field (r >> 3 disk scale lengths). This is particularly important: one cannot say much about dark matter in regions where there are no kinematic data! (This is a real problem in studying dwarf galaxies.)

The estimates of the halo parameters given in Table 1 should not be taken too literally. Uncertain internal absorption corrections, the unknown structure of the halo (including the lack of selfconsistency of the halo in the gravitational field of the halo plus the disk), and some remaining uncertainty about whether the maximum disk solutions are appropriate, will all affect the values of the halo parameters.

Table 1: Parameters for dark halos in spirals.

	N3198	N6503	N3109	U2259	DDO127
M_B	-19.8	-19.3	-17.7	-17.1	-14.5
r_c[kpc]	12		10	7	2.8
σ[km s^{-1}]	100		45	57	28
$\rho_o[M_\odot pc^{-3}]$	0.008		0.002	0.013	0.017
R_{lim}[kpc]	30	23	10	7	3
R_{lim}/h_R	11	14	8	6	7
M_{halo}/M_{disk}	4		7	2	8
M/L_B	12	10	8	15	12
Source	[5]	[6]	[4]	[7]	[8]

For the smaller dwarf irregulars, it is more difficult to study their dark matter distributions. For these systems, the rotational velocities are low, support from random gas motions becomes significant, the contribution of the gas itself to the potential field increases, and the galaxies are probably not flat (as inferred from their observed V/σ values). Their HI velocity fields are irregular, and the kind of rotational decomposition that was done for the larger spirals becomes very uncertain. Most of the estimates of their total masses are just HI virial masses. Here are some examples of fainter dwarf irregulars ($M_B > -14.5$):

Holmberg I (TULLY et al. [9]). This system has M_B = -14.4. It shows a well defined rotation field of amplitude ±10 km s^{-1}, and a virial mass of (2 to 6).10^8 $M_\odot$. The ratio of HI mass to total mass, M_{HI}/M, is 0.2 to 0.6. The M/L ratio is 3 to 7. (The wide range in parameters comes from the uncertain inclination.)

I ZW 36 (VIALLEFOND and THUAN [10]). This blue compact galaxy system has M_B = -14.5, and shows ordered motion with a total velocity spread of 40 km s^{-1}. The HI distribution is clumpy. Its virial mass is 2.10^8 $M_\odot$ and the ratio M_{HI}/M is about 0.2. The M/L ratio is 2. Near-IR photometry suggests that an old population can account for the virial mass if the distribution of this old population is more extended than that of the young population.

Leo A (ALLSOPP [11]; SARGENT and LO [12]). This dwarf irregular has M_B = -12.7 and shows rotation of amplitude ±3 km s^{-1} for an inclination of 54°. Its virial mass is 3.10^7 $M_\odot$ and the ratio M_{HI}/M is 0.7. Its M/L = 1.5.

GR8 ([12]). This faint dwarf irregular has M_B = -11.5. Its velocity field is irregular, with some evidence for minor axis rotation. The HI distribution is clumpy. Its virial mass is 1.2.10^8 $M_\odot$, its M_{HI}/M ratio is only 0.05, and the M/L ratio has the high value of 17.

M81 dwA ([12]). This faint system has M_B = -11.0, and its velocity field shows rotation with an amplitude of ±2.5 km s^{-1}. The virial mass is 4.10^7 $M_\odot$, M_{HI}/M = 0.3 and its M/L has again a relatively high value of 11.

Table 2 gives the absolute magnitude, virial mass [$M_\odot$], M_{HI}/(virial mass) ratio, M/L ratio and the HI profile width at 50% peak intensity [km s^{-1}] for a selection of dIrr galaxies covering the range $-15.9 < M_B < -9.4$. Most of the data in Table 2 comes from the compilation [12]. The very faint system LGS 3 may be intermediate between the dIrr and dSph galaxies: its brightest stars are red (CHRISTIAN and TULLY [13]), so its most recent episode of star formation probably occurred several Gyr ago. The systems shown in Table 2 have varied stellar populations. Some have massive stars and HII regions (I Zw 36, GR8), while others do not (Peg, LGS 3). Some of them have very low HI content.

Table 2: Data for dwarf irregular galaxies.

	$-M_B$	M_{virial}	M_{HI}/M	M/L_B	ΔV_{50}
DDO125	15.9	6.10^8	0.2	2	30
Ho I	14.4	3.10^8	0.4	4	
IZw36	14.5	2.10^8	0.2	2	47
U4483	13.9	$1.4.10^8$	0.4	3	
Peg	12.7	$1.6.10^8$	0.01	9	
Leo A	12.7	$2.6.10^7$	0.7	1.5	
DDO187	12.1	$2.3.10^8$	0.05	21	34
GR8	11.6	$1.2.10^8$	0.05	17	30
CVn dwA	11.5	$1.8.10^8$	0.2	31	
DDO210	11.0	$6.5.10^7$	0.04	12	26
M81 dwA	11.0	$4.0.10^7$	0.3	11	19
Sag dw	10.5	$3.0.10^7$	0.3	11	
LGS3	9.4	$2.2.10^7$	0.01	27	

2.1 Summary for Dwarf Irregulars and Spirals

For systems brighter than $M_B \approx -14.5$, there is good direct evidence for dark matter from the HI rotation curves, for at least some galaxies.

Dwarf irregulars with $-14.5 < M_B < -12$ have generally low M/L ratios. This says nothing strong about dark matter, because the HI is often not much more extended than the light. It is possible that the dark matter has a much more extended distribution (as it does in the brighter spirals) and it need not contribute much to the gravitational field in the optical/HI region.

All of the dwarf irregulars fainter than $M_B \approx -12$ have high M/L ratios (10 to 30). These high ratios are not due to the HI content, because the ratio $M_{HI}/M < 0.3$ for all of these systems in Table 2. These systems appear to be promising candidates for significant amounts of dark matter, as if at $M_B \approx -12$ the (optical + HI) component becomes non-selfgravitating.

However, we should note that, for the faintest dwarfs ($M_B > -12$), the HI velocity dispersions are small. They are similar to the HI random motions seen in the interstellar medium of larger spirals. The amplitudes of the random motions in the spirals are set by local energy balance processes in the interstellar medium. Therefore, it may be that in some of these dIrr systems, the HI velocity dispersions are not virial, and therefore the virial masses are overestimates. An example of such a system is the irregular NGC 1705. This is a starburst system with $M_B \approx -15$; it has split optical emission lines with a spread of about 100 km s^{-1}, resulting from outflow, and a similar HI velocity width. These velocities are clearly not virial (MEURER et al. [14]). Of the systems in Table 2, with larger M/L ratios, we note that GR8 has HII regions and massive stars, while Peg and LGS 3 do not (HOESSEL and MOULD [15], [13]).

We conclude that at least some of the high M/L values for the faintest dIrr systems are probably real and indicate the presence of dark matter. The nature of this dark matter is, of course, as yet unknown. However we recall the argument of MELNICK and TERLEVICH [16] that the faintest dIrr systems are the most metal weak and may therefore have the flattest mass functions, so their dark matter may simply be in the form of stellar remnants.

3. Dwarf Spheroidal Galaxies

One of the first suggestions that some dwarf spheroidal galaxies may contain significant amounts of dark matter came from FABER and LIN [17], who attempted to estimate the masses of the nearby dSphs from their tidal radii. Hodge's (see [18]) star counts for these systems suggested that they were tidally truncated by the Galaxy. The tidal radius r_t derived from the star counts is given approximately by

$$r_t = R_{peri} \{M_d/(3+e)M_G\}^{1/3} \qquad (1)$$

where R_{peri} and e are the perigalactic radius and eccentricity of the dwarf of mass M_d, and the (point) mass of the Galaxy is M_G. As a test, this relation can be used to estimate the masses for the outer galactic globular clusters, assuming that R_{peri} is the present galactocentric distance R of the cluster. For clusters with R > 10 kpc, the mean M/L ratio for the clusters is 1.3 ± 0.4, which is very close to the M/L values derived dynamically from the internal velocity dispersions of individual clusters. For the nearby dSph galaxies, a similar procedure gives the following M/L_V ratios:

system	For	Scl	LeoI	LeoII	Dra	UMi	Car
M/L_V	1	7	0.2	0.3	13	126	54

However, the tidal approach has its problems. (1) Although the star count profiles appear to be tidally truncated when plotted in the log(n)-log(r) plane (n and r are the star count surface density and the radius), the profiles appear approximately linear in the log(n)-r plane. This suggests that the dSph galaxies may be structurally related to the dIrr systems (LIN and FABER [19]), and that their density distributions are not tidally truncated. If this is so, then the tidal estimates for the masses of the dSph systems would be too small. (2) Tidal truncation is known to take several orbital periods to become established. Then, even if the dSph systems are tidally truncated, they are probably not fully truncated because their orbital periods are so long. Therefore the tidal estimates of their masses would be too large.

We now turn to estimates of the masses of dSph galaxies from internal velocity dispersions. These have their own problems. However I would like to make two preliminary points: (1) Small velocity dispersions can be measured; for example, the measured velocity dispersion of M67 is about 1 km s^{-1} (MATHIEU [20]). (2) There are well behaved stellar systems whose internal kinematics are exactly as expected; for example, the globular cluster ω Cen has the structure of a King model and its observed velocity dispersion - radius relation follows very closely the expected relation for that King model, all the way from the cluster center to the tidal radius. Its M/L ratio is 2.6, which is close to the epected value from stellar population models (SEITZER and FREEMAN [21]).

Table 3 summarises the available velocity dispersion data for five dSph systems. The rows give the absolute V magnitude, the central surface brightness [V mag arcsec^{-2}], the core radius [kpc], the ratio of tidal radius to core radius, the velocity dispersion estimates [km s^{-1}] with the kind of objects used given below each estimate (cl, C*, K* denote clusters, carbon stars and K giants respectively), and finally the M/L_V value.

Table 3: Data for dwarf spheroidal galaxies

	For	Scl	Car	Dra	UMi
$-M_V$	12.6	11.1	9.4	8.5	8.5
$I_{o,V}$	23.3	23.9	24.9	25.4	26.1
r_c	0.5	0.2	0.2	0.15	0.15
r_t/r_c	6	6	3	3	6
σ	7.5±3 (3cl)	6±2.5 (3C*)	5.5±1.5 (6C*)	11±2 (2C*,11K*)	10±2 (10C*)
	6.5±2 (5C*)	6.5±1 (16K*)	6±2 (5C*)		
	5.0±0.9 (40K*)				
M/L_V	2	5	8	60	80

The last velocity dispersion entry for Fornax is for 40 K giants at about one core radius from the center of the system ([22]: see below). Sources for the other entries can be found in Marc Aaronson's review. We note that the parameters given in the first four rows of Table 3 may also have significant uncertainties. The mass estimator used in Table 3 is

$$M = 167\mu\sigma_{obs}{}^2[\mathrm{km}^2\mathrm{s}^{-2}]\, r_c[\mathrm{pc}] \qquad (2)$$

where μ is a parameter which depends on the concentration of the King model (see KING [23]). This mass estimator underestimates the mass for highly truncated King models, and we should now discuss the procedures for estimating the M/L ratio from the observed velocity dispersion and surface density profile.

If we assume that the mass is distributed like a King model, with a distribution function

$$f(E) = \exp(-E/\sigma^2) - \exp(-E_o/\sigma^2) \quad (3)$$

where E_o is the cutoff energy and σ is the scale velocity dispersion, then the expression for the mass is similar to that given in (2):

$$M = 167\mu\sigma^2[\mathrm{km^2s^{-2}}]r_c[\mathrm{pc}] \quad (4)$$

except that the velocity dispersion in (4) is the scale dispersion σ rather than the observed dispersion σ_{obs}. Therefore, to use (4), the scale velocity dispersion σ must be derived from the observed velocity dispersion σ_{obs}. In a King model, the true dispersion $\sigma(r)$ decreases with radius r, and $\sigma(r) < \sigma$ everywhere because of the truncation of the distribution function. So it is necessary to correct σ_{obs} for the effects of truncation, projection and the mean radius of the stars. This correction can be large. For example, for a King model with $r_t/r_c = 3.16$ (appropriate to the Fornax dSph), $\sigma_{obs}(0) = 0.49\sigma$, so the use of σ_{obs} in (2) underestimates the mass by a factor of 4.

Another approach is core fitting (KORMENDY [24], RICHSTONE and TREMAINE [25]). Here one uses the stars as test particles, and assumes an isotropic velocity dispersion. Then, from hydrostatic equilibrium,

$$(1/\rho)d(\rho\sigma^2)/dr = -GM(r)/r^2 \quad (5)$$

where ρ and σ are the density and velocity dispersion for the test particles. Therefore the LHS of the equation pertains to the test particles and the RHS is the total gravitational field. If the total matter and the test particles have similar distributions, then

$$M/L = 9\eta\sigma^2_{obs}(0)/(2\pi GI(0)R_{1/2}) \quad (6)$$

where $R_{1/2}$ is the half brightness radius, and $\eta \approx 1.0$ for a wide range of systems. All the quantities on the RHS are observables.

If the total matter and the test particles have different distributions, then it is still possible to estimate the total matter density at the center, $\rho_{total}(0)$, somewhat less directly from

$$[(1/4\pi Gr^2)(d/dr)\{(r^2/\rho)d(\rho\sigma^2)/dr\}]_{r=0} = \rho_{total}(0) \quad (7)$$

which will lead to

$$M/L = 9\eta\sigma^2_{obs}(0)/(2\pi GI(0)R_{1/2}) \quad (8)$$

where η now depends on $\rho(r)$ and $\sigma(r)$ for the test particles.

I would like now to present some preliminary new results by PALTOGLOU et al. [22] on the internal kinematics of the Fornax dSph system. This work began as an attempt to measure the rotation of this system. Fornax has an apparent axial ratio of about 0.67. Say it was once a dIrr, which faded or was stripped. We estimated the amplitude of its rotation before stripping (from the Tully-Fisher relation) and then after stripping. We calculated that the present observable velocity difference along the major axis of Fornax, due to rotation, could be as large as 10 to 20 km s^{-1}, if Fornax is now a disk of intrinsic axial ratio 0.2 and inclination of about 50°.

The star count profile from our new star counts is very well represented by a King model with $\log(r_t/r_c) = 0.5$ and $r_c = 17'.7 = 720$ pc, in excellent agreement with HODGE's [18] earlier estimates. From new photographic photometry, we selected two samples of about 20 K-giants near the major axis of Fornax, at about one core radius to the NE and one core radius to the SW. Their radial velocities were measured simultaneously, using the AAT fiber system.

The rotation is small: the velocity difference $\langle V\rangle_{NE} - \langle V\rangle_{SW} = 3.4 \pm 2.4$ km s^{-1}. The raw observed velocity dispersion, including measuring errors, is 7.4 km s^{-1}. The measuring error is 5.5 km s^{-1}, derived from many repeated observations. The intrinsic projected velocity dispersion at about one core radius from the center of the system is therefore 5.0 ± 0.9 km s^{-1}. For a King model, we would then expect the projected velocity dispersion at the center of Fornax to be about 6.5 km s^{-1}, which is in fair agreement with the values given in Table 3. Our value of the velocity dispersion leads to a mass of 6.10^7 $M_\odot$ and a M/L ratio of 3.6, assuming that the total matter is distributed in the same way as the star counts. This M/L value suggests that the dark matter does not contribute significantly to the gravitational field in the inner parts of the system.

More work would give the shape of the velocity dispersion profile $\sigma(r)$, and would show whether $\sigma(r)$ follows the King model $\sigma(r)$ (as is the case for ω Cen). Such an observation would suggest that the dark matter is not important anywhere in this system. This result would not be conclusive, however, because we do not know the core radius of the possible dark matter distribution, or how the shape of the velocity ellipsoid (ie the anisotropy of the velocity dispersion) changes with radius.

2.3 Summary for Dwarf Spheroidal Galaxies

There is no compelling evidence for dark matter in the Fornax, Sculptor and Carina systems: it could be there, but it does not appear to be dynamically significant in the inner parts of these systems.

For Ursa Minor and Draco, the M/L ratios are about 60 to 80. This appears to be strong evidence for dark matter, and it is difficult to find a way out of this conclusion (see Scott Tremaine's talk).

It is interesting to compare the inferred total central densities for Ursa Minor and Draco with that for the dark matter halo of the dwarf spiral DDO127 (KORMENDY [24]):

	UMi	Dra	DDO127
$\rho_o[M_\odot\ pc^{-3}]$	0.42 ± .23	0.80 ± .43	0.02 to 0.04

These values for UMi and Draco were calculated with velocity dispersions that were uncorrected for the effects of truncation etc; correction would further increase their central densities. If the core radii for their dark matter distribution were much larger than for their luminous matter, then these central density values would be too large, but only by a factor of about two. We are left with the apparent conclusion that the central densities of the dark matter in UMi and Draco are about an order of magnitude larger than in DDO127 (or in any of the other spiral systems that were discussed earlier). Also, as Kormendy pointed out, the dark matter density in UMi and Draco exceeds their luminous matter density by about an order of magnitude, so the luminous matter could not be selfgravitating and it may be difficult to understand how stars could form.

Acknowledgements

I am very grateful to Marc Aaronson, K. Begeman, Albert Bosma, John Kormendy and Wal Sargent, who kindly allowed me to use unpublished material in this talk, and to Gregg Rowley who computed some models.

References

1. A. Dekel, J. Silk: Astrophys.J. 303, 39 (1986)
2. K. Freeman: in Dark Matter in the Universe, ed. by J. Kormendy (Reidel, Dordrecht 1986), in press.
3. A. Kalnajs: in Internal Kinematics and Dynamics of Galaxies, ed. by E. Athanassoula, (Reidel, Dordrecht 1983) p. 87
4. C. Carignan, K. Freeman: Astrophys.J. 294, 494 (1985)
5. T.S. van Albada, J.N. Bahcall, K. Begeman, R. Sancisi: Astrophys.J. 295, 305 (1985)
6. K. Begeman: unpublished
7. C. Carignan, R. Sancisi, T.S. van Albada; unpublished
8. A. Bosma, J. Kormendy: unpublished
9. R. Tully, L. Bottinelli, J.R. Fisher, L. Gouguenheim, R. Sancisi, H. van Woerden: Astron.Astrophys. 63, 37 (1978)
10. F. Viallefond, T.X. Thuan: Astrophys.J. 269, 444 (1983)
11. N.J. Allsopp: Mon.Not.R.astr.Soc. 184, 379, (1978)
12. W.L.W. Sargent, K-Y. Lo: in Star Forming Dwarf Galaxies, ed. by D. Kunth, T.X. Thuan, J.T.T. Van (Editions Frontieres, Gif sur Yvette, 1986)
13. C. Christian, R.B. Tully: Astron.J. 88, 934, (1983)
14. H.G. Meurer, C. Cacciari, K. Freeman: unpublished
15. J.G. Hoessel, J.R. Mould: Astrophys.J. 254, 38 (1982)
16. J. Melnick, J. Melnick: Observatory 106, 69 (1986)
17. S.M. Faber, D. Lin: Astrophys.J. 266, L17 (1983)
18. P. Hodge: Ann.Rev.Astron.Astrophys. 9, 35 (1971)
19. D. Lin, S.M. Faber: Astrophys.J. 266, L21 (1983)
20. R. Mathieu: in Dynamics of Star Clusters, ed. by P. Hut and J. Goodman (Reidel, Dordrecht, 1985)
21. P. Seitzer, K. Freeman: unpublished
22. G. Paltoglou, K. Freeman, J.E. Norris: unpublished
23. I.R. King: Astron.J. 71, 64 (1966)
24. J. Kormendy: in Dark Matter in the Universe, ed. by J. Kormendy (Reidel, Dordrecht 1986), in press.
25. D. Richstone, S. Tremaine: Astron.J. 92, 72 (1986)

Dark Matter in Early-Type Galaxies

S. MICHAEL FALL

1. Introduction

The quest for dark matter in galaxies has been an active field of research for more than a decade. (See, for example, the classic review by FABER and GALLAGHER [1] and the proceedings of IAU Symposium 117, edited by KORMENDY and KNAPP [2].) Most of the results, however, pertain to late-type galaxies and only in the last year or two has evidence for dark matter in early-type galaxies begun to emerge. This imbalance has hindered our understanding of the formation and evolution of galaxies. Many questions regarding the initial collapse, chemical enrichment, tidal interactions, and so on are affected by whether early-type galaxies do or ever did possess massive halos. Another important problem concerns the number and splitting angles of gravitational lenses, which depend sensitively on the projected density in all forms of matter. The evidence for dark matter in early-type galaxies comes from polar rings, shells, and X-rays. Unfortunately, only a few galaxies have been studied in detail and the observations are not always free of ambiguity. My purpose in this article is to collect together the fragmentary results that are now available.

Before considering early-type galaxies, it may be useful to give a brief summary of the evidence for dark matter in late-type galaxies. In most cases, at least half the mass within the Holmberg radius can be accounted for by the luminous components alone. (For reference, the Holmberg radius, defined by the isophote at 26.5 B mag/arcsec^2, is 4.5 scale lengths for an exponential disk with a central surface brightness of 21.65 B mag/arcsec^2.) However, when rotation curves can be measured beyond the Holmberg radius, they are usually found to be flat or gently rising and therefore provide unambiguous evidence for dark matter. This is often modelled as a distinct spheroidal component – the halo – although the arguments for such an interpretation are not yet conclusive. In the best cases, the rotation curves extend out to 2-3 Holmberg radii and the enclosed mass-to-light ratios are 5-15. (See VAN ALBADA and SANCISI [3] and references therein. All distance-dependent quantities in this article are based on a Hubble constant of 50 km s^{-1} Mpc^{-1} and all luminosities are in the B-band.) The fact that the rotation curves of spiral galaxies are flat over a large range of radii, in regions where they are influenced to various degrees by the bulge, disk and halo components, is a "conspiracy" that requires some cosmological explanation.

Velocity dispersion and surface brightness profiles have been measured for a large number of elliptical and SO galaxies. With some assumptions about the amount of isotropy in the velocity distributions, these data can, through the equations of stellar hydrodynamics, be used to place constraints on the mass distributions in the galaxies. EFSTATHIOU, ELLIS, and CARTER [4] have applied this method to the bright elliptical galaxy NGC 5813. Their measurements extend to the effective radius r_e and can be fitted with constant M/L and a velocity distribution that is isotropic. In most cases, the kinematic data to not reach beyond $0.5r_e$ as a result of the rapid fall off in the surface brightness. JEDRZEJEWSKI [5] has studied about a dozen elliptical galaxies and finds that most of them can be fitted by models with constant M/L and isotropic velocity distributions. For the others, radial variations in M/L are not required if modest degrees of anisotropy are allowed. The available data are therefore consistent with the notion that the inner parts of elliptical galaxies are self-gravitating, with mass-to-light ratios in the range 5-10, and that any dark halos they possess must have core radii exceeding the effective radii of the luminous components. To explore the distribution of matter at larger radii, we must turn to other methods.

2. Some Recent Results

A small fraction of early-type galaxies have polar rings. These are thought to originate in the transfer of material during close encounters with other galaxies or in the complete capture of much smaller galaxies. In some cases, the rings are quite broad, resembling the rings of Saturn. The preferential alignment with the minor axes of the galaxies is not fully understood but polar rings are probably easier to discover than equatorial rings and more stable against precession than tilted rings. At any rate, they are ideal for determining the mass distributions of early-type galaxies because they consist of gas and stars on nearly circular orbits. WHITMORE, MCELROY, and SCHWEIZER [6] have obtained photometric and kinematic data for three faint SO galaxies with polar rings: NGC 4650A, ESO 415-G26, and A0136-0801 (see also SCHWEIZER, WHITMORE, and RUBIN [7]). In terms of the effective radii of the bulges r_e or the exponential scale lengths of the disks α^{-1}, the rings extend to large radii r_{max}, namely 55 r_e, 13 r_e and 45 r_e, or equivalently, 12.5 α^{-1}, 4.3 α^{-1}, and 13.7 α^{-1}. The rotation curves are flat or gently rising out to r_{max} and the enclosed mass-to-light ratios are 4.5, 9.3, and 25. The situation for this small sample of SO galaxies is therefore reminiscent of that for spiral galaxies. Another quantity of interest is the ratio of the circular velocity at r_{max} to the central velocity dispersion of the bulge, v_c/σ_o, which, for the galaxies studied by Whitmore, et al., takes the values 1.6, 1.5, and 2.5.

Many early-type galaxies have faint arclike features, and in favorable cases, the spacing of these "shells" can be used to infer the mass distributions at large radii. The method is based on the idea that the shells represent the debris of a small galaxy that merged with and was tidally disrupted by a larger galaxy (QUINN [8]). A process of phase wrapping then ensures that more of the liberated material is seen at the turning points of orbits with commensurate periods. It follows that the radii of the shells r_n, their ranks n (when counted inward from the outermost one), and the corresponding orbital periods p are related by the expression

$$p(r_n)/p(r_1) = 3/(2n+1). \tag{1}$$

This can be inverted to derive the gravitational field and hence the mass distribution of a galaxy. The method is only reliable when the shells are distinct and have a high degree of symmetry resulting from the capture of a single satellite with a small internal velocity dispersion on a radial orbit. To date, the example that has given the most convincing results is the bright elliptical galaxy NGC 3923, which is the subject of a detailed study by HERNQUIST and QUINN [9]. They find that the circular velocity rises gradually to the outmost shell at 17 r_e, that the enclosed mass is about 40 times the mass of the luminous component, and that the corresponding mass-to-light ratio is 100-200. The absolute value of the circular velocity at large radii is not given directly by this method, but its value relative to that at small radii implies $2.4 \lesssim v_c/\sigma_o \lesssim 3.6$.

The discovery of extended X-ray emission from early-type galaxies raised hopes that their mass distributions could be inferred through the equations of hydrostatic equilibrium. If the X-rays originate in a gaseous corona with a density profile $\rho_g(r)$ and a temperature profile $T_g(r)$, the total mass enclosed within any radius r is given by

$$M(r) = \frac{-rkT_g(r)}{G\mu m_H}\left(\frac{dlog\ \rho_g}{dlog\ r} + \frac{dlog\ T_g}{dlog\ r}\right). \tag{2}$$

The density gradient can often be determined reliably but the temperature and especially its gradient are usually major sources of uncertainty. The most notable exception is the giant elliptical galaxy M87, for which the results are $M/L \approx 200$ at $r \approx 10r_e$ and $2.3 \lesssim v_c/\sigma_o \lesssim 2.8$ (FABRICANT and GORENSTEIN [10]). Although dark matter is definitely present, there is no way to tell whether it "belongs" to M87 or to the Virgo Cluster. The only other galaxy with a temperature gradient that can be constrained by the observations is NGC 4472 (E1/SO) and in this case the results are $M/L \approx 20$ at $r \approx 3r_e$ and $1.4 \lesssim v_c/\sigma_o \lesssim 1.9$ (FORMAN, JONES, and TUCKER [11]; TRINCHIERI, FABBIANO, and CANIZARES [12]). Again, dark matter appears to be present but it may not be required given the large uncertainties in the data. Unfortunately, these results cannot be checked by other means. For the Sa galaxy NGC 4594 (the Sombrero), however, the circular velocity inferred from X-ray observations, assuming the gas is isothermal, exceeds 600 km s^{-1} whereas the value derived from HI observations is 380 km s^{-1} (BAJAJA, et al. [13]). This difference can be reconciled if in fact the temperature gradient is large and positive: $dlog\ T_g/dlog\ r \gtrsim 0.9$. Other possibilities include a substantial contribution to the X-ray emission from discrete sources or a non-solar abundance of heavy elements in the gas, either of which could lead to systematic errors in the temperature.

A lower limit on the total mass of a galaxy can be derived from X-ray observations without knowledge of the temperature gradient (FABIAN, THOMAS, FALL, and WHITE [14]). The method rests on the assumption that the gas (a) is in hydrostatic equilibrium, (b) is convectively stable, and (c) has a negative pressure gradient. If (a) were not satisfied, impossibly large inflow or outflow rates would be required to explain the observed mass of gas, and if (b) were not satisfied, the resulting convection would ensure that the gas found a stable equilibrium within a few sound-crossing times. Assumption (c) is not as easy to justify, but it certainly holds for M87 and NGC 4472 and is generally indicated by constraints on the confinement of the gas by any intergalactic medium. The lower limit on the total mass of a galaxy is then

$$M(r_\infty) \geq \frac{5r_o kT_o[1-(p_\infty/p_o)^{2/5}]}{2G\mu m_H(1-r_o/r_\infty)}, \tag{3}$$

where T_o and p_o are the observed temperature and pressure at some radius r_o and p_∞ is the pressure, assumed negligible, at a much larger radius r_∞. When applied to a dozen or so early-type galaxies with the most reliable X-ray data, the mass-to-light ratios are typically found to exceed 70, implying large amounts of dark matter. The main limitation of this method is that it provides no information on how the mass is distributed. The possibility of systematic errors in the temperatures is worrisome but not likely to change the mass-to-light ratios derived from (3) by more than a factor of two.

3. Conclusions and Comparisons

The results discussed in the previous section indicate that early-type galaxies have dark matter in significant quantities. This conclusion is based on circular velocities that are constant or rise gradually out to large radii and on global values of M/L that are higher than the local values within the luminous components. One can quibble about individual cases, but the evidence for dark matter in at least some and perhaps most early-type galaxies now seems compelling. What is less certain is whether we should think of the dark matter as belonging to a distinct halo of the kind often associated with spiral galaxies. The composition and shape of this component are hardly constrained at all by present observations. Nevertheless, there are some indications, to be described next, that whatever picture turns out to be correct for late-type galaxies, may also apply to early-type galaxies.

A useful way of comparing the bulges of spiral galaxies with their halos has been discussed by WHITMORE, KIRSHNER, and SCHECHTER [15]. They plot v_c/σ_o, the ratio of the circular velocity in the disk to the central velocity dispersion in the bulge against L_{bul}/L_{tot}, the ratio of the luminosity of the bulge to that of the whole galaxy. The results, as modified and extended by WHITMORE and KIRSCHNER [16], are shown as the dots in Fig. 1. The median value of v_c/σ_o is 1.8 and the full range is from 1.3 to 2.6. For reference, a non-rotating stellar system with a space density $\rho(r)$ and velocity dispersions $\sigma_r(r)$ and $\sigma_t(r)$ in the radial and tangential directions has, whether or not self-gravity is important,

$$\left(\frac{v_c}{\sigma_r}\right)^2 = -\frac{d\log\,\rho}{d\log\,r} - 2\,\frac{d\log\,\sigma_r}{d\log\,r} - 2\left[1-\left(\frac{\sigma_t}{\sigma_r}\right)^2\right]. \tag{4}$$

Since bulges have Hubble profiles down to small radii, values of v_c/σ_o near $\sqrt{3}$ are entirely plausible although some deviations are expected as the result of velocity anisotropies or non-zero gradients in the velocity dispersions. Equation (4) is of course a means of comparing v_c and σ_r at the same radius whereas the observed quantities generally pertain to very different radii. Some of the scatter in Fig. 1 may therefore be caused by rotation curves that are not perfectly flat. Similarly, the weak (anti-) correlation of v_c/σ_o with L_{bul}/L_{tot} might be explained by the fact that the circular velocities often increase outward and that they are usually measured at smaller radii relative to r_e in bulge-dominated galaxies.

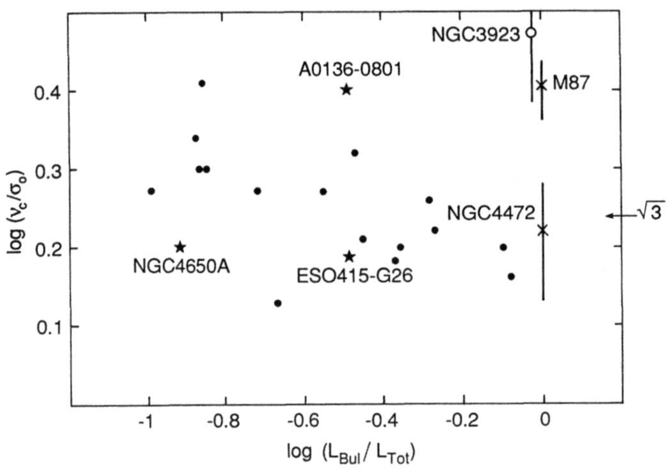

Figure 1. log (v_c/σ_o) against log (L_{bul}/L_{tot}) for early- and late-type galaxies.

The results collected in this article for early-type galaxies are plotted in Fig. 1 as the labelled points. The median value of v_c/σ_o is 2.1 and the full range is from 1.5 to 3.0. These are slightly higher than the corresponding quantities for late-type galaxies but probably not significantly so considering the small samples. Moreover, the galaxies with the highest values of v_c/σ_o are generally those with measurements at the largest radii relative to r_e, which, as mentioned previously, could lead to a small bias in the comparison with other galaxies. The bulges of spiral and SO galaxies have several basic properties in common with elliptical galaxies; they have similar surface brightness profiles, and at a given luminosity, they have similar velocity dispersions. Figure 1 suggests that the dark matter bears roughly the same relation to the bulges in early-type galaxies as it does in late-type galaxies. Clearly, more work is needed to test this conjecture and to explore its consequences. My guess is that the similarity in the values of v_c/σ_o for early- and late-type galaxies is closely related to the conspiracy that produces nearly flat rotation curves over large ranges in radii.

I am grateful to Pepi Fabbiano, Jacqueline van Gorkom, Peter Quinn, and Brad Whitmore for helpful discussions of the material in Section 2 of this article.

References

1. S. M. Faber and J. S. Gallagher: Ann. Rev. Astr. Ap. **17**, 135 (1979)
2. J. Kormendy and G. R. Knapp, eds.: IAU Symposium 117, Dark Matter in the Universe, (Reidel, Dordrecht) in press
3. T. S. van Albada and R. Sancisi: Phil. Trans. Roy. Soc. Lond., in press
4. G. Efstathiou, R. S. Ellis, and D. Carter: Mon. Not. Roy. Astr. Soc., **201**, 975 (1982)
5. R. Jedrzejewski: Ph.D. Thesis, University of Cambridge (1985)
6. B. C. Whitmore, D. B. McElroy, and F. Schweizer: Ap. J., in press
7. F. Schweizer, B. C. Whitmore, and V. C. Rubin: Astr. J., **88**, 909 (1986)
8. P. J. Quinn: Ap. J. **279**, 596 (1984)
9. L. Hernquist and P. J. Quinn: Ap. J., in press
10. D. Fabricant and P. Gorenstein: Ap. J., **267**, 535 (1983)
11. W. Forman, C. Jones, and W. Tucker: Ap. J., **293**, 102 (1985)
12. G. Trinchieri, G. Fabbiano, and C. R. Canizares: Ap. J., in press
13. E. Bajaja, G. van der Burg, S. M. Faber, J. S. Gallagher, G. R. Knapp, and W. W. Shane: Astr. Ap., **141**, 309 (1984)
14. A. C. Fabian, P. A. Thomas, S. M. Fall, and R. E. White: Mon. Not. Roy. Astr. Soc., **221**, 1049 (1986)
15. B. C. Whitmore, R. P. Kirshner, and P. L. Schechter: Ap. J., **234**, 68 (1979)
16. B. C. Whitmore and R. P. Kirshner: Ap. J., **250**, 43 (1981)

What is the Cosmological Density Parameter Ω_0?

AMOS YAHIL

1. Introduction

It has been clear for more than a decade that the deceleration of the universe can not be determined from the magnitude-redshift relation for first-ranked elliptical galaxies, which suffer from strong evolution, dominating the cosmological effects. Attention has therefore shifted to alternate methods which relate peculiar velocities to the density inhomogeneities which have given rise to them. These fall under three categories:

- The study of well defined structures, primarily the local Virgo supercluster (VSC) of galaxies and its surrounding.
- The use of a statistical relation between the positional correlation functions of galaxies and their velocity dispersion, known as the "cosmic virial theorem" (CVT).
- Direct comparison between the observed density and velocity distributions of galaxies and N-body simulations, in which Ω_0 is an input parameter.

In a review two years ago [1], I argued that the observational evidence was converging toward a baryon dominated open universe, with Ω_0=0.1-0.2. A brief summary of the arguments is given in §2. The main objection was the theoretical disposition in favor of a flat universe with Ω_0=1, which led to the idea of biased galaxy formation, §3. A few additional dissonances remained in the interpretation of the observations, §4, but they were not deemed major. However, a serious reassessment occurred in 1985 regarding the peculiar gravitational field in the VSC, and doubts were raised about the previous interpretation of the observations. This review describes these new developments, §5, and the resulting ongoing research, which is aimed at resolving the seeming contradictions, §6.

2. Observational Consensus (1984)

2.1 Virgocentric Flow

There was, and still is, general agreement that the peculiar velocity of the Local Group (LG) with respect to Virgo is u=250±50 km s^{-1} [2-8]. With a Virgo mean recession velocity v=1000 km s^{-1}, this gives the dimensionless velocity perturbation $u/H_o r = u/(u+v) = 0.2 \pm 0.04$. The mean number density of galaxies in a sphere centered on Virgo, with the LG at its periphery, was evaluated to be in the range $\langle \delta D/D_o \rangle = 2$ [9,10] to $\langle \delta D/D_o \rangle = 3$ [2,11]. If it can be assumed that the number density of galaxies, D, is proportional to the total mass-energy density, ρ, at least on a scale ~10 Mpc, then the above data can be combined with the relation [1]

$$u/H_o r = \frac{1}{3}\Omega_o^{0.6} \langle \delta\rho \rangle/\rho_o \left(1 + \langle \delta\rho \rangle/\rho_o\right)^{-0.25} \tag{1}$$

to determine that Ω_o=0.1-0.2.

In order to reconcile the above data with Ω_o=1, one would have to establish one of the following:

- Increase $u/H_o r$ to 0.5-0.7, which requires u to be in the unacceptable range 1000-2400 km s^{-1}.
- Reduce $\langle \delta D \rangle/D_o$ to 0.56-0.80, which again seems difficult.
- Abandon the assumption that galaxies trace the matter and $\langle \delta\rho \rangle/\rho_o = \langle \delta D \rangle/D_o$.
- Modify the Virgocentric flow model (VFM).
- A combination of the above effects, in which each one of them is less severe.

2.2 Cosmic Virial Theorem

The CVT is really a statistical statement of local hydrostatic equilibrium [12]. It is assumed that the mean peculiar acceleration between pairs of galaxies is balanced by a pressure gradient. Beyond a minimum separation, this acceleration is dominated by the 3-point correlation function, while the pressure is derived from a product of the 2-point correlation function and the pairwise velocity dispersion. Applications of the CVT have differed on both the observed velocity dispersion, and on details of the usage of the 3-point correlation function (see §4.2 and §4.3), but they agree that $\Omega_o \ll 1$ [13-15].

2.3 N-Body Simulations

The conclusion from the CVT, that the low observed velocity dispersion of galaxies is indicative of an open universe, is supported by N-body simulations [e.g., 16-18]. To the extent that these simulations are realistic (cf. §4.4), they provide direct evidence for an open universe, that is independent of the assumptions of the CVT.

3. Biased Galaxy Formation

Elementary particle theories of baryogenesis [19-22] and inflation [23-26] both imply $\Omega_0=1$. These theories have so far not been confirmed or refuted by laboratory experiments, so currently their only testing ground is in observational cosmology. Since in 1984 the bulk of the observational evidence seemed to suggest $\Omega_0<<1$, this conclusion was called into question by undermining the assumption that galaxies trace the mass. Theories have been proposed in which galaxy formation is assumed to be more efficient in regions that have higher densities, even though at the epoch of galaxy formation their overdensity may still have been very small [27-29]. As a result of this biased formation process, galaxies are no longer fair tracers of the total mass-energy density, and $<\delta\rho>/\rho_0 \neq <\delta D>/D_0$. Details of the biasing physics are still sketchy [30].

4. Other Dissonances

4.1 Galaxy Pairwise Velocity Differences: Isolated Pairs

The velocity differences of isolated pairs of galaxies, presumed to be binaries, have been determined by many authors [13,31-34]. Subtraction of the effect of optical pairs must be performed carefully [13], although the problem is less severe if a median Δv is used, instead of the r.m.s. [34]. In any case, there is agreement between the different investigations: $\Delta v=100$ km s^{-1} (one dimensional r.m.s. velocity difference), independent of separation and luminosity. Binary galaxies therefore do not behave like point masses. Moreover, the total mass of a system in which a binary resides seems to be uncorrelated with luminosity, unlike the situation inside single galaxies [35,36].

4.2 Galaxy Pairwise Velocity Differences: All Pairs

There is some question whether the background due to optical pairs has been correctly subtracted in the case of the general pairwise velocity dispersion of all galaxies (nonisolated and not necessarily each other's nearest neighbors). Estimates have ranged from $\Delta v=100$ km s^{-1} [13] to $\Delta v=250$ km s^{-1} [14,15]. The investigations have differed in both the datasets and the methods of background subtraction. A cross analysis is therefore needed, in which all datasets are subjected to all analyses.

4.3 Cosmic Virial Theorem

In addition to the observational difficulty, there is a problem with the calculation of the mean pairwise peculiar gravitational acceleration. The integral over the 3-point correlation function is not rapidly converging, and there is some concern that it may overestimate the acceleration, leading to an underestimate of Ω_o [13,14]. The integral was tested in the output of an N-body simulation [17], and found to be reasonably well satisfied [37]. The 2-point correlation function in that simulation, however, was significantly different from the observed one, so there remains the question of the general validity of the integral.

Complete redshift samples make it possible to attack the problem in another way, by determining the gravitational acceleration directly, without recourse to the 3-point correlation function. Appropriate assumptions need to be made in order to deal with projection effects, but these are not much different from the ones encountered in converting the angular 3-point correlation function to the spatial one.

A very important additional parameter that needs to be examined is the dependence on density. After all, the CVT is really only local hydrostatic equilibrium, so it can be applied separately in different regions. If biased galaxy formation is important, and the number (or luminosity) density of galaxies is not a good measure of the total mass-energy density, then the effect should show up in tests of local hydrostatic equilibrium as a function of density.

4.4 Extended Nature of Galaxies

Most of the nonlinear dynamical models at the basis of mass estimates assume galaxies to be point masses. But galaxies are known to have extended massive halos [38,39], and simulations have shown that encounters between extended systems are highly inelastic [e.g., 40-42]. The bulk orbital energy of galaxies is constantly used to disrupt individual halos, and produce common halos. The timescale for this process is of the order of the characteristic dynamical timescale, usually much shorter than the Hubble age of the universe. The interactions between the unseen extended halos around galaxies therefore have serious effects on the majority of galaxy encounters, and thus affect the dynamics of most binaries, small groups, and clusters of galaxies [43].

The only tractable way realistically to model the nonlinear clustering of extended bodies is through N-body experiments. Most cosmological simulations to date have assigned one particle per galaxy. Although these simulations have modeled extended galaxies to some degree by using a softened r^{-2} force law, the force has remained a conservative one, thus making collisions between galaxies elastic. Given adequate initial conditions, such simulations should be able reasonably to model the largescale structure of the universe, but they become suspect on the smaller scales in which "virialized" systems are usually investigated. In particular, single particle galaxies have been unable to reproduce the kinematic properties of isolated binary galaxies. The point-like interaction between galaxies leads to relative velocities which show a Keplerian fall-off with separation, instead of remaining constant as observed [44].

The obvious course to take would be to model galaxies as swarms of particles, thus increasing the number of degrees of freedom per galaxy, and allowing interchange of energy between bulk and internal motions. Such an approach has recently been used in simulating the dynamical evolution of small groups of galaxies possessing massive dark halos [45,46], as well as the universe at large [47]. It was found that peculiar velocities of galaxies were reduced significantly, but not enough for the data to be consistent with a closed universe.

4.5 N-Body Tests of the Virgocentric Flow Model

The validity of the Virgocentric flow model has been questioned on the basis of N-body simulations, which seem to show both scatter and biasing in the values of Ω_0 "derived" from "superclusters" in the simulations [48-50]. The implication is that the value deduced from the observations may be incorrect. It was pointed out, however, that the analysis of the simulations did not always closely follow that of the observations [51]. Moreover, it did not take into account the effect of the tidal field due to neighboring structure (see §5.1).

5. The Local Gravitational Field

The velocity of the LG relative to the microwave background radiation (MBR) is 600 km s^{-1} in a direction that is 45° away from the direction of Virgo [52,53]. The vector difference between that velocity and the Virgocentric velocity of 250 km s^{-1} is 500 km s^{-1}. It is most plausibly attributed to the net gravitational attraction and repulsion due to superclusters and holes outside the VSC.

5.1 Tidal Field in the Virgo Supercluster

The same outside structures also exert a *tidal* field inside the VSC. Since they are typically 2-3 times farther from us than Virgo, the quadrupolar tidal velocity field which they generate between us and Virgo should be 2-3 times smaller than the bulk velocity which they impart to the entire supercluster, i.e., ~200 km s^{-1}. That is comparable to the Virgocentric velocity, and may therefore not be neglected. A revised fit to a flow model, including this tidal field, has shown that this is indeed the case [8]. Fortuitously, the component of the tidal field in the direction of Virgo is small, and the estimate of Ω_o from the revised flow model is not seriously modified. This need not have been so, and is probably not the case in a typical N-body simulation. That could be a partial explanation for the bias and scatter found in the analyses of the N-body simulations, which have so far failed to take this effect into account (cf. §4.5).

5.2 IRAS Dipole Anisotropy

The advent of the IRAS catalog has made it possible to search for the density perturbations outside the VSC. The catalog is ideal for this purpose, because it is calibrated homogeneously over almost the entire sky, ranges in distance far beyond the limits of present redshift surveys, and is unaffected by extinction. Two independent teams quickly discovered a dipole anisotropy in the sky distribution of the IRAS galaxies, aligned within the errors with the velocity of the LG relative to the MBR [54,55]. The interpretation has been that the density structure responsible for the bulk motion of the VSC has been identified.

Most of the IRAS galaxies do not yet have redshifts, but initial surveys have provided the infrared luminosity function [56-58]. This was to work out the dipole moment of the peculiar gravitational field [54]. When coupled via eq. (1) to the peculiar velocity, assumed to be equal to the velocity of the LG relative to the MBR, the estimate $\Omega_o \approx 1$ was obtained, in seeming contradiction with the previous "consensus".

6. Ongoing Investigations

The emergence of a clearcut paradox often results in great strides forward, because old assumptions need to be reexamined, and previous conclusions re-evaluated. This section describes ongoing research that is aimed at resolving the current difficulties and contradictions in the determination of Ω_o.

6.1 IRAS Dipole Moment

The main *observational* argument for $\Omega_o=1$ comes from the interpretation of the dipole anisotropy of the IRAS galaxies. The first question is whether the dipole moment was correctly determined. The flux calibration of all sources, and the effect of underestimating the fluxes of extended sources, need to be re-evaluated. The effect of "cirrus" contamination can be overcome by identifications.

6.2 Largescale Velocity Fields

It has recently been reported that elliptical galaxies with $v \lesssim 6000$ km s^{-1}— overlapping with that part of the volume covered by the IRAS galaxies which is responsible for the dipole anisotropy— partake in a largescale bulk peculiar velocity relative to the MBR [59]. This calls into question the use of the peculiar velocity of the LG relative to the MBR to determine Ω_0, although there is then no explanation for the directional coincidence between the IRAS and MBR anisotropies. Careful evaluation of the reported bulk velocity of the ellipticals must await publication of the data. It can be noted from the velocity plots already presented, however, that the velocity field is by no means homogeneous, and there is a considerable turbulent tidal component. (In fact, for $v<3000$ km s^{-1} this tidal field resembles the one found in the VSC [8].) The determination of the bulk flow velocity might possibly be affected by this tidal field, which has hitherto not been included in the analysis.

6.3 Do Galaxies Trace the Matter?

It is well known that the frequencies of early and late type galaxies are functions of density [60-62], but this does not seem to bias the determination of the largescale density structure in the VSC [11]. It is important to see whether a similar effect, a form of biased galaxy formation, might result in different determinations of density from IR (spiral) and optical (spiral and elliptical) galaxies. A comprehensive comparison of the two is not yet possible, because of the lack of redshifts for the IRAS galaxies. However, in the small area of the sky surveyed [56], 90% of the IRAS galaxies with $S_{60} \geq 0.85$ and $v<3000$ km s^{-1} are already included in the CfA catalog. (Of course the CfA catalog also contains many galaxies which are not IRAS galaxies, e.g., elliptical galaxies.) If this is true in all the overlap region between the IRAS and CfA surveys, then their density structures can be compared directly within the limited redshift range $v<3000$ km s^{-1}. Such a comparison clearly shows the two density structures to be similar [63]. It also shows that the mean cosmological density baseline defined by the IRAS galaxies is higher than that determined in the optical studies [2,9-11]. A possible interpretation might be that the existing optical surveys do not extend deep enough, and the volume covered is not a fair sample of the universe. This possibility needs to be examined carefully, both by re-evaluating the existing density determinations [64], and by extending both the optical and IR redshift surveys.

6.4 IRAS Redshift Survey

A full understanding of the density structure of the IRAS galaxies can come only from a complete redshift catalog. Such a catalog will enable us to:

- Redetermine the dipole gravitational field using individual distances for each galaxy, instead of fluxes and a broad luminosity function. This will greatly increase the reliability of the result, and minimize biases.
- Determine higher moments of the gravitational field. The quadrupole moment, in particular, should be compared with the quadrupolar tidal velocity field [8].
- Establish the cosmological baseline density to use in determining the density perturbation in the VSC (and elsewhere). This will allow a more reliable evaluation of Ω_o from the peculiar velocity field.
- Obtain other quantitative descriptions of the density structure around us, such as correlation functions, hole distribution, and continuous largescale structures.
- Compare the density structure of IR selected galaxies with optical galaxies, to obtain some understanding of possible biased galaxy formation.
- Test for biased galaxy formation using the principle of superposition, by determining the effect of introducing density-dependent weighting on the resultant gravitational field [65].

A collaboration (Davis, Huchra, Strauss, Tonry, Yahil) is already underway to obtain a redshift catalog complete to $S_{60} \gtrsim 2$ Jy for $|b|>10^o$. The flux limit was chosen in order to make the project manageable, with the aim of completing it within two years. The broad luminosity function [56-58] guarantees that the sample will extend to large distances. The eighty percentile of the redshift distribution will be at 10,000 km s^{-1}, but there will be adequate sampling even beyond that range. A British team (Efstathiou, Ellis, Frenk, Hewitt, Kaiser, Rowan-Robinson) is undertaking the complimentary approach of sparse sampling the catalog down to $S_{60} \gtrsim 0.5$.

6.5 What If $\Omega_o=1$?

All the pieces of the puzzle must fit, and none can be left out. If the largescale structure ($\gtrsim$5000 km s^{-1}), seen in the IRAS and other galaxy catalogs, continues to point toward $\Omega_o=1$, then the existing evidence for $\Omega_o<<1$ on smaller scales ($\lesssim$1000 km s^{-1}) needs to be explained. This involves both theoretical and observational work on the nonlinear growth of gravitational perturbations. In essence, mass determinations become coupled to the dynamical understanding of galaxy formation and clustering. Things to consider include:

- The validity of the Virgocentric flow model.
- The validity of the CVT, and its dependence on density, as a test of biased galaxy formation
- N-body simulations with inelastic galaxy collisions, to understand the velocity distribution of galaxy pairs.
- The survival of galaxies in merged halos, and possible observable effects that give a clue to their dynamics.

Acknowledgments

This research was supported in part by USDOE grant DE-AC02-80ER10719 at the State University of New York. The hospitality of the Research Institute for Fundamental Physics, Kyoto University, where this work was completed, is gratefully acknowledged.

References

1. A. Yahil: "Dynamics and Evolution of the Virgo Supercluster", in O.G. Richter, B. Binggeli, eds.: *The Virgo Cluster of Galaxies*, pp. 359-373, ESO, Munich 1985.
2. A. Yahil: *Ann. N.Y. Acad. Sci.*,, **375**, 169 (1981)
3. M. Aaronson, J. Huchra, J.R. Mould, P.L. Schechter, R.B. Tully: *Astrophys. J.*, **258**, 64 (1982)
4. A. Dressler: *Astrophys. J.*, **281**, 512 (1984)
5. R.B. Kraan-Korteweg: "On the Infall Velocity Towards Virgo", O.G. Richter, B. Binggeli, eds.: in *The Virgo Cluster*, pp. 397-406, ESO, Munich 1985.
6. G.A. Tammann, A. Sandage: *Astrophys. J.*, **294**, 81 (1985)
7. M. Aaronson, G. Bothun, J.R. Mould, J. Huchra, R.A. Schommer, M.E. Cornell: *Astrophys. J.*, **302**, 536 (1986)
8. P.B. Lilje, A. Yahil, B.J.T. Jones: *Astrophys. J.*, **307**, 91 (1986)
9. M. Davis, J. Tonry, J. Huchra, D. Latham: *Astrophys. J. (Letters)*, **238**, L113 (1980)
10. M. Davis, J. Huchra: *Astrophys. J.*, **254**, 437 (1982)
11. A. Yahil, A. Sandage, G.A. Tammann: *Astrophys. J.*, **242**, 448 (1980)
12. P.J.E. Peebles: *Astrophys. J. (Letters)*, **205**, L109 (1976)
13. A.R. Rivolo, A. Yahil: *Astrophys. J.*, **251**, 477 (1981)
14. M. Davis, P.J.E. Peebles: *Astrophys. J.*, **267**, 465 (1983)
15. A.J. Bean, G. Efstathiou, R.S. Ellis, B.A. Peterson, T. Shanks: *Mon. Not. Roy. Astr. Soc.*, **204**, 615 (1983)
16. J.R. Gott, E.L. Turner, S.J. Aarseth: *Astrophys. J.*, **234**, 13 (1979)
17. G. Efstathiou, J.W. Eastwood: *Mon. Not. Roy. Astr. Soc.*, **194**, 503 (1981)
18. M. Davis, G. Efstathiou, C.S. Frenk, S.D.M. White: *Astrophys. J.*, **292**, 371 (1985)
19. A.D. Sakharov: *JETP Letters*, **5**, 24 (1967)
20. M. Yoshimura: *Phys. Rev. Letters*, **41**, 381 (1978)
21. S. Dimopoulos, L. Susskind: *Phys. Rev.*, **D18**, 4500 (1978)
22. S. Weinberg: *Phys. Rev. Letters*, **42**, 850 (1979)
23. K. Sato: *Mon. Not. Roy. Astr. Soc.*, **195**, 467 (1981)
24. A. Guth: *Phys. Rev.*, **D23**, 347 (1981)

25. A. Albrecht, P.J. Steinhardt: *Phys. Rev. Letters*, **48**, 1220 (1982)
26. A.D. Linde: *Phys. Letters*, **114B**, 431 (1982)
27. N. Kaiser: *Astrophys. J. (Letters)*, **284**, L9 (1984)
28. R. Schaeffer, J. Silk: *Astrophys. J.*, **292**, 319 (1985)
29. J.M. Bardeen, J.R. Bond, N. Kaiser, A.S. Szalay: *Astrophys. J.*, **304**, 15 (1986)
30. M. Rees: this volume.
31. E.L. Turner: *Astrophys. J.*, **208**, 20, 304 (1976)
32. S.D. Peterson: *Astrophys. J. Suppl.*, **40**, 527 (1979); *Astrophys. J.*, **232**, 20 (1979)
33. I.D. Karachentsev: *Astrofizika*, **17**, 135, 238, 675, 693 (1981)
34. S.D.M. White, J, Huchra, D. Latham, M. Davis: *Mon. Not. Roy. Astr. Soc.*, **203**, 701 (1983)
35. R.B. Tully, J.R. Fisher: *Astr. Astrophys.*, **54**, **661** (1977)
36. S.M. Faber, R.E. Jackson: *Astrophys. J.*, **204**, 668 (1976)
37. A.E. Evrard, A. Yahil: *Astrophys. J.*, **296**, 310 (1985)
38. J.P. Ostriker, P.J.E. Peebles, A. Yahil: *Astrophys. J. (Letters)*, **193**, L1 (1974)
39. J. Einasto, A. Kaasik, E. Saar: *Nature*, **250**, 309 (1974)
40. S. Tremaine: "Galaxy Mergers", in S.M. Fall, D. Lynden-Bell, eds.: *The Structure and Evolution of normal Galaxies*, pp. 67-84, Cambridge University Press, Cambridge 1981.
41. S.D.M. White: "Simulations of Galaxy Mergers", in E. Athanassoula, ed.: *Internal Kinematics and Dynamics of Galaxies*, pp. 337-345, Reidel, Dordrecht 1983.
42. S. Alladin, K. Narasimhan: *Phys. Rep.*, **92**, 339 (1983)
43. S.D.M. White: *Mon. Not. Roy. Astr. Soc.*, **184**, 185 (1978)
44. A.E. Evrard, A. Yahil: *Astrophys. J.*, **296**, 299 (1985)
45. J. Barnes: *Mon. Not. Roy. Astr. Soc.*, **203**, 223 (1983); *ibid*, **208**, 885 (1984)
46. C.S. Frenk, S.D.M. White, G. Efstathiou, M. Davis: *Nature*, **317**, 595 (1985)
47. A.E. Evrard: *Astrophys. J.*, in press (1986); *ibid*, submitted (1986)
48. H. Bushouse, A. Melott, J. Centrella, J. Gallagher: *Mon. Not. Roy. Astr.* Soc., **217**, 7p (1985)
49. H. Lee, Y. Hoffman, C. Ftaclas: *Astrophys. J. (Letters)*, **304**, L11 (1986)
50. J.V. Villumsen, M. Davis: *Astrophys. J.*, **308**, 499 (1986)
51. A. Yahil: "Is the Criticism of the Virgocentric Flow Model Justified?", in B.F. Madore, R.B. Tully, eds.: *Galaxy Distances and Deviations from Universal Expansion*, pp. 143-145, Reidel, Dordrecht 1986.
52. P.M. Lubin, G.L. Epstein, G.F. Smoot: *Phys. Rev. Letters*, **50**, 616 (1983)
53. D.J. Fixsen, E.S. Cheng, D.T. Wilkinson: *Phys. Rev. Letters*, **50**, 620 (1983)

54. A. Yahil, D. Walker, M. Rowan-Robinson: *Astrophys. J. (Letters)*, **301**, L1 (1986)

55. A. Meiksin, M. Davis: *Astr. J.*, **91**, 191 (1986)

56. A. Lawrence, D. Walker, M. Rowan-Robinson, M. Penston, K. Leech: *Mon. Not. Roy. Astr. Soc.*, **219**, 687 (1986)

57. B.T. Soifer, D.B. Sanders, G. Neugebauer, G.E. Danielson, C.J. Lonsdale, B.F. Madore, S.E. Persson: *Astrophys. J. (Letters)*, **303**, L41 (1986)

58. B. Smith, S. Kleinmann, J. Huchra: in preparation.

59. A. Dressler, S.M. Faber, D. Burstein, R. Davies, D. Lynden-Bell, R. Terlevich, G. Wegner: *Astrophys. J. (Letters)*, submitted (1986)

60. A. Dressler: *Astrophys. J.*, **236**, 351 (1980)

61. B. Binggeli: this volume.

62. M. Haynes: this volume.

63. A. Yahil: "The Local Gravitational Field", in B.F. Madore, R.B. Tully, eds.: *Galaxy Distances and Deviations from Universal Expansion*, pp. 143-145, Reidel, Dordrecht 1986.

64. P.B. Lilje, A. Yahil: in preparation.

65. J.E. Gunn: "Conference Summary", in G. Knapp, J. Kormendy, eds.: *Dark Matter in the Universe*, in press, Reidel, Dordrecht 1986.

Fundamental Physics and Dark Matter

KATHERINE FREESE

For the past ten years or so there has been a lot of speculation about the existence and behavior of nonbaryonic dark matter. In this talk I would like to re-examine the evidence for a non-baryonic dark component, contrasting the theoretical prejudice that $\Omega = 1$ with the Big Bang Nucleosynthesis constraints on the amount of baryonic matter, $\Omega_b \leq 0.2$. Assuming then that such a component does exist, I will present a list of candidates from particle physics to explain this dark matter, and illustrate what I find to be a most exciting development, the fact that all the most likely candidates stand to be ruled out or even detected in the next five to ten years.

I promised a review of the arguments why cosmologists believe $\Omega = 1$. Under the assumptions of isotropy and homogeneity for the universe, Einstein's equations take the form

$$H^2 = (\dot{R}/R)^2 = 8\pi G\rho/3 - k/R^2 \tag{1}$$

where R is the scale factor, the Hubble parameter H describes the expansion of the universe, ρ is the energy density of the universe, and k is the spatial curvature. Different values of the spatial curvature divide the universe into three types of models:

1) $k < 0$, negatively curved spatial sections, open universe, $\Omega < 1$.

2) $k = 0$, flat spatial sections, open universe, $\Omega = 1$.

3) $k > 0$, positively curved spatial sections, closed universe, $\Omega > 1$,

where $\Omega = \rho/\rho_c$, the ratio of the energy density of the universe to the critical density required to close the universe. The evolution of the universe can then be parameterized in the following way:

$$\Omega(t) = \frac{1}{1 - x(t)} \tag{2a}$$

where

$$x(t) = \frac{k/R^2}{8\pi G\rho/3} \propto \begin{cases} R^2(t) & \text{radiation dominated} \\ R(t) & \text{matter dominated} \end{cases} \tag{2b}$$

is an increasing function of time. Plotting Ω as a function of x, or equivalently as a function of time (see Fig. 1), one can see that for $k = 0$, Ω stays exactly equal to 1 forever, but for any other value of k, Ω diverges away from 1 rapidly. More quantitatively, since we know $\Omega = \mathcal{O}(1)$ today, at the Planck time it must have satisfied $|\Omega(10^{-43}\text{ sec}) - 1| \simeq \mathcal{O}(10^{-60})$. Thus unless $k = 0$, it is very difficult to explain the energy density we see today. The

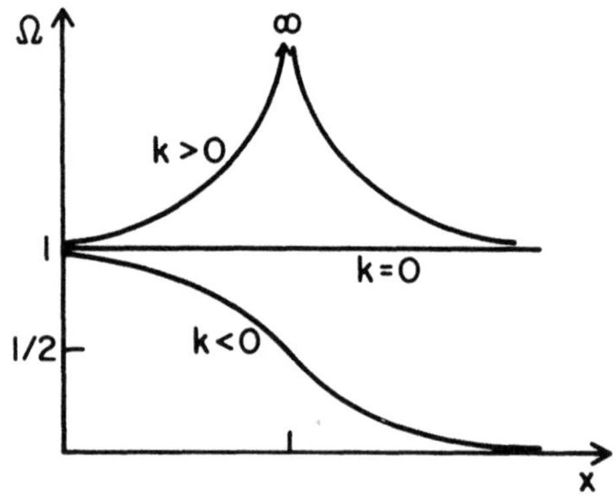

Figure 1: The energy density of the universe as a function of x as defined in Eq. (2b) for different values of the spatial curvature k.

selection of this particular value of the curvature cannot be explained in the standard model, but can be solved in an inflationary scenario. In an inflationary model, as the universe passes through a phase transition in the early universe, the universe suddenly finds itself in a *false* vacuum characterized by a large vacuum energy density $\mathcal{O}(M^4)$ where M is the mass scale associated with the phase transition. During the slow rollover phase to the true vacuum, this vacuum energy dominates the energy density of the universe,

$$H \sim \left[\frac{8\pi V(0)}{3m_{\rm pl}^2}\right]^{\frac{1}{2}} \sim \mathcal{O}\left[M^2/m_{\rm pl}\right] \tag{3}$$

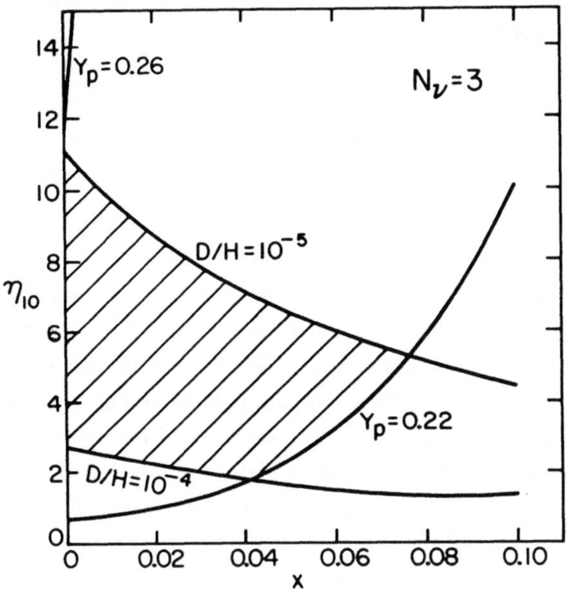

Figure 2: Element abundances as a function of vacuum energy $x = \Omega_{vac}$ and η_{10} (the baryon-to-photon ratio at nucleosynthesis in units of 10^{-10}) for three neutrino species. Primordial ^{4}He satisfies $0.22 \leq Y_p \leq 0.26$ and the ratio of deuterium to helium satisfies $10^{-5} \leq D/H \leq 10^{-4}$. Cross-hatching indicates the allowed region.

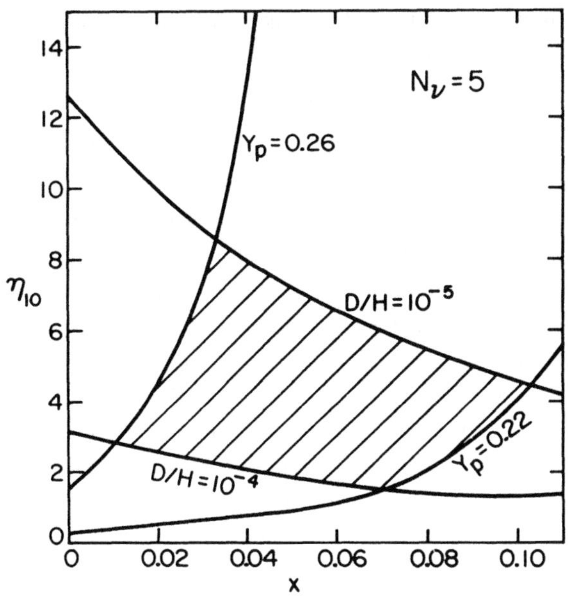

Figure 3: Element abundances as a function of vacuum energy for 5 neutrino species (for notation see Fig. 2). Cross-hatching indicates the allowed region; n.b. $x \neq 0$ can give consistency with $N_\nu > 4$.

where $m_{\rm pl} \simeq 10^{19}$ GeV is the Planck mass, $V(0)$ is the free energy of the false vacuum, and the universe expands exponentially:

$$R \sim e^{Ht}. \tag{4}$$

If the expansion persists, e.g. for 100 e-folds, R expands by e^{100}, and from Eq. (2b) $x_{\rm after} = e^{200} x_{\rm before}$. From Eq. (2a) we see that this predicts $\Omega = 1.0000\cdots$ (Discussion taken from TURNER [1]).

Contrasted with the theoretical arguments for $\Omega = 1$, abundances of various elements from Big Bang Nucleosynthesis including ^{4}He, ^{3}He, D, and ^{7}Li constrain the amount of baryonic matter in the universe (subscript b) to lie in the range $0.01 \leq \Omega_b \leq 0.2$ [2]. This is the origin of the claim that the "missing mass" must be nonbaryonic and has led to the study of various exotic dark matter candidates. In the light of recent observations, however, although the Ω corresponding to objects of various sizes does indeed increase with scale (from stars to galaxies to clusters) it seems to plateau at a value of 0.2 or 0.3 [3]. Hence an observer might question the need for invoking any dark matter beyond what is baryonic, and argue that $\Omega_b = 0.2$ is sufficient to explain the observed properties of the universe.

Aside from the theoretical prejudice that Ω should be 1, what are the arguments in favor of a nonbaryonic component? First, there is the issue of perturbations leading to galaxy formation. In a universe dominated by baryons, perturbations in the baryon density can start to grow only after recombination at a redshift $z = 10^3$. [Before this time baryons are tied to photons which diffuse out of any overdense region.] Since the perturbations grow as $\delta\rho/\rho \propto \frac{1}{1+z}$, in order to reach nonlinearity by today the value at recombination must have been $\left.\frac{\delta\rho}{\rho}\right|_b (t_{\rm rec}) \simeq 10^{-3}$. Since for adiabatic perturbations the anisotropy in the microwave background $\frac{\delta T}{T} \simeq \left.\frac{\delta\rho}{\rho}\right|_b (t_{\rm rec})$ this value is in conflict with observations of the microwave background $\frac{\delta T}{T} \leq 10^{-4}$. In a universe with nonbaryonic dark matter, on the other hand, the perturbations in the dark matter can begin growing sooner, whenever the nonbaryonic component starts to dominate the energy density of the universe. The baryon perturbations remain small until recombination, and then fall into the gravitational potential wells provided by the dark matter. If the nonbaryonic species becomes dominant, e.g. at $z = 10^4$, then one obtains $\left.\frac{\delta\rho}{\rho}\right|_b (t_{\rm rec}) \simeq 10^{-4}$, consistent with observations of the microwave anisotropy [4]. Second, OLIVE and HEGYI [5] have argued that the dark matter in galactic haloes probably must be nonbaryonic. They have considered various possible baryonic candidates, including snowballs, gas, low mass stars and Jupiters, high mass stars, and high metallicity objects such as rocks or dust. By various different arguments they find that all of these classes are unlikely to solve the dark matter of the halo (with the exception of a possible loophole: low mass stars with a different initial mass function) and argue for a nonbaryonic component.

In an attempt to resolve the discrepancy between the theoretically preferred $\Omega = 1$ and the observed $\Omega \simeq 0.2$, several groups have proposed various smooth components, *i.e.*, unclustered matter that would not show up in surveys of large scale structure. These include failed galaxies [6], relativistic particles produced by decaying weakly interacting massive particles [7], and a vacuum energy component that can be either constant [8,9] or decaying in time. I will briefly discuss the last of these possibilities FREESE *et al.* [10]. In the presence of a decaying vacuum energy ρ_{vac}, Einstein's equations take the form

$$(\dot{R}/R)^2 = \frac{8\pi G}{3}(\rho + \rho_{\text{vac}}) \qquad (\text{for } k = 0) \tag{5}$$

where the vacuum decays to radiation and in the radiation dominated era $\rho_{\text{vac}} \propto \frac{1}{t^2}$. The cosmology of this model is somewhat different from the standard one: e.g. the relationship between temperature and time in the early universe is changed, and there is significant entropy generation due to vacuum energy decay throughout Nucleosynthesis: the baryon-to-photon ratio

$$\frac{n_b}{n_\gamma} \propto T^{\frac{3\Omega_{\text{vac}}}{1-\Omega_{\text{vac}}}}. \tag{6}$$

We considered the effects of a vacuum component on element abundances in Big Bang Nucleosynthesis, as illustrated in Fig. 2 for three neutrino species. Although we find that the vacuum component must be small, $\Omega_{\text{vac}} \leq 0.1$ at nucleosynthesis and $\Omega_{\text{vac}} \leq 10^{-5}$ today, and cannot provide the missing mass of the universe, it can still have important consequences. Because of the altered temperature-time relation, the time at which dark matter perturbations can begin to grow is pushed backward, thus allowing more time for galaxy formation. Also, a nonzero vacuum energy can accommodate five or more neutrino species (see Fig. 3), which are ruled out in the standard model. If more than four neutrinos or equivalent light particle species are found to exist, then one may be driven to such a scenario.

Assuming for the remainder of the talk that a nonbaryonic component does exist, I will briefly run through some of the candidates from particle physics. The standard model of particle physics has as its fundamental constituents the following particles:

i) Three families of quarks and their associated leptons:

$$\text{quarks} \quad \begin{pmatrix} u \\ d \end{pmatrix} \begin{pmatrix} c \\ s \end{pmatrix} \begin{pmatrix} t \\ b \end{pmatrix}$$

$$\text{leptons} \quad \begin{pmatrix} e \\ v_e \end{pmatrix} \begin{pmatrix} \mu \\ v_\mu \end{pmatrix} \begin{pmatrix} \tau \\ v_\tau \end{pmatrix}$$

ii) Gauge bosons which mediate the fundamental interactions (γ mediate electromagnetism; $W^{\pm}, Z^0$ mediate weak interactions; gluons mediate strong interactions; $X, Y \cdots$ mediate suppressed processes such as proton decay).

But the standard model does not appear to be complete since it leaves many questions unanswered: why are there three families, how does one explain quark and lepton masses, the hierarchy problem, how can one incorporate gravity, etc. Several attempts at more fundamental theories have been proposed, each of them predicting the existence of new particles which serve as dark matter candidates.

Table I gives a list of various dark matter candidates, their masses, and the models in which they arise. I have put a check next to those candidates I consider most likely, and have also indicated the cases where detectors to look for the particle are either presently running or proposed for the future. A very exciting development is the fact that all three of the most likely candidates stand to be either discovered or ruled out as the dark matter within the next decade. The three most likely candidates are light massive neutrinos, axions, and GeV mass particles (also known as WIMPs, weakly interacting massive particles). Tritium β-decay experiments as well as searches for neutrino oscillations have already been running to probe the existence of a light neutrino species, perhaps the most likely candidate since neutrinos are actually known to exist. Claims of the discovery of a neutrino mass have been made but are extremely controversial [11]. Proposed detection schemes for axions [12], the very light pseudoscalar particles necessary for the solution of the strong CP problem in theories of the strong interaction, exploit the coupling of axions to two photons to convert halo axions into photons. Several groups are designing detectors using similar ideas.

I will focus on recent ideas for detecting particles in the GeV mass range. To remind you why this mass range is interesting for dark matter candidates, consider the energy density of a nonbaryonic species (in this case neutrinos) as a function of the particle's mass. Requiring that $\Omega_\nu < 2$ in order not to overclose the universe, one sees that two interesting mass ranges remain: $m_\nu < 200$ eV or $m_\nu > 1.5$ GeV [13]. The light neutrinos would provide hot dark matter, and the heavy ones cold dark matter. I will concentrate on particles in this GeV

TABLE I

DARK MATTER CANDIDATES

particle	mass	model	most likely	detector
axion	10^{-5} eV	QCD	√	proposed
light neutrino	25 eV	GUTs	√	running: tritium β-decay, ν oscillations
$\tilde{\gamma}$, $\tilde{\nu}$, $\tilde{H}^0$ heavy neutrino shadow matter	GeV	SUSY	√	proposed
preon bound states	(20–200) TeV	composite models		
magnetic monopoles	10^{16} GeV	GUTs		running
pyrgons	$\geq 10^{19}$ GeV	higher dimensional theories		
primordial black holes	$\geq 10^{15}$ gm			

mass range. Various supersymmetric particle candidates also exhibit similar behavior, and again could have masses $\mathcal{O}$(GeV). Supersymmetry predicts higher mass partners for known particles: for every fermion a boson and vice versa. Conservation of a new quantum number in many models, R-parity, implies that the lowest mass supersymmetric particle must be stable against decay and may therefore provide the dark matter. Candidates include the photino, the supersymmetric partner of the photon, the scalar neutrino or sneutrino, the partner of the neutrino, and the Higgsino, the partner of the Higgs. I wanted to point out also that exactly this class of GeV mass particles has recently been proposed as a solution of the solar neutrino problem [14,15,16], although existing candidates may not work because the particles annihilate with one another in the sun [17].

The class of cold dark matter candidates in the GeV mass range can be probed in two ways:

1) Their behavior in the earth, sun, or galactic halo might give rise to observable consequences. (cost = 0 dollars)
2) Direct detection of these particles [$\mathcal{O}$(1 million bucks)].

Several detectors have been proposed differing significantly in design yet all based on the same principle. A particle going through the detector scatters elastically off of nuclei; the detector must be sensitive to a nuclear recoil energy of a few keV, a very small energy deposit. Proposals in the U.S. include those of DRUKIER and STODOLSKY [18]; CABRERA, KRAUSS and WILCZEK [19]; and SADOULET. For the numbers presented in the talk I will assume that the particles are present with a halo density $\rho \simeq 0.007\ M_{\odot}/pc^3 = 0.4\ \mathrm{GeV/cm^3}$ and a Maxwellian velocity distribution characterized by a velocity dispersion of 300 km/s.

One way to probe these dark matter candidates is to consider their behavior in, e.g. the earth [20,21]. Massive Dirac or scalar neutrinos with masses (1-20) GeV are captured in the earth with an efficiency $10^{-10} - 10^{-7}$ relative to the number entering the earth. In spite of this tiny efficiency for capture, a sufficient number are captured, sink to the center of the earth, and start annihilating with one another to give an observable signal of massless neutrinos at the surface of the earth. The flux at the earth's surface is sufficiently above atmospheric background that a signal should have been seen in a proton decay detector. Angular acceptance due to the fact that we expect this signal to be coming from the center of the earth can be further used to reject background. In conclusion, Dirac or scalar neutrinos more massive than about 10 GeV cannot be the dominant component of the halo (we cannot say anything about less massive ones because they could have evaporated off the high-energy tail of the Boltzmann distribution). A similar analysis in the sun [22,23,24] can be used to show that photinos more massive than 3 GeV are marginally possible as the dark matter of the halo.

A more exciting way to probe this class of dark matter candidates is to directly detect them. I will focus on the particular detection scheme of superheated superconducting colloids first proposed by DRUKIER and STODOLSKY [18] as a detector for neutrino astronomy. GOODMAN and WITTEN [25] then noticed that it could be used as a dark matter detector. In a paper by DRUKIER, FREESE and SPERGEL [26] we worked out the actual count rates one could expect for various particles. The total mass of the detector is 1 kg, with micronsize grains in a dielectric filling material. The superconducting grains are placed in a magnetic field and maintained in a superheated state. When a halo particle passes through the detector, it interacts with a nucleus inside a grain, the nucleus recoils, energy is deposited in the grain, the temperature of the grain goes up, and the grain goes normal. This change of state can be detected by the Meissner effect. Penetration of the magnetic flux into the grains produces an observable signal on the readout electronics. The flipping of a grain a few microns in size produces a change of flux of typically 0.1 ϕ_0 (flux quanta) in a few centimeter superconducting loop encircling the detector. SQUIDs can detect this flux change. Demanding a signal-to-noise ratio of at least 10, we calculated the count rates in the detector for various particles. We expect 10^4 counts/day for sneutrinos and Dirac neutrinos and $\mathcal{O}(1)$ count/day for photinos more massive than a few GeV.

Advantages over conventional particle detectors for GeV mass particles are the sensitivity to heavy nucleus recoils with energy < 1 KeV and good background rejection. The detector is small (1 kg) so one can use anticoincidence or shielding effectively, e.g. to eliminate cosmic rays. Also the signal from such small energy deposits will be the flip of one and only one grain. Most backgrounds such as charged particles will flip many grains, e.g. a 0.5 MeV electron with range 0.5 gm/cm^2 must pass through 100 grains to lose its energy. The worst background for high energy threshold is natural radioactivity (^{40}K, U, Th) and for low energy threshold is solar neutrinos. These backgrounds as well as Compton effect backgrounds may contribute $\mathcal{O}$(1 count/day), well below the count rate for neutrinos or sneutrinos, but troublesome for photinos. Even if one obtains a signal, how can a real signal be distinguished from the systematics of the detector? One can take advantage of a modulation effect to help reject background. The motion of the earth around the sun produces a modulation in the dark matter count rate. When the earth's velocity relative to the halo is highest, there are more particles above threshold. The maximum count is predicted in May when the earth's velocity is parallel to the sun's motion around the galactic center. This modulation gives rise to a 10–20% effect over the course of a year. Taking advantage of this type of effect for background rejection one can hope to detect or rule out GeV mass dark matter candidates within the next decade. Once this type of detector is perfected to even lower energy sensitivity it may be used eventually as a neutrino telescope to study, e.g. solar neutrinos and supernovae, the purpose for which it was originally proposed.

In conclusion, fundamental particle physics offers a host of dark matter candidates. Detectors have been proposed to explicitly look for several of the most likely classes of these. A new type of observational astrophysics is attracting experimentalists from many branches of physics.

REFERENCES

1. M.S. Turner: "Cosmology and Particle Physics," in *Proceedings of the 1st and 2nd Jerusalem Winter Schools*, eds. T. Piran and S. Weinberg, 101 (1986).
2. D.N. Schramm and R.V. Wagoner: *Ann. Rev. Nucl. Sci.* **27**, 37 (1977).
3. J.P. Huchra: in *Inner Space Outer Space* ed. Kolb, Turner, Lindley, Olive, and Seckel (Chicago: Univ. of Chicago Press), 65 (1986).
4. A.G. Doroshkevich, M. Yu. Khlopov, R.A. Sunyaev, A.S. Szalay and Ya. B. Zel'dovich: in *10th Texas Symposium on Relativistic Astrophysics,* 32 (1981).
5. K.A. Olive and D.J. Hegyi: in *Inner Space Outer Space* ed. Kolb, Turner, Lindley, Olive, Seckel (Chicago: Univ. of Chicago Press), 112 (1986).
6. N. Kaiser: *Ap. J.* **284**, L9(1984).

7. M.S. Turner, G. Steigman, and L.K. Krauss: *Phys. Rev. Lett.* **52**, 2090 (1984); R. Flores, G. Blumenthal, A. Dekel, and J.R. Primack, *Nature*, in press (1986).

8. P.J.E. Peebles: *Ap. J.* **284**, 439 (1984).

9. J.R. Gott, J.E. Gunn, D.N. Schramm, and B.M. Tinsley: *Ap. J.* **194**, 543 (1974).

10. K. Freese, F.C. Adams, J.A. Frieman, and E. Mottola: ITP preprint "Cosmology with Decaying Vacuum Energy" (1986).

11. V. Lubimov, E. Novikov, V. Nozik, E. Tretyakov, and V. Kosik: *Phys. Lett.* **94B**, 266; *JETP* **54**, 616 (1981).

12. P. Sikivie: P. *Phys. Rev. Lett.* **51**, 1415 (1983).

13. B.W. Lee and S. Weinberg: *Phys. Rev. Lett.* **39**, 165 (1977).

14. G. Steigman, C.L. Sarazin, H. Quintana, and J. Faulkner: *Ap. J.* **83**, 1050 (1978).

15. D.N. Spergel and W.H. Press: *Ap. J.* **294**, 663 (1985).

16. J. Faulkner and R. Gilliland: *Ap. J.* **299** (1985).

17. L.K. Krauss, K. Freese, D.N. Spergel, and W.H. Press: *Ap. J.* **299** (1985).

18. A. Drukier and L. Stodolsky: *Phys. Rev.* **D30**, 2295 (1984).

19. B. Cabrera, L.K. Krauss, and F. Wilczek: *Phys. Rev. Lett.* **55**, 25 (1985).

20. K. Freese: *Phys. Lett.* **167B**, 295 (1986).

21. L.K. Krauss, M. Srednicki, and F. Wilczek: *Phys. Rev.* **D33**, 2079 (1986).

22. J. Silk, K.A. Olive, and M. Srednicki: *Phys. Rev. Lett.* **55**, 257 (1985).

23. K. Greist and D. Seckel: UC Santa Cruz preprint SCIPP 86/60 (1986).

24. T.K. Gaisser, G. Steigman, and S. Tilav: Bartol Research Institute preprint BA-86-42 (1986).

25. M.W. Goodman and E. Witten: *Phys. Rev.* **D30**, 3059 (1985).

26. A.K. Drukier, K. Freese, and D.N. Spergel: *Phys. Rev.* **D33**, 3495 (1986).

Session 9

Galaxies Before Recombination

J. Primack, Chair
July 31, 1986

Inflationary Universe Models and the Formation of Structure

ROBERT H. BRANDENBERGER

Abstract

I briefly review the main features of inflationary universe models. Inflation provides a mechanism which produces energy density fluctuations on cosmological scales. In the original models it was not possible to obtain the correct magnitude of these fluctuations without fine tuning the particle physics models. I discuss two mechanisms, "chaotic inflation", and a "dynamical relaxation" process, by which inflation may be realized in models which give the right magnitude of fluctuations.

1. Introduction

Until recently, most cosmological models were based on coupling classical general relativity as the theory of space-time with an ideal gas as a matter source. In the most popular of these models, the "big bang" theory, the universe starts out at an initial time t=0 with infinite temperature and energy density. Obviously, at very high temperatures, i.e. very early times, the classical description of matter as an ideal gas will be very misleading.

In the new cosmological theories the ideal gas as matter source for classical gravity is replaced by a description of matter which will be valid until much higher temperatures, namely in terms of quantum fields. In inflationary universe models matter is taken to be given by a grand unified theory. This may lead to deviations from the standard big bang model at very early times.

The standard big bang model leads to various puzzles for cosmologists, the first of which is the horizon problem. The universe looks homogeneous and isotropic on scales much greater than the causal horizon at the time the features we observe were established. The prime example is the homogeneity and isotropy of the microwave background radiation (Figure 1). If we look back at the last scattering surface along two light rays subtending an angle of more than 1°, we are looking at points which were causally disconnected at last scattering. Why should the microwave temperature be the same?

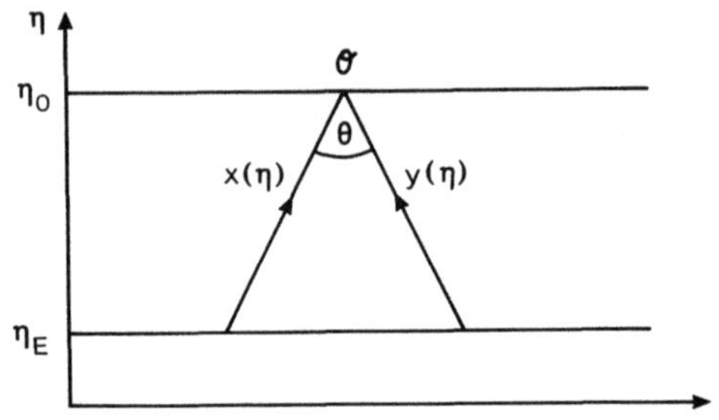

Figure 1 A conformal space-time diagram showing two light rays $X(\eta)$ and $Y(\eta)$ seen by the observer σ and subtending an angle θ. The last scattering surface is $\eta = \eta_E$. η_0 is the present conformal time.

The second puzzle is the flatness problem. We observe no spatial curvature of the universe. Hence the energy density today must be equal to the critical density (the density which is required for a spatially flat solution of Einstein's equations) to within one order of magnitude. In an expanding universe flat space is unstable in the sense that any deviation from the critical energy density grows in time as T^{-2} (T is the temperature). In order to explain the present day flatness of the universe, the energy density ρ would have had to be equal to the critical energy density ρ_0 so one in 10^{55} at a temperature of 10^{17} Gev, a temperature at which we imagine setting up initial conditions. This is a completely unnatural fine tuning of initial conditions.

Finally, the standard big bang model does not explain why inhomogeneities such as galaxies and clusters of galaxies are correlated on scales which initially were outside the causal horizon.

The key to resolving these cosmological puzzles is to change the expansion rate of the universe at very early times. In the standard model the space-time metric is given by

$$ds^2 = -dt^2 + a(t)^2 d\underline{x}^2 \tag{1}$$

a(t) is the scale factor and determines the physical distance between two points at rest. The time dependence of a(t) is determined by the equation of state of matter. For relativistic matter the pressure p equals $1/3\rho$ (ρ is the energy density), and $a(t) \sim t^{1/2}$. For non-relativistic matter p=0 and $a(t) \sim t^{2/3}$ In both cases the Hubble radius

$$H^{-1}(t) = \left[\frac{\dot{a}(t)}{a(t)}\right]^{-1} \tag{2}$$

increases faster than the distance between two points at rest. $H^{-1}(t)$ is the maximal distance microphysical forces can act at time t and can thus be called the causal horizon.

What changes if there is a period in the early universe in which the scale factor increases exponentially (hence the physical size of the universe expands exponentially, i.e. inflates)? Now the Hubble radius is constant while physical scales increase. Distances outside the causal horizon at later times were initially inside the Hubble radius, and thus a solution of the horizon problem is at hand. Provided the period of exponential expansion lasted sufficiently long, our entire observed universe started out inside the causal horizon (Figure 2). In addition, correlations of energy density fluctuations can be created inside the Hubble radius in the exponential phase (the de Sitter phase), expand to be far outside the causal horizon at the beginning of the radiation-dominated Friedmann-Robertson-Walker (FRW) period, and reenter the Hubble radius later as the fluctuations we observe on the scales of galaxies and clusters of galaxies. Finally, the flatness problem is resolved. Our observed universe stems from an exponentially small patch P of the universe at the beginning of the de Sitter phase. Any reasonable curvature initially on scales of the Hubble radius is exponentially small compared to the scale of P. The inflationary universe predicts $\Omega = |\rho - \rho_0| \rho_0^{-1} \simeq 1$.

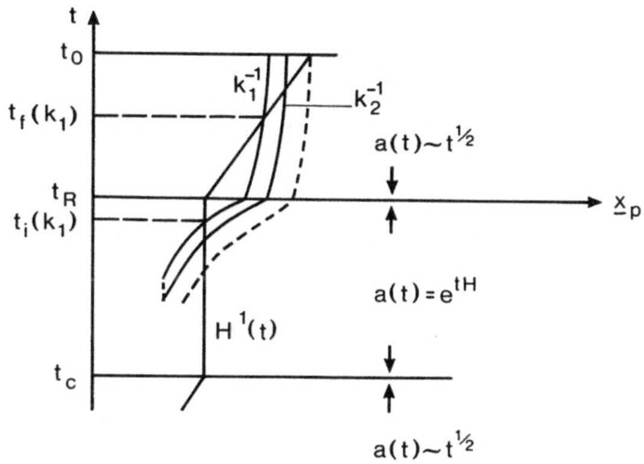

Figure 2 Evolution of the comoving scale corresponding to the present day ($t = t_0$) horizon (dotted line) in the inflationary universe. The solid line $H^{-1}(t)$ is the Hubble radius which is proportional to t in the radiation dominated phases $t < t_c$ (t_c the time when inflation starts) and $t > t_R$ (t_R is the reheating time). Also shown are the scales corresponding to fluctuations of wave numbers k_1 and k_2.

The crucial observation by Guth [1] was that an equation of state $p = -\rho$ for matter which gives rise to an exponentially increasing scale factor emerges naturally when considering a class of grand unified field theories in the early universe. In section 2 I will describe the basic mechanism of inflation (see Ref 2 for more detailed expositions). I then discuss the causal mechanism by which energy density fluctuations required to form galaxies and clusters of galaxies are generated. The mechanism is very successful in explaining the qualitative features of the spectrum of density perturbations, but in the initial models [3] fails to produce the correct amplitude [4,5]. In section 4 I outline two new mechanisms by which inflation may be realized in models with the correct amplitude of fluctuations. (See also Ref. 6.)

2. The New Inflationary Universe

To illustrate the main ideas I will consider a simple toy model field theory, a double well $\lambda\phi^4$ model with potential (Figure 3)

$$V(\phi) = \frac{\lambda}{4}(\phi^2 - \sigma^2)^2 \tag{3}$$

σ is the scale of symmetry breaking, and is of the order 10^{15} Gev for grand unified theories. It is very natural to consider such a potential, since in any grand unified theory in which the symmetry is spontaneously broken there will be a Higgs field with a similar potential.

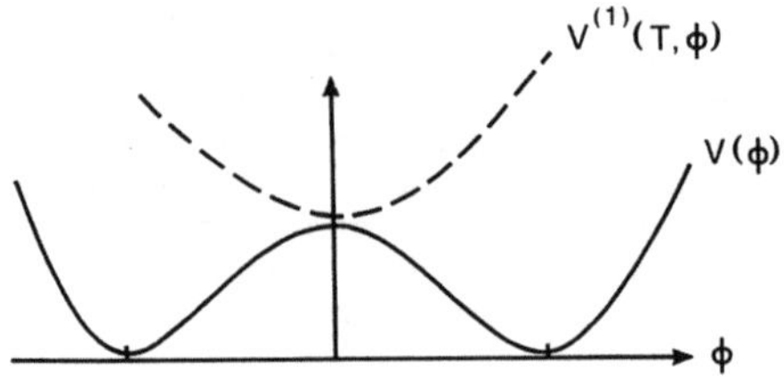

Figure 3 A typical potential $V(\phi)$ for the new inflationary universe $V^{(1)}(T,\phi)$ is the finite temperature one-loop effective potential, drawn at a temperature T much larger than the critical temperature T_c.

The equation of state for the theory with Lagrangian

$$L(\phi) = \tfrac{1}{2}\partial_\mu\phi\partial^\mu\phi - V(\phi) \tag{4}$$

is

$$\rho = \tfrac{1}{2}\dot{\phi}^2 + \tfrac{1}{2}a^{-2}(t)(\underline{\nabla}\phi)^2 + V(\phi) \tag{5}$$

$$p = \tfrac{1}{2}\dot{\phi}^2 - \tfrac{1}{6}a^{-2}(t)(\underline{\nabla}\phi)^2 - V(\phi)$$

We consider a modified big bang model in which there is ordinary matter plus the scalar field ϕ. The energy-momentum tensor is

$$T_{\mu\nu} = T_{\mu\nu}(\phi) + T_{\mu\nu}(\text{rad}). \tag{6}$$

$T_{\mu\nu}(\text{rad})$ is the energy-momentum tensor of ordinary matter and gives an equation of state $p = \frac{1}{3}\rho$. $T_{\mu\nu}(\phi)$ is the contribution to $T_{\mu\nu}$ from the scalar field.

The standard lore of new inflation is as follows. After the big bang $\phi(\underline{x},t) = 0$. At high temperatures $T_{\mu\nu}$ is dominated by radiation, the effective equation of state is $p = 1/3\rho$ and the scale factor grows as $a(t) \sim t^{1/2}$. Once the radiation gas has cooled to below the critical temperature T_c determined by

$$\lambda\sigma^4 = \frac{\pi^2}{30} NT_c^4 \tag{7}$$

(where N is the number of spin degrees of freedom at temperature T_c). $T_{\mu\nu}(\text{rad})$ becomes unimportant and the equation of state is given by the scalar field. If $\phi(\underline{x},t) = 0$ then $p = -\rho$, and by the Einstein equations

$$\left[\frac{\dot{a}}{a}\right]^2 = \frac{8\pi G}{3}\rho \tag{8}$$

$$\dot{\rho} = -3H(t)(\rho + p)$$

we see that a(t) expands exponentially

$$a(t) = e^{Ht} \ , \ H = \left[\frac{8\pi G}{3}\rho\right]^{1/2} . \tag{9}$$

The equation of motion for a spatially homogeneous scalar field configuration in an expanding universe is

$$\ddot{\phi} + 3H(t)\dot{\phi} = -\partial V/\partial\phi \tag{10}$$

$\phi(t) = 0$ is an unstable fixed point; any deviations from it will grow in time. Thus, starting near $\phi(\underline{x}) = 0$, the scalar field will slowly roll down the potential slope (Figure 3). As long as the rolling is slow, the equation of state is still approximately $p = -\rho$, and inflation continues. When $\phi(t)$ approaches $\phi = \pm\sigma$, the scalar field will begin to roll fast, it will oscillate about $\phi = \pm\sigma$ and its energy is converted into thermal energy. In this period, which generally is much shorter than a Hubble expansion time H^{-1}, the universe reheats to close to the critical temperature T_c. The phases of the new inflationary universe are sketched in Figure 4.

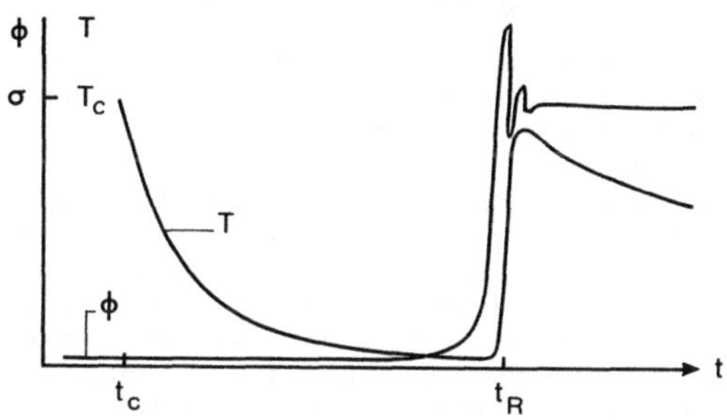

Figure 4 Sketch of the phases in the new inflationary universe. During most of the inflationary (de Sitter) phase ϕ remains close to the origin and T decreases exponentially. At reheating ϕ increases to its ground state value σ and the universe reheats to almost T_c.

Why does the scalar field start out with $\phi(\underline{x},t(T)) = 0$ for $T > T_c$? The answer depends on whether the scalar field ϕ is weakly or strongly coupled to other fields. Consider first the case in which ϕ is strongly coupled to a second field ψ, which for simplicity we take to be a second scalar field. The interaction Lagrangian is $\frac{1}{2}\tilde{\lambda}\phi^2\psi^2$ and will give rise to an effective mass term in the equation of motion for ϕ

$$\ddot{\phi} + 3H\dot{\phi} = -\frac{\partial V}{\partial\phi} - \tilde{\lambda}\psi^2(T)\phi \tag{11}$$

At high temperatures the mass term will drive $\phi(\underline{x})$ towards $\phi = 0$. An alternate way to see this effect would be to compute the one loop finite temperature effective potential, which will give a similar mass term [7] (see Figure 3). The case in which ϕ is only very weakly coupled to other fields will be considered in Section 4. (See also Ref. 8.)

3. Origin of Energy Density Fluctuations

Probably the most astonishing success of the inflationary universe is that it provides a mechanism which in a causal way generates a spectrum of primordial energy density fluctuations necessary to produce galaxies and clusters of galaxies. The first crucial observation is that in inflationary universe models all scales of interest today originate inside the causal horizon in the de Sitter period [9] (see Figure 2).

We can also derive the shape of the perturbation spectrum on very general grounds. Consider the two scales indicated in Fig 2 and assume that there is some mechanism which produces perturbations at all times on a fixed physical scale. Then the evolution of the perturbations k_1 and k_2 between when they were formed and when they left the Hubble radius at time $t_i(k)$ are related by time translation. Hence the magnitude of the perturbation is independent of k when evaluated at time $t_i(k)$:

$$\frac{\delta M}{M}(k,t_i(k)) = \text{const} \tag{12}$$

$\delta M/M(k,t)$ is the r.m.s. mass excess in a ball of radius k^{-1} at time t. While the perturbation is outside the Hubble radius microphysical processes can not influence it. Hence its physical size will not change until it reenters the horizon at time $t_f(k)$. Therefore,

$$\frac{\delta M}{M}(k,t_f(k)) = \text{const} \tag{13}$$

This shape of spectrum is called scale-invariant or Harrison-Zel'dovich spectrum. It was postulated a long time ago [10] as a reasonable spectrum for galaxy and cluster formation.

Now the second crucial observation [4,5]. Quantum fluctuations in the scalar field $\phi(\underline{x})$ can produce classical energy density perturbations. This problem has recently been studied semiclassically [11], i.e. without quantizing gravity, and in linearized quantum gravity [12,13]. The main idea in the semiclassical analysis is very simple. In quantum mechanics, we all know that the expectation value of the operator Q^2 in a state $|\psi\rangle$ of a simple harmonic oscillator can be interpreted as the square of its classical amplitude q in state $|\psi\rangle$

$$q^2 = \langle\psi|Q^2|\psi\rangle \tag{14}$$

Translating this procedure to field theory we can define a classical field

$$\phi(\underline{x},t) = \phi_0(t) + \delta\phi(\underline{x},t) \tag{15}$$

(ϕ_0 is homogeneous, $\delta\phi(\underline{x})$ has vanishing spatial average) in terms of the expectation values of the operators $\Phi^2(\underline{x})$ and $\tilde{\Phi}^2(\underline{k})$ ($\tilde{\Phi}$ is the Fourier transform of Φ) [15]

$$\phi_0(t) = \langle\psi|\Phi^2(\underline{x})|\psi\rangle_{ren} \tag{16}$$

$$\delta\tilde{\phi}(\underline{k}) = \langle\psi|\tilde{\Phi}^2(\underline{k})|\psi\rangle ,$$

where the subscript ren indicates a renormalized expectation value (see Ref 11).

If we choose the state $|\psi\rangle$ to be the state empty of particles at the beginning of the de Sitter phase, then its wave functional is the product of the ground state wave functions for each Fourier mode of ϕ. The expectation values can easily be evaluated. For a free, massless, minimally coupled scalar field we obtain

$$\phi_0(t) = (2\pi)^{-1}H^{\frac{3}{2}}t^{1/2} \tag{17}$$

$$\delta\tilde{\phi}(\underline{k},t) = V^{1/2}(2\pi)^{\frac{3}{2}}a^{-\frac{3}{2}}(t)H^{-1/2} ,$$

where V is the cutoff volume. It is now easy to evaluate the r.m.s. mass perturbations at initial Hubble radius crossing $t_i(k)$. For spectra with reasonable k dependence we can express $(\delta M/M)(k)$ in terms of the Fourier transform of the energy density perturbation

$$\left[\frac{\delta M}{M}\right]^2(k) = V^{-1}k^3\left[\frac{\delta\tilde{\rho}}{\rho}\right]^2(k) \qquad (18)$$

Since $T_{\mu\nu}$ for a free scalar field is well known, we can immediately compute the mass perturbations

$$\left[\frac{\delta M}{M}\right](k,t_i(k)) = O(1)\frac{H^4}{\rho} \sim \lambda^{-1}\left[\frac{H}{\sigma}\right]^4 \sim \lambda\left[\frac{\sigma}{m_{pl}}\right]^4 \qquad (19)$$

For typical grand unified theories $\sigma \sim 10^{15}$ Gev and hence the mass perturbation at initial Hubble radius crossing is of the order 10^{-16}. The third crucial point was the observation that the amplitude of these fluctuations is amplified by a large factor as a consequence of the change in the equation of state from de Sitter-like $(p = -\rho)$ to radiation dominated $(p = 1/3\rho)$.

In linear perturbation theory fluctuations with different wavelengths do not couple and hence can be analysed independently. While the wavelength of a given perturbation is inside the Hubble radius, microphysical processes determine the amplitude. Once the wavelength is greater than the Hubble radius, microphysical processes can no longer act coherently, and the evolution of the perturbation is determined by gravity alone.

The analysis of the evolution of perturbations outside the Hubble radius is not straightforward. The main problem is to separate the physical degrees of freedom from pure gauge modes. For a detailed explanation, we refer the reader to [5] and [14]. The main point is the following: if we restrict our attention to scalar modes, we can form a gauge invariant variable Φ_H by combining the components of the tensor of the metric perturbation in a clever way [15]. When the wavelength equals the Hubble radius (at time t_H), Φ_H is essentially equal to the energy density perturbation (evaluated in comoving coordinates, which we indicate by a subscript c)

$$\Phi_H(t_H) = \alpha\left[\frac{\delta\rho}{\rho}\right]_c(t_H) \qquad (20)$$

Here α is a constant of the order 1. The next step is to consider the linearised Einstein equations and to transform them into an equation of motion for Φ_H. The upshot of a lot of algebra is a fairly simple equation

$$\ddot{\Phi}_H + (4+3c_s^2)H\dot{\Phi}_H + 3(c_s^2-w)H^2\Phi_H = I(t) \qquad (21)$$

where $w = p/\rho$ and $c_s^2 = \dot{p}/\dot{\rho}$ determine the equation of state and its change in time, and $I(t)$ is a combination of matter source terms which is negligible for scales outside the Hubble radius.

If the equation of state does not change in time, then Φ_H is constant, as can easily be verified from (21). In inflationary universe models, the equation of state changes during reheating. Given that $\dot{\Phi}_H$ vanishes before and after reheating, Eq (21) can be recast into the form

$$\frac{\Phi_H}{1+w} = \text{const} \tag{22}$$

This and (24) allows us to determine the amplitude of the energy density fluctuations when they enter the Hubble radius at time $t_f(k)$ in the radiation dominated period in terms of the amplitude at initial Hubble radius crossing $t_i(k)$ and the net change in the equation of state

$$\left[\frac{\delta M}{M}\right](k,t_f(k)) = \frac{1+w(t_f)}{1+w(t_i)}\left[\frac{\delta M}{M}\right](k,t_i(k)) = \frac{4\rho}{3\dot{\phi}_0^2(t_i(k))}\left[\frac{\delta M}{M}\right](k,t_i(k)) \tag{23}$$

using in the final step

$$1+w(t) = \frac{\dot{\phi}_0^2}{\rho} \tag{24}$$

for the equation of state of a homogeneous scalar field.

If we evaluate (23) for a scalar field $\phi(\underline{x},t)$ slowly rolling down the slope of the potential in Figure 3 we find

$$\left[\frac{\delta M}{M}\right](k,t_f(k)) = O(1)\frac{H\delta\dot{\phi}}{\dot{\phi}_0} \tag{25}$$

For a potential which near the origin can be approximated by [3]

$$V(\phi) = -\lambda\phi^4 \tag{26}$$

we find

$$\left[\frac{\delta M}{M}\right](k,t_f(k)) \sim \lambda^{1/2} \tag{27}$$

This brings us to an embarrassing problem. The amplification of the perturbations during inflation is too efficient. Unless λ is fine tuned to a value smaller than 10^{-8} the primordial perturbations produced by inflation are too large. They would have generated observable anisotropies in the microwave background radiation. It looks like cosmological fine tuning has been replaced by particle physics fine tuning. There exist, however, many models which give sufficiently small energy density perturbations. Many of these are supergravity models [16], many others employ a new Higgs singlet to produce inflation [17].

4. Two new mechanisms : Chaotic Inflation and Dynamical Relaxation

In the new inflationary universe it was assumed that the scalar field ϕ takes on the value 0 at all points in space at the time of the phase transition. At the end of Section 2 we justified this assumption in the case of a scalar field which is strongly coupled to other fields.

However, in models which satisfy the fluctuation constraint $\lambda < 10^{-8}$, ϕ is usually also weakly coupled to other fields. Hence the confining forces are very small, not large enough to produce a homogeneous field configuration by the time of the phase transition. The standard picture of inflation must be modified.

There are two mechanisms by which inflation can be realized in models in which ϕ is only weakly coupled to other fields. The first, championed in particular by Linde [6], is called chaotic inflation. The basic idea is simple. We consider a general stable potential e.g.

$$V(\phi) = \frac{\lambda}{4}\phi^4 + \beta\phi^3 + \gamma\phi^2 + \delta\phi + \epsilon \tag{28}$$

At a high initial temperature T_0 when initial conditions are set up (for simplicity we shall take $T_0 = m_{pl}$), all values of ϕ with $V(\phi) < \frac{\pi^2}{30} m_{pl}^4$ should arise.

Consider now a region of space in which the scalar field takes on a value close to its maximal value ϕ_{max}. If βm_{pl}, γm_{pl}^2, δm_{pl}^3 and ϵm_{pl}^4 are small, then

$$\phi_{max} \sim \lambda^{-1/4} m_{pl} \tag{29}$$

If $\phi(\underline{x})$ is fairly homogeneous so that the spatial gradient term in the equation of motion for ϕ is negligible compared to the nonlinear force term $\partial V/\partial\phi$, and if $\dot{\phi}$ is small initially, then the equation of motion can be approximated by

$$3H\dot{\phi} = -\frac{\partial V}{\partial \phi} = -\lambda\phi^3 \tag{30}$$

with the Hubble "constant"

$$H(t) = \left[\frac{2\pi\lambda}{3}\right]^{1/2} \frac{\phi^2}{m_{pl}} \tag{31}$$

The solution of (34) is [18]

$$\phi(t) = \exp\left\{-\left[\frac{\lambda}{6\pi}\right]^{1/2} m_{pl} t\right\} \phi(0) \tag{32}$$

where we have taken the initial time to be t=0. As a consistency check, we note that neglecting $\ddot{\phi}$ is self consistent until $\phi(t) \simeq (6\pi)^{-1/2} m_{pl}$.

It can easily be verified [18] that the spectrum of energy density fluctuations is again given by Eq (27). The equation of state is inflationary as long as (32) is valid. We conclude that provided $\lambda < 10^{-8}$ chaotic inflation works in the sense that we have a sufficient period of inflation, and a spectrum of energy density fluctuations consistent with the bounds on microwave background anisotropies.

Chaotic inflation works for fairly general potentials, which is an improvement over the new inflationary universe. However, we have to be in a special region of the universe for the mechanism to work. We need an initial region of several Hubble volumes in which ϕ is very large and homogeneous.

Chaotic inflation has recently received considerable attention. It has been shown (using "naive" [19] and "sophisticated" [20] measures) that almost all initial configurations for the scalar field lead to inflation. Again, spatial inhomogeneities have not been considered in these analyses.

The second mechanism does not work for arbitrary potentials. We need a potential of double well shape. In particular the potential energy at $\phi = 0$ must be large and positive. The mechanism, however, allows for large spatial fluctuations in the initial scalar field configuration.

I will call this mechanism "dynamical relaxation of ϕ". Consider the equation of motion for the scalar field

$$\ddot{\phi} + 3H(t)\dot{\phi} - a^{-2}(t)\nabla^2\phi = -\partial V/\partial\phi \tag{33}$$

The spatial gradient term leads to oscillations of $\phi(\underline{x},t)$, the Hubble term will damp the oscillation, thus producing a configuration $\phi(\underline{x}t) \approx 0$. In contrast, the nonlinear force $\partial V/\partial\phi$ will tend to drive $\phi(\underline{x})$ towards $\pm\sigma$ and thus produce a domain structure. For a scalar field theory with weak self coupling, the Hubble damping force will be much stronger than the nonlinear force, and inflation will occur [21].

To make the above arguments more precise, we shall for a moment replace the nonlinear force by a small mass term :

$$\frac{\partial V}{\partial\phi} \rightarrow \frac{1}{6}R\phi \tag{34}$$

where R is the Ricci scalar. With this substitution, Eq (33) is the equation of motion of a conformally coupled free scalar field. Any solution $\phi(\underline{x},t)$ can be obtained by conformal transformation from a solution $\tilde{\phi}(\underline{x},t)$ of the flat space-time Klein-Gordon equation for a free massless scalar field.

$$\phi(\underline{x},t) = a^{-1}(t)\tilde{\phi}(\underline{x},\tau) , \tag{35}$$

where τ is conformal time given by $dt^2 = a^2(t)d\tau^2$. In flat space-time the solution $\tilde{\phi}$ will be an oscillating wave with constant amplitude. Hence in an expanding universe ϕ will be a wave oscillating in conformal time with amplitude damped as $a^{-1}(t)$.

At this stage the result is no surprise, since we have thrown away the domain forming force. For weak coupling we can analyse the effects of $\partial V/\partial\phi - \frac{1}{6}R\phi$ using a perturbative Green function method [9]. We Fourier expand $\phi(\underline{x},t)$ and study the equation of motion for each Fourier mode $q_K(t)$. Demanding that the effect of the nonlinear force be small for a period of time long enough for sufficient inflation to occur gives an estimate of the maximal value of the coupling constant λ. We can guess this value by comparing the damping force and the nonlinear force at an initial temperature $T \approx \sigma$ and demanding that the damping force dominates. This gives

$$\lambda < \frac{8\pi^3}{90} N\left[\frac{\sigma}{m_{pl}}\right]^2 \tag{36}$$

where m_{pl} is the Planck mass. This result has been confirmed in numerical work [22].

5. Conclusions

The inflationary universe has led to a major change and improvement in cosmological models. The description of matter as an ideal gas can never be the complete answer, and whatever may happen to particular particle physics models, quantum field theoretical processes in the early universe will have to be considered.

I would like to stress two main achievements of inflationary models. Firstly, quantum field theory effects in the early universe can lead to a period in which the universe expands exponentially. In this case the entire observable universe starts out inside the causal horizon in the de Sitter period. This solves the horizon and flatness puzzles of the standard big bang model.

Secondly, inflation provides a mechanism which from first principles in a causal way gives rise to primordial energy density perturbations required for the formation of galaxies and clusters of galaxies.

Inflation is clearly not the answer to all questions. It does not address the question why the cosmological constant today is so small. From a mathematical point of view inflationary models are not consistent in that they treat matter quantum mechanically while retaining the classical description of space-time. This inconsistency leads to the necessity of an ad-hoc prescription when computing the spectrum of energy density perturbations : at some point it is necessary to make the transition between quantum mechanical expectation values and classical quantities. I tried to convince you of the plausibility of this prescription, but it is still only a prescription. Another weak point of the technical analysis of perturbations is that it was necessary to prescribe an initial state for the quantum field. Finally, a certain amount of fine tuning in the particle physics sector is required in order to make inflation work at all.

References

1. A. Guth, Phys. Rev. D23, 347 (1981).

2. A. Linde, Rep. Prog. Phys. 47, 925 (1984);
   R. Brandenberger, Rev. Mod. Phys. 57, 1 (1985).

3. A. Linde, Phys. Lett. 108B, 389 (1982);
   A. Albrecht and P. Steinhardt, Phys. Rev. Lett. 48, 1220 (1982);
   S. Hawking and I. Moss, Phys. Lett. 110B, 35 (1982).

4. A. Guth and S. Y. Pi, Phys. Rev. Lett. 49, 1110 (1982);
   S. Hawking, Phys. Lett. 115B, 295 (1982);
   A. Starobinsky, Phys. Lett. 117B, 175 (1982).

5. J. Bardeen, P. Steinhardt and M. Turner, Phys. Rev. D28, 679 (1983).

6. A. Linde, Phys. Lett. 129B, 177 (1983).

7. L. Dolan and R. Jackiw, Phys. Rev. D9, 3320 (1974);
   S. Weinberg, Phys. Rev. D9, 3357 (1974);
   C. Bernard, Phys. Rev. D9, 3313 (1974).

8. G. Mazenko, R. Wald and W. Unruh, Phys. Rev. D31, 273 (1985).

9. W. Press, Physica Scripta 21, 702 (1980);
   K. Sato, Mon. Not. R. Astron. Soc. 195, 467 (1981);
   V. Lukash, JETP 52, 807 (1980);
   G. Chibisov and V. Mukhanov, JETP Lett. 33, 532 (1981).

10. E. Harrison, Phys. Rev. D1, 2726 (1970);
    Ya. Zel'dovich, Mon. Not. R. Astron. Soc. 160, 1p (1972).

11. R. Brandenberger, Nucl. Phys. B245, 328 (1984).

12. J. Halliwell and S. Hawking, Phys. Rev. D31, 1777 (1985).

13. W. Fischler, B. Ratra and L. Susskind, Nucl. Phys. B262, 159 (1985).

14. R. Brandenberger and R. Kahn, Phys. Rev. D29, 2172 (1984).

15. J. Bardeen, Phys. Rev. D22, 1882 (1980);
R. Brandenberger, R. Kahn and W. Press, Phys. Rev. D28, 1809 (1983).

16. A. Linde, Phys. Lett. 132B, 137 (1983);
A. Linde, JETP Lett. 37, 724 (1983);
D. Nanopoulos, K. Olive, M. Srednicki and K. Tamvakis, Phys. Lett. 123B, 41 (1983) and ibid 124B, 171 (1983);
B. Ovrut and P. Steinhardt, Phys. Lett. 133B, 161 (1983);
R. Holman, P. Ramond and C. Ross, Phys. Lett. 137B, 343 (1984).

17. Q. Shafi and A. Vilenkin, Phys. Rev. Lett. 52, 691 (1984);
S. Y. Pi, Phys. Rev. Lett. 52, 1725 (1984).

18. R. Kahn and R. Brandenberger, Phys. Lett. 141B, 317 (1984).

19. V. Belinsky, L. Grishchuk, I. Khalatnikov and Ya. Zel'dovich, Phys. Lett. 155B, 232 (1985);
T. Piran and R. Williams, Phys. Lett. 163B, 331 (1985).

20. G. Gibbons, S. Hawking and J. Stewart, DAMTP preprint (1986).

21. A. Albrecht and R. Brandenberger, Phys. Rev. D31, 1225 (1985).

22. A. Albrecht, R. Brandenberger and R. Matzner, Phys. Rev. D32, 1280 (1985).

Formation and Evolution of Cosmic Strings

ANDREAS ALBRECHT

1. What is a Cosmic String?

First of all, cosmic **strings** are *not* superstrings. Superstring theory is an attempt to supplant field theory as the most fundamental language with which to describe physics. Standard field theories would then appear as a "low energy" limit of superstring theory. The strings I shall discuss today are just particular field configurations that can appear in "ordinary" field theories, regardless of whether there is an underlying superstring theory. Their **existence** was first demonstrated by Nielsen and Olesen [1] and these strings are often called Nielsen-Olesen strings.

In order to describe Nielsen-Olesen strings let us consider the simplest field theory in which they can occur, a theory with one complex scalar field $\Phi(x)$. This theory assigns to every point in space (x) two real numbers, $Re(\Phi(x))$ and $Im(\Phi(x))$. At every point the potential energy density corresponding to the value of $\Phi(x)$ is given by $V(\Phi)$ which we choose to be

$$V(\Phi) = \lambda(|\Phi|^2 - \eta^2)^2. \tag{1}$$

This choice is none other than the famous "Mexican hat" potential which is illustrated in Fig. (1). An important feature of V is that the value of Φ for which $V(\Phi)$ takes on its minimum value ($V = 0$) is not unique. In fact V is minimized for any Φ such that

$$|\Phi|^2 = \eta^2 \tag{2}$$

(corresponding to the circle at the bottom of the Mexican hat).

To construct the Nielsen-Olesen string field configuration I will first define the field boundary conditions. Consider a cylinder with a very large radius. I will define cylindrically symmetric boundary conditions so I only need to define them on some circle that goes around the cylinder perpendicular to the axis of the cylinder. One then gets the conditions on the rest of the cylinder by translating up and down the cylindrical axis. Since I wish to take the radius of the cylinder to be arbitrarily large $V(\Phi)$ must be constrained to be zero on the cylinder to prevent a divergence of the energy. As we saw above, however, this constraint does not uniquely determine $\Phi(x)$. At this point, instead of just fixing $\Phi(x)$ at some random value which satisfies (2) I will let $\Phi(x)$ vary from point to point. In fact, I will choose $\Phi(x)$ so that as one moves completely around the circle in space, the corresponding value of $\Phi(x)$ describes

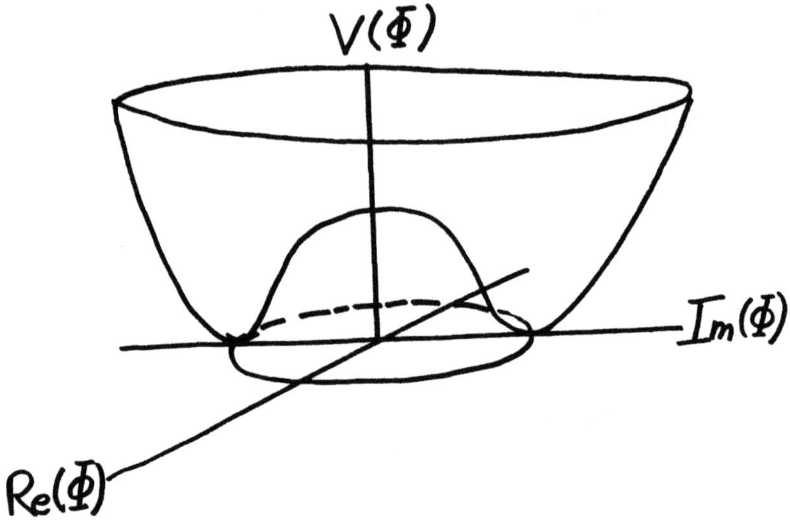

Figure 1: The "Mexican hat" potential

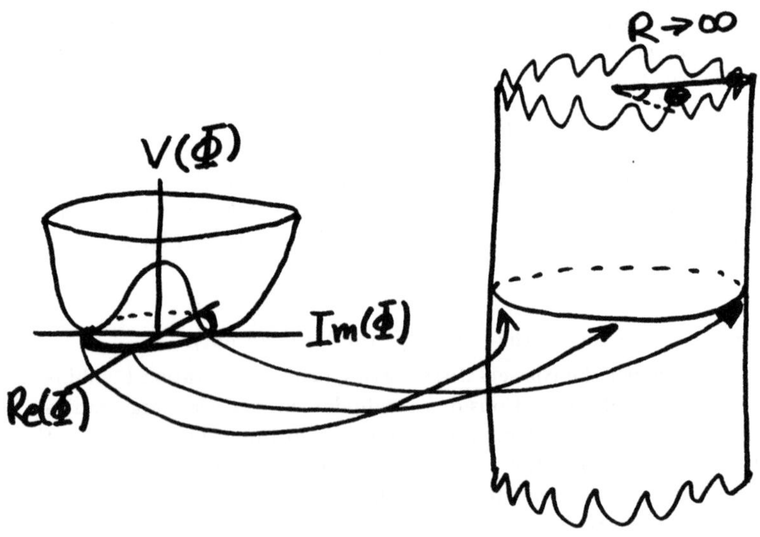

Figure 2: The correspondence of Φ on the cylinder with the bottom of the Mexican hat ($\Phi(R,\theta) = \eta e^{i\theta}$)

Figure 3: At high temperatures $\Phi(x)$ varies wildly. When the universe cools Φ must settle toward a minimum of V.

a complete circle around the bottom of the Mexican hat, as illustrated in Fig. (2). One then chooses the field inside the cylinder so as to minimize the energy, subject to the given boundary conditions. What one finds is that even though Φ is chosen so that the energy is zero on the boundaries, the minimum energy configuration has non-zero energy. The gradient energy due to the variation of Φ around the circle in Fig. (2) was negligible, due to the large size of the circle. As one moves toward the center of the cylinder, however, Φ must vary in phase over a smaller and smaller distance and the gradient energy becomes appreciable. Depending on the details of the model this gradient energy can be compensated in various ways, but the minimum energy state is always left with a string of non-zero energy density running down the center of the cylinder. This is the Nielsen-Olesen string.

2. Formation of Strings in the Early Universe

Now let us look at how Nielsen-Olesen strings could form in the early universe. According to the big bang model the universe was once very hot, and has been expanding and cooling up until the present day. The effects of this on Φ are represented in Fig. (3). High temperatures mean high energy densities, so initially Φ would be expected to be varying wildly, all around the Mexican hat. As the universe cools, the energy density decreases so Φ settles down. When the temperature becomes much less than

η (in units where $\hbar = k_B = 1$) Φ approaches a value which minimizes V. But which value does it choose? As we have just discussed, there are many values of Φ which meet this criterion. Causality arguments tell one that Φ will not choose the same value everywhere in space, but rather a domain structure will appear in which Φ takes on different values (consistent with (2)) in different regions and varies smoothly from one region to another.

With this sort of domain structure an interesting situation can arise. Since the orientation of Φ is uncorrelated over large distances, it is likely that regions appear where when one moves around a circle in space the value of Φ makes a complete circle around the bottom of the Mexican hat. As we saw above, the minimum energy configuration inside such a circle must have a Nielsen-Olesen string running through it. That means that as the universe expands and cools, Nielsen-Olesen strings will form. What is happening here is that a true minimum energy state is not chosen because distant regions cannot agree on which minimum energy state to choose. Instead, cosmic strings appear as "defects" in the resulting domain structure. To iron out these defects would require shifting $\Phi(x)$ over the top of the Mexican hat in large regions of space, a process which rapidly becomes energetically impossible as the universe cools.

The above discussion probably looks familiar to those of you who remember similar discussions about magnetic monopoles [2,3,4]. Magnetic monopoles are also non-trivial minimum energy field configurations which arise when special boundary conditions are imposed. Likewise, in field theories which contain magnetic monopoles the necessary "boundary conditions" are likely to arise in numerous regions as the universe cools from high temperatures, resulting in the formation of monopoles. However, monopoles are not string-like, but point-like and they require a special twist to be put into the field configuration on a large sphere rather than on a cylinder. These differences mean that different fields with different potentials are required to construct monopoles rather than Nielsen-Olesen strings. Consequently, a field theory in which monopoles exist may or may not have strings, and vice versa. One reason magnetic monopoles have received so much attention is that it is hard to construct a grand unified model that does not have magnetic monopoles. On the other hand, the case for Nielsen-Olesen strings is different. It is equally easy to construct grand unified models with or without Nielsen-Olesen strings.

Given our present ingnorace of physics at very high energies, we are not compelled to consider Nielsen-Olesen strings in cosmology. (Nielsen-Olesen strings do not appear in any model which has been completely tested.) On the other hand, if the existence of strings had consequences which contradict observation, some models of high energy physics could be **ruled out.** Recall, however, that magnetic monopoles were once thought to have unacceptable consequences. It was felt that monopoles were almost guaranteed to dominate the universe shortly after their production, contradicting virtually every observation ever made. It is currently popular to get rid of the "monopole problem" by arranging for a period of inflation [5,6,7] to occur after the monopoles are produced. In this way the monopoles are diluted to acceptable levels. One might ask, then, whether the inflation required for other reasons may also dilute the strings, making the study of cosmic strings uninteresting. This argument is not valid for several reasons. First of all, cosmic strings may well be produced *after* inflation has finished. Secondly, if strings are produced *before* inflation they may affect the onset of inflation in an important way. Lastly, our understanding of the

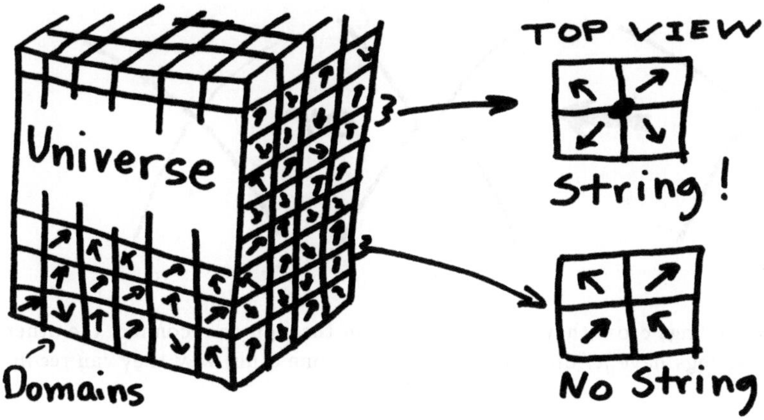

Figure 4: To calculate the initial string configuration the universe is taken to be a cube, with the domains as smaller cubes stacked inside. The arrows represent a random choice of Φ orientation (in the bottom of the Mexican hat). Given the Φ orientations it can be determined which domain edges have string on them and the resulting network can be constructed.

very early universe is simply not solid enough to be sure of of inflation, monopoles or many other things. It may turn out that there is no inflation, no monopoles, but that there *are* cosmic strings.

3. The Initial String Network

The first step in studying cosmic strings is understanding the type of network that forms when the strings are first produced. This issue was first studied in detail by Vachaspati and Vilenkin [8]. Turok and I [9] essentially repeated their work to provide initial conditions for our evolution simulations. As illustrated in Fig. (4), the universe is treated as a large cube (with periodic boundary conditions). The domains that form as Φ approaches different minima of V are represented by smaller cubes which are stacked into the large one. In each domain the value of Φ which is approached as the universe cools is chosen at random. A string will form on a domain edge if when one moves around the four adjacent domain cubes the corresponding Φ values move completely around the bottom of the Mexican hat. Otherwise there is no string there. With this information the entire initial string network can be constructed.

The strings considered here can have no ends. If one domain edge has string on it, there is sure to be string on one of the domain edges connecting up to it (no matter how Φ is chosen in each domain). What we find (in agreement with [8]) is that almost 20% of the resulting string network is in small loops, made up of four domain edges. There are some slightly larger loops, but most of the remaining string is what we call "infinite" string. That is, 80% of the string goes clear across the large box, no matter how large (in domain lengths) the box is taken to be. The long string is far from straight. If r is the **spatial** separation of two points on a long string, then

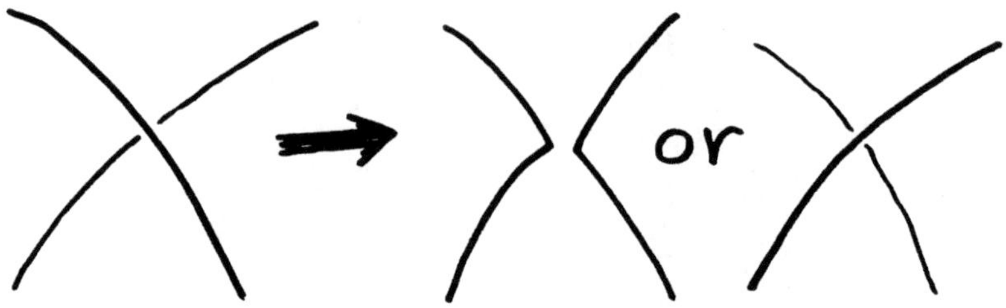

Figure 5: When strings cross they can either pass on through or reconnect the other way. Since the strings have a "direction" there is only one other way they can reconnect

the length of string (l) that connects them is given by

$$l = r^2/\xi \tag{3}$$

where ξ is the length of a domain edge. Because (3) is also obeyed for Brownian walks the long strings are often called "Brownian" although they do not resemble Brownian walks in every respect.

4. Calculating the String Evolution

Calculating the initial string network, of course, is only the beginning. One then would like to understand how the strings evolve. It is the pursuit of this understanding which drove Neil Turok and me to perform the large numerical simulations that I will report here [9,10]. The first step is to determine what dynamical equations the Nielsen-Olesen strings obey. One is aided by the fact that soon after their production the width of the cosmic strings is less than any other length scale in the problem. In that "thin string" limit the equations of motion for Nielsen-Olesen strings are just the Nambu equations for relativistic strings [10]. The strings would be produced in the radiation dominated era so a radiation dominated gravitational background was coupled to the Nambu equations.

There is actually one circumstance where the thin string approximation is not valid. That is when two bits of string cross one another. The Nambu equations would have the strings pass on through without interacting, but these strings are really configurations of interacting fields and are bound to interact in some way. There are two ways the string can emerge from such a crossing, as illustrated in Fig. (5). They can either break and reconnect the other way (intercommuting) or, despite the interaction, they can emerge connected the same as before the crossing. Considerable effort is going into understanding exactly which of these two types of string crossing occurs under what circumstances [11,12,13]. So far, indications are that intercommuting is highly likely, but when current research is complete, this issue should be much clearer. In the mean time we treat the string interactions "by hand" in our simulations. As the strings are evolved the code watches for string

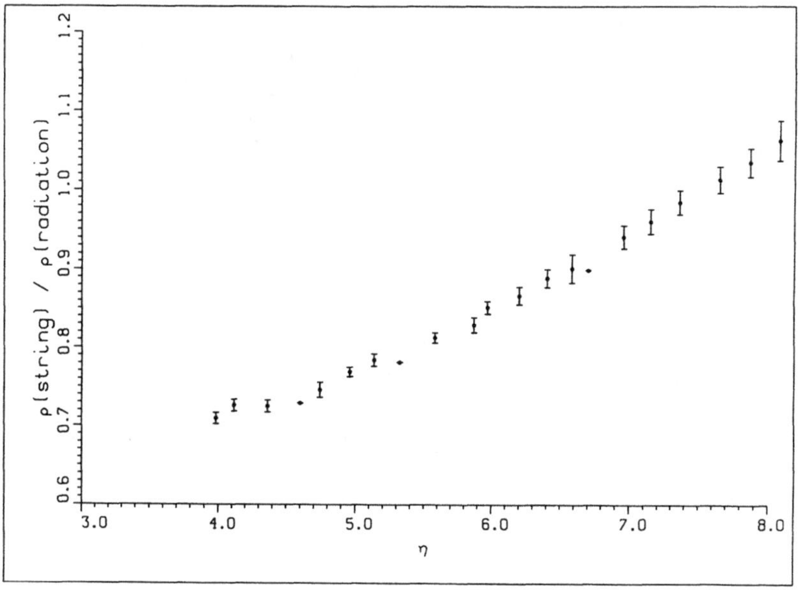

Figure 6: The ratio $\rho_{string}/\rho_{radiation}$ as a function of conformal time η with $p_{cross} = 0$. Both axes are in arbitrary units

crossing events. The strings are then either intercommuted or not according to an intercommuting probability p_{cross} which is set at the start of each run.

The first issue our calculations can address is the evolution of the energy density in strings. If the energy density in strings decreases more slowly than the radiation energy density as the universe expands then the strings will come to dominate. If the strings had energies typical of grand unification string domination would be a complete disaster, just like monopole domination. Figure (6) depicts the ratio of string energy density to radiation energy density for the case where $p_{cross} = 0$. It is clear from Fig. (6) that the energy in strings is coming to dominate over the radiation. Furthermore, since conformal time (η) goes as $t^{1/2}$, the extent to which the curve in Fig. (6) is straight is the extent to which the energy density in strings scales as "dust" (proportional to the scale factor to the minus three). If this were the end of the story, grand unified strings would be completely inconsistent with observation and would have to either simply not exist, or have been diluted by the likes of inflation.

What is there that could save cosmic strings? The only parameter available to fiddle with is p_{cross}. What is needed is a mechanism by which energy can be transferred out of strings and into radiation. Fortunatly, intercommuting strings can provide just such a mechanism. The additional ingredient necessary is the fact that loops of string oscillate relativistically and radiate [14,8]. Through this process a loop of string can decay entirely into radiation. The intercommuting is necessary to break loops off the long strings (which initially amount to 80% of all string). Figure (7) illustrates the energy transfer mechanism. Intercommuting can also cause loops to reattach to long strings before much energy is radiated. If there is to be

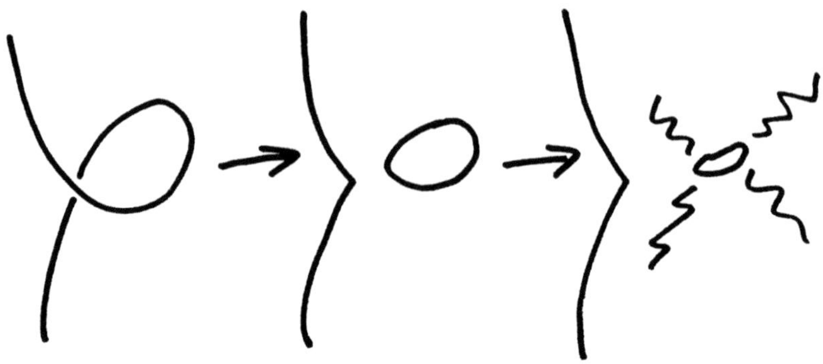

Figure 7: Energy transfer from strings to radiation: First a loop breaks off due to intercommuting. Then the loop decays into radiation.

much energy flow out of long strings the loop production must dominate over loop reattachment. To see whether the loop production is sufficient we turn again to the numerical simulations. Figure (8) shows samples of the strings at different stages of the numerical evolution where $p_{cross} = 1$. There does appear to be loop production, but one can not tell from the picture how much energy flow is occurring [15]. Figure (9) plots the ratio of energy density in long strings to the energy density in radiation as a function of conformal time, this time with $p_{cross} = 1$. This plot shows that the loop production is sufficient to keep the string energy under control. (For a discussion of how this result depends on the ratio of the initial Hubble length to domain length see reference [9].)

Upon concluding that (with sufficient string interaction) strings will not come to dominate, one is allowed to further consider cosmologies in which strings are present. Cosmic strings can have a variety of interesting effects on the evolution of the universe. Most notably, gravitational forces will cause matter to cluster around the strings, giving rise to structure in the universe. Neil Turok's talk in Friday's session will address the exciting possibility that strings might be responsible for the formation of galaxies, clusters of galaxies, and voids. In order to adress this possibility more detailed information is needed about the string network. In principle, most of the necessary information is available from the computer simulations. In practice, the hard work of extracting this information is still in its preliminary stages. Nonetheless, a definite picture of the nature of the string network has been developing, based partly on the simulations and partly on physical intuition.

The string network is believed to approach a particular ("scaling") form rather independently of the initial conditions. In this scaling network of order one loop will break off the long strings per Hubble volume per Hubble time. The length of these loops is of order the Hubble length. One can define a differential number density of loops, $n(r)$ such that $n(r)drdV$ is the number of loops with radius between r and $r + dr$ in volume dV. In the scaling network one finds

$$n(r) \sim r^{-5/2}t^{-2/3} \tag{4}$$

with a small r cutoff due to the eventual decay of the loops.

It is important to stress that our present understanding of the string network is rather rough. There is much more to be learned and the present rough

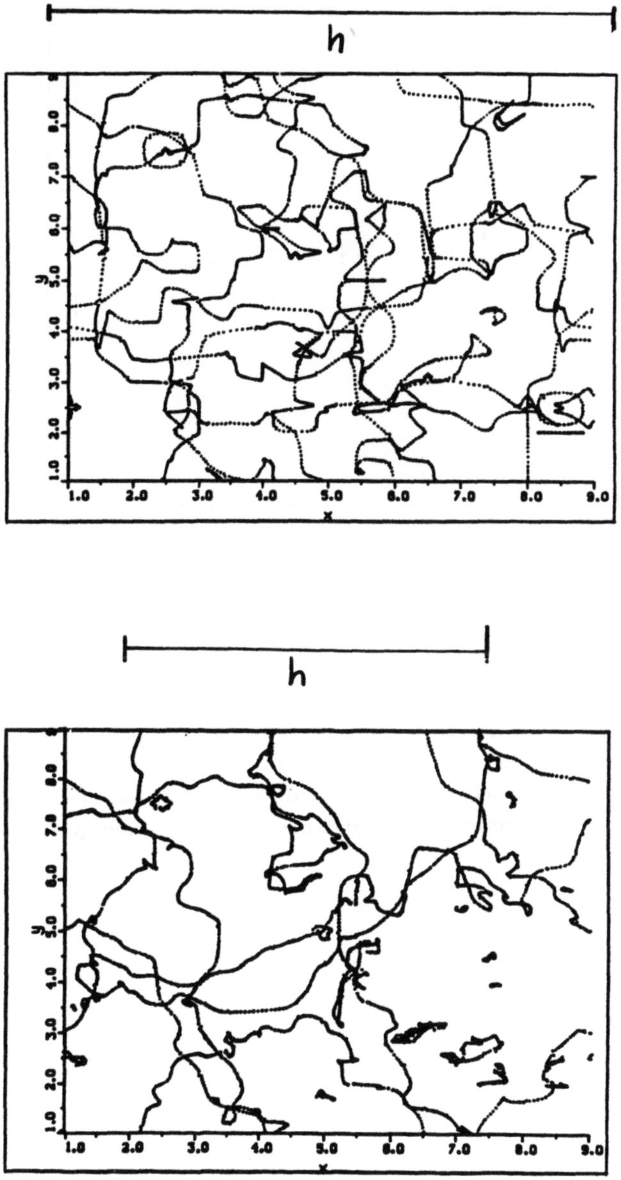

Figure 8: Strings from numerical simulations. These picures are projections into the x-y plane and the coordinates are co-moving (in units of initial domain length). Time evolution is indicated by the Hubble length (h) marked at the top of each picture ($h \sim t$).

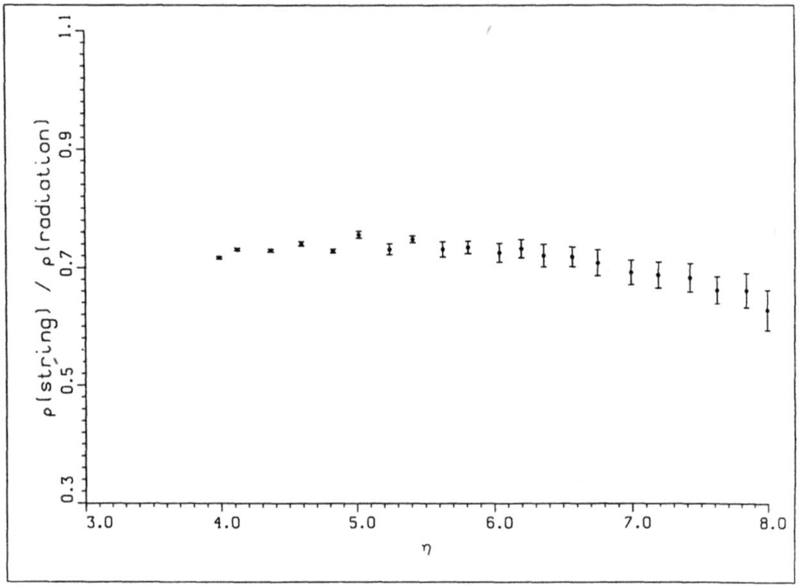

Figure 9: Energy density in long string over energy density in radiation versus conformal time for $p_{cross} = 1$. Again, units are arbitrary.

understanding could be upset by more careful calculations (some of which are already underway). In principle, however, there is nothing standing between us and a clear understanding of the cosmic string network except hard work. The promise strings appear to hold as seeds of structure in the universe makes the situation quite exciting.

References

[1] H.B. Nielsen and P. Olesen. *Nucl. Phys.*, B61:45, 1973.

[2] J. P. Preskill. *Phys. Rev. Lett.*, 43:1365, 1979.

[3] Ya. B. Zeldovich and M. Y. Khlopov. *Phys. Lett.*, 79B:239, 1978.

[4] T. W. B. Kibble. *J. Phys*, A9:1387, 1986.

[5] A. H. Guth. *Phys. Rev. D*, 23:347, 1981.

[6] A. D. Linde. *Phys. Lett.*, 108B:389, 1982.

[7] A. Albrecht and P. Steinhardt. *Phys. Rev. Lett.*, 48:1220, 1982.

[8] T. Vachaspati and A. Vilenkin. *Phys. Rev. D*, 30:2036, 1984.

[9] A. Albrecht and N. Turok. Evolution of cosmic strings. *Phys. Rev. Lett.*, 54:1868, 1985.

[10] A. Albrecht and N. Turok. In preperation.

[11] P. Shellard. PhD thesis, Cambridge University, 1986.

[12] E. Copeland and N. Turok. Cosmic string interactions. 1986. Imperial College Preprint.

[13] A. Albrecht and R. Matzner. Numerical studies of local Nielsen-Olesen strings. in preperation.

[14] N. Turok. *Nucl. Phys.*, b242:520, 1984.

[15] The results presented in this paper are from runs where the large box had 16 domain edges on a side. The pictures in Fig. (8) are from smaller runs for easier viewing.

The Quark-Hadron Phase Transition and Primordial Nucleosynthesis

CRAIG J. HOGAN

This talk deals with two different but related topics. In the first part I summarize the current understanding of processes occurring during the cosmological transition from "quark soup" to normal hadron matter. In the second part I describe what happens to cosmological nucleosynthesis in the presence of small-scale baryon inhomogeneities—of which the QCD phase transition is one plausible (if not the only) source. I reach the happy conclusion that there is perhaps after all something qualitatively new to learn about the early universe from looking at nucleosynthesis products in greater detail.

QCD (for "quantum chromodynamics") is now the standard model of the strong interactions [1,2]. The fundamental Fermions of the theory are called quarks, which interact with each other via a "color force" carried by intermediate particles called gluons. The theory is constructed by analogy with electromagnetism, where color plays the role of charge and gluons play the role of photons. Instead of one charge (and its anti-charge), which corresponds to a simple phase or U(1) symmetry in an internal space, there are now three charges (often called red, green, blue) which represent an internal SU(3) symmetry; instead of one type of photon, there are eight types of gluons whose properties, like those of the photon, are determined when this internal symmetry is made "local", or subject to arbitrary basis transformations at each point in space-time. An important difference arising from the properties of the symmetry group is that the gluons themselves carry color charge, unlike the photon which is electrically neutral.

This last property has the interesting consequence that two quarks which get closer to each other tend to polarize the vacuum in the sense that the strength of their interaction becomes weaker, a property called "asymptotic freedom." A plasma of quarks at high temperature or high density thus approaches an ideal gas, sometimes called "quark soup." The corollary is that as quarks are separated their interaction becomes stronger, until at some point their interaction energy is sufficient to create new quark pairs out of the vacuum, leading to the confinement of color charge at low temperature inside of hadrons or other particles composed of several quarks. It is not known for certain whether the transition from quark soup to normal or hadron phase of matter is first-order or second-order. Analytical understanding of the QCD phase transition is confined to the limit of infinite or zero quark mass [3,4] or various other approximations [5,6]. A very active field at present involves studying the equilibrium phases of QCD on a numerical lattice [7,8]. Several excellent conference proceedings are available which summarize the current state of the art [9].

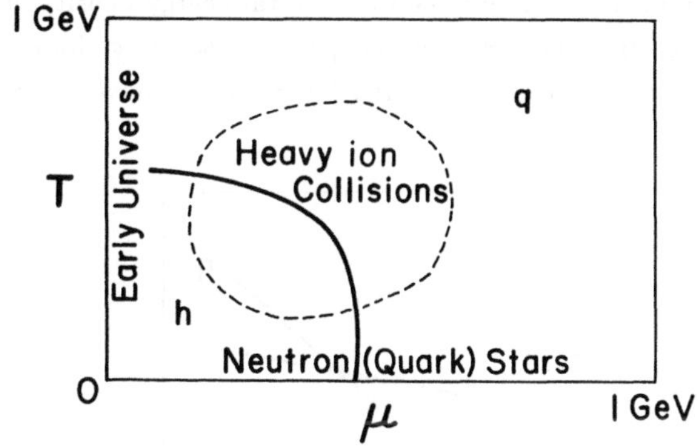

Figure 1: Schematic phase diagram for QCD as a function of temperature T and chemical potential μ. Regions corresponding to early universe, heavy ion collisions, and neutron (quark) stars are shown. The solid curve represents the transition between hadron (h) and quark soup (q) phases. Stable strange quark matter [10,11], if it exists, would coincide here with quark stars, if they exist [12].

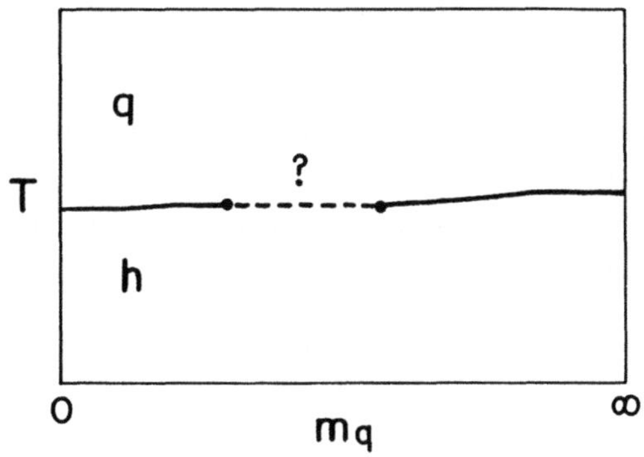

Figure 2: Schematic phase diagram for QCD as a function of quark mass m_q and temperature T (for $\mu = 0$). Solid lines represent known first-order phase transitions. The discontinuity at right represents the confinement transition; that at left is the chiral transition. Numerically it is found that these two order parameters always undergo simultaneous discontinuity, and there may be no true distinction [5]. Physical QCD probably lies in the left-hand part of the figure [8].

Many current studies indicate that the physical transition is actually first-order [8], and if this is so, then one expects macroscopic cavitation in the early universe. A first-order transition is characterized by a discontinuous change in thermodynamic properties of the equilibrium state of matter at a certain critical temperature T_c. At T_c, the two phases can coexist in pressure and temperature equilibrium, separated by a surface interface with surface tension. The transfer of matter from q phase to h phase releases a latent heat comparable to the total energy density of the medium. The h phase, at the same temperature and pressure as the q phase, has a much lower density. Qualitatively, the density and latent heat behave like quark = water, hadron = ice.

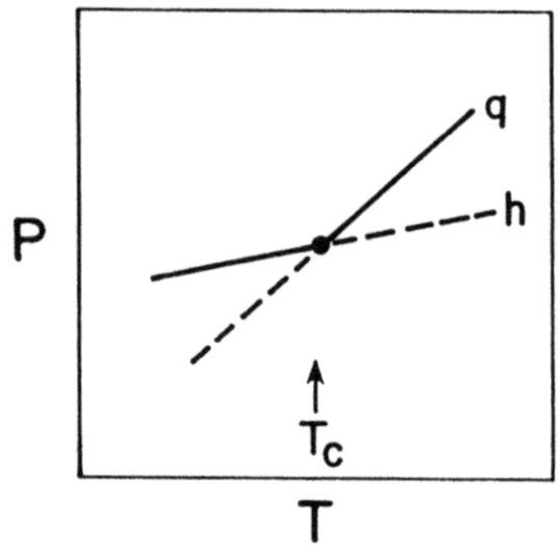

Figure 3: Schematic equation of state for quark and hadron phases near the critical temperature T_c. At other temperatures, the equilibrium phase is that with the higher pressure (solid lines).

Several important thermodynamic features of the transition can be described using a simple "bag model" [14-22], wherein the gluon interactions of the quark soup are simply incorporated into a uniform negative pressure, cosmological-constant-like term, the "bag constant" B:

$$\begin{aligned} p_q &= \frac{1}{3}N_q(aT^4) - B \\ \rho_q &= N_q(aT^4) + B \\ p_h &= \frac{1}{3}N_h(aT^4) \\ \rho_h &= N_h(aT^4) \end{aligned}$$

and the thermal pressure and density of the particles are incorporated in the usual way as a certain number of "effective photon degrees of freedom" N_q, N_h of a relativistic ideal gas. Because of the large number of quark and gluon degrees of freedom, $N_q \gg N_h$. The critical temperature is determined by B from $p_q(T_c) = p_h(T_c)$. This model, while it does not do justice to the complexities of QCD and does not include surface tension, enables one to go beyond the study of equilibrium states (lattice work) and investigate the actual kinematics of the transition.

The kinematics of the cosmological transition are in some ways considerably simpler than those of heavy-ion collisions [19-20,23-36], because the number of particles is very large and the timescale for expansion is very long compared to the microscopic timescales. It is useful to describe the cosmic quark-hadron transition in various stages (Fig. 4).

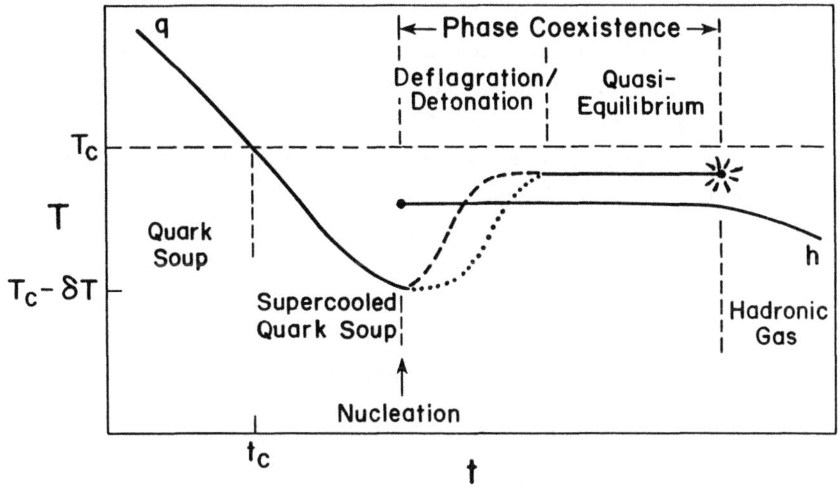

Figure 4: Representation of temperature of two phases during various stages of transition outlined in the text. Dashed and dotted lines indicate temperature history at different points which are overtaken by deflagration or detonation fronts at different times. The amount of supercooling before nucleation δT is determined by the effective surface tension between the phases.

(1) Quark soup. As T approaches T_c from above, the gluon interactions (bag constant) assume greater importance and the pressure falls faster than that of an ideal gas. When the temperature falls below T_c, the system is no longer in its ground state, which would include an admixture of h phase. But it remains for awhile as pure metastable supercooled q matter because of the difficulty of nucleating new phase.

(2) Nucleation [27-31]. Thermal fluctuations or quantum tunneling events at random points throughout the medium cause the formation of h bubbles large enough to undergo unstable growth [bubbles below a certain critical size are stabilized by surface tension and shrink; above this size, the energy available from latent heat drives bubble growth]. More supercooling increases the probability of nucleation.

(3) Nonequilibrium phase coexistence [13,17,18,22]. Around each "successful" nucleation site a bubble forms which begins self-similar deflagration or detonation, eating its way into the q phase, converting it into hadrons and releasing latent heat. For realistically small supercooling, the deflagration solutions are applicable and the bubbles resemble classical chemical burning fronts. The sphere of influence of each bubble travels at the sound speed, and the release of heat warms the medium. When the bubbles overlap, they heat up the remaining medium enough to inhibit further nucleation. At this point the matter is still mostly q phase with widely separated bubbles of h phase. The separation between the bubbles is typically about $10^{-2}H^{-1}$, based on simple but generic models of nucleation and deflagration [31].

(4) Quasi-equilibrium phase coexistence [10,21,22]. The medium is now warmed almost back to T_c and further transfer of matter between phases is regulated by the removal of heat by the cosmic expansion. The universe expands in this stage by about a factor of 2 in volume (depending on the latent heat) and gradually changes from being predominantly q to predominantly h and from having isolated hadron bubbles to isolated quark bubbles. In general the bubbles are big enough that surface tension is not strong enough to enforce sphericity in the quark bubbles, so the distribution is likely to be highly irregular; under some circumstances it may even undergo various nonspherical dendritic-type instabilities [32]. The h phase is now slightly cooler than q and at slightly lower pressure, driving a hydrodynamic flow across the phase interface such that the latent heat released just matches the pdV work done by the expansion. The temperature difference between phases adjusts itself to be just sufficient to drive the requisite flux across the interface.

(5) Finally, there is so little q left that the interface remaining is unable to perpetuate quasiequilibrium, and T starts to fall away from T_c. Before it falls very far, there is no q left at all, and the universe is once again a simple single-phase system.

However, it is not a homogeneous system. The expansion factor at the time each fluid element makes the transition varies from point to point, and temperature gradients ensure that entropy is conducted between fluid elements during the transition by neutrinos; thus, perturbations are created in density and in entropy per baryon. The spatial coherence length of these perturbations is fixed by the nucleation scale [31] and their amplitude is determined by transport processes. One consequence of the coexistence of two lumpy phases of relativistic matter of different density is a significant gravitational wave background [10,33]; a consequence of entropy inhomogeneities on these scales is an interesting perturbation to cosmic element abundances [21], to which we turn below.

I would like to advertise a point here to do with the transfer of material between phases, which was brought up by Applegate and myself [21]. If the temperature gradient driving the transfer is small, then the transport of energy across the phase boundary is predominantly hydrodynamical—and the flow carries net baryon charge with it, in spite of the fact [10] that forming a massive ~1 GeV baryon at T_c ~200 MeV carries a substantial thermodynamic penalty; this penalty is "paid" by the flow itself, which is dominated by non-baryonic particles and driven by latent heat. An astrophysicist would say that baryons are entrained in a wind. The non-fluid energy flux (e.g. optically-thin conduction by neutrinos) for a temperature difference δT is of order $(\delta T/T)$ times the blackbody flux whereas the fluid flux, the flux of material with subrelativistic bulk velocity, is (v_{flow}/c) times the blackbody flux. The temperature gradient drives velocities with kinetic energy of order the free energy available, $(v/c)^2 \simeq (\delta T/T)$, so the conductive flux is always of order $\sqrt{\delta T/T}$ times the fluid flux and this is the characteristic magnitude of entropy perturbations created by the transition. If the system is optically thick to neutrinos, the amplitude is smaller. Thus the formation of primordial "quark nuggets" and other cold exotica requires very low entropy regions to start with (although this is not these days an outlandish possibility [34,35]), and even the more modest nonlinearities which perturb nucleosynthesis (described below) probably require some ingredient in addition to a quiescent, mildly supercooled transition.

I now back off from the specifics of the QCD transition and consider the effects of entropy perturbations on later events, particularly cosmic nucleosynthesis. Just as temperature fluctuations with heat conduction between fluid elements implies entropy perturbations, so too perturbations in entropy per baryon with differential diffusion of neutrons and protons inevitably leads to fluctuations in neutron-to-proton ratio. In the case where entropy fluctuations are nonlinear and have a coherence scale larger than the proton diffusion length, but smaller than the neutron diffusion length at nucleosynthesis, cavities in the baryon distribution fill with neutrons but not protons, and some nucleosynthesis occurs in a very unconventional neutron-rich environment. In the remainder of this talk I will summarize some recent work of Applegate, Scherrer and myself [36] where we analyze big-bang baryon diffusion and the consequent neutron-rich nucleosynthesis. An excellent recent review of standard nucleosynthesis and observed abundances is given by Boesgaard and Steigman [37].

Fig. 5 shows the rms comoving distance travelled by a baryon up to time t (temperature T), for two values of the baryon-to-photon ratio η. Up to $T \sim 800$ keV, a baryon is continually intertransmuted between neutron and proton, so its diffusion is primarily limited by the neutron interactions—magnetic dipole interactions with e^+e^- plasma, and strong interaction scattering with protons. After 800 keV a baryon is forever "stuck" as a proton, or as a neutron with a normal zero-temperature half-life to beta decay. The protons remain where they were at $\simeq$800 keV, while the neutrons diffuse away, limited mainly by their interaction with protons. In an inhomogeneous medium where a neutron finds itself in a region with less than the mean proton density, its diffusion will carry it even farther than indicated in fig. 5.

By the time of nucleosynthesis, $T \sim 0.1$ MeV, the neutrons have diffused at least 100 times farther than the protons. If one had started with baryon voids with diameters anywhere between these two scales, they would by nucleosynthesis have filled with neutrons but not with protons. It is interesting that the range of comoving scales for which this occurs encompasses distances from $H^{-1}/30$ at 100 GeV to $H^{-1}/100$ at 100 MeV—just the range of plausible scales from cavitation caused by first-order transitions associated with QCD, electroweak symmetry breaking, or supersymmetry breaking (see fig. 5). Physically, at .1 MeV the proton diffusion length for $\eta = 10^{-8}$ is $\approx$300 m and the neutron diffusion length $\simeq$30 km; this corresponds to baryon masses $\simeq 10^{11} - 10^{17}$g. (Note that spatial variations in abundance on larger scales are washed out as $N^{-1/2}$, so no residual inhomogeneity in primordial abundance would be detectable.)

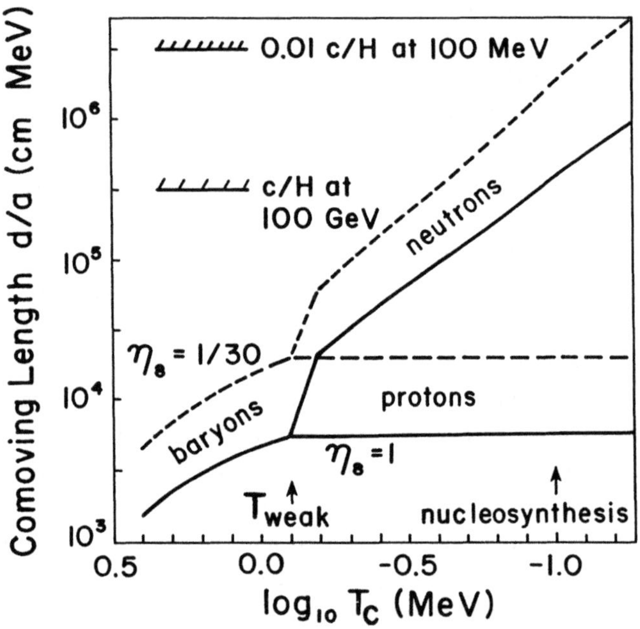

Figure 5: Diffusion of baryons in homogeneous big bang plasma. Comoving rms diffusion length is shown as a function of T for two values of $\eta_8 = 3\Omega_b h^2$, the baryon-to-photon ratio times 10^8.

Nucleosynthesis in neutron-rich regions proceeds in the usual way, except that protons are only available from the decay of neutrons. This has profound effects on abundances, the most important of which is a dramatically increased production of deuterium. This arises from the availability of free protons and neutrons at a low temperature when, in the standard model, all D, p, n would have burned already to He; here, D can form when T is low enough that the rate for further burning is reduced. Another surprising consequence is a nonnegligible abundance of heavy elements—carbon and beyond. We do not yet know the distribution of abundance between the various heavy elements, but this is under investigation.

To explore the contact with observation we have adopted a specific simplified model wherein a fraction f_v of the volume is involved in pure-neutron nucleosynthesis and a fraction f_n of the free neutrons usually present at 800 keV (freeze-out) participate in nucleosynthesis in these regions. The products of the pure-neutron regions are evaluated using a modified version of Wagoner's [38] nucleosynthesis code. A sample of our results is given in table 1, where we have compared abundances of a standard (homogeneous) model with a segregated model, adopting $H_o = 50h_{50}$ km sec^{-1} Mpc^{-1}. The "standard" values come from Wagoner [38]. By choosing f_v and f_n to range from zero up to a few tenths, probably any value of $\Omega_b h_{50}^2$ up to about one can fit current observations. The "observations" of light element abundances shown in the table are indicative values distilled from various reviews [37,39,40], while that for $A \geq 12$ comes from measurements of the metal-poorest star known, CD-38°425 [41].

Table 1. Primordial Element Mass Fractions

Element	Standard Model		Segregated Model	"Observed(?)"
	$\Omega_b h_{50}^2 = 1$	$\Omega_b h_{50}^2 = 0.1$	$\Omega_b h_{50}^2 = 1$	
^{4}He	.26	.24	$.26(1 - f_n/2)$	0.24 ± 0.01
^{2}He	10^{-8}	4×10^{-5}	$1.2 \times 10^{-4} f_v$	$\gtrsim 3 \times 10^{-5}$
^{3}He	4×10^{-6}	2×10^{-5}	$7.2 \times 10^{-6} f_v$ F	$\lesssim 6 \times 10^{-5}$
^{7}Li	1.7×10^{-8}	1.2×10^{-9}	$3.5 \times 10^{-8} f_n$	7×10^{-9} Pop I 8×10^{-10} Pop II
A$\geq$ 12	$< 10^{-12}$	$< 10^{-12}$	$10^{-6} f_n$	$\lesssim 10^{-6.5}$

At present the most clear-cut observational signatures of these are (1) a high primordial Lithium content, which would lead to high ^{7}Li/^{6}Li in population I objects; and (2) a significant primordial enrichment of heavy elements, which could manifest itself in distinctive element abundance ratios, in models of synthesis of "secondary" nuclei, and in a minimum "floor" abundance below which no stars could be found. In a model with $\Omega_b h^2 = 1$, this floor is expected not far below the enrichment of CD-38°425.

This work was supported by an Alfred P. Sloan Research Fellowship, and by NASA Theoretical Astrophysics Grant NAGW-763 at the University of Arizona.

References

1. I. J. R. Aitchsion, A. J. G. Hey: "Gauge Theories in Particle Physics: a Practical Introduction", Adam Hilger, 1982
2. C. Quigg: "Gauge Theories of the Strong, Weak and Electromagnetic Interactions", Benjamin Cummings, 1983
3. K. Kajantie, C. Montonen, E. Pietarinen, Z.: Z. Phys. *C9*, 253 (1981); B. Svetitsky, L. Yaffe: Nucl. Phys. *B210*, 423 (1982); T. Celik, J. Engels, H. Satz: Phys. Lett. *125B*, 411 (1983)
4. R. D. Pisarski, F. Wilczek: Phys. Rev. *D29*, 338 (1984)
5. J. Kogut *et al.* : Phys. Rev. Lett. *50*, 393 (1983)
6. T. Banks, A. Ukawa: Nucl. Phys. *B225*, 455 (1983); T. A. De Grand, C. E. De Tar: Nucl. Phys. *B225*, 590 (1983)
7. T. Celik, J. Engels, H. Satz: Phys. Lett. *133B*, 427 (1983); J. Polonyi *et al.* : Phys. Rev. Lett. *53*, 644 (1984); H. Satz: Nucl. Phys. *A418*, 447c (1984); J. Kogut: Nucl. Phys. *A418*, 381c (1984); F. Fucito, S. Solomon: Phys. Rev. Lett. *55*, 2641 (1985); A. Ukawa, M. Fukugita: Phys. Rev. Lett. *55*, 1854 (1985)
8. M. Fukugita, A. Ukawa: Kyoto Preprint RIFP-642 (1986)
9. K. Kajantie, ed.: *Proceedings of Quark Matter '84*, Springer-Verlag, 1985
10. E. Witten: Phys. Rev. D *30*, 272 (1984)
11. E. Farhi, R. L. Jaffe: Phys. Rev. D *30*, 2379 (1984)
12. C. Alcock, E. Farhi, A. Olinto: MIT Preprint, CTP #1329 (1986)
13. T. DeGrand, K. Kajantie: Phys. Lett. *147B*, 273 (1984)
14. B. Friman, K. Kajantie, P. V. Ruuskanen: Nucl. Phys. *B266*, 468 (1986)
15. E. von Gersdorff, M. Kataja, L. McLerran, P. V. Ruuskanen: Fermilab Preprint 86/13-T (1986)
16. K. Kajantie, M. Kataja, L. McLerran, P. V. Ruuskanen: Helsinki Univ. Preprint HU-TFT-86-6 (1986)
17. M. Gyulassy, K. Kajantie, H. Kurki-Suonio, L. McLerran: Nucl. Phys. *B237*, 477 (1984)
18. H. Kurki-Suonio: Nucl. Phys. *B255*, 231 (1985)
19. L. Van Hove: Z. Phys. *C1*, 93 (1983)
20. L. Van Hove: Z. Phys. *C27*, 135 (1985)
21. J. H. Applegate, C. Hogan: Phys. Rev. D *31*, 3037 (1985)
22. K. Kajantie, H. Kurki-Suonio: Preprint, Austin, Texas (1986)
23. J. D. Bjorken: Phys. Rev. D. *27*, 140 (1983)
24. L. Van Hove: Phys. Lett. *118B*, 138 (1982)
25. K. Kajantie, R. Raitio, P. V. Ruuskanen: Nucl. Phys. B *222*, 152 (1983)
26. G. Baym, B. Friman, J.-P. Blaizot, M. Soyeur, W. Czyz: Nucl. Phys. A. *407*, 541 (1983)

27. R. E. Peierls, K. S. Singwi, D. Wroe: Phys. Rev. *87*, 46 (1952)
28. S. Coleman: Phys. Rev. D *15*, 2929 (1979)
29. A. H. Guth, E. Weinberg: Phys. Rev. D *23*, 876 (1981)
30. C. J. Hogan: Astrophys. J. *252*, 418 (1982)
31. C. J. Hogan: Phys. Lett *133B*, 172 (1983)
32. J. S. Langer: Rev. Mod. Phys. *52*, 1 (1980)
33. C. J. Hogan: Mon. Not. Roy. astr. Soc. *218*, 629 (1986)
34. I. Affleck, M. Dine: Nucl. Phys. B *249*, 361 (1985)
35. M. Fukugita, V. A. Rubakov: Phys. Rev. Lett. *56*, 988 (1985)
36. J. H. Applegate, C. J. Hogan, R. J. Scherrer: submitted to Phys. Rev. D. (1986)
37. A. Boesgaard, G. Steigman: Ann. Rev. Astr. Ap. *23*, 319 (1985)
38. R. V. Wagoner: Astrophys. J. *179*, 343 (1973)
39. B. Pagel: in *Inner Space/Outer Space*, E. W. Kolb *et al.* , ed., University of Chicago Press. p. 72 (1986); B. J. Pagel: Phil. Trans. Roy. Soc. London A *307*, 19 (1982)
40. J. Yang, M. S. Turner, G. Steigman, D. N. Schramm, K. Olive: Astrophys. J. *281*, 493 (1984)
41. M. S. Bessel, J. Norris: Astrophys. J. *285*, 622 (1984)

Testing Cosmic Fluctuation Spectra

J. RICHARD BOND

Abstract

Recent observations of large scale streaming velocities and the rich cluster correlation function indicate that extra power exists on large scales over that of the CDM spectrum with $\Omega = 1$ and scale-invariant (Zeldovich) initial conditions. The extra power required to explain the reported observations is explored, and it is demonstrated that the two problems could be solved by adding an $n = -1$ ramp to the density fluctuation spectrum between wavenumbers $k^{-1} \sim 5$ and $\sim 300\ h^{-1}Mpc$. The cluster-galaxy correlation function also comes out at the right level and shape. Low Hubble constant models, hybrid models with hot and cold dark matter, models with both adiabatic and isocurvature modes present and decaying dark matter models are shown not to work. Models that are open ($\Omega \sim 0.2$) or have a large cosmological constant ($\Omega_{vac} \sim 0.8$) do better, but still fail. The easiest way to add large scale power is to make use of features that can appear at the horizon scale at recombination when the density contribution of baryons is similar to that in the dark matter ($\Omega_B \approx \Omega_X$). A vacuum energy dominated model with $\Omega_{vac} = 0.8$, $\Omega_B = \Omega_X = 0.1$ does reproduce the large scale results, but it has a low redshift of galaxy formation. Deviations of the initial spectrum from the scale-invariant form are another possibility, but only if a characteristic scale is built into the fluctuation generation mechanism. Angular anisotropies of cluster-cluster and cluster-galaxy correlation functions are predicted to be large for $r < 25\ h^{-1}Mpc$.

1. Introduction

The theory of structure formation from Zeldovich initial conditions with $\Omega = 1$ and $\Omega_B < 0.1$, cold dark matter, and biased galaxy formation (DAVIS *et al.*[1], BARDEEN *et al.*[2]) is quite successful in explaining the galaxy-galaxy correlation function and the formation of galaxy halos, and even appears to reproduce voids and superclusters (WHITE *et al.*[3]). However, observations of large scale streaming velocities and the rich cluster correlation function do not agree with the predictions [3,4], and indicate that extra power exists on large scales $k^{-1} > 5\ h^{-1}Mpc$ over that of the CDM spectrum with $\Omega = 1$.

The extra power could be obtained by adding power to the Zeldovich spectrum in the initial conditions, though this involves building in such a scale to the fluctuation generation mechanism, and there seems to be no natural way in which this arises (§2). The other method is to modify the transfer function which takes the initial fluctuation spectrum into an output one which describes the perturbation spectrum from the post recombination epoch to the onset of nonlinearity. In §3, I consider the effect of varying the various input cosmological parameters such as $h \equiv H_0/(100\ km\ s^{-1})$ and the following density parameters: total Ω_{tot}, baryon Ω_B, stable cold dark matter (CDM) Ω_X, stable hot dark matter Ω_ν, vacuum energy $\Omega_{vac} \equiv \Lambda/(3H_0^2)$ and decaying cold dark matter Ω_{nrd} with its relativistic decay products Ω_{erd}.

In the spirit of this investigation, all possible combinations of these parameters is allowed, though many of the hybrid models are not compelling from the particle physics viewpoint.

These many model universes are passed through the tests of the large scale streaming velocities, the rich cluster correlation function, the cluster-galaxy correlation function and the angular anisotropies in the microwave background. The first three tests can be made from the fluctuation spectrum alone without resort to N-body calculations. These test results presented in §4 are summaries of work done in collaboration with Bardeen and Efstathiou (BARDEEN, BOND and EFSTATHIOU [5]). In §5, these fluctuation spectra are shown to imply significant cluster alignments, at the 30% level at $\sim 20\ h^{-1}Mpc$. This work is reported in more detail elsewhere (BARDEEN, BOND and SZALAY [6]).

2. Fluctuation Initial Conditions

2.1 **Statistics**: I shall assume fluctuation generation results in a homogeneous and isotropic Gaussian random field so that the statistics are fully specified initially by a power spectrum $P(k)$ which is a function only of the magnitude $|\vec{k}|$ of the wavenumber required to specify the modes. This simple assumption is the one predicted in inflationary models, and follows more generally from a wide class of possible generation models as a result of the central limit theorem. A counterexample is the cosmic string theory requiring full numerical evolution of a complex string network to determine the probability distribution.

2.2 **Perturbation Mode**: Linearity of the fluctuations as evidenced by the high degree of isotropy of the microwave background suggest a perturbative treatment will be valid. Perturbations can be decomposed into spherical tensor harmonics. The vector (vortex), tensor (gravitational wave) and higher order spherical tensor perturbations do not couple directly to density, and consequently are expected to have little effect on structure formation. I therefore consider only scalar modes, which do couple to density. These modes may be the usual adiabatic modes for which the total energy density varies ($\delta\rho_{tot} \neq 0$), or they may be isocurvature for which $\delta\rho_{tot} = 0$ initially. Isocurvature fluctuations have perturbations in some matter density component (baryons or cold dark matter), with compensating fluctuations of opposite sign in the radiation (photons, quark-antiquark pairs, gluons etc.) at the epoch of generation. Possibilities include isocurvature baryon perturbations (the old isothermal perturbations) or isocurvature CDM perturbations, with the CDM being in the form of a pseudo-Goldstone boson such as the axion or the venerable original CDM candidate, primordial black holes.

2.3 **Initial Spectral Shape**:

2.3.1 **Adiabatic Mode**: For the adiabatic mode, there really has been only one case discussed that arises naturally: the scale-invariant Zeldovich spectrum, with the power in initial gravitational potential fluctuations per decade of wavenumber being constant, $k^3\langle\Phi_H^2(k)\rangle = constant$. Here, Φ_H is a gauge-invariant measure of scalar metric fluctuations satisfying a Poisson-Newton equation, $\nabla^2\Phi_H = 4\pi G(\delta\rho)_{com}$, where $(\delta\rho)_{com} = \delta(U_aT^{ab}U_b)$ is the fluctuation in the timelike eigenvalue of the total stress-energy tensor (the density measured with respect to the comoving hypersurfaces). Thus, $\langle|((\delta\rho)_{com}/\rho)(k,t_i)|^2\rangle \sim k_i^{n}$, $n_i = 1$. Such a spectrum, modified only by monotonic very weak functions of k, is predicted in standard inflationary models. If there are two inflationary epochs, finely tuned, it may be possible to get structure in an astrophysically interesting regime of k-space though this seems to give more power on smaller scales than on larger (SILK and TURNER [7]). To get specific features built into the initial conditions will require building in such a scale in the mechanism for fluctuation generation in some way.

2.3.2 **Isocurvature Mode**: For the isocurvature mode, two cases have been discussed. The first arises in inflationary models and is a scale-invariant spectrum, with the power in now the initial CDM density fluctuations per decade of wavenumber being constant, $k^3\langle|((\delta\rho)_X/\rho_X)(k,t_i)|^2\rangle = const.$ Such a spectrum can arise if the CDM is in the form of axions. For then a scale invariant spectrum of zero-point quantum mechanical oscillations of the pseudo-Goldstone boson field in the de Sitter space vacuum is generated during inflation. Upon boson mass generation at t_i, the CDM (X) mass density perturbations which arise take this scale-invariant form. For axions, t_i corresponds to the epoch just after the quark-hadron phase transition. This spectrum is the analogue of the Zeldovich one in the adiabatic case. By finely tuning the horizon scale when Peccei-Quinn symmetry breaking occurs to be an astrophysically interesting length, it is possible to have scale-invariant spectra with a long wavelength cutoff beyond it (KOFMAN and LINDE [8]), but this diminishes large scale power. Artful choices of the interaction potential between two scalar fields may lead to a small wavelength cutoff as well [8]. Adding this to a scale-invariant adiabatic spectrum could give the extra large scale power required, though then the complaint of overly fine-tuning the inflationary model is justified. No good particle physics inspired models currently exist for the generation of the isocurvature baryon mode.

Another isocurvature spectrum which is possible is a seed model with white noise initial conditions $n_i = 0$ for the CDM fluctuations down to the horizon scale at the epoch of seed generation, probably smaller than a parsec, the horizon scale at the quark hadron phase transition. Spectra like this have been discussed for primordial black holes. They have very little power on large scales, so it is unlikely they will solve the large scale power problem, and they are not discussed further in this paper.

3. Fluctuation Evolution and Characteristic Scales

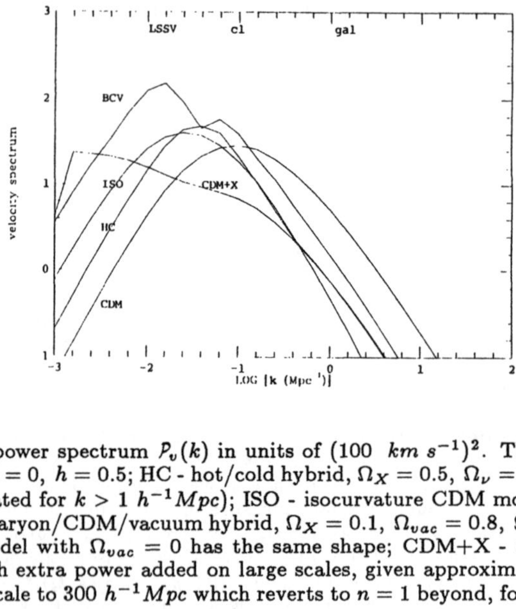

Figure 1. Velocity power spectrum $\mathcal{P}_v(k)$ in units of $(100\ km\ s^{-1})^2$. The cases shown are: CDM - $\Omega_X = 1$, $\Omega_B = 0$, $h = 0.5$; HC - hot/cold hybrid, $\Omega_X = 0.5$, $\Omega_\nu = 0.4$, $\Omega_B = 0.1$, $h = 0.5$ (curve extrapolated for $k > 1\ h^{-1}Mpc$); ISO - isocurvature CDM mode, $\Omega_X = 1$, $\Omega_B = 0$, $h = 0.4$; BCV - baryon/CDM/vacuum hybrid, $\Omega_X = 0.1$, $\Omega_{vac} = 0.8$, $\Omega_B = 0.1$, $h = 0.75$, though an open model with $\Omega_{vac} = 0$ has the same shape; CDM+X - a phenomenological CDM spectrum with extra power added on large scales, given approximately by an $n = -1$ ramp from cluster scale to 300 $h^{-1}Mpc$ which reverts to $n = 1$ beyond, for an $\Omega_X = 1$, $\Omega_B = 0$, $h = 0.4$ model.

Our fluctuation spectra must be compatible with the observed large scale streaming velocities (LSSV), distributed according to

$$P(v)dv = (54/\pi)^{1/2}\ exp[-3v^2/(2\sigma_v^2)]v^2dv/(\sigma_v^3),$$
$$\sigma_v^2(R_f) = \int \mathcal{P}_v(k)\ e^{-k^2R_f^2}\ d\ ln(k). \tag{3.1}$$

$\sigma_v(R_f)$ is the *rms* 3-dimensional streaming velocity smoothed on the Gaussian filtering scale R_f. Velocity spectra, $\mathcal{P}_v \equiv k^3\langle\vec{v}(k)\cdot\vec{v}(-k)\rangle/(2\pi^2)$, are plotted in Fig. 1 for a number of model universes.

If the local power law index for the density fluctuations is n, then the velocity spectrum $\sim k^{n+1}$. Thus the initial Zeldovich spectrum for the adiabatic cases $\sim k^2$, behaviour which is retained on very large scales. The critical index $n = -1$ marks the boundary between a rising and falling velocity curve. It is an important feature of the CDM spectrum that it peaks around cluster scale, for then the flows on this scale are relatively coherent, not greatly disturbed by the nonlinear motions generated as smaller scale regions in the hierarchy collapse. This coherence can be seen in the flow patterns of MELOTT's N-body simulations [10]. I believe this peak is an essential ingredient for the CDM universe to generate relatively coherent voids of scale $2\pi k_c^{-1} \sim 30\ h^{-1}Mpc$. For $h = 0.5$, the CDM spectrum flattens to $n \approx -2$ on galactic scales and $n \approx -3$ on smaller scales.

An obvious way to increase streaming velocities on large scales is to extend the flat $n = -1$ portion of the CDM spectra in Fig. 1 to larger scales. Such a ramp, indicated by the curve CDM+X, gives σ_v's large enough to agree with the data (Table 1). The mass autocorrelation function $\sim r^{-2}$ if $n = -1$, indicating this spectrum may give the observed shape of the cluster-cluster and cluster-galaxy correlation functions as well. This is demonstrated in Figs. 2 and 3.

The shape of the transfer functions describing the deviations from k^2 initial conditions is characterized by the following length scales:

$$\begin{aligned}
k_{Heq}^{-1} &= 10\ \theta^{1/2}(\Omega_{nr}h_{50})^{-1}\ h^{-1}Mpc = \tau_{eq}/\pi, \quad \theta \equiv \rho_{er}/(1.69\rho_\gamma) \\
k_{\nu damp}^{-1} &= 6\ (\Omega_{nr}\Omega_\nu h_{50}^2)^{-1/2}(g_{m\nu}/2)^{1/2}\ h^{-1}Mpc \\
k_{Hrec}^{-1} &= 31\ (\Omega_{nr})^{-1/2}\ h^{-1}Mpc = \tau_{rec}(z = 1500)/\pi \\
k_{Silk}^{-1} &= 1.3\ (\Omega_B^{-1/2}\Omega_{nr}^{-1/4})^{10/9}h_{50}^{-2/3}\ h^{-1}Mpc \\
k_{JBrec}^{-1} &= 0.0016\ (\Omega_X)^{-1/2}\ h^{-1}Mpc \\
k_{curv}^{-1} &= 3000(1-\Omega)^{-1/2}\ h^{-1}Mpc.
\end{aligned} \tag{3.2}$$

Only the first 3 are of interest in obtaining extra large scale power. This suggests how to systematically vary the parameters describing the global properties of the cosmological models to push these scales into the 'LSSV' range of k-space shown in Fig. 1.

The CDM spectra with $\Omega_B \ll \Omega_X$ form a one parameter of curves characterized by k_{Heq}^{-1}. It can be increased by lowering h, by raising θ (= 1 with the canonical three massless neutrino species present), or by lowering Ω_{nr}, the energy density in nonrelativistic dark matter present at the epoch of matter-radiation equality, but there is a price to pay: the spectrum significantly flattens between galaxy scale, $k_g^{-1} \equiv R_g \approx 0.35\ h^{-1}Mpc$, and cluster scale, $k_{cl}^{-1} \equiv R_{cl} \approx 5\ h^{-1}Mpc$. ($R_g$ and R_{cl} are Gaussian filtering radii chosen to roughly give the mass of the objects.) This flattening lowers the difference in time between the epoch of galaxy formation and that of cluster formation and for normalizations of the spectrum based on light traces mass, or on biasing so light is more clustered than the mass, galaxy formation occurs too late.

In Fig. 1, the curve ISO shows the effect of changing the mode to isocurvature (and also decreasing h), HC the effect of varying $k_{\nu damp}$, and BCV of getting extra power out beyond the scale k_{Hrec}. The latter clearly looks more promising in trying to obtain a spectrum similar to the phenomenological one CDM+X.

The *rms* level of the fluctuations in the density at time t, $\sigma_0(R_f, t)$, has a power spectrum $\mathcal{P}_\rho \propto k^2 \mathcal{P}_v$. The linear fluctuation growth law is used to extrapolate this *rms* level into the nonlinear regime; for $\Omega_{nr} = 1$ universes, the redshift at which the *rms* fluctuations of scale R_f reach nonlinearity is $1 + z_{nl}(R_f) = \sigma_0(R_f,\ now)$. The $n = -1$ ramp also raises σ_0 on scales much larger than R_{cl}, increasing the frequency of rare large scale collapse events.

There remains one overall *normalization amplitude*, parameterized by b, the *biasing factor*, defined so that mass traces light for $b = 1$. The technical definition of b relates a statistical average of mass density fluctuations to a volume average of fluctuations in the density of bright galaxies

$$\langle \frac{\Delta M}{M}(< r = 10\ h^{-1}Mpc)\frac{\delta\rho}{\rho}(0)\rangle \equiv b^{-2}\ \langle \frac{\Delta N_g}{N_g}(< 10\ h^{-1}Mpc)\frac{\delta n_g}{n_g}(0)\rangle$$
$$\equiv b^{-2}\ 3J_{3g}(r)/r^3 \approx 0.81/b^2. \qquad (3.3)$$

The numerical value for J_{3g} is the result from the CfA redshift survey. This normalization basically fixes the spectrum at cluster scale to have $b\sigma_0(5\ h^{-1}Mpc) \approx 0.75$. If $b > 1$, mass is *less* clustered than light, the preferred possibility in the biased galaxy formation theories considered here, and if $b < 1$ mass is *more* clustered than light. $b < 1$ might arise naturally in neutrino-dominated models (SZALAY, BOND and SILK [9]).

Dense cosmic structures form at the smoothed peaks of the the density field [3]. The height of a peak is characterized by $\nu \equiv (\delta\rho/\rho)/\sigma_0$. A selection function must be chosen for peaks which are to form a given class of cosmic objects. The mathematical framework used to determine the properties of the peaks is the theory of the statistics of peaks of Gaussian random fields. The relevant tools are developed in BBKS and a summary is given in [4].

Following KAISER [11] and BBKS, clusters are identified with $\sim 10^{15}\ M_\odot$ peaks which have collapsed by the present: $\nu\sigma_0(R_{cl},\ now) > f_c$. The threshold ν_t above which collapse has occurred is obtained by equating the spacing of peaks with $\nu > \nu_t$ to 55 $h^{-1}Mpc$, the spacing of richness ≥ 1 clusters according to BAHCALL and SONEIRA [12]. ν_t is independent of dynamics. However, the value of the collapse factor f_c must be large enough to ensure that a compact Abell cluster core has formed. The value required needs to be checked by comparison with N-body results. $f_c = 1.69$ corresponds to a uniform spherical mass of cold particles in an $\Omega = 1$ universe having completely collapsed (concentrated at a point in the center). Turn-around of the spherical perturbation occurs at 1.06, hence a value in between may be more appropriate. If f_c were determined, the collapse criterion could be used to fix the spectrum normalization, though it would also be sensitive to the choice of R_{cl}.

In the simplest version of the biased galaxy formation theory, 'bright' galaxies are assumed to form only above some global threshold value $\nu_t \sim 3$, found by equating the spacing of peaks of higher height to the density field smoothed on a galactic scale to the spacing of bright galaxies as determined in the CfA redshift survey, 4.6 $h^{-1}Mpc$ (DAVIS and HUCHRA [13]). The biasing factors quoted in specific models in Table 1 below are derived assuming this spacing, unless I adopt $b = 1$, light traces mass if biasing doesn't work. For the hot DM model, fixing the nonlinear epoch z_{nl} for the fluctuations is used to determine b through $\sigma_0(now, R_f = 0) = 1 + z_{nl}$.

4. Direct Tests of Fluctuation Spectra

4.1 **Microwave Background Anisotropies**: Though the anisotropies $\Delta T/T(\theta)$ are direct tests of the fluctuations in the linear regime, they require detailed evolution of the radiation field through recombination. In Table 2, I only quote results that could lead to significant constraints on the models, using results derived from the work of BOND and EFSTATHIOU [14,15,16,17].

4.2 **Large Scale Streaming Velocities**: Provided the filtering scale R_f is large enough so that the velocities on that scale are linear, the distribution given by eq. (3.1) is appropriate. There is only a 5% probability of finding a bulk speed above $1.6\sigma_v$, and a 5% probability of finding one below $\sigma_v/2.97$. Table 4.1 gives the range of velocities interior to these values within which the speed lies 90% of the time for a number of cosmological models. The Gaussian filtering radii are chosen to roughly correspond to the observational results: the Local Group motion relative to the CMB rest frame (LUBIN and VILLELA [18]), with $R_{TH} \approx$ 5 $h^{-1}Mpc$; the AARONSON *et al.*[19] cluster motion, corrected by $\sqrt{3}$ for going from a 1D to a 3D dispersion and by $\sqrt{2}$ since peaks (clusters) have smaller velocity dispersions than σ_v, which applies to field points, by roughly this factor (BBKS), though, for example, the model

CDM+X only has a 1.1 multiplicative factor; the elliptical galaxy results of BURSTEIN *et al.*[20] (BDDFTLW); and the Rubin-Ford sample results of COLLINS *et al.*[21]. $R_f = 0.6R_{TH}$ has been used to relate Gaussian to top hat radius where required.

Table 1: Large Scale Streaming Velocities for Models

NAME	$\Omega, \Omega_X(\Omega_\nu), \Omega_B$	Dipole	Cluster	BDD	Collins	σ_{0g}	σ_{0cl}
	Ω_{vac}, H_0, b	Lubin	AHM	FTLW	etal		
	$R_f\ h^{-1}Mpc$	3.2	5	25	40	0.35	5
	OBS $km\ s^{-1}$	~ 600	~ 500	~ 700	970 ± 300		
	CDM STD						
CDM	1,1,0,0,50,1.7	136-654	110-570	53-250	35-180	2.9	.42
	$\Delta k_{Heq}, \Delta H_0$						
C40	1,1,0,0,40,1.8	134-643	119-573	55-266	40-192	2.4	.39
VAC/C	1,.2,0,.8,50,1	-	-	115-550	94-450	2.7	.75
OP/C	.2,.2,0,0,50,1	-	-	108-516	88-422	2.7	.75
	$\Delta MODE$						
ISO/C	1,1,0,0,50,1	294-1410	273-1310	151-726	114-547	2.6	.73
ISO/AD (2::1)	1,1,0,0,50,1.7	148-709	132-634	64-309	47-226	2.5	.42
	$\Delta k_{\nu damp}$						
HOT (ν)	1,(1),0,0,50,.53	594-2850	558-2680	419-1350	200-958	2.0	1.4
HC	1,.5(.4),.1,0,50,1.7	185-889	174-835	91-435	65-311	1.5	.44
	$\Delta\Omega_B,\ k_{Hrec}$						
C+B	1,.8,.2,0,40,1.8	149-714	136-651	71-342	52-251	2.2	.40
B+C	1,.5,.5,0,50,1.7	208-1000	198-950	119-573	87-419	1.9	.44
BCV	1,.1,.1,.8,75,1	232-1120	228-1090	186-894	160-771	2.4	.72
BCO	.2,.1,.1,0,75,1	218-1050	214-1030	174-839	151-724	2.4	.72
	PHENOM						
CDM+X	1,1,0,0,40,1.8	280-1331	273-1298	233-1106	214-1016	2.2	.41

4.3 Rich Cluster Correlation Functions

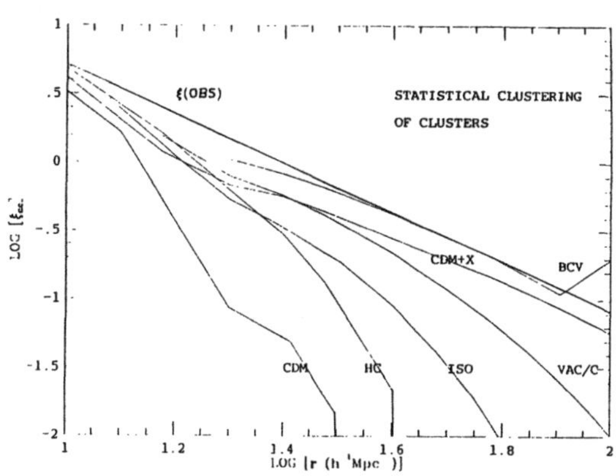

Figure 2. The statistical part of the correlation function of rich cluster peaks for some of the models of Table 1 are compared with the correlation function $\xi(OBS)$ [12]. The dynamical contribution due to the motion of the clusters towards each other typically multiplies these results by ~ 1.4 for the biased cases, and by ~ 1.8 for the unbiased cases. This brings the CDM+X case close to the reported values.

High ν peaks of the density field are already correlated in the initial conditions. This correlation function for the ***statistical clustering*** depends upon the normalized mass correlation function $\psi = \xi_{\rho\rho}/\sigma_0^2$ and its first few derivatives, and the local structural properties defining the character of the peaks. It is independent of the dynamical amplitude σ_0. If the orientation of the principal axes of one of the two peaks is fixed, then there are nonzero higher multipole moments of the correlation function describing anisotropies in the clustering. Fig. 3 shows the $L = 0$ part of $\xi_{pk,pk}$ for cluster-scale peaks above the threshold chosen to fix the cluster abundance. It is evaluated in a linear approximation to the full expression which is exact in the far field ($r > 4R_f \sim 20\ h^{-1}Mpc$), and is not wildly off in the near field. The smoothing function chosen determines the shape in the interior anyway. Inclusion of the terms involving derivatives of ψ is an improvement over the BBKS approximations. These terms make a significant difference from the BBKS approximation only when ξ is small, as in the CDM case. For models with significant power on large scales such as CDM+X the derivative corrections are small in the far field. The equations for $\xi_{pk,pk}$ are given in [22]. There is also a dynamical contribution to the clustering, which I crudely estimate to be the values quoted above using the BBKS approximation $\xi^{tot}_{pk,pk} \approx [(\xi^{stat}_{pk,pk})^{1/2} + (\xi_{\rho\rho})^{1/2}]^2$: the biasing factor describing the extent to which clusters do not trace the mass is then ~ 6 compared with ~ 2 for galaxies.

Though the amplitude for the CDM case in Fig. 2 is not far off at $r = 20\ h^{-1}Mpc$, difficulties in shape and amplitude begin beyond this point. How seriously one views the failure of a model not having a correlation function which stays positive out to 100 or even out to 50 $h^{-1}Mpc$ depends upon one's attitude as to the reliability of the Abell catalogue. However, for ξ_{cl} to plunge as the CDM model indicates at the scale 25 $h^{-1}Mpc$ where we can see groups of clusters seems unlikely.

4.4 Cluster-Galaxy Correlation Function:

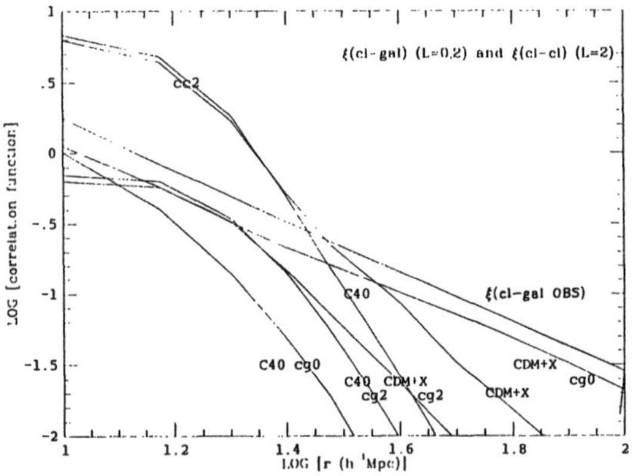

Figure 3. The cluster-galaxy correlation function determined by SELDNER and PEEBLES [23] ($\xi(cl - gal\ OBS)$) is compared with the results cg0 (the $L = 0$ part of the correlation function) of the models C40 and CDM+X of Table 1. The pure CDM model clearly does not work. The $L = 2$, $m = 0$ parts of the cluster-cluster (cc2) and cluster-galaxy (cg2) correlation functions if the clusters are oriented along their principal axes illustrate the strong alignment effects already present in the initial conditions.

The cross-correlation function of the cluster-scale peaks with the mass density smoothed on cluster scale $\xi_{pk,\rho}$ can also be evaluated. It depends not only upon ψ and its derivatives and the peak properties but also linearly on σ_0. There are $L = 0$ and $L = 2$ components in the spherical harmonic expansion if the clusters are oriented. The $\xi_{pk,\rho}$ results do not apply in the immediate neighbourhood of the cluster, for then the correlation function is measuring the collapse profile of the clusters, dynamics which is not included. To evaluate the cluster-galaxy correlation function, I assume the number density of galaxies at any point is a nonlinear function of the local smoothed mass density in the 'peak-background split' approximation of BBKS. The simple far field approximation used in Fig. 3, $\xi_{pk,g} = b\xi_{pk,\rho}$, where b is the biasing factor, follows, demonstrating the enhanced clustering around clusters of the galaxies compared with the mass.

The model CDM+X matches both ξ_{cc} and ξ_{cg}. Current observations are compatible with $\xi_{cg}/\xi_{cc} \approx 0.3$ being r-independent in the far field. With biasing this ratio is ~ 0.4, and without biasing it is ~ 0.3, almost model independent provided both ξ's are positive.

5. Alignments and Anisotropy in the Cluster Neighbourhood

The $L = 2$, $m = 0$ correlation functions are given in Fig. 3: cc2 is ξ_{cc20} which multiplies the Legendre polynomial $P_2(cos\theta)$, with θ the angle from the short axis of the oriented cluster to the other cluster; cg2 is $1.5\xi_{cg20}$ which multiplies $P_2(cos\theta)$, with θ now the angle from the long axis of the oriented cluster to the galaxy, following a convention used in [26]. In either case, ξ_{20} measures the asymmetry in the plane containing the long and short axes of the cluster. The $L \neq 0$ correlation functions only involve derivatives of ψ, and, for $\psi \sim r^{-\gamma}$, $\xi_{20} \sim r^{-(\gamma+2)}$ in the far field, rapidly approaching nonmeasurability. The effects are predicted to be large in the near field. BINGGELI [24] and PEEBLES and STRUBLE [25] have both reported that alignment effects do occur in clusters, although such effects may be weak [25]. ARGYRES *et al.*[26] report relatively large anisotropies in the angular correlation function w_{cg} in the Shane-Wirtanen catalogue, and suggested a crude model which would fit these results for $r < 25\ h^{-1}Mpc$, $1.5\xi_{cg20} \approx 2\epsilon\xi(cl - gal\ OBS)$, with $\epsilon = 0.46 \pm 0.10$. Over this limited range the C40 and CDM+X results compare favourably with these values.

Therefore one does not have to resort to galaxy and cluster formation in gaseous pancakes or to dubiously strong tidal forces to get large alignment effects up to $\sim 20\ h^{-1}Mpc$. They already exist in the initial conditions due the clusters being high ν peaks: the waves which must come together to construct such an oriented peak preferentially lie along the long axis of the peak, leading to large asymmetries in the surrounding field. Galaxies and clusters will therefore be more likely to be found along the long rather than the short axis. Further, the other clusters tend to have their long axes pointing toward the oriented cluster, measured by a correlation function varying with position angle as well as distance which also has an $L = 2$, $m = 0$ component in relative position angle; its amplitude is $0.5\xi_{cc20}$ in the long-short plane.

Unfortunately the amplitude where deviations indicating extra power occur in ξ_{Lm} is too small to be easily measurable, so alignment effects will not be a good probe of the large scale fluctuation spectrum. However, the absence of such effects over the range of $10 - 25\ h^{-1}Mpc$ would be a devastating blow to the viability of any of the models discussed here, since they all predict about the same amplitude (provided clusters are indeed associated with peaks).

Finally, in Table 2, I present grades for the various models considered assuming the observational results hold up exactly as stated. The LSSV and ξ_{cc} hurdles are remarkably difficult to satisfy with believable variations. Results for decaying particle models were not presented because the effect of having a large $\Omega_{nrd} \sim 3$ prior to recombination in order to have $\Omega_{erd} \sim 0.8$ now to bring Ω up to 1 is to decrease k_{Heq}^{-1}, thereby decreasing the large scale power. It fails badly on the first two tests. Possible interpretations of Table 2 are: (1) It is evidence for large $\Omega_B \sim 0.5$ (B+C), signalling the breakdown of the standard primordial nucleosynthesis story [27]. (2) It is evidence for a large vacuum energy (BCV). (3) It is evidence that the primordial fluctuation generator does not create scale-invariant spectra. To preserve the successes of the CDM spectrum, the generator should preserve the flat spectrum at high k, give an $n = -1$ rise from cluster to well beyond the LSSV scale, then return to the flat spectrum, either by levelling off to a plateau or by dropping down to the high k level (mountain). The plateau cannot be too high or else the intermediate angle $\Delta T/T$ constraints are violated. CDM+X satisfies all of these criteria. A mountain was chosen but it makes little difference to the results if a plateau is chosen, except $\Delta T/T$ nears the 6° limit. (4) It is evidence against inflation, since only scale-invariant spectra are 'natural'. (5) The observations must be wrong. This work was supported by a Sloan Foundation Fellowship, NSERC and a Canadian Institute for Advanced Research Fellowship.

Table 2: Test Results for Models

NAME	LSSV	ξ_{cc}	$\Delta T/T$ $< (3-6)\times 10^{-5}$ at 4.5′ $< 4\times 10^{-5}$ at 6°, 10°	z_{gf}
CDM	F	F	P (2.5×10^{-6} at 4.5′)	P?
C40	F	F	P	P?
VAC/C	F	F	P (2×10^{-5} at 4.5′)	F? (F with biasing)
OP/C	F	F	F (7×10^{-5} at 4.5′)	P
ISO/C	F	F	F (1×10^{-4} at 6°, 10°)	F? (F with biasing)
ISO/AD	F	F	P (just OK on 6° and 10°)	P?
HOT, $z_{nl}=1$	F	F	P	F
HC	F	F	P	F?
C+B	F	F	P	P?
B+C	F	F	Incomplete (P likely)	P??
BCV	P	P	P ($\sim 10^{-5}$ at 4.5′)	F? (F with biasing)
BCO	P	P	F (8×10^{-5} at 4.5′)	P?
CDM+X	P	P	P (just below 6° constraint if in IC)	P

References

1. M. Davis, G. Efstathiou, C. Frenk, S. White: *Astrophys. J.* **292**, 371 (1984).
2. J. M. Bardeen, J. R. Bond, N. Kaiser, A. S. Szalay (BBKS): *Astrophys. J.* **304**, 15 (1986).
3. S. White, C. Frenk, M. Davis, G. Efstathiou, preprint (1986)
4. J. R. Bond. "Large Scale Structure in Universes Dominated by Cold Dark Matter", in B. F. Madore and R. B. Tully, eds. *Galaxy Distances and Deviations from the Hubble Flow.* Reidel, Dordrecht 1986
5. J. M. Bardeen, J. R. Bond, G. P. Efstathiou, in preparation.
6. J. M. Bardeen, J. R. Bond, A. S. Szalay, in preparation.
7. J. I. Silk, M. S. Turner, preprint (1986)
8. L. A. Kofman, A. D. Linde, Tallinn preprint (1986)
9. A. S. Szalay, J. R. Bond, J. I. Silk. "Pancakes, Hot Gas and Microwave Distortions", in J. Audouze, ed. *Proc. 3rd Rencontre de Moriond.* Reidel, Dordrecht 1986
10. A. L. Melott. "Some Like It Hot", in B. F. Madore and R. B. Tully, eds. *Galaxy Distances and Deviations from the Hubble Flow.* Reidel, Dordrecht 1986
11. N. Kaiser: *Astrophys. J. Lett.* **284**, L49 (1984).
12. N. Bahcall, R. Soneira: *Astrophys. J.* **270**, 70 (1983).
13. M. Davis, J. Huchra: *Astrophys. J.* **254**, 437 (1982).
14. J. R. Bond, G. P. Efstathiou: *Astrophys. J. Lett.* **285**, L45 (1984).
15. G. P. Efstathiou, J. R. Bond: *Mon. Not. Roy. Astron. Soc.* **218**, 103 (1986).
16. J. R. Bond, G. P. Efstathiou, CITA preprint (1986)
17. J. R. Bond, in W. Unruh, ed. *The Early Universe.* Reidel, Dordrecht 1987
18. P. Lubin, T. Villela, in B. F. Madore and R. B. Tully, eds. *Galaxy Distances and Deviations from the Hubble Flow.* Reidel, Dordrecht 1986
19. M. Aaronson, G. Bothun, J. Mould, J. Huchra, R. A. Schommer, M. E. Cornell: *Astrophys. J.* **302**, 536 (1984).

20. D. Burstein, R. Davies, A. Dressler, S. Faber, D. Lynden-Bell, R. Terlevich, G. Wegner, in B. F. Madore and R. B. Tully, eds. *Galaxy Distances and Deviations from the Hubble Flow.* Reidel, Dordrecht 1986

21. C. A. Collins, R. D. Josephs, and N. A. Robertson, *Nature* **320**, 506 (1986)

22. J. M. Bardeen, J. R. Bond, L. G. Jensen, A. S. Szalay, in preparation.

23. M. Seldner, P.J.E. Peebles: *Astrophys. J.* **215**, 703 (1977).

24. B. Binggeli, *Astron. Astrophys.* **107**, 388 (1982).

25. P. J. E. Peebles, M. F. Struble, *Astron. J.* **90**, (4)582 (1986).

26. P. C. Argyres, E. J. Groth, P. J. E. Peebles, M. F. Struble, *Astron. J.* **91**, (3)471 (1986).

27. J. Yang, M. S. Turner, G. Steigman, D. N. Schramm, K. A. Olive: *Astrophys. J.* **281**, 493 (1984).

Session 10

Galaxies After Recombination

G. Blumenthal, Chair
August 1, 1986

Models of Protogalaxy Collapse and Dissipation

GEORGE R. BLUMENTHAL

Abstract

When protogalaxies collapse, the cooling and infall of what will become the visible galactic component affects the mass distribution of dissipationless dark matter particles which constitute the halo. For spiral galaxies, the adiabatic approximation shows that flat rotation curves arise only when about 10% of the protogalaxy's mass is dissipational, the initial core radius is large, and the visible matter falls in by roughly a factor of ten. For spirals, the amount of infall is directly related to the initial protogalaxy angular momentum. Observed variations in rotation curves may be due to several factors including variations in angular momentum, core radius, and bulge-to-disk ratio.

1. Introduction

If the universe is dominated by nonbaryonic dark matter, perturbations on galactic scales must have a very schizophrenic history. They are characterized by two "personalities": baryons and dark matter. For adiabatic perturbations produced at very early times, the ratio of dark to baryonic matter is a constant throughout the universe. However, when perturbations of galactic scale cross the horizon, the effects of pressure and diffusion decrease baryonic perturbations relative to those of dark matter, thereby increasing the ratio $\rho_{\text{dark}}/\rho_{\text{baryon}}$ within a perturbation. Once recombination occurs, this separation or schizophrenia is temporarily "cured" as baryons fall into the gravitational potential wells of the dark matter until the ratio of their densities is again a constant on scales above the baryonic Jeans mass. After the perturbations go nonlinear, the baryons (which dissipate energy radiatively) continue to fall inward leaving behind a dissipationless halo of dark matter particles [1,2]. This infall of baryons is halted either by angular momentum or star formation (producing spirals or ellipticals) and leads to a bright visible galaxy imbedded within a large dissipationless dark matter halo. Note that in some biasing schemes, such as explosions, (see paper by DEKEL in this volume) the baryons can be separated, at least temporarily, from the dark matter.

What would an ideal calculation of galaxy collapse entail? First, one must choose initial conditions which include the density of the universe Ω_o, the cosmological constant Λ, the type of dark matter (baryons, cold dark matter, massive neutrinos, *etc.*), and the type of initial density fluctuations (*e.g.* adiabatic, isocurvature, scale invariant, strings). Next, the spectrum of fluctuations after recombination must be calculated, as has been done for many of the above cases [3-5]. Then the nonlinear development of a galaxy-sized perturbation must be calculated including (a) a correct treatment of gravity, (b) cooling and infall of visible matter, (c) star formation and evolution, and (d) the feedback of star formation and evolution on the remaining gas. To date, no calculations have treated all four effects.

Both analytic [2] and N–body [6] simulations suggest that a universe dominated by cold dark matter shows considerable agreement with observations on galactic scales. For example, consider a simple spherical top hat model for perturbations on a galactic scale. If at some very early redshift z_i the relative density perturbation was $\delta\rho/\rho = \Delta_i$, then the perturbation begins to collapse when its density is [4]

$$\rho_{\min} = \frac{3\Omega_o {H_o}^2}{8\pi G}\left[1 - {\Omega_o}^{-1} + \frac{5}{3}(1+z_i)\Delta_i\right]^3 . \tag{1}$$

Subsequently, the perturbation collapses and virializes so that it achieves a velocity dispersion $\langle v^2 \rangle \sim GM/R_{\text{virial}}$. If these perturbations had an initial rms spectrum of the form

$$\Delta(M) \propto (1+z)^{-1} M^{-(3+n_{eff})/6}, \tag{2}$$

then they virialize with

$$M \propto v^{12/(1-n_{eff})}. \tag{3}$$

For cold dark matter, $n_{eff} \approx -2$ on galactic scales, so for constant mass to light ratio, (3) just reduces to the standard Tully-Fisher and Faber-Jackson relations. In addition, for protogalaxies between about 10^7 and $10^{13} M_\odot$, the baryons are able to cool and fall toward the center on a timescale of order the dynamical time for the system, forming a visible core within a large dark matter halo [2].

Recent N–body simulations also show that the cold dark matter spectrum leads to roughly the observed density of spiral galaxies [6]. They also show that for spiral galaxies, the mass distribution relaxes with a core radius $\approx$ 25 kpc [7]. For further details of N–body simulations, see the paper by FRENK in this volume.

When the baryons in a protogalaxy dissipate energy and fall inwards, the changing gravitational potential can affect the distribution of dissipationless halo material, even if only a small fraction of the mass is dissipational. In a sense, this is analogous to the tail wagging the dog. This effect has been studied using both analytic approximations and N–body simulations (with dissipation included either by letting nearby particles collide inelastically or by assuming a final distribution for the visible matter). Simulations (with sticky particles) of a single elliptical galaxy have shown that dissipation may lead to a deVaucouleurs distribution of visible matter and an r^{-2} distribution for the halo density [8]. CARLBERG [9] has found that as ellipticals collapse, dissipation transports angular momentum outward as the visible matter falls in. For spiral galaxies, it has been found using both N–body [10,11] and analytic methods [11,12] that flat rotation curves will result if (1) the core radius of the protogalaxy is large in the absence of dissipation, (2) the ratio F of dissipational to total mass lies within a factor of two of 0.1, and (3) the visible matter falls in by roughly a factor of ten. When these conditions are met, there is roughly comparable amounts of visible and dissipationless matter within an optical radius, in agreement with the observations.

The following sections describe some recent analytic work on spirals [13] done by BLUMENTHAL, FLORES, FABER and PRIMACK. These calculations show the relation between the amount of dissipational infall and the initial galactic angular momentum. They also explain how different forms of rotation curves may arise.

2. Analytic Model for Dissipation

A convenient analytical model, which has been extensively checked by numerical simulations, can be used to calculate the radial redistribution of the dissipationless halo matter of a protogalaxy when its dissipational (visible) matter radiates energy and falls in toward the center [11-14]. For a particle moving in a periodic orbit, $\oint p dq$ is an adiabatic invariant, where p is the canonical conjugate momentum of the coordinate q. Thus, for particles moving in circular orbits about a spherically symmetric mass distribution, the adiabatic invariant is $rM(r)$ provided that $M(r)$, the mass inside the orbital radius r, changes slowly within an orbital time. For purely radial orbits, $r_{\max}M(r_{\max})$ is also an adiabatic invariant if $M(r)$ varies in a self-similar fashion.

Consider a spherically symmetric protogalaxy consisting of a fraction $F \ll 1$ of dissipational baryons and a fraction $1 - F$ of dissipationless particles which will constitute the halo. Both types of particles are assumed to be well mixed initially. Define a truncation radius R beyond which no dissipation occurs. This is a rough approximation to the final shape of a protogalaxy, which is certainly not spherical although its inner regions ($r < R$) may be nearly so.

Since there is more phase space for nearly circular orbits than for nearly radial orbits, the simplest approximation is that the dark matter particles move in circular orbits about the protogalaxy center with almost randomly oriented angular momentum vectors. (The initial angular momentum of the protogalaxy is assumed to be very small.) As the baryons dissipatively cool and fall in to the center forming a disk, a dark matter particle initially at radius r_i will move in to radius $r < r_i$. The adiabatic invariant for such a particle implies that

$$r\left[M_{\text{disk}}(r) + M_{\text{halo}}(r)\right] = r_i M_i(r_i), \tag{4}$$

where $M_i(r_i)$ is the initial total mass distribution, $M_{\text{disk}}(r)$ is the final mass distribution of dissipational baryons, and $M_{\text{halo}}(r)$ is the final distribution of dissipationless halo particles. If the orbits of the halo particles do not cross, then

$$M_{\text{halo}}(r) = (1 - F)M_i(r_i), \tag{5}$$

so (4) and (5) can be used to calculate the final radial distribution of the halo particles once $M_i(r_i)$ and $M_{\text{disk}}(r)$ are given. If the dissipational mass fraction $F \ll 1$, then for a halo particle not too near the center of the protogalaxy, the mass interior to its orbit will undergo a small fractional change in one orbital period, even if dissipation occurs in an orbital time. Therefore, the adiabatic invariant given in (4) is expected to be a good approximation for all but the innermost halo particles.

The initial protogalaxy will be assumed to relax to an isothermal sphere with core radius $r_{\text{core}} = 3\langle\sigma^2\rangle/4\pi G\rho_o$, where σ is the one-dimensional velocity dispersion and ρ_o is the central density. This initial state is a good approximation (in the absence of dissipation) to the equilibrium radial mass distribution of an isolated, expanding protogalaxy that consists of a large number of "clouds" initially in pure Hubble flow and distributed inside a truncated sphere with a Poisson distribution. In reality, one expects some dissipation to occur before a protogalaxy has fully virialized, but this does not significantly change the results [11,15].

Large values of the core radius, $r_{\text{core}}/R \sim 0.4$, are needed for baryonic infall to produce mass distributions with flat rotation curves [11]. Such large values would arise naturally if protogalaxies had a substantial amount of kinetic energy at maximum expansion, which is the case for the isolated expanding protogalaxy. In addition, simulations of a large expanding region from which collapsed protogalaxies can be extracted by FRENK, WHITE, EFSTATHIOU and DAVIS [6] for a universe with a cold dark matter spectrum indeed show large values for the core radius, $r_{\text{core}}/R \sim 0.2 - 0.5$. Furthermore, QUINN, SALMON and ZUREK [7] find r_{core}/R to be rather sensitive to the power spectrum, with $r_{\text{core}}/R \sim 0.1$ (0.4,1.0) for a $n = -1$ (-2,-3) power law perturbation spectrum of the form $|\delta_k|^2 = k^n$.

For a spiral galaxy, the final radial mass distribution of the baryons is constrained by the initial angular momentum of the protogalaxy. Here, the final mass distribution of the disk is assumed to be

$$M_{\text{disk}}(r) = \mathcal{M}_{\text{disk}}\left[1 - (1 + r/b)e^{-r/b}\right], \tag{6}$$

which is the radial mass distribution of a thin disk of mass $\mathcal{M}_{\text{disk}}$ whose surface density $\mu_{\text{disk}}(r) = \mathcal{M}_{\text{disk}}\exp(-r/b)/(2\pi b^2)$ decreases exponentially with scale length b.

The adiabatic approximation is not expected to accurately describe the dissipationless matter distribution near the center, where dissipating particles dominate the mass [10]. However, N–body simulations confirm that the total $M(r)$, and therefore the rotational velocity is quite well described by the adiabatic approximation for this case [16].

3. Relation of Infall to Initial Angular Momentum

It is usual to describe the angular momentum, **J**, of a protogalaxy in terms of a dimensionless quantity

$$\lambda \equiv \frac{J|E|^{1/2}}{G\mathcal{M}^{5/2}}, \tag{7}$$

where $\mathcal{M}$ is the total mass of a protogalaxy with total energy (gravitational plus kinetic) E. Simple theoretical arguments show that if protogalaxies receive most of their angular momentum through tidal torques, then one expects the population of protogalaxies to have a mean λ of ~ 0.07. This has been confirmed by N–body simulations which find $\langle\lambda\rangle \approx 0.07$ with a width for the distribution $\Delta\lambda \approx 0.03$ [17,18].

FALL and EFSTATHIOU [1] first calculated how the amount of infall of the disk material is related to the initial angular momentum λ of a protogalaxy containing both dissipational and dissipationless material. They assumed that the formation of a disk involved no transfer of angular momentum between the disk and halo components. They also took a specific form for the halo density distribution, which they assumed was unaffected by the dissipational infall of the disk material. Using the adiabatic invariant (4), it is straightforward to relax the latter assumption [13]. In the absence of a halo, a thin exponential disk has roughly half of its angular momentum in the outer 25% of its mass, so the adiabatic invariant assumption should be quite good.

Consider a protogalaxy with core radius $r_{\rm core}$, truncation radius R, and dissipational baryonic fraction F. If the dissipational material settles to a disk with mass distribution given by (6), then the angular momentum of the thin disk is given by

$$J_{\rm disk} = 2\pi \int_0^\infty r dr \mu_{\rm disk}(r) v_{\rm rot}(r). \tag{8}$$

The rotation velocity $v_{\rm rot}(r)$ is given by

$$v^2(r) = r\left[g_{\rm disk}(r) + g_{\rm halo}(r)\right]. \tag{9}$$

The inward acceleration of an exponential disk is given by

$$g_{\rm disk}(r) = \frac{G\mathcal{M}_{\rm disk}}{b^2}\left(\frac{r}{2b}\right)\left[I_0(r/2b)K_0(r/2b) - I_1(r/2b)K_1(r/2b)\right], \tag{10}$$

where I_n and K_n denote modified Bessel functions of order n [19]. The acceleration due to the halo is $g_{\rm halo}(r) = GM_{\rm halo}(r)/r$, where $M_{\rm halo}(r)$ is determined by iteratively solving equations (4) and (5). If there is no transfer of angular momentum between disk and halo particles, then $J_{\rm disk}/\mathcal{M}_{\rm disk} = J_{\rm halo}/\mathcal{M}_{\rm halo}$, where $J_{\rm halo}$ and $\mathcal{M}_{\rm halo}$ are the halo angular momentum and mass out to the maximum radius of infall R, and the total angular momentum of the protogalaxy is $J = J_{\rm disk}/F$ out to that radius. Since the form of (7) suggests that λ is not very sensitive to the cutoff radius, one can therefore simply relate λ for the protogalaxy to the disk scale length b.

Figures 1 and 2 show how the amount of dissipative infall b/R is related to the initial angular momentum, λ. The dependence of the b/R versus λ relation on the baryonic fraction F is shown in Fig. 1. Here, smaller values of F lead to less infall essentially because for a given b/R smaller F produces less halo infall and therefore less disk angular momentum per unit disk mass. In the limit that $F \to 0$ the curve is expected to approach a limiting relation. Figure 2 shows how the relation depends on the initial protogalaxy core radius $r_{\rm core}/R$. Here, larger core radii produce less infall because the large core radius case gives less contribution of the halo to the disk rotation velocity and angular momentum. As $r_{\rm core}$ becomes large, the initial protogalaxy density is nearly constant for $r < R$, and the curves become indistinguishable. A stiff halo which does not respond to the dissipative infall would lead to a larger value for the disk scale length b, about 27% larger for λ=0.07 and F=0.1 [13].

These results are consistent with the conclusions of earlier work [11] that to within a factor of 2, flat rotation curves require $r_{\rm core}/R \approx 1/2$, $F \approx 0.1$, and

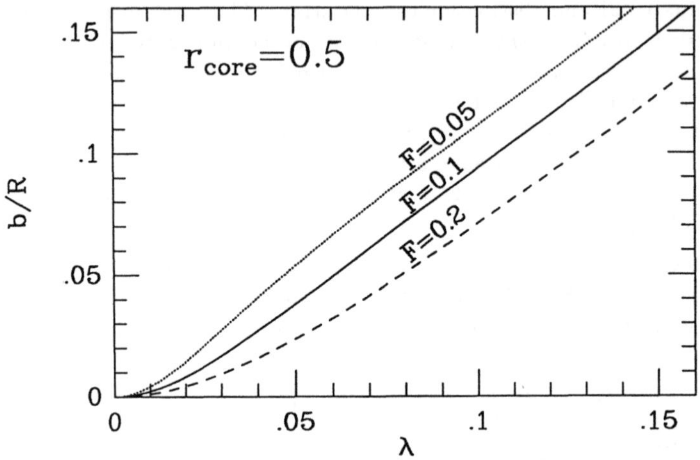

Figure 1: Plot of the amount of dissipational infall b/R versus the dimensionless initial angular momentum λ for $r_{\rm core}/R = 0.5$ and $F = 0.05$, 0.1, and 0.2.

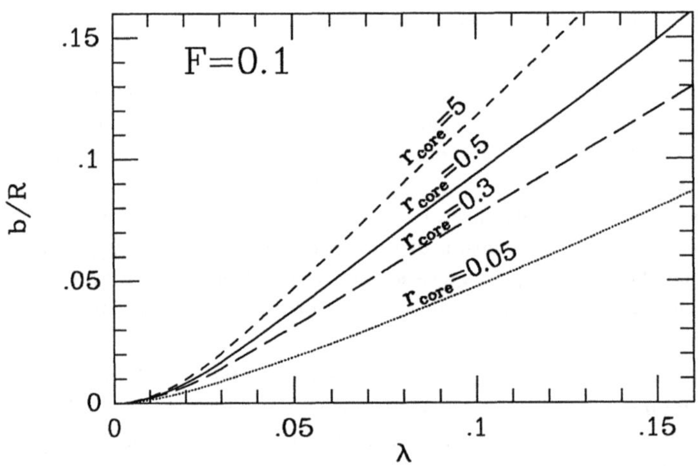

Figure 2: Plot of the amount of dissipational infall versus the dimensionless initial angular momentum λ for $F = 0.1$ and $r_{\rm core}/R = 0.05$, 0.3, 0.5 and 5.0.

$b/R \approx 0.06$. Figure 2 shows that these values are entirely consistent with a value of $\lambda = 0.07$ which is implied by the tidal torque theory of angular momentum in galaxies.

4. Systematics of Rotation Curves

BURSTEIN and RUBIN [20] have found that the rotation curves of spiral galaxies can be separated into three different classes which correlate much more strongly with environment than with Hubble type [21]. It is therefore of interest to calculate theoretical rotation curves using the adiabatic invariant to set limits on the parameters consistent with relatively flat rotation curves out to large distances and to determine which parameters may be responsible for the various forms of rotation curves which are observed [13].

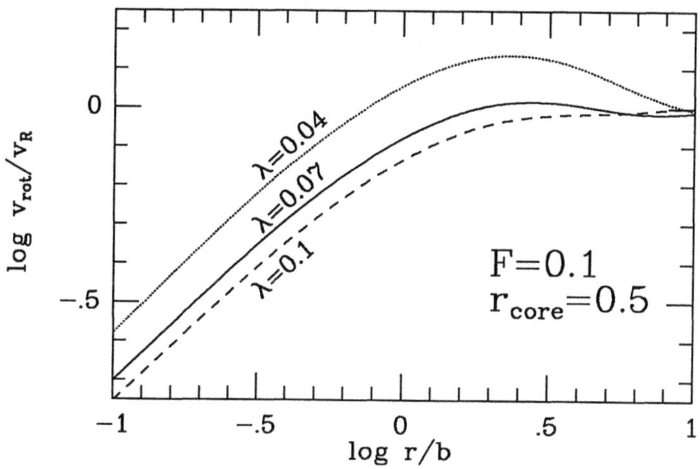

Figure 3: Rotation curves $\log v_{rot}/v_R$ versus $\log r/b$ for various values of the angular momentum λ. v_R is the rotation velocity at $r = R$.

Figure 3 contains a log–log plot of the rotation velocity versus radial distance measured in units of the scale length, r/b. This figure shows that both rising and falling rotation curves are expected beyond a couple of disk scale lengths for typical values of the angular momentum parameter λ. At the Holmberg radius of a galaxy, which roughly corresponds to $R_H = 4.5b$, the rotation curve can show a substantial decline for small λ because small λ leads to greater infall (smaller b/R) and therefore a very centrally condensed mass distribution.

Figures 4 and 5 show the rotation profiles of the mass distributions that result after baryonic infall, for various values of F and r_{core}/R. One can see that flat rotation curves out to large radii are possible only for $F \sim 0.1$ and $r_{core}/R \sim 0.4$. Furthermore, for $F \gtrsim 0.15$ only falling rotation profiles ($v_{rot}(4.5b) < v_{rot}(3b)$) are possible for $0.04 \leq \lambda \leq 0.1$ and any value of r_{core}/R. Likewise, for $F \lesssim 0.03$ only rising rotation profiles arise because of the absence of a central mass concentration. Figure 5 also illustrates the point that small initial core radii inevitably lead to falling rotation curves because of the large central mass concentration.

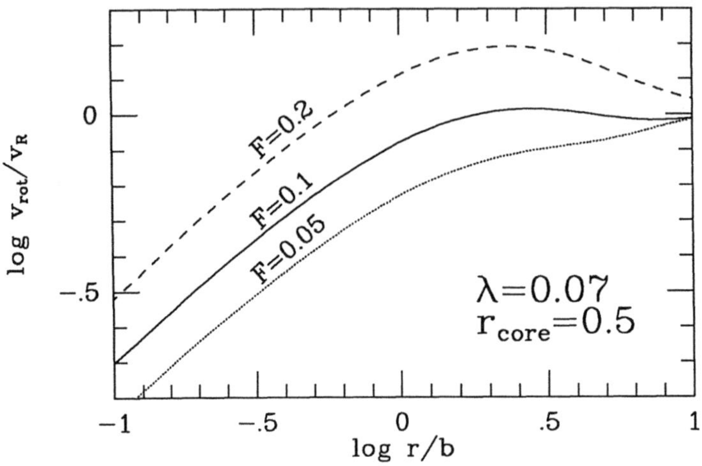

Figure 4: Rotation curves for various values of F.

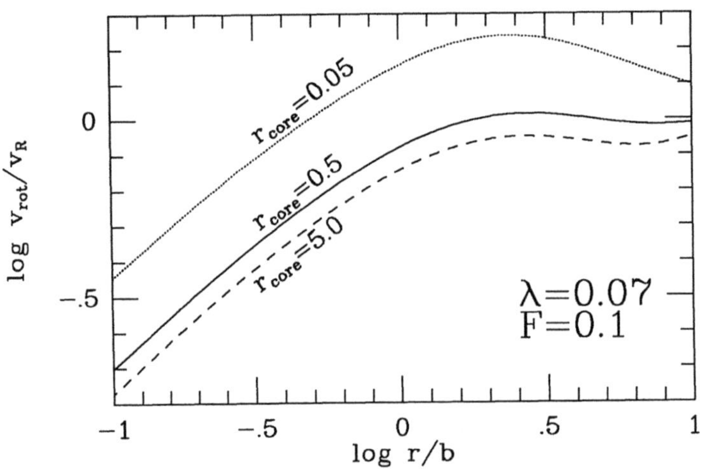

Figure 5: Rotation curves for various values of r_{core}/R.

Rather flat rotation curves arise naturally for $r_{core}/R \gtrsim 0.2$ and for values of λ resulting from tidal torques. This point is illustrated in Fig. 6, which shows the relative increase in rotation velocity between $r = 3b$ and $r = 4.5b$, roughly the Holmberg radius. In this region, the adiabatic approximation should be quite accurate. The figure further illustrates that for very small values of λ, flat rotation curves would not be observed in this range of r.

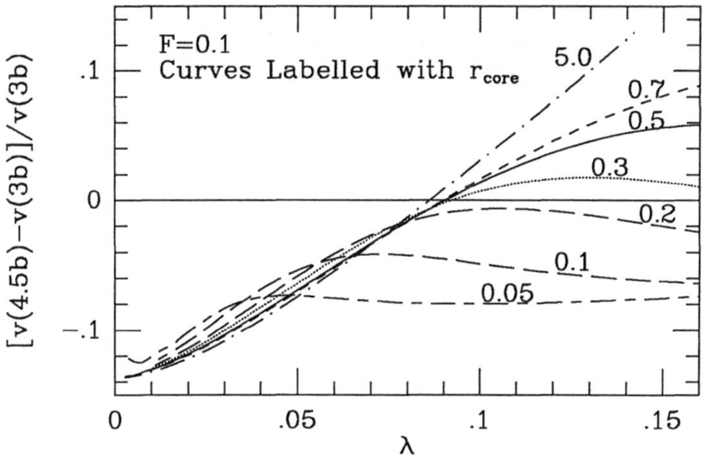

Figure 6: Relative increase in rotation velocity between $r = 3b$ and $r = 4.5b$ versus λ for $F = 0.1$ and several values of r_{core}/R. For comparison, a Keplerian system with $v_{\text{rot}} \propto r^{-1/2}$ has $[v(4.5b) - v(3b)]/v(3b) = -0.18$.

One notable aspect of Figs. 3–5 is that no feature in the rotation profile separates the inner, disk-dominated region from the outer, halo-dominated region [11,13]. The coupled motion of the two components, as the baryons fall into the center, avoids any noticeable separation of the two components. This is a direct consequence of the response of the halo to the infall of the visible matter. Observationally, the absence of such a separation into two components has been noted by several groups [20,22,23]. It has been observed that disk and halo matter contribute about equally to the rotational velocity at the optical radius of a galaxy, R_H.

Figure 7 shows the ratio of the rotational velocity due to the disk to that due to the halo matter at $r = 4.5b$. For $F = 0.1$, this ratio is indeed close to unity for $0.04 < \lambda < 0.1$ and for all values of $r_{\text{core}}/R \gtrsim 0.05$. For large F, the disk gives a relatively greater contribution to the rotational velocity because of the relatively larger central mass concentration. Therefore, the fact that $v_{\text{disk}} \approx v_{\text{halo}}$ at $r = 4.5b$ is a consequence of the fact that $F \approx 0.1$ and $\lambda \gtrsim 0.04$ for galaxies.

5. Discussion

One conclusion that can be drawn is that flat rotation curves out to many disk scale lengths severely constrain the parameters of protogalaxies. In particular, it requires that the baryonic mass fraction $0.05 \lesssim F \lesssim 0.2$, that the protogalaxy angular momentum $0.04 \lesssim \lambda \lesssim 0.1$, and that the protogalaxy core radius $r_{\text{core}}/R \gtrsim 0.2$.

What is responsible for the various forms of rotation curves seen in spiral galaxies? An obvious candidate is the bulge-to-disk ratio. Figure 8 shows how the bulge-to-disk ratio affects the rotation curve assuming all other parameters remain constant and assuming a deVaucouleurs law with $r_{\text{eff}} = b/2$ for the mass distribution of the bulge [13]. Clearly, different forms for the rotation curve arise. However, observations show no significant correlation between the form of the rotation curve

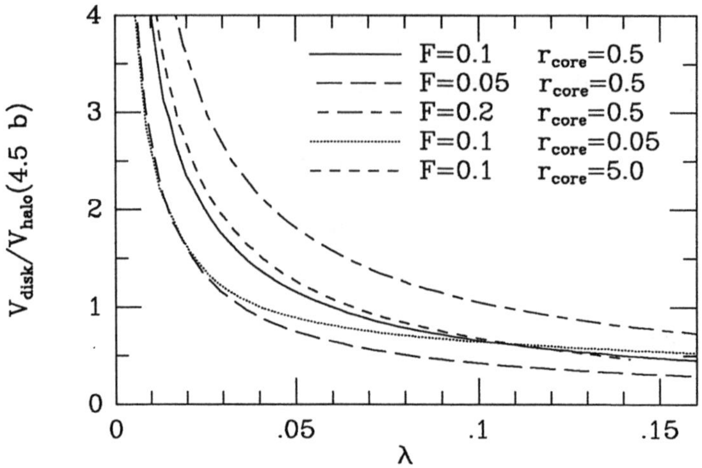

Figure 7: Ratio of the disk to halo contributions to the rotational velocity at $r = 4.5b$ versus the angular momentum parameter λ for various values of the dissipational fraction F and the initial core radius $r_{\rm core}/R$.

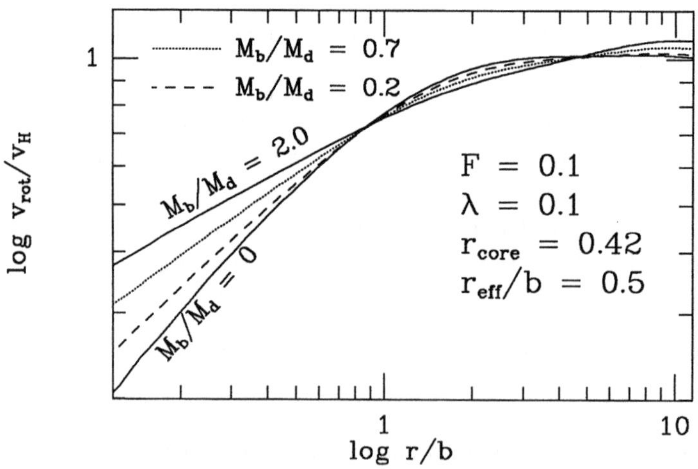

Figure 8: Rotation curves showing the contribution of a bulge for typical values of F, $r_{\rm core}$, and λ. Here, $\log v_{\rm rot}/v_H$ is plotted against $\log r/b$, where $v_H = v_{\rm rot}(r = R_H = 4.5b)$. The bulges are assumed to have deVaucouleurs profiles with $r_{\rm eff}/b = 0.5$ and various bulge to disk mass ratios M_b/M_d.

and either Hubble type or bulge-to-disk ratio [21].

Alternatively, Figs. 3 and 5 suggest that variations in the initial angular momentum λ or in the core radius r_{core}/R from galaxy to galaxy may lead to various forms for the mass distribution, assuming that F is a universal constant. Indeed, N–body simulations show a substantial range in λ among galaxies, and Fig. 3 suggests that galaxies with a higher value of λ will show a more steadily rising rotation curve. If field spirals have systematically higher initial angular momenta λ than spirals in clusters, then one would expect field spirals to show more steadily rising rotation curves than cluster spirals, as has been observed [21]. However, more detailed N–body simulations (including the effects of dissipation) will be needed to confirm whether variations in λ, r_{core}/R, or bulge-to-disk ratio or correlations among these quantities are responsible for the differences among observed rotation curves.

Acknowledgements

I gratefully acknowledge many stimulating interactions with my colleagues Ricardo Flores, Sandra Faber, and Joel Primack, with whom most of the work reported here was done. This research was supported by NSF grant PHY84-15444 and by a faculty research grant at the University of California, Santa Cruz.

References

1. S. M. Fall and G. Efstathiou, *M.N.R.A.S.* **193**, 189 (1980)
2. G. Blumenthal, S. Faber, J. Primack, and M. Rees, *Nature* **311**, 517 (1984); also *Nature* **313**, 72 (1985)
3. P. J. E. Peebles *Ap. J.* **258**, 415 (1982)
4. J. R. Primack and G. R. Blumenthal in *Clusters and Groups of Galaxies*, eds. F. Mardirossian, G. Giuricin and M. Mezzetti, Reidel, Dordrecht, 1984, p. 435. Also G. R. Blumenthal and J. R. Primack, preprint (1986)
5. J. R. Bond and A. S. Szalay, *Ap. J.* **274**, 443 (1984)
6. C. Frenk, S. White, G. Efstathiou and M. Davis, *Nature* **317**, 595 (1985)
7. P. Quinn, J. Salmon, and W. Zurek, preprint (1986)
8. R. Carlberg, G. Lake and C. Norman, preprint (1986)
9. R. Carlberg, preprint (1986)
10. J. Barnes, in this volume (1986)
11. G. R. Blumenthal, S. M. Faber, R. Flores and J. R. Primack, *Ap. J.* **301**, 27 (1986)
12. B. Ryden and J. Gunn, *Bull. AAS* **16**, 487 (1984)
13. G. R. Blumenthal, R. Flores, S. M. Faber and J. Primack, in preparation (1986)
14. Ya. B. Zeldovich, A. Klypin, M. Yu. Khlopov and V. Chechetkin, *Sov. J. Nucl. Phys.*, **31**, 664 (1980)
15. R. Flores, G. Blumenthal, A. Dekel and J. Primack, *Nature,* in press (1986)
16. K. S. Oh, in preparation (1986)
17. G. Efstathiou and B. J. T. Jones, *M.N.R.A.S.* **186**, 133 (1979)
18. J. Barnes and G. Efstathiou, preprint (1986)

19. K. Freeman, *Ap. J.* **160**, 811 (1970)

20. D. Burstein and V. Rubin, *Ap. J.* **297**, 423 (1985)

21. D. Burstein, V. Rubin, W. Ford and B. Whitmore *Ap. J. Letters* **305**, L11 (1986)

22. J. Bahcall and S. Casertano, *Ap. J. Letters* **293**, L7 (1985)

23. T. S. van Albada and R. Sancisi, preprint (1986)

Unstable Dark Matter and Galaxy Formation

RICARDO A. FLORES

1. Introduction

Inflation has become a very attractive cosmological model because it can naturally resolve major cosmological puzzles that the standard Big Bang Cosmology leaves unanswered, such as the flatness, isotropy and homogeneity of the Universe and the origin of density perturbations; inflation naturally produces the scale-invariant, Harrison-Zeldovich spectrum of density perturbations as well. Because of the tremendous expansion that the Universe undergoes at very early times in the inflationary scenario, it is predicted to be extremely flat at present; if the cosmological constant vanishes, this implies that the mean energy density of the Universe ρ, in units of the critical density ρ_c, is $\Omega \equiv \rho/\rho_c = 1$ to a very high accuracy. Observations, on the hand, indicate that the clustered matter (including dark matter) on scales of up to a few megaparsecs amounts only to $\Omega_{cl} = 0.1 - 0.3$ [1,2]. Several solutions to this puzzle have been proposed, such as a relic cosmological constant Λ [3] such that $\Omega + \Lambda/8\pi G\rho_c = 1$, or "biased" galaxy formation [4], in which it is postulated that galaxies are rare events, resulting only from large fluctuations of the density field −thus they would not be good tracers of mass in the Universe.

Here we consider another solution [5-7], which postulates that the non-radiative decay of a heavy elementary particle species, after it has driven the formation of galaxies and clusters, provides a smooth, undetected background of relativistic particles that at present contribute Ω_r to the total energy density of the Universe. In the original model [5,6] ("type I" models) the non-relativistic matter contributes $\Omega_{nr} = \Omega_{cl}$ to the total energy density and $\Omega_r = 1 - \Omega_{nr} \sim 0.8$. Another possibility is that the Universe becomes dominated by a primordial, stable non-relativistic dark matter (DM) species after decay of the unstable species [8]; in these models ("type II") the decay of a heavy species unbinds large quantities of a primordial, stable light species causing it to stream away from clusters and larger structures, thus providing a smooth background of non-relativistic DM. Here $\Omega_{cl} \sim 0.2$ reflects only the component of the DM that remains clustered on small scales, while the true Ω_{nr} is ~ 0.8.

The redshift corresponding to the epoch of decay, z_d, is constrained by several astrophysical considerations. The isotropy of the microwave background requires that the decay happen sufficiently late, $1+z_d \lesssim 5$ [9-11], because fluctuations can only grow until the epoch of decay in type I models. Also, $1+z_d \lesssim 5$, for the cores of rich clusters to remain in virial equilibrium [12]; if the decay occurs on a time scale shorter than the dynamical time, a cluster would become unbound after losing 50% of its mass. A similar upper bound is implied if normal galaxies are assumed to be the source of gravitational lensing [13]. On the other hand, the decay must happen sufficiently early for our infall velocity towards the Virgo cluster not to be too large. A linear theory analysis implies $1 + z_d \gtrsim 20$ for type I models [12], but this bound is exponentially sensitive to uncertainties in the observational parameters and is affected by the non-sphericity [14] and non-linearity [15] of the infall, so a value as low as $1 + z_d \sim 5$ is not excluded. The large streaming velocities that seem to exist on large scales and the reported large amplitude of the rich–cluster correlation function would rule out type I models (see the article by BOND in this volume) as well as any cold– or hot–DM model of formation of structure in the universe. These observations, however, remain controversial (see the article by GUNN in this volume).

We discuss here the constraints imposed on cosmologies with unstable DM by the observed rotation curves of spiral galaxies [16] if galaxies are assumed to have formed via dissipative collapse inside a gravitationally induced protogalaxy consisting of a homogeneous mixture of dissipationless DM and a small fraction of baryonic material [17]. We discuss first a simple analytic model based on adiabatic invariants; one finds that the calculated rotation curves are sharply peaked if the fraction of stable DM is very small. Numerical models are discussed next: they relax many of the assumptions made in the analytic model, but they are still limited by numerical effects in the very inner parts of the galaxies; it is found that the analytic model works fairly well. The analytic model is then used to compute rotation curves, which are compared with the observational rotation curves of spirals; agreement is found only if at least 50% of the DM is stable. Finally, we discuss the implications of this constraint for both type I and type II models and summarize the conclusions.

2. Analytic Model

Consider a spherically symmetric protogalaxy of radius R that consists initially of a mass fraction F_i of baryons and $1-F_i$ of DM particles that are well mixed initially (i.e. their density ratio is F_i throughout the protogalaxy). This is a rough approximation to the final shape of an expanding protogalaxy which will cut itself out of the general expansion with some characteristic size although there is no reason to expect it to be exactly spherical. Two processes change the amount of mass inside a given radius: baryonic infall due to dissipation and mass loss due to DM decay. The astrophysical constraints discussed above imply that the DM must decay on a time scale $\tau \gtrsim 10^9$ years, and since the mass fraction in baryons at the present must be small [1], a DM particle's orbit about the protogalaxy is expected to change adiabatically during dissipation and decay. This allows one to calculate the effect of these processes on a galaxy halo with a simple analytic model [16].

For particles moving in circular orbits about a spherically symmetric mass distribution, $rM(r)$ is an adiabatic invariant provided that $M(r)$, the mass inside the orbital radius r, changes slowly with time (see the article by BLUMENTHAL in this volume, and references therein). For purely radial orbits, $r_{max}M(r_{max})$ is an adiabatic invariant as well, provided that $M(r)$ varies in a self-similar fashion. Since there is more phase space available for nearly circular orbits than for nearly radial orbits, we shall make the simplifying assumption that the orbits of DM particles are circular.

As the baryons cool, they fall into the center to a final mass distribution $M_b(r)$ which, in the case of a spiral galaxy, is constrained by the initial angular momentum (it will be assumed that no baryons fall inside R from beyond R). After the DM decay is complete, the baryonic mass fraction is $F \equiv \mathcal{M}_b/\mathcal{M}_f$, where $\mathcal{M}_f$ is the final total mass and $\mathcal{M}_b$ is the total mass in baryons. Since F must be small [1], $F_i \ll 1$, and the mass interior to a DM particle's orbit will not show a large fractional change during one orbital period even if dissipation occurs rapidly. provided the particle is far enough from the galaxy center. This assures that the orbits of all but the innermost halo particles change adiabatically under baryonic infall. There is always some inner region of the halo that changes adiabatically under DM decay as well, since the mass inside the orbit of a halo particle with orbital time $t_{orb} \ll \tau$ changes slowly. For τ much longer than the system's dynamical time, t_{dyn}, the orbits of all (except, perhaps, the outermost) halo particles change adiabatically. Thus, after dissipation and decay are complete, a halo particle initially at radius r_i is at radius r, such that $rM(r) = const.$:

$$r\,[M_b(r) + M_{DM}(r)] = r_i\, M_i(r_i). \tag{1}$$

If shells of DM do not cross each other, $M_{DM}(r) = f(1 - F_i)M_i(r_i)$, where f is the mass fraction of stable DM. Thus (1) can be used to find the DM mass distribution $M_{DM}(r)$ once the initial distribution of total mass $M_i(r_i)$ and the final baryon mass distribution $M_b(r)$ are given. The initial mass distribution $M_i(r_i)$ must be an equilibrium configuration and will be taken to be that of an isothermal sphere with core radius $a \equiv 3v^2/4\pi G\rho_o$, where v is the one-dimensional velocity dispersion and ρ_o is the central density. The final radial mass distribution of the baryons is assumed to be $M_b(r) = \mathcal{M}_b\,(1 - (1 + r/b)exp(-r/b))$, which describes the mass distribution of a thin disk whose surface density falls off exponentially with scale length b. (When using (1) the baryonic mass is actually treated as if it were distributed spherically rather than in a disk. See second ref. of [17].)

Figure 1 shows rotation curves resulting from dissipation and decay within an initial isothermal sphere whose core radius is 42% of R. The two solid lines are the rotation curves for two values of f, and typical values of F and the collapse factor b/R. The ratio b/R is constrained by the initial angular momentum. Theoretical analysis and N-body simulations of the tidal torque theory of angular momentum estimate the mean value of the spin parameter $\lambda \equiv J|E_i|^{1/2}/GM_i^{5/2}$ to be 0.07 [18]; here J, E_i and M_i are respectively the total angular momentum, energy and mass of a system (for reference, a rotationally suported sphere has $\lambda \sim 0.25$, and a uniform density disk

$\lambda \sim 0.5$). Unlike F, which is assumed to be a spatial constant (as expected if most of the DM is non-baryonic), λ varies among protogalaxies with a dispersion about the mean of ~ 0.03. This implies that $0.02 \lesssim b/R \lesssim 0.1$ with a mean value of 0.05 [17,19]. The effect of this dispersion on rotation curves is shown in the two dotted lines, which are rotation curves for $f = 0.25$ with collapse factors that correspond to b/R at its upper (3) and lower (5) limit. There are several observational and theoretical constraints on F that place it in the range $0.05 \lesssim F \lesssim 0.2$ [1]. Decreasing F produces less peaked rotation curves, as shown by the thick dotted line in Fig. 1, which is the rotation curve for $f = 0.75$ with $F = 0.05$ and $b/R = 0.1$.

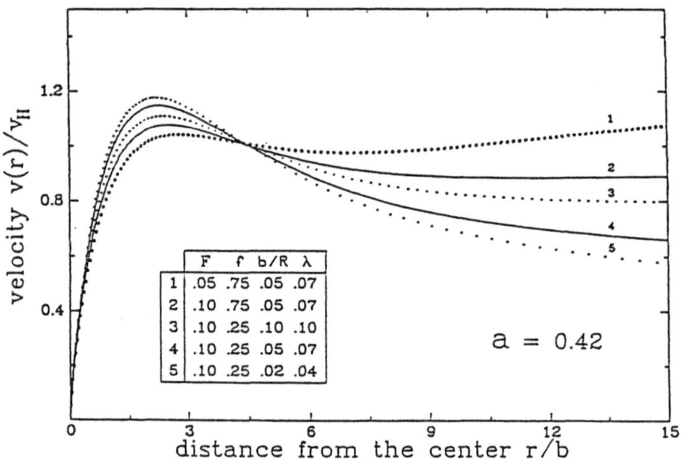

Fig. 1: Rotation curves calculated with the analytic model: $v(r)$ is the rotational velocity, in the plane of the disk, needed by a particle to remain in circular orbit at distance r/b from the center and $v_H \equiv v(r = R_H \equiv 4.5b)$.

The main result of Fig. 1 is that velocities decrease at large distances for small f. This can be understood as follows: after the baryons have cooled to an exponential disk (dissipation and decay do not commute since λ is invariant under decay in the adiabatic approximation; we assume that by $t \simeq \tau \gtrsim 10^9$ years the luminous part of galaxies have collapsed and dissipated, since their dynamical and cooling times are shorter than τ), the DM decay hardly changes $v(r)$ at small radii because the baryons dominate there. At large radii, where the DM dominates, $v(r)$ decreases substantially as a result of the mass loss. Thus, a lower bound on f is anticipated from the constraint that observed rotation curves do not fall with distance even well outside the optical radius.

3. Numerical Model

The analytic model just described assumes circular or purely radial orbits, spherical symmetry, an equilibrium starting configuration, no shell crossing, and does not include expansion. These assumptions can be relaxed and more general cases can be studied using N-body simulations. We carried out extensive numerical work to test the analytic model, using a dissipative N-body code based on the code developed by S. J. Aarseth[20]. Some of this work is discussed in more detail in Ref. 16; here we briefly outline its features and the conclusions one can draw from it.

The simulations follow a large number of particles that interact via a Newtonian potential, softened on small scales in order to suppress spurious two-body relaxation effects due to the relatively small number of bodies in the simulation. Integration time steps were chosen so that the fractional change in the total energy is less than 1% over a whole simulation, and the integration continued for several dynamical times after the system reached approximate equilibrium ($t_{dyn} \equiv GM_i^{5/2}/|2E_i|^{3/2}$). The system was chosen to have total mass $M_t = 10^{12}M_\odot$, and to consist of $N_{1-f} + N_f$ dissipationless particles (the DM) and 500 dissipational particles (the baryons). We chose the number of stable (N_f) and unstable (N_{1-f}) DM particles large enough to have adequate statistics. For example for $f = 0.4$ we took $N_f = 500$ and $N_{1-f} = 750$. In all simulations, the baryonic particles were initially mixed uniformly with the DM particles.

In simulations with dissipation, the baryonic particles were assumed to have a collisional cross section σ , which was kept constant throughout a simulation. A collision occurred if two particles came within a distance $(\sigma/\pi)^{1/2}$ of each other, in which case they were merged to form a single particle located at their center of mass, moving with its velocity and mass. Their relative energy was thus dissipated (σ controls the rate of baryonic infall; for larger σ the baryons fall in faster). We chose σ such that the total energy loss rate due to inelastic collisions equals the physical cooling rate (assuming fully ionized H + He). For more discussion see [17].

The decay of the unstable DM was mimicked by randomly selecting N_{1-f} dissipationless particles and assigning them a time $t_{decay} = -\tau \log\theta$, where θ is a random number in the interval (0,1) chosen for each particle. At time t_{decay} the particle was eliminated from the simulation.

An example of such simulations is shown in Fig. 2, where the results of a "Hubble start" model are presented. The model starts with an expanding, homogeneous sphere with pure Hubble flow, except for a small amount of rigid-body rotation with $\lambda = 0.07$, and the final fraction of baryons is $F = 0.1$. At t = 0 the Hubble constant is chosen so that the protogalaxy expands by a factor of ~ 2.5 before collapse. The lower solid line is the resulting "velocity" profile, $(M(r)/r)^{1/2}$ vs. r, after a fraction $1 - f = 0.4$ of the DM has decayed with a lifetime $\tau = 5t_{dyn}$, which corresponds to $1 + z_d \sim 4$, and no dissipation has taken place. The upper solid line is the resulting profile when the baryons are allowed to dissipate. The dotted line is the result of the analytic model for the same values of F and f, and a collapse factor $b/R = 0.05$ corresponding to the chosen values of λ and F: the initial state needed was taken to be the profile resulting from an N-body simulation without dissipation or decay. The agreement between the two models is fairly good and we regard it as confirming the validity of the analytic model.

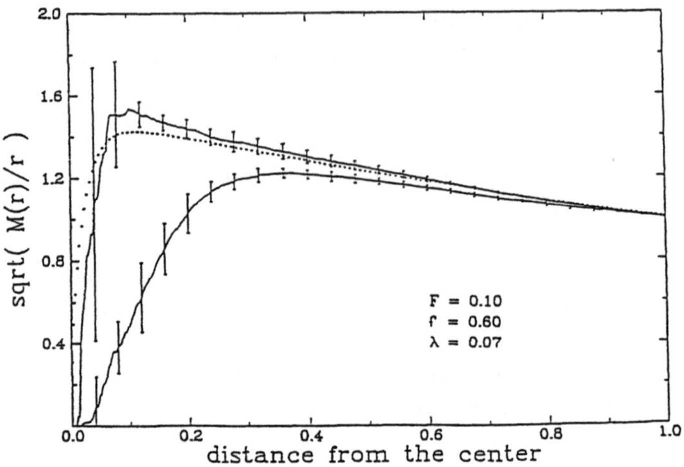

Fig. 2: "Rotation" curves for the numerical model. The y- axis is $(M(r)/r)^{1/2}$, where $M(r)$ is, at fixed r, averaged over several dynamical times after the system reached approximate equilibrium. The bars are one standard deviation fluctuation levels. The x-axis is the distance from the center in units of the radius that contains 80% of the final total mass (edge effects probably affect the results of the numerical models outside this radius). The dotted line is the result of the analytic model for $(M(r)/r)^{1/2}$.

4. Discussion

The remarkably flat or rising velocity profiles of spiral galaxies [21,22] are a strong constraint on the models being studied. Figure 1 shows that for typical values of F and λ the final velocity profile is rather sensitive to the value of f and not every value of f yields acceptable rotation curves. The numerical simulations showed that the analytic model works well and, therefore, we can use it to study the theoretical rotation curves and compare them to observations.

The parameters of the analytic model are the final baryon fraction F, the collapse factor b/R, the core radius a/R of the initial isothermal sphere and the fraction of stable DM f. As Fig. 1 shows, if for a given value of f one obtains a falling velocity profile for F and R/b taken at their lowest values and a/R $\sim$ 0.4, then only falling rotation curves can be produced for higher values of F and R/b. However, Fig. 3 shows that this is so for any value of the core radius for $f \lesssim 0.3$. The figure shows a measure of the flatness of the rotation curve, $\Delta v/v \equiv [v(4.5b) - v(3.5b)]/v(3.5b)$, ploted as a function of the core radius of the initial configuration for the lowest values of F and R/b (higher values would only lower the curves).

The rotation curves of spiral galaxies are observed to be remarkably flat or rising inside the optical radius, with only a small fraction $\sim$ 20% of the Burstein-Rubin [21] sample observed to have falling velocity profiles. For $f \lesssim 0.3$, however, Fig. 3

shows that the theoretical rotation curves decrease ($\Delta v/v < 0$) inside the optical radius $R_H \equiv 4.5b$ for all values of a/R, even for F and R/b at their lowest values. Furthermore, velocity profiles are observed to be flat or rising outside the optical radius as well [22] and this provides further constraints on f. Because we assume f to be a spatial constant, one can use the data of individual galaxies to constrain it. For example, no value of the initial core radius can yield a flat rotation curve out to $r = 11b$ (like that of NGC 3198 [23]) unless $f \gtrsim 0.4$ nor can a profile rising out to $r \sim 8b$ (like that of NGC 3109 [24]) be obtained if $f \lesssim 0.5$. These bounds are, again, for F and R/b taken at their minimum values.

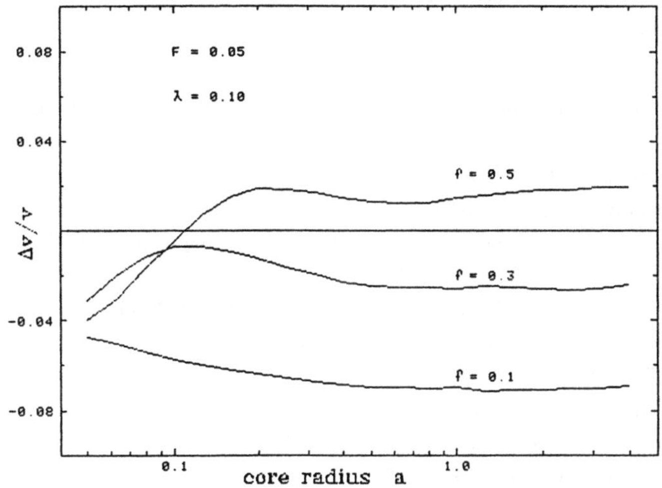

Fig. 3: $\Delta v/v$ as a function of of the core radius of the initial isothermal sphere for $F = 0.05$ and $b/R = 0.1$ (e.g. $\Delta v/v \sim -0.1$ for a self-gravitating exponential disk).

5. Conclusions

The rotational profiles of spiral galaxies require that the fraction of stable DM $f \gtrsim 0.5$. One can relate f to the relative contribution by relativistic particles to the energy density:

$$\Omega_r/\Omega_{nr} \leq f^{-1}(1-f)(1-F)(1+z_d)^{-1} \tag{2}$$

(the equality is satisfied in the instantaneous decay approximation [3]). Since $F \ll 1$, for $f \geq 0.5$ one gets $\Omega_r/\Omega_{nr} \leq (1+z_d)^{-1}$. Thus, the Universe could not be dominated by weakly interacting relativistic particles. In the type I model, the relativistic contribution to the energy density required for $\Omega = 1$ amounts to $\Omega_r/\Omega_{nr} \sim 4$. Therefore, even for a decay epoch as recent as $1 + z_d \sim 5$ Eq. 2 requires $f \sim 0.05$, a factor of 10 smaller than the bound obtained here. Thus, with the assumptions we have made, the type I model is ruled out. Only if a substantial fraction of the baryonic gas contracts and dissipates after the decay might one lower the upper bound on f, although it is not clear that such a scenario can lead to flat rotation curves. Type II models[8], however, do not require that the Universe be radiation dominated at present; in these models $\Omega_r/\Omega_{nr} \sim 0.25$, which is only marginally consistent with the upper bound of Eq. 2 even for a decay epoch as recent as $1 + z_d \sim 4$. However, an upper bound on f must come from the requirement that on large scales Ω be unity with Ω_{cl} being only ~ 0.2. We are trying to quantify this constraint by studying in detail the formation of clusters and large scale structure in type II models [25].

Acknowledgements

It is a pleasure to thank G. Blumenthal, A. Dekel, S. Faber and J. Primack, with whom most of the work reported on here was carried out, for much fun and enlightment throughout our collaborations. This work has been supported by DOE contract DE-AC02 -76ER03230 at Brandeis.

References:

1. G. Blumenthal, S. Faber, J. Primack and M. Rees: Nature 311, 517 (1984) and references therein
2. P. J. E. Peebles: Nature, 321, 27 (1986)
3. P. J. E. Peebles: Astrophys. J. 284, 439 (1984)
4. J. Bardeen, in E. W. Kolb *et. al.* eds.: Inner Space/ Outer Space, p. 212, U. of Chicago Press, 1986. N. Kaiser: *ibid.*, p. 258
5. M. S. Turner, G. Steigman and L. Krauss: Phys. Rev. Lett. 52, 2090 (1984)
6. G. Gelmini, D. Schramm and J. Valle: Phys. Lett. 146B, 311 (1984)
7. M. S. Turner: Phys. Rev. D31, 1212 (1985)
8. K. Olive, D. Seckel and E. Vishniac, Astrophys. J. 292, 1 (1985)
9. J. Silk and N. Vittorio: Phys. Rev. Lett. 54, 2269 (1985)
10. M. S. Turner: Phys. Rev. Lett. 55, 549 (1985)
11. E. W. Kolb, K. Olive and N. Vittorio: preprint, Fermilab-PUB-86/40-A, 1986
12. G. Efstathiou: M. N. R. A. S.213, 29 (1985)
13. A. Dekel and T. Piran: Weizmann preprint WIS-86/30-June Ph
14. M. Davis and J. V. Villumsen: Berkeley preprint, 1985
15. Y. Hoffman: U. of Pennsylvania preprint UPR-0291-T(Jan. 1986)
16. R. Flores, G. Blumenthal, A. Dekel and J. Primack: UCSC preprint (April, 1986). (submited to Nature)
17. G. Blumenthal, S. Faber, R. A. Flores and J. Primack: Astrophys. J. 301, 27 (1986); UCSC preprint, in preparation
18. G. Efstathiou and B. J. T. Jones: M. N. R. A. S. 186, 133 (1979)
19. S. M. Fall and G. Efstathiou, M. N. R. A. S. 193, 189 (1980)
20. S. J. Aarseth, in J. U. Brackbill and B. I. Cohen eds.: Multiple Times Scales, p. 377, Academic Press, 1985
21. D. Burstein and V. C. Rubin: Astrophys. J. 297, 423 (1985) and references therein
22. A. Bosma: Astron. J. 86, 1791 (1981); *ibid.* 86, 1825 (1981)
23. T. S. van Albada, J. N. Bahcall, K. Begman and R. Sanscisi: Astrophys. J. 295, 305 (1985)
24. C. Carignan: Astrophys. J. 299, 59 (1985)
25. G. Blumenthal, A. Dekel, R. Flores and J. Primack, in progress.

Halos and Angular Momentum Generation

C. S. FRENK

1. Introduction

A major goal of Cosmology is to arrive at specific models for the origin of cosmic structures that can be tested against observations. Much recent effort has focused on the clustering properties of galaxies on large scales where non-linear effects are relatively weak and gravity is usually (although not always) assumed to be the primary interaction. Detailed models of galaxy formation have proved more elusive because of the attendant complexities introduced by non-linear gravitational and dissipative effects. (For recent reviews see [1] and [2]).

From the confrontation of theory with observations of galaxy clustering, the cold dark matter (CDM) cosmogony has emerged as the clear frontrunner [3,4] . In its standard form, this scheme assumes that: (i) the Universe is flat (Ω=1), (ii) the dark matter consists of "cold" weakly interacting elementary particles, (iii) primordial fluctuations are adiabatic, have random phases and a scale-invariant power spectrum and (iv) galaxies form only near high peaks of the smooth linear density fluctuation field. This model is completely specified by only four "free" parameters. Two determine the normalization of the primordial fluctuation spectrum: an amplitude which should, in principle, be calculable from a theory of the early Universe, and a linear scaling which depends only on the Hubble constant $H_o (L \alpha (\Omega H_o^2)^{-1})$. The other two describe a simple model for the bias in the sites of galaxy formation: the scale of the smoothing filter applied to the linear density field and the threshold height for peaks identified with galaxies [4-7]. (In practice these two parameters can be thought of as the abundance of galaxies and the strength of the bias.) Thus, not all of these four "free" parameters are freely adjustable and as I will discuss below the biasing parameters may not even be required.

Empirically, the four parameters of the CDM model can be fixed by comparing the model predictions with the galaxy two-point correlation function. In an earlier paper we compared N-body simulations of clustering with the Center for Astrophysics redshift survey [4]. We found that by choosing H_c=50km s^{-1}Mpc^{-1}and plausible values for the remaining parameters, reasonable agreement is obtained not only with the observed two-point correlation function but also with the three-point function and the peculiar relative velocity of galaxy pairs. With no further free parameters we subsequently carried out a comparison of the model with observations on larger scales [8,9]. We showed that the model is consistent with the luminosity function, abundance, mass and mass-to-light ratio of Abell clusters, with our inferred infall velocity towards Virgo, with voids as large as the one in Bootes, with filaments similar to the Perseus-Piscis chain, with the "bubbles" and general morphology seen in recent redshift surveys, and with our observed motion through the microwave background. Significant discrepancies were found only on the largest scales where the model may not produce a sufficient level of superclustering.

On galactic scales, the CDM cosmogony is also an attractive model. BLUMENTHAL et al., have given scaling arguments to explain the characteristic masses and luminosities of galaxies [3] and FRENK et al. have shown that galactic halos with flat rotation curves are formed with the observed abundance and potential depth [10]. In this paper I present new results which extend the latter study and raise the possibility that galaxies in a CDM universe may be naturally biased towards high density regions without requiring additional mechanisms to suppress galaxy formation elsewhere [11, 12]. I also discuss the generation of angular momentum in this model. Most of the material described here is part of an on-going collaboration with M. Davis, G. Efstathiou and S.D.M. White whom I thank for permission to present our new results prior to joint publication.

2. Methods

This study is based on high resolution N-body simulations carried out with the P^3M code [13]. We followed the evolution of 9 small cubic regions of a CDM universe, of present size $2(1+z_i)$ Mpc, from time $t_i=7.2\times10^8$ yrs (corresponding to $z_i=6$ for $\Omega=1$) to the present day. (Throughout this paper $H_0=50$ km $s^{-1}Mpc^{-1}$, the value required by our previous study of galaxy clustering [4]). The resolution is determined by the softening parameter and corresponds to $\sim$2kpc at the start of the calculations and $\sim$14 kpc at the end; thus, these models are able to resolve the internal regions of galaxies. Three of the calculations (ensemble EFLAT) have the mean cosmological density, three (ensemble EOPEN) have a density slightly below average and the remaining three (ensemble ECLOSED) have a density slightly above average. By varying the density, we incorporate in an approximate way the modulating effect of wavelengths larger than the box size; such wavelengths are not expected to affect significantly the internal structure of small clumps, but are important in determining the characteristic mass and distribution of collapsed objects. So long as these large wavelengths remain linear, a region of higher than average density will evolve as a closed universe, whereas a region of smaller than average density will evolve as an open universe. More specifically we considered regions of density $\rho=\rho_{crit}(1\pm\Delta)$ at t_i, where ρ_{crit} is the closure density and Δ is the expected rms density fluctuation in a 2 Mpc cube at this time, $\Delta=0.094$. The expansion factors (7, 8.1, and 5.5 for EFLAT, EOPEN and ECLOSED respectively) were chosen so that all regions have the same age; Ω varies from 0.85 to 0.41 in EOPEN and from 1.17 to 4.79 in ECLOSED. The simulations all have 32768 particles corresponding to a mass per particle of 5.8, 3.9, and 14.3(10^9 $M_\odot$) in EFLAT, EOPEN and ECLOSED respectively. To investigate the spatial distribution of galactic halos and as a check on the abundances predicted in these ensembles we also a ran a larger simulation in a box of comoving size 32 Mpc, with the critical density and 110592 particles; the resolution in this calculation is reduced to $\sim$32 kpc at the present day.

3. The formation of galactic halos

In the CDM cosmogony galaxy formation is predicted to be a recent and protracted process which continues until the present [10]. Before $z \sim 3$ the mass distribution in our simulations is quite smooth but by $z=2.5$ a few tens of clumps with circular speeds exceeding 100 km s^{-1} have formed near high peaks of the initial density field. Between $z=2.5$ and the present the clumps undergo considerable evolution. Some of them merge while others remain in relative isolation during extended periods, gradually accreting surrounding material. By $z=0$ almost all the clumps present at $z=2.5$ have been disrupted and incorporated into larger systems. The final halos are smooth, centrally concentrated and, in most cases, significantly triaxial. Figure 1 shows mass distributions at $z=1$ and at the present day in one realization of each ensemble. Most of the large halos visible at $z=0$ had precursors which merged with at least one other similar sized clump since $z=2.5$. The merging activity is at its peak prior to $z=1$ and only about 30% of the final halos undergo a major merger between $z=0.4$ and the present. About half the halos evolve from $z=1$ to their final state by accretion of material onto their outer parts, increasing their mass by at least a factor 2 during this process. Thus in the CDM cosmogony violent dynamical activity is expected even at fairly recent epochs and galaxy mergers are expected to be frequent at moderate redshifts. The merger rate is somewhat larger in high density regions where the largest clumps are formed. All three panels of Fig. 1 show halos with a wide range of masses, but the plot from ECLOSED has a few very large objects which have no counterpart in EOPEN.

The formation paths of two halos are shown in the first and third rows of Fig. 2; these display the time evolution of all the particles that end up in the two most massive clumps, identified at the final time at a density contrast of $\sim$ 500, in one of the models of EFLAT (see Fig. 1). The second and last rows correspond to the dense regions of these clumps, identified at a density contrast of $\sim$ 10000, and will be discussed later. Since the masses of these halos are very large, these evolutionary paths are not necessarily representative, but I have chosen them here because they are particularly interesting. The first halo forms through a complicated series of merger events involving 6 or 7 clumps, originally distributed on sheets, which by $z=0.4$ have coalesced onto a centrally concentrated core. The second is originally a binary with a few outlying satellites which has almost completely merged at the end of the simulation after extensive tidal interactions.

The rotation curves of halos at $z=0$ in all three ensembles are shown in Fig. 3. Plotted here is the average circular velocity, $V_c(r)=(GM(r)/r)^{\frac{1}{2}}$, of halos binned in logarithmic intervals of width 0.1 in the asymptotic circular velocity, V_{max}; this is defined as the value of V_c at the radius of a sphere of overdensity 500 and only halos with $V_{max}>63$ km s^{-1} are included. The most striking feature of this plot is the flatness of most rotation curves at radial distances beyond a few tens of kiloparsecs. This behaviour is already apparent at $z=2.5$, shortly after the first clumps in the simulations have collapsed; as halos grow through mergers and accretion, their rotation curves increase in amplitude. In EOPEN growth freezes out at around $z=1$ but in ECLOSED it continues until the present reflecting the higher merger rate in this ensemble. The flat rotation curves in Fig. 3 resemble the rotation curves measured in the outer parts of spiral galaxies. These tend to remain flat to smaller radii than we can resolve in the calculations; however, the gravitational effects of the luminous material collecting onto the centre are likely to extend a flat rotation curve into the inner regions of the galaxy [14-16]. The deepest potential wells in ECLOSED have rotation curves which continue

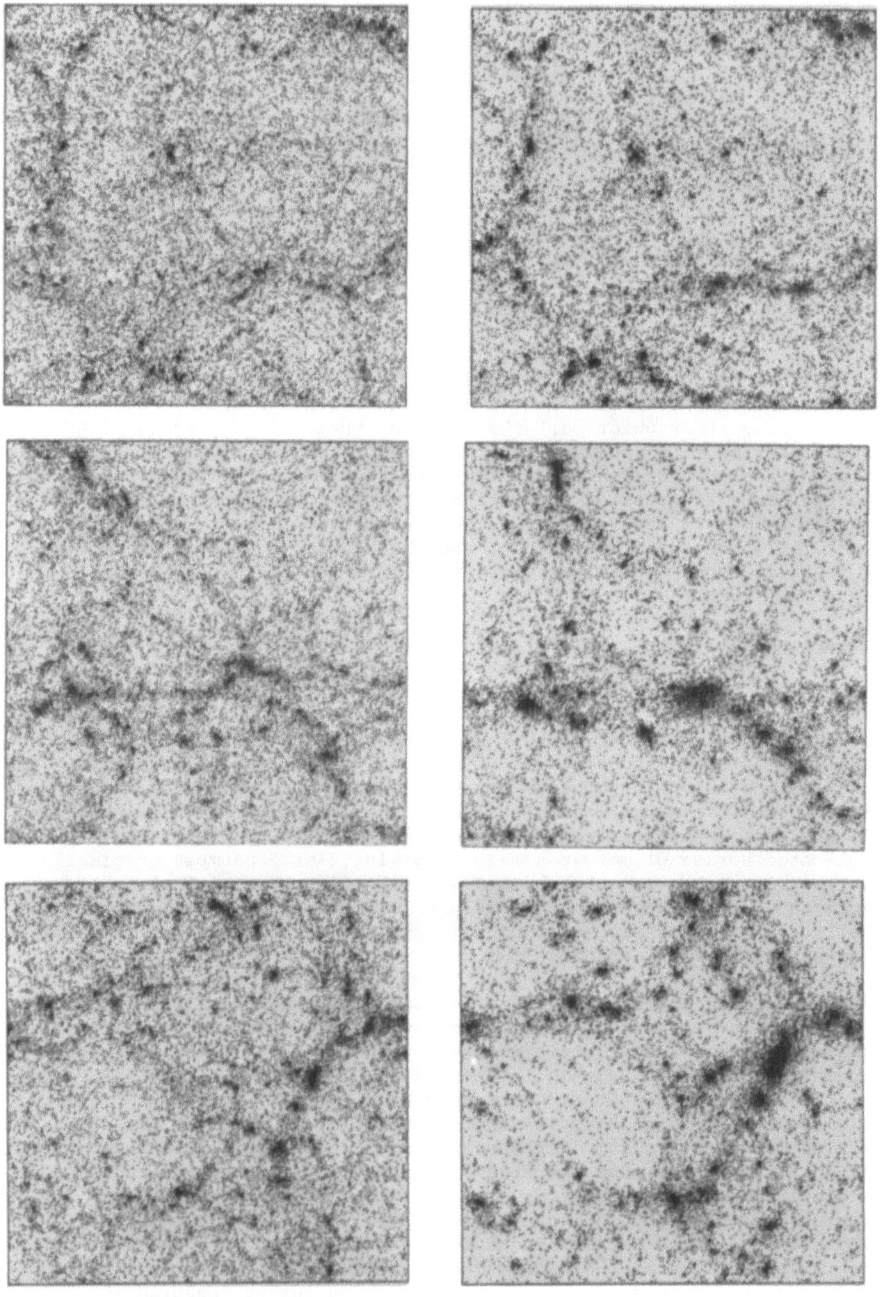

Figure 1: Mass distributions at z=1 (left) and z=0 (right) in a simulation of EOPEN (top), EFLAT (middle) and ECLOSED (bottom). Co-moving coordinates are used and the box sizes are 16.5 Mpc/(1+z) (top), 14 Mpc/(1+z) (middle) and 11.2 Mpc/(1+z) (bottom).

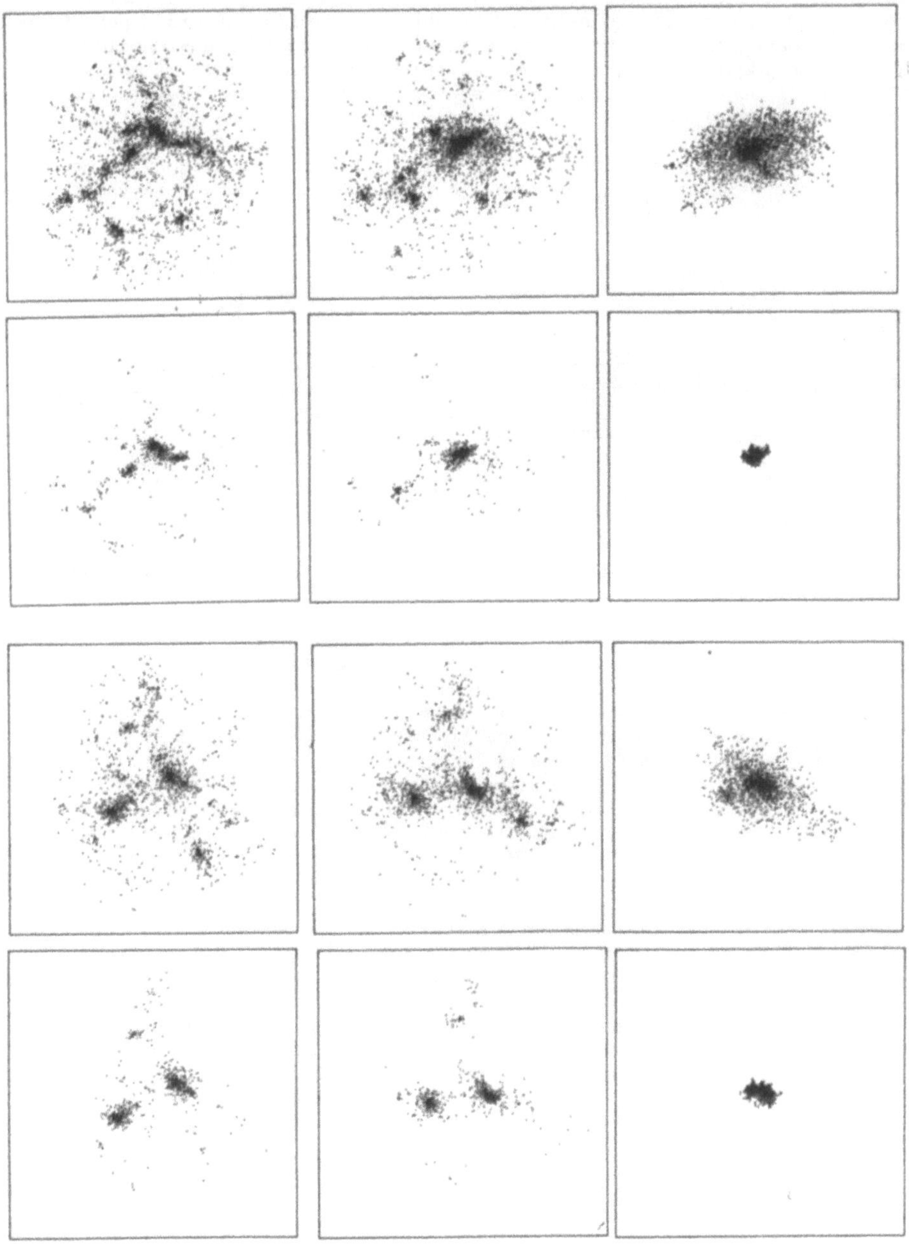

Figure 2: Positions at z=1 (left), z=0.4 (middle) and z=0 (right) of particles that end up in 2 halos (rows 1 and 3) and in their inner regions (rows 2 and 4). Physical coordinates are used and the box sizes are: 3.6 Mpc (top 2 rows) and 2.8 Mpc (bottom 2 rows). The angular momentum in arbitrary units of the particles shown from left to right are: 24.6, 31.6, 32.4 (row 1); 4.0, 1.8, 1.3 (row 2); 12.3, 18.2, 22.9 (row 3); 11.8, 14.8, 6.3 (row 4).

to rise steeply beyond ~100 kpc and have amplitudes well above those inferred for the halos of most spiral galaxies. While some of them may correspond to the massive halos of bright ellipticals, others appear to be much too massive to be associated with any single galaxy and should perhaps be identified with the merged halos of small groups of galaxies. This cannot be decided with our present simulations since they do not include the dissipational physics required to separate baryonic from dark material. Small galaxy groups, however, are observed in the real Universe and our results suggest that for the cold dark matter cosmogony to be viable, the galaxies must merge on a much longer timescale than their individual halos [17].

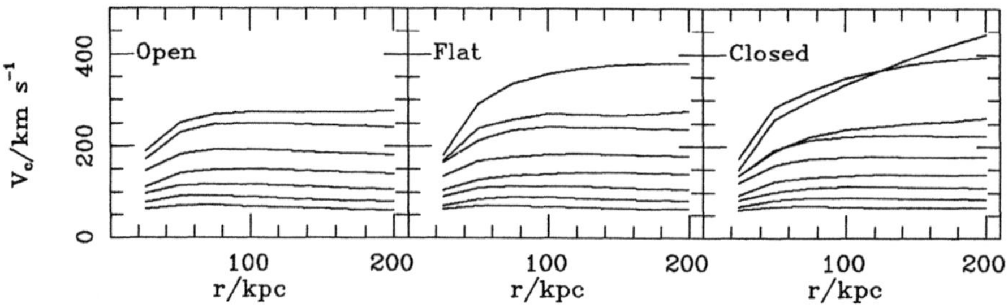

Figure 3: Rotation curves for halos in the three ensembles at the end of the simulations.

From the simulations we can readily calculate the predicted number density of halos as a function of characteristic velocity V_{max} and of mean background density. No such data exist for real galaxies, but their mean abundance as a function of V_{max} can be derived from the Tully-Fisher relation and the galaxy luminosity function. The latter is fairly well established for L_B near the characteristic luminosity L_*, but there is a considerable spread of luminosities at a given rotational velocity. For example, using FELTEN'S recommended luminosity function [18] and the M_B-V_{max} relation obtained by RUBIN et al. [19], the density of galaxies brighter than the median luminosity corresponding to V_{max}=150 km s^{-1} is $n_{150} \sim 2.1 \times 10^{-3}$ Mpc^{-3} with a spread of about a factor 2 due to the spread in luminosity whereas the density of galaxies brighter than the median luminosity corresponding to V_{max}=250 km s^{-1} is $n_{250} \sim 3.2 \times 10^{-4}$ Mpc^{-3} with a spread of about a factor 5. In the simulations, averaging over all the ensembles, we find $n_{150} = (4.3 \pm 0.4) \times 10^{-3}$ Mpc^{-3} and $n_{250} = (7.7 \pm 0.7) \times 10^{-4}$ Mpc^{-3}, where the error reflects the variance in the total number of halos found. This agreement with observations is remarkable, especially since there are no adjustable parameters in the calculations. At any given V_{max}, the number density of halos is larger in the high density models than in the low density ones, the difference ranging from a factor 5 for $V_{max} >$ 150 km s^{-1} to a factor 14 for $V_{max} >$ 250 km s^{-1}. No halos with $V_{max} >$275 km s^{-1} were formed in EOPEN, whereas a small number of halos with $V_{max} >$ 400 km s^{-1} and number density $n_{400} = (9.5 \pm 4.7) \times 10^{-4}$ Mpc^{-3} was found in ECLOSED.

4. Biased galaxy formation

We have seen that large halos tend to form preferentially in regions where the local mass density is high. This bias is of dynamical origin and is imposed by the different rates at which non-linear gravitational clustering proceeds in regions of different density. It raises the attractive possibility that the "ad hoc" bias in the galaxy distribution that has been postulated to reconcile a flat universe with observations may actually arise naturally in the CDM model. To test this possibility we compare the abundance and clustering properties of galactic size halos in our simulations with those predicted by the high-peak model. In this model, there is a global threshold for galaxy formation which in our clustering study we estimated to be 2.5 σ, where σ is the rms fluctuation of the linear density field filtered with a Gaussian of width 1.08 Mpc [4]. In high density regions, such as those of ensemble ECLOSED, the global threshold appears lower than in low density regions, such as those of ensemble EOPEN. The effective thresholds for these two ensembles, corresponding to a global 2.5σ value, are 2.48σ' and 2.94σ' respectively, where σ' is the rms fluctuation contributed by the waves present in the simulation. From the formalism of BARDEEN et al. [7], we find that 8.5 more peaks above threshold are expected in ECLOSED than in EOPEN. It is worth emphasizing that this bias refers exclusively to the bright galaxies employed in clustering studies which typically have $V_{max} \gtrsim 220$ km s^{-1}. The bias in the populations of dark halos found in our high and low density simulations depends on V_{max}; it is 2.1 for $V_{max} > 200$ km s^{-1}, 4.3 for $V_{max} > 250$ km s^{-1} and formally infinite for $V_{max} > 275$ km s^{-1}. These values are lower limits to the expected bias in the population of bright galaxies since, as discussed above, the largest halos must correspond to groups. For example, groups of $\sim$5 galaxies in all halos with $V_{max} > 400$ km s^{-1} would account for the required factor of 8.5.

To examine the spatial correlations of halos we ran the large box model CFLAT. We found that the clustering length of halos with $V_{max} > 200$ km s^{-1} is 5.8 Mpc, an enhancement by a factor 1.4 relative to the mass, whereas we had previously found that in the high peak model the correlations of bright galaxies are enhanced by a factor of 2.4 [4]. Again there is an ambiguity in this comparison because underestimating the number of galaxies in groups will depress their correlations. We have made a crude attempt to correct for this effect in the data by smoothing the galaxy counts within a velocity difference of 300 km s^{-1} and projected separation of 1 and 2 Mpc in the CfA catalogue. By replacing galaxies within such groups by the group centres we find that the correlation length drops from 10 Mpc to 6.4 and 5 Mpc respectively. These group selection scales correspond roughly to the size of the largest halos in the simulations. Thus, provided these contain small galaxy groupings, the natural bias seems to be enough to account for the requirements of an $\Omega=1$ CDM universe. In this case, less luminous galaxies would be expected to be less clustered than brighter galaxies.

5. The generation of angular momentum

In the gravitational instability theory, protoclumps acquire angular momentum during their linear growth phase as a result of tidal torques exerted by neighbouring objects [20,21]. In general, however, the angular momentum of a collapsed clump will be modified by non-linear mode-mode couplings. WHITE has shown that in linear theory the angular momentum of a protogalaxy grows to first order, in proportion to t. Using N-body simulations, he also showed that the angular momenta of non-linear clumps falls below the extrapolation of linear theory, with considerable variation from clump to clump [22]. In an extensive study, also based on N-body simulations, BARNES & EFSTATHIOU found that tightly bound objects depart from the linear behaviour earlier and end up with less angular momentum than loosely bound objects [23]. Departures from linear theory are due to 2 different mechanisms. Tidal torquing is most effective when objects attain their maximum size; after collapse, their coupling to the tidal field decreases and the transfer of angular momentum to them becomes ineffective [21]. In addition, as a result of the large asymmetries present, angular momentum is transferred outwards as substructure is destroyed and the inner regions of merged remnants tend to be rotating slowly [10,23].

The second effect is illustrated in Fig. 2. The first and third rows show the evolution of two different perturbations in an N-body simulation of a CDM universe with the behaviour of the most tightly bound particles shown in the rows directly beneath. The angular momentum of the particles displayed is given in the figure caption. At z=1, both perturbations are near their maximum expansion; prior to this time, their angular momenta accurately obeyed the linear growth law. At z=0.4 the perturbations have become fully non-linear; their angular momenta remain virtually frozen until the final time, z=0. Now consider the evolution of the tightly bound particles, rows 2 and 4. At z=1, there is still considerable substructure in the inner regions of the first perturbation, but by z=0.4 much of it has been destroyed; the angular momentum of these particles, most of which was originally invested in the orbital motions of the subclumps, drops by more than a factor of 2 as they merge. There is a further transfer of angular momentum to the outer halo as the last of the substructure is erased between z=0.4 and the present. The same process operates in the inner regions of the second perturbation. As the components of the original binary merge between z=0.4 and z=0, 60% of their angular momentum is given up to surrounding material and the merger remnant shows little rotation. This mechanism is likely to play an important role in galaxy formation. In particular, the formation paths illustrated here may explain the high luminosities and slow rotation velocities of bright ellipticals.

An unfortunate consequence of the importance of non-linear effects is that linear theory, which is tractable and relatively simple, has little predictive power. BARNES & EFSTATHIOU [23] have examined correlations between rotational properties and various clump parameters such as mass, initial overdensity or distance to the nearest neighbour. Contrary to the expectations of linear theory, these correlations are weak or non-existent. The only property that seems to correlate with final angular momentum is the substructure of the progenitors. Thus, linear theory can provide no more than an order of magnitude estimate of the rotation of collapsed clumps.

6. Conclusions

From N-body simulations of the growth of fluctuations in an $\Omega=1$ universe dominated by cold dark matter we have found that:

(i) Galactic halos with rotation curves that are flat beyond a few tens of kiloparsecs (the resolution limit of the simulations) are formed. The rotation curves become flat shortly after the halos collapse, at $z\sim3$. At early times there is a good correspondence between halos and high peaks of the linear density field, but at late times several peaks are submersed in merger remnants.

(ii) Halos grow by mergers and accretion of surrounding material. The merger rate peaks at around z=1, and about 30% of present halos have suffered a major merger since z=0.4.

(iii) The predicted abundance of galactic halos with different rotation curve amplitudes agrees remarkably well with observations. However, a few large halos with characteristic velocities in excess of 400 km s-1 form in high density regions. For the model to be viable these objects must correspond to the merged halos of small galaxy groups but the present simulations cannot resolve this issue because they only include the dissipationless component.

(iv) There is a natural bias in the spatial distribution of halos relative to the distribution of mass, in the sense that galactic size halos form

preferentially in high density regions. This is a purely dynamical effect which arises from the rate at which non-linear gravitational clustering proceeds from initial conditions with a CDM spectrum. Our results strongly suggest that this effect may account for the requirements of an $\Omega=1$universe, but they cannot provide a definitive answer because of the ambiguity in identifying individual galaxies within merged halos.
(v) Angular momentum grows during the linear phases of protogalactic evolution in accordance with the predictions of the tidal torque theory. However after clumps collapse non-linear effects become dominant and large amounts of angular momentum are transferred from the orbital motions of subclumps to surrounding material as substructure is erased.

The successes of the CDM model discussed here are remarkable, not least because there are no free adjustable parameters. The calculations presented refer exclusively to the formation of halos but nevertheless suggest possible formation paths for galaxies of different morphological types. Mergers play a central role and may explain the high luminosity, slow rotation and enhanced clustering of bright ellipticals. Disks, on the other hand can only form during extended periods of quiescent evolution and the predicted fraction of halos which remain relatively undisturbed since a redshift of 1 is similar to the observed fraction of spirals. Galaxy formation is predicted to be a recent but protracted process; violent dynamical activity is expected at low redshifts and may provide a direct observational test of the CDM cosmogony.

References

1. C.S. Frenk: Phil. Trans. Roy. Soc., in press (1986)
2. S.D.M. White: In Inner Space/Outer Space, ed. by E.W. Kolb, M.S. Turner, D. Lindley, K. Olive, D. Seckel (Univ. Chicago Press 1986) p.228
3. G.R. Blumenthal, S.M. Faber, J.R. Primack, M.J. Rees: Nature 311, 517 (1984)
4. M. Davis, G. Efstathiou, C.S. Frenk, S.D.M. White: Astrophys. J. 292, 371 (1985)
5. J.M. Bardeen: In Inner Space/Outer Space, ed. by E.W. Kolb, M.S. Turner, D. Lindley, K. Olive, D. Seckel (Univ. Chicago Press 1986) p.212
6. N. Kaiser: In Inner Space/Outer Space, ed. by E.W. Kolb, M.S. Turner, D. Lindley, K. Olive, D. Seckel (Univ. Chicago Press 1986) p.258
7. J.M. Bardeen, J.R. Bond, N. Kaiser, A.S. Szalay: Astrophys. J. 304, 15 (1986)
8. S.D.M.White, C.S. Frenk, M. Davis, G. Efstathiou: Astrophys. J. in press (1986)
9. S.D.M. White: this volume.
10. C.S. Frenk, S.D.M. White, G. Efstathiou, M.Davis: Nature 317, 595 (1985)
11. M.J. Rees: Mon. Not. Roy. Astron. Soc. 213, 75p (1985)
12. J.I. Silk: Astrophys. J. 297, 1 (1985)
13. G. Efstathiou, M. Davis, C.S. Frenk, S.D.M. White: Astrophys. J. Suppl. 57, 241 (1985)
14. J. E. Gunn: In Dark Matter in the Universe, ed G. Knapp (Reidel, Dordrecht 1986) in press
15. G.R. Blumenthal, S.M. Faber, R.A. Flores, J. Primack: Astrophys. J. 301, 27 (1986)
16. J. Barnes: this volume
17. S.D.M. White, M.J. Rees: Mon. Not. Roy. Astron. Soc. 183, 341 (1978)
18. J.E. Felten: Comm. Astrophys. Sp. Sci. 11, 53 (1985)
19. V.C. Rubin, D. Burstein, W.K. Ford, N. Thonnard: Astrophys. J. 289, 81 (1985)

20. F. Hoyle: In Problems of Cosmical Aerodynamics (Central Air Documents Office, Dayton, Ohio 1949)
21. P.J.E. Peebles: Astrophys. J. 155, 393 (1969)
22. S.D.M. White: Astrophys. J. 286, 38 (1984)
23. J. Barnes, G. Efstathiou: preprint, Univ of Cambridge (1986)

Cosmic Strings and the Formation of Galaxies and Clusters of Galaxies

NEIL TUROK

1. Introduction

As many speakers have discussed at this meeting, recent observations of the universe on very large scales have revealed a great deal of structure. Large voids, filaments and sheets with dimensions of 50-100 h^{-1}Mpc stand out in the observed distribution of galaxies. Rich clusters of galaxies are significantly clumped on scales of at least 100 h^{-1}Mpc. And most recently observations apparently show large scale coherent streaming velocities of ~700 kms^{-1} on scales of about 60 h^{-1}Mpc.

What is most significant about these observations is that they provide us with a direct 'window' into what the very early universe was like. There is very strong evidence that the structure we see has not been much disrupted since the big bang. Random motion of galaxies relative to the observed structures is no more than a few hundred kms^{-1}. This means that they have moved no more than a few h^{-1}Mpc since the big bang! The large scale structure is therefore telling us quite directly about conditions in the very early universe.

Different theories of high energy physics make different predictions for the fluctuations present in the early universe, so large scale observational astronomy therefore provides a unique and direct testing ground for fundamental particle physics theories.

In this talk I will discuss one class of theories which have provoked a lot of interest recently - theories predicting cosmic strings. It was realised as early as 1973 that strings could occur in unified theories, but Kibble in 1976 was the first to suggest they might have interesting consequences in cosmology[1]. Subsequently Kibble, Zel'dovich and Vilenkin suggested they might play a role in seeding the formation of galaxies[2]. Over the last two years there has been a great deal of progress with this idea.

As Albrecht has discussed at this meeting, numerical simulations of the formation and evolution of strings have clarified many of the initial speculations[3]. They have also shown a remarkable agreement between the correlation of string loops and Abell clusters[4], and in the near future will have a lot more to say about the predicted large scale structure. The simulations have already enabled a series of more detailed calculations from which the theory has so far emerged very successfully[5-7].

2. Cosmic Strings - Formation and Evolution

An attractive feature of the string theory is that there is only one free parameter, μ, the mass per unit length of the string. For grand unified theories producing strings the dimensionless number $G\mu = (m_{gut}/m_{planck})^2$ is typically about 10^{-4}- 10^{-10} , neatly bracketing the values required for galaxy formation, as we shall see. Strings with $G\mu = 10^{-6}\mu_6$ have a mass per unit length of $2.10^7\mu_6 M_\odot$ parsec^{-1} (about 10^{21} kg m^{-1}!). More importantly however, the dynamics of the string network is completely independent of μ. The characteristic velocity of motion for the string under its own tension is the velocity of light - the detailed motion is determined solely by the string configuration and the background metric. This gives the theory a lot of predictive power.

Let me quickly review the formation and evolution of a string network[8,3]. Strings are formed in a phase transition as 'defect lines' in the orientation of a Higgs field. The string forms a tangled Brownian network of loops and 'infinite' strings (i.e. strings which stretch right across the universe) with a fixed number of segments of length ξ per volume ξ^3. ξ is the initial correlation length after the phase transition, $\xi \sim T_c^{-1}$ where T_c is the temperature of the phase transition. The strings form at a time $t \sim 10^{-34}$s and ξ is very small - it corresponds to a comoving scale of about a kilometre today. At formation, most of the string is in infinite strings and the rest is in closed loops, with a fixed number of loops of size R per volume R^3.

Interactions between strings are crucial in determining the subsequent evolution of the network. Numerical calculations and analytical arguments[9,10] show that when two strings cross one another, they almost always reconnect the other way. This allows the infinite strings to slowly grind themselves up into little loops as the universe expands. In fact they do this at nearly the fastest possible rate - by chopping off about one loop of radius ~t per expansion time t per horizon volume $\sim t^3$, the long strings (strings longer than the horizon) straighten themselves out on a scale ~t.

Loops chopped off the network of long strings undergo a brief period of chopping themselves up into smaller loops but then settle into non-self-intersecting motion for about a factor of about $(G\mu)^{-1}$ in time before finally disappearing due to gravitational radiation.

One of the main open questions which we hope to resolve in the near future is the question of exactly how a "primary" loop chopped off the network subsequently chops itself up. As we will see below this is crucial for the "Ω =1 problem". It is particularly important to see whether a large loop produces a great many much smaller loops.

One possible model for loop self-intersection is the following. Assume that there is always a fixed probability x for any loop to self-intersect, and that if it does so it will break into two roughly equal pieces. Now imagine starting with N "primary" loops of length L. After one oscillation period (a loop's motion is always periodic in its centre of mass frame with period one half of its length[10]) $(1-x)N$ loops with length L are left and these will survive for many oscillation periods[11]. Similarly there will be $2x(1-x)N$ loops of length L/2 and so on. After a total time L the chopping is over and there are $(2x)^n(1-x)N$ loops of length $L/2^n$. Note that if $x > 1/2$ the number of loops increases monotonically with n and in fact the total number diverges. However the energy in loops always decreases monotonically with n. Treating the length as continuous, the number of loops with length between l and l+dl is $n(l) = N(1-\alpha)(L/l)^{\alpha}dl/l$ where $\alpha = \log(2x)/\log(2)$, $-\infty<\alpha<1$. We shall use this model later.

At any time t there is a network of long strings straightened out on a scale ~t and a distribution of loops of all sizes within the horizon. In the radiation dominated period, the number of loops n(R) with radii between R and R+dR is given by

$$n(R)dR = \nu dR(R+\gamma G\mu t)^{-5/2}t^{-3/2} \qquad o < R < t \qquad (1)$$

and the mass of a loop of size R is $\beta\mu R$. According to numerical simulations , $\nu \simeq 10^{-2}$ and $\beta \simeq 9$. γ is a constant which depends on the efficiency of gravitational radiation for loops[12,13]. Typically $\gamma \sim 5$. The density in long strings is given by $\rho \sim 3\mu\ t^{-2}$.

3. Accretion of Matter by Loops

Let us now turn to the accretion of matter onto loops. For simplicity I will assume an $\Omega = 1$ cold dark matter dominated universe, although as I discuss later there are several reasons why a neutrino-dominated universe may be more attractive. Unless explicitly stated, h is taken to be 0.5.

Only a small amount of accretion onto loops takes place before the time of equal matter and radiation density t_{eq}. This will change our normalisation of $G\mu$ by at most 2 or 3.

One of the nice features of the string theory is that the loops' radii are very much smaller than their mean separation, because they have moved apart with the Hubble flow since they were produced. They therefore provide unambiguous peaks in the density distribution with which similarly discrete objects like galaxies can be identified. This contrasts favourably with the usual picture where the definition of what constitutes a galaxy is less clear.

Density perturbations grow around loops in a fairly simple way, well approximated by a spherical model on scales much larger than the loop radius and smaller than the mean separation of loops (at least if we ignore their peculiar velocities - see later). Larger loops produce larger potential wells, accreting proportionately larger masses containing smaller loops as well as matter about them. Thus a hierarchy of objects, smaller ones in bigger ones, is produced which we shall identify with galaxies and clusters of galaxies[4,5].

Now imagine drawing a sphere of radius r and moving it through space. Neglecting correlations between loops, r.m.s. fluctuations in the mass inside r due to loops of radius R are given by[12]

$$\delta M/M = (N)^{1/2}(\mu R)/(4\pi\rho r^3/3) \sim G\mu r^{-3/2}R^{1/4}t^{5/4} \qquad (1)$$

with ρ the background density and N the mean number of loops in the sphere, for $N \gg 1$. These fluctuations are dominated by the largest loops for which $N \sim 1$ i.e. $R \sim r^2/t$. Now imagine looking for regions containing much greater mass excess than this typical fluctuation. These are places where the sphere contains a much larger loop of radius R with $r > R \gg r^2/t$. In fact this selection procedure will 'find' all loops of radius bigger than some cutoff such that their mean separation is equal to the mean separation of the regions so selected[4].

This selection procedure is in fact precisely the one employed observationally by George Abell to define rich clusters of galaxies over 25 years ago[15]. He defined them as regions containing more than 50 bright galaxies within a radius $1.5h^{-1}$ Mpc. For comparison, the mean separation of bright galaxies[16] is $5h^{-1}$ Mpc so Abell clusters are exceptionally dense clusters of galaxies. Their mean separation d_c is now known[17] to be $55h^{-1}$ Mpc. As mentioned in the intoduction, on scales as large as this there has been hardly any gravitationally induced motion of matter since perturbations started to grow. The positions of Abell clusters should thus accurately trace the positions of the loops that produced them.

Now Abell himself noticed that the two dimensional distribution of Abell clusters on the sky is in fact highly non random[15], but it was not until quite recently that enough redshift data became available for a precise determination to be made of a good statistic for measuring this, the two point correlation function $\xi(r)$ (the excess probability over random of finding two objects at a separation r). This was done in 1983 by Bahcall and Soneira,[17] and Klypin and Kopylov,[18] who found that the two point correlation function for clusters was similar in form, but 18 times larger than that for galaxies. It was also observed to be positive out to distances of more than $100h^{-1}$ Mpc[17].

This suggested a very clean test of the cosmic string theory[4]. The correlations of loops may be measured in simulations such as those described above. The two point correlation function $\xi(r)$ of loops of mean separation d (i.e. loops bigger than some cutoff radius chosen so that their mean separation is d) produced in the radiation-dominated era must in fact be a universal dimensionless function of r/d. This is because all loops are produced in the same way, and at the time loops of a given size are formed, there is only one relevant scale, the horizon scale. At formation therefore, and measuring distances in terms of the horizon scale, all loops show identical correlations. Subsequent expansion of the universe simply stretches all separations by the same amount, so $\xi(r/d)$ does not change. Furthermore, there are no free parameters in $\xi(r/d)$ at all, the dynamics of string being independent of μ, the mass per unit length, and independent of the cosmological parameters h or Ω. $\xi(r/d)$ thus provides a very good test of the cosmic string theory[4].

As is well known by now, the numerical simulations show that the loop-loop correlation function matches that of Abell clusters remarkably closely[4]. This is a remarkable success, the more so since theories based on Gaussian perturbations with a Zel'dovich spectrum, get the correlations of Abell clusters completely wrong. Since then more evidence has accrued for such a scale-free correlation function on very large scales. The same correlation function $\xi(r/d)$ appears to hold for poor clusters[19], with $d = 30h^{-1}$Mpc as well as for superclusters[20] where $d = 100h^{-1}$Mpc. Of course the two point function is only a very partial measure of the full probability distribution and much remains to be done in comparing higher correlations as well as "filamentariness" and other statistics. It is also not yet clear that galaxies will turn out to have the correct correlations. The galaxy-galaxy correlation function has the same shape as that for clusters but (written as a function of r/d) has an amplitude 4 times larger[21]. This is consistent with correlations having been enhanced on small scales due to gravitationally induced clumping and in Ref.5 it was argued that at least in a simplified model of this process the slope would be preserved. It is important that this be checked in N-body simulations.

Having identified loops with the separation of Abell clusters with Abell clusters and loops with the separation of galaxies with galaxies one can then demand that they be massive enough to have accreted objects with the appropriate masses by today. This determines μ, the mass per unit length of the string, in a very direct way and provides a useful consistency check[5].

According to the spherical model[22], a seed mass m laid down at time t_i in an otherwise unperturbed $\Omega = 1$ matter dominated universe has turned around (i.e. reduced the relative velocity of the shell with respect to the origin to zero) at time t a mass

$$M = (4/3\pi)^{2/3} m (t/t_i)^{2/3} \tag{2}$$

At this time the overdensity inside the shell turning round is $(3\pi/4)^2$. The shell then collapses by a factor of 2 in radius before virialising, multiplying its density by 8. In addition to this the background density continues to fall as t^{-2}. Putting this together, an object with overdensity δ today turned around at a redshift

$$1 + Z_{turn} = (2\delta/9\pi^2)^{1/3} \tag{3}$$

For Abell clusters for example, observations of the rotation velocities of galaxies about the core show that the mass inside an Abell radius $r = 1.5\ h^{-1}$Mpc is $\sim 3\times10^{15} M_\odot$, so using $\rho_c = 2.8\times10^{11}\ h^2 M_\odot\ \mathrm{Mpc}^{-3}$, $\delta \sim 170$ and $1 + Z_{turn} \sim 1.6$.

For galaxies rotation curves show that inside $r \sim 50$ kpc the mass is $\sim 10^{12} M_\odot$ so $\delta \sim 3\times10^4$ and $1 + Z_{turn} \sim 9$. This tells us when galaxies and clusters formed.

Now from (2) we can calculate the seed mass required to form galaxies and clusters;

$$m = (3\pi/4)^{2/3}\ M\ (1 + Z_{turn})/(1 + Z_{eq}) \qquad (4)$$

Using $1+Z_{eq} = 6250$ (the value for h=.5 and 3 light neutrinos) we find that for galaxies the seed mass $m_g \simeq 2.6\times10^9 M_\odot$ while for clusters $m_c \simeq 6\times10^{11} M_\odot$.

However equation (1) uniquely relates the size of loops and therefore their mass to their mean separation; loops of mean separation dh^{-1} Mpc today had size

$$R = (2/3\nu)^{2/3}\ (d/2000)^2 \cdot 6250\ t_{eq} \qquad (5)$$

For galaxies, the mean separation is $5h^{-1}$ Mpc whereas for Abell clusters it is $55h^{-1}$ Mpc. Thus the loops giving rise to galaxies had $R_g \sim 1.4\times10^{-3}\ t_{eq} \simeq 11$ pc whereas the loops giving rise to clusters had $R_c \sim 1.7\times10^{-1}\ t_{eq} \sim 1.4$ kpc. Using $\mu = 2\times10^7\ \mu_6\ M_\odot\ pc^{-1}$ we find these loops had masses $M_g = \beta\mu\ R_g \simeq 2.2\times10^9\ \mu_6 M_\odot$ and $M_c = \beta\mu\ R_c \simeq 2.8\times10^{11}\ \mu_6\ M_\odot$. Thus galaxies and clusters come out with the right masses for $\mu_6 = 1-2$. Including the small amount of growth in the radiation dominated era[22] we would find $\mu \simeq .5-1$.

Thus for $G\mu \approx 10^{-6}$ and with cold dark matter the cosmic string theory predicts (in gross terms at least) the right mass spectrum of objects, provided the larger loops produce clusters of galaxies rather than very large single galaxies. We shall see in the next section why this is justified. The two point correlation function for poor clusters, rich clusters and superclusters also comes out right.

Where are the loops now? The most numerous loops around today have radii $R \sim \gamma G\mu t_o \sim 10\ \mu_6\ h^{-1}$kpc and would be in or near large galaxy clusters. Galaxy forming loops decayed long ago.

4. The Shapes of Galaxies, Clusters and Superclusters

Let us turn now to the shapes and density profile of the objects accreted around loops. The spherical model gives a reasonable approximation on large scales but as we shall see on small scales the predicted profile is far from spherical. This is because the loops have appreciable peculiar velocities and in fact move a good fraction of the distance to neighbouring loops of similar sizes during the lifetime of the universe. A loop's peculiar velocity $a\dot{\underline{x}}$ decays as a^{-1} (where a is the scale factor and $\underline{x}$ comoving coordinates) and the total comoving distance moved by a loop turns out to be a fixed fraction of the horizon distance when it was formed:

$$z = \int_{t_i}^{t} v_i (t_i / t')^{4/3} \, dt' = 3 v_i \, t_i (1-(t_i / t)^{1/3}) \equiv d(1-a^{-1/2}) \tag{6}$$

where v_i is its velocity at initial time t_i, when we take a to be unity.

In linear perturbation theory, the density perturbation grows in proportion to the initial perturbation in comoving coordinates. Thus to determine the shapes of objects produced by loops we need the shape they trace out in comoving coordinates. A loop's comoving radius shrinks like a^{-1} : so a loop traces out a 'trumpet': a roughly axially symmetric object with radius

$$r = R_i (1-z/d)^2 \tag{7}$$

where R_i is the initial loop radius and z runs from 0 to d. Rewriting this in terms of z' = d-z which measures the comoving distance from the apex of the trumpet we find something interesting:

$$r = R_i (z'/d)^2 \tag{8}$$

Now a loop's peculiar velocity decays like a^{-1} after it is produced so since loops of radius R are produced at a time $t \propto R$ we have, for all loops produced before t_{eq},

$v_i = v_o(R/t_{eq})^{1/2}$ with v_o a constant of order unity. Since $d \propto v_i$ we see in (8) that the shape of the 'trumpet' produced by all loops is the same - smaller loops just produce shorter lengths of it. This will be very important.

What about the density profile? In linear perturbation theory the perturbation $\delta = \delta\rho(\underline{x},t)/\rho_b$ with $\rho_b = 1/6\pi G t^2$ obeys

$$\ddot{\delta} + (4/3t)\dot{\delta} - (2/3t^2)\delta = 4\pi G\, \delta\rho_s(\underline{x},t) \qquad (9)$$

where $\delta\rho_s(\underline{x},t)$ is the source perturbation for a moving loop

$$\delta\rho_s(\underline{x},t) = (1/\pi r(t)^2)\Theta(r(t)-r)(m/a^3)\delta(z-z(t)) \qquad (10)$$

with r(t) given by (7) in terms of z(t) in (6).

Using a Greens function and ignoring the decaying mode we easily find that the linear mass density along the z axis is given by

$$d\delta M/dz = \delta M_{tot}(2/d)(1-z/d) \qquad (11)$$

$$= \delta M_{tot}\,(2z'/d^2)$$

with the total accreted mass $\delta M_{tot} = 3/5m(t/t_i)^{2/3}$, the usual result. Thus the line density is greatest where the loop started and least where it ends up. However the actual density

$$\delta\rho(z) = (1/\pi r^2)d\delta M/dz = 2(\delta M_{tot}/\pi R_i^2)(d/z')^2 \qquad (12)$$

is greatest where the loop ends up. (11) and (12) are also independent of R_i for loops produced before t_{eq}.

The universal profile predicted is very important - it says that if we ask for the greatest density perturbation produced by loops on a given fixed length scale, less than d for all loops considered, then all loops will produce the same. Thus 'cluster' loops produce no higher density on a galactic scale than 'galaxy' loops. Peculiar velocities of loops smear out the mass distribution on small scales.

Whilst linear perturbation theory gives us some idea of what goes on around loops and is valid on surfaces within which $\delta M/M < 1$, it clearly breaks down on small scales. Let us now look at the shape of the 'turnaround' surface around loops - inside this surface we know that complicated nonlinear dynamics will determine the structure.

We can calculate the turnaround radius in linear perturbation theory - if we assume the collapse to be mainly radial a comoving radius r turns around when

$$\frac{d}{dt}(ar) = 0 \qquad \text{i.e. } \dot{r} = -\dot{a}/a\ r \tag{13}$$

However in linear perturbation theory outside the region where $\delta\rho_s(\underline{x},t)$ is non zero the density is uniform: $\delta(\underline{x},t)$ is identically zero. Thus mass conservation yields for the radius of a shell initially at r_i

$$r^2 = r_i^{\,2} - (d\delta M/dz)/\pi\rho_i \tag{14}$$

where we know $d\delta M/dz$ from (11), and ρ_i is the initial density. Differentiating and using (13) we find

$$r^2 = (d\delta M/dz)/\pi\rho_i \tag{15}$$

for the radius just turning around at time t at distance z along the z axis.

Putting (15) and (11) together, we find a paraboloid

$$r^2 = (36/5)Gm\ (t_i/d)^2(t/t_i)^{2/3}\ z' \tag{16}$$

for the turnaround radius. Once more for loops produced before t_{eq} it is independent of loop radius. A similar result (ignoring the finite size of the loop) has been reported by Bertshinger at this meeting using a slightly different argument[23].

Inside r given by (16) nonlinear effects are important and the 'trumpet' picture breaks down. Clearly near $z' = 0$ (16) is most important. As z' increases however, the

paraboloid $r \propto \sqrt{z'}$ crosses the 'trumpet' $r \propto z'^2$ and our turnaround surface calculation breaks down.
This happens at

$$z' = ((324/5)G\mu\beta v_o^2 a)^{1/3}\ t_i \tag{17}$$

which is the comoving distance travelled by loops of radius

$$R = ((12/5)G\mu\beta\ a/v_o)^{2/3}\ t_i \tag{18}$$

Using $v_o \simeq 1$ we find $R \simeq .2\ t_i$. Roughly speaking therefore for loops smaller than those giving rise to Abell clusters, the turnaround effect is more important; for larger loops the 'trumpet' structure due to the loops' finite size should be visible.

First we discuss the smaller loops. Equation (16) gives an estimate of the ellipticity of the structure - the diameter at half-length is given by

$$2r = ((8/15)\ \beta G\mu v_o^{-3}\ (t_i/R)^{1/2} a)^{1/2}\ d \tag{19}$$

$$\simeq .18\ (t_i/R)^{1/4}\ d$$

So for Abell clusters $2r/d \simeq .3$ whereas for galaxies $2r/d \simeq 1$ (which means of course that our cylindrical symmetry has broken down).

Cosmic strings therefore predict roughly spherical galaxy halos but quite prolate clusters. There is in fact some observational evidence that indeed clusters are more prolate than galaxies[24] - the best test would be to look at the largest clusters, where the prolateness should be stronger in the cosmic string theory.

This shape of rare peaks provides a useful comparison to theories with Gaussian perturbations. As is well known, with Gaussian perturbations higher peaks are predicted to be more and more spherical[25]. More observations on the shapes of very massive clusters may well be able to check this.

What about larger loops?. Here we should be able to see the effect of the 'trumpet' $r \propto z'^2$ shape. Of course such objects will be rare - the mean separations would be larger than for Abell clusters but the structures should be visible as giant superclusters. Note that the 'trumpet' shape $r \propto z'^2$ would be observed in projection onto the sky whatever the orientation of the object relative to us.

An obvious candidate is the Perseus-Pisces supercluster, usually referred to as a giant filament[26,27]. However as may be seen in pictures of it (for example in Oort, reference 26 p.399), a 'trumpet' - like shape is clearly visible. The dense tip is at Right Ascension 02h and declination 37°, and the structure extends some 50 h^{-1}Mpc, with Right Ascension 23h declination 20° marking the centre of the wide end of the 'trumpet'. The ratio of the base radius to the length is about .4.

This is of course rather speculative, but it does give an idea as to where we should look for surviving large loops. If the loop that produced this structure had radius $R = \alpha\ t$ at the time t when it was produced, we find the redshift when it was produced $1 + Z \simeq 10^4\ \alpha^2$. Taking $\alpha \sim 0.1$ for example we find $1 + Z \simeq 100$, and for the radius of the loop $R = 10^{-6}\alpha^{-2}t_o$ $= 4\ \alpha^{-2}$ kpc $\simeq 0.4$ Mpc. The loop would still be near the dense tip of the structure. Obviously all this is terribly sensitive to α. However, the total mass accreted is actually independent of α. It is given, from linear perturbation theory as $M \approx (1+Z)\beta\mu\ R \approx 10^{-2}\ \beta\mu t_o \simeq 6\times10^{15}\ \mu_6\ M_\odot$ which is certainly reasonable for $\mu_6 \sim 1\text{-}2$ since observations indicate the total mass is $10^{16}\text{-}10^{17}\ M_\odot$.

5. Peebles' Problems

So far I have mainly mentioned successes of the cosmic string theory. Not surprisingly there are also some problems. In a recent preprint, Peebles has discussed some of these[28]. He lists five potential problems: small loops, big loops, angular momentum, $\Omega = 1$ and very large scale structure - sheets, filaments and voids.

The small loop problem is that there are too many small loops around at t_{eq}. As equation (1) shows, at t_{eq} the number density of loops cuts off at radii $r \approx \gamma G\mu\ t_{eq} \approx 4\times10^{-2}\ \mu_6$pc. These loops had masses $8\times10^5\ \mu_6{}^2\ M_\odot$ and would today have mean separations of 0.5 Mpc. In the cold dark matter scenario they would have accreted objects of mass $3\times10^{9}\ \mu_6^2\ M_\odot$. Such objects (dwarf galaxies) are seen but not as frequently as predicted.

Equation (1) predicts $n(m)\,dm \propto m^{-5/2}dm$ for the number of objects with masses between m and m+dm. However observations show that the number of objects with luminousity between luminousity ℓ and $\ell + d\ell$, $n(\ell)\,d\ell \propto \ell^{-1.3}d\ell$ and unless luminousity is a very steep function of seed mass indeed it is hard to reconcile the two.

There do however appear to be several ways out of this problem. First, we know that nonlinear dynamics has been important on scales smaller than a few Megaparsecs and it is conceivable that the dwarf galaxies formed around small loops were stripped by interactions with larger galaxies. Presumably they would in this case be visible in voids. A second possibility would be to take $\mu_6 \simeq 3$ or so and begin accretion at a later redshift (this would be the case with neutrinos - see below). If accretion only starts at decoupling, as would happen for "galaxy" loops in a neutrino - dominated universe, $t_{dec} \simeq 8\ t_{eq}$, and the cutoff in the loop distribution would correspond to loops of mean separation $\sim2\mu_6^{1/2}$ Mpc. With $\mu_6 \simeq 3$ this is close to the mean separation for galaxies. Note that the cutoff would not be sharp - below the threshold mass n(m) tends to a constant acording to (1). As loops lose an appreciable fraction of their length due to gravitational radiation, one would also expect them to intersect occasionally and split into smaller loops. This would increase n(m) at smaller m. Whilst this is far from clear cut, it should also be said that theories based on Gaussian perturbations are nowhere near predicting the correct form for n(m).

The large loop problem is that above the galactic luminosity $L_* = 4\times10^{10}\ L_\odot$ $n(\ell)$ decreases exponentially, as $e^{-\ell/L_*}$. However as I have discussed above, the density perturbation produced by large loops on small scales is reduced by their peculiar velocities. Provided very small peculiar velocities for loops are strongly suppressed (which seems reasonable), there will be no "supergiant" galaxies produced. Hence the large loop problem is not really a problem.

The angular momentum problem is that galaxy-forming loops are very much smaller than the regions that eventually form galaxies when they turn around. For example the galaxy halo shell now at 50 kpc turned around at $z \simeq 10$ when its radius was ~ 100 kpc. The loop that formed the galaxy was only 11 pc in radius however - so very little angular momentum could be transferred to the matter (except perhaps in some very violent process). Here again however the peculiar velocities of loops come to the rescue. The comoving scale loops move after t_{eq}, $d_m = 3v_i t_{eq} = 3v_o(Rt_{eq})^{1/2}$, is not very much smaller than their mean separation, $d = (2/3\nu)^{-1/3}\ (Rt_{eq})^{1/2} \simeq 5(Rt_{eq})^{1/2}$ if $v_o \simeq 1$. Thus the paraboloid masses they accrete (which from equation (19) are quite strongly aspherical at $z \simeq 10$ have considerable quadrupole moments. A calculation along the line of Thuan and Gott[29] shows that the tidal torquing between neighbouring loops can give, at least in order of magnitude, the right amount of angular momentum. In this respect the cosmic string theory does not seem worse than other theories. Another solution to the angular momentum problem involving splitting of loops into clouds of smaller loops has been suggested by Zurek[30].

The $\Omega = 1$ problem is probably the most serious one for the cosmic string theory at present. From a theoretical point of view, $\Omega = 1$ is the only natural value. However when one measures the mass per galaxy in clusters and multiplies up by the total space density of galaxies one finds $\Omega \simeq 0.2 - 0.3$. This can be seen crudely from the numbers we used above. The mass overdensity $\delta M/M \sim 170\ \Omega^{-1}$ is measured from the dynamics of clusters. However the mean separation of galaxies is $5\ h^{-1}$ Mpc, so one would only expect to find $4\pi/3\times(1.5/5)^3 \simeq 0.11$ galaxies in an Abell radius $1.5\ h^{-1}$ Mpc. We see more than 50, so the overdensity in galaxies is >440. If the density of galaxies is proportional to the mass density, this means $\Omega <$ 0.4. Thus if $\Omega = 1$ the mass per galaxy in clusters must be less than the mass per galaxy elsewhere.

Possibly related to this is the problem of morphology segregation - why galaxies in clusters are mainly ellipticals but galaxies outside clusters are mainly spirals.

One suggestion which has been widely adopted is 'biasing', the assumption that galaxy formation is very sensitive to environment and that the small density enhancement in the neighbourhood of a cluster is enough to trigger the formation of many more galaxies than elsewhere. We could adopt this solution here too, but since no convincing explanation of the origin of such biasing has emerged, we prefer not to.

There is one possiblity in the cosmic string theory that may resolve the problem. If a large loop typically chops off very many small loops the loops that gave rise to clusters might actually be sitting inside a cloud of galaxy-forming loops. Clearly most of the galaxies in the cluster would have to have been formed by 'progeny' of this large loop - so at least 30-50 small loops (massive enough to accrete a galaxy) would have to be produced. This may in fact occur. The mass of an Abell cluster-forming loop is over a hundred times larger than the mass of a galaxy forming loop. So one "parent" loop could have split into one or more large loops and lots of smaller loops. According to the model for loop splitting discussed above, the fraction of energy of a loop going into loops of size ℓ or smaller is $(\ell/L)^{1-\alpha}$ so if $\alpha \approx 0.8$ for example, about 40% of the original loop's energy goes into loops of size L/100 or less. In this case about 40 galaxy-forming loops would be produced. Checking this only requires a moderate increase in the resolution of the numerical simulations - in the near future we should be able to check whether it does in fact occur.

If the angular momentum of galaxies is due to tidal torquing then one would expect a change in the local density of galaxies to produce some morphology segregation, although the details of this process are far from clear.

Peebles' last problem is the observations of sheets, filaments and voids - very large scale structure. Vachaspati recently suggested[31] that the wakes formed behind long strings could form sheets in the distribution of galaxies such as those apparently observed in the CfA survey. However Stebbins, Veeraraghavan, Brandenberger, Silk and myself[32] have recently calculated the effects of wakes in more detail, and find that wakes are unlikely to have accreted more than about 20% of the galaxies onto them. Thus the sheets would not stand out very strongly. A more likely explanation for sheets is that they are due to large scale coherent velocity fields, which can easily produce sheets perpendicular to the line of sight, as many appear to be. Brandenberger, Kaiser, Shellard and myself are presently calculating the magnitude of these in the cosmic string theory.

I have already suggested an explanation for filaments - their being formed by large moving loops.

Lastly, voids are really not a problem in any theory with a form of 'biasing' i.e. where the galaxies do not trace the mass exactly. I will discuss this in a forthcoming preprint.

Finally let me turn to the question of what the dark matter is. As is well known, the standard nucleosynthesis calculations constrain $\Omega_{\text{baryonic}} < 0.12\ \Theta^3\ h_{.5}^{-2}$ where Θ is the microwave background temperature in units of 2.7°K, although as Hogan has argued at this meeting depending on the details of the QCD phase transition these may be open to question. Accepting this bound requires that most matter in the universe be non-baryonic.

Unfortunately cosmic strings do not help here. They provide only a small fraction of the mass in the universe. Probably the best motivated dark matter candidate is a massive neutrino - it is not hard to construct models where a single neutrino has a mass of 25 eV required to close the universe.

Cosmic strings and neutrinos may be an attractive scenario - the string loops preserving perturbations on small scales and forming galaxies out of the baryons. This would require μ_6 to be slightly larger (by a factor of 2-3) than with cold dark matter since one would lose the growth during the radiation dominated era - the neutrino Jeans length falls to the comoving radius corresponding to a cluster mass at around Z_{eq}. This larger value of μ_6 probably helps however in producing larger streaming velocities on large scales and in solving the "small loop problem".

6. Other Effects of Cosmic Strings

I will not review here the other possible methods that have been suggested for detecting cosmic strings. The principal ones are their distinctive imprint on the microwave background, the gravitational radiation they emit which can in principle be detected in timing the millisecond pulsar, and lensing by string[33]. Suffice it to say that all known constraints are consistent with values of $G\mu < 5\times10^{-6}$ or so.

7. Conclusions

In the near future many of the main issues in the cosmic string theory will be resolved. Large scale string simulations can in principle give clear predictions for the very large scale structure of the universe. They will soon be "plugged in" to N-body codes and subjected to the same tests as other models have been. Perhaps the greatest virtue of the theory is the lack of freedom in interpreting it, at least for the simplest version I have outlined here. If it is wrong we shall know pretty soon!

Acknowledgement

I would like to acknowledge many helpful discussions with A. Albrecht, D. Bennett, R. Brandenberger, E. Copeland, T. Kibble, J. Peebles, D. Schramm, A. Stebbins, T. Vachaspati, W. Zurek and the hospitality of the Aspen center for Physics where most of the work described above was done. I also acknowledge the SERC for support through an Advanced Research Fellowship.

References

1. T.W.B. Kibble, J. Phys A9 (1976) 1387.
2. T.W.B. Kibble, Phys. Rep. 67 (1980) 183;
   Ya. B. Zel'dovich, Mon. Not. Roy. Ast. Soc., 192 (1980) 663;
   A. Vilenkin, Phys. Rev. Lett. 46 (1981) 1169, 1496 (E).
3. A. Albrecht and N. Turok, Phys. Rev. Lett. 54 (1985) 1868, and in preparation (1986).
4. N. Turok, Phys. Rev. Lett. 55 (1985) 1801.
5. N. Turok and R. Brandenberger, Phys. Rev. D33 (1986) 2175;

R. Brandenberger and N. Turok, Phys. Rev. D33 (1986) 2182;
J. Traschen, N. Turok and R. Brandenberger, Phys. Rev. D (1986) in press;
R. Brandenberger, A. Albrecht and N. Turok, Nucl. Phys. B D (1986) in press.
6. H. Sato, Mod. Phys. Lett. A1 (1986) 9.
7. A. Stebbins, Ap. J. Lett. 303 (1986) L21
8. T. Vachaspati and A. Vilenkin, Phys. Rev. D30 (1984) 2036.
9. E.P.S. Shellard, DAMTP preprint (1986) and E. Copeland, E.P.S. Shellard and N. Turok, in preparation (1986).
10. T.W.B. Kibble and N. Turok, Phys. Lett. 116B (1982) 141.
11. E. Copeland and N. Turok, Phys, Lett. B173 (1986) 129.
12. N. Turok, Nucl. Phys. B242 (1984) 520.
13. T. Vachaspati and A. Vilenkin, Phys. Rev. D30 (1985)
14. N. Turok, Phys. Lett. B126, 437.
15. G.O. Abell, Ap. J. Suppl. 3 (1958) 211, Ap. J 66 (1961) 607.
16. M. Davis and J. Huchra, Ap. J. 254 (1982) 437.
15. N.A. Bahcall and R.M. Soneira, Ap. J. 270 (1983) 20.
17. A.A. Klypin and A.I. Kopylov, Sov. Astr. Lett. 9 (1983) 41.
18. S. Schectman, Ap. J. Suppl. 57 (1985) 77.
19. N. Bahcall and W. Burgett, Ap. J. 300 (1986) L35.
20. A. Szalay and D. Schramm, Nature, 1985.
21. P.J.E. Peebles, 'The Large Scale Structure of the Universe', Princeton University Press, 1980.
22. E. Bertshinger, Berkeley preprint, 1986 and talk at this meeting.
23. B. Bingelli, Astron. Astrophys. 107 (1982) 338.
24. J.M. Bardeen, J.R. Bond, N. Kaiser and A.S. Szalay, Ap. J. 304 (1986) 15.
25. J. Oort, Ann. Rev. Astron. Astrophys. 21 (1983) 373.
26. R. Giovanelli and M.P. Haynes, Astron J. 87 (1982) 1355.
27. P.J.E. Peebles, Princeton preprint (1986).
28. T.X. Thuan and J.R. Gott, Ap. J. 216 (1977) 194.
30. W. Zurek, Los Alamos preprint (1986)
31. T. Vachaspati, Phys. Rev. Lett. 57 (1986) 1655.

32. A. Stebbins, S. Veeraraghavan, R. Brandenberger, J. Silk and N. Turok, Berkeley preprint, (1986).
33. For a discussion and references, see for example N. Turok, in 'The Very Early Universe', ed. G. Lazarides and Q. Shafi, North Holland, (1986).

Large Scale Drift and Peculiar Acceleration as Cosmological Tests

N. VITTORIO AND R. JUSZKIEWICZ

Introduction

The observations of large scale fluctuations in the Hubble flow and in the spatial distribution of galaxies can be used to constrain cosmological scenarios. Here we confront some model predictions with recent observations by Collins et al. (1986) and Burstein et al. (1986), suggesting that the Hubble expansion on scales $\sim 50h^{-1}Mpc$ (h$\equiv$ Hubble constant / 100 km s^{-1} Mpc^{-1}) is distorted by matter currents of surprisingly large amplitude. Confirmation of this will have grave consequences for scenarios with random phase scale invariant primeval mass inhomogeneities (Vittorio, Juszkiewicz, Davis, 1986).

We also discuss the theoretical conclusions that may be drawn from the analysis of the IRAS Point Source Catalogue, that provides us with a uniformly selected galaxy sample with nearly full sky coverage (Yahil et al., 1986; Meiksin and Davis, 1986). These observations are extremely important for our understanding of the large scale distribution of galaxies. However, we show that the IRAS data alone are not sufficient to make a reliable and model independent estimate of the density parameter Ω_0.

The Large Scale Drift

The best experimental evidence for our peculiar motion relative to the comoving frame is provided by the cosmic microwave background (CMB) dipole anisotropy. This implies a Local Group peculiar velocity of $|\mathbf{V}_{LG}| = 600$ km s^{-1} (see Lubin and Villela 1986 for a recent review). This velocity is the result of the infall of the Local Group onto the Virgo Cluster ($|\mathbf{V}_{inf}| \sim 250 \pm 50$ km s^{-1}) and the motion of the Virgo Cluster as a whole relative to the comoving frame ($|\mathbf{V}_{VC}| \sim 470 \pm 70$ km s^{-1}). Galaxy formation scenarios can be tested by checking if the expected gravitational field of "typical" mass concentrations is sufficient to generate $\mathbf{V}_{VC}$.

To compare the observations with the model predictions it is important to determine the minimum depth of a sample of galaxies necessary to define the comoving frame. If these galaxies are unperturbed tracers of the Hubble flow , the peculiar velocity of the Local Group relative to the sample should be equal to the velocity of the Local Group relative to the CMB. If the two velocities are different, a coherent motion relative to the CMB of all the matter inside the volume is implied.

Rubin et al. (1976) reported evidence of anisotropy of the Hubble flow on very large scales. This result has been recently confirmed by Collins et al. (1986), who concluded that the original Rubin et al. sample has a large amplitude

($\sim 1000 kms^{-1}$) bulk motion relative to the comoving frame. A lower but still large amplitude velocity ($\sim 700 kms^{-1}$) for a sample of elliptical galaxies has been recently found by Burstein et al. (1986). Both the Rubin et al. and Burstein et al. samples probe a spherical region of radius $R \sim 50 h^{-1} Mpc$.

The expected magnitude of the drift of a galaxy sample of depth R is (e.g., Vittorio and Turner, 1986),

$$v_{rms}^2(R) = \frac{\Omega_0^{1.2} H_0^2}{2\pi^2} \int_0^\infty dk |\delta_k|^2 \, exp(-k^2 R^2) \tag{1}$$

where $|\delta_k|^2$ is the density fluctuation power spectrum. Clearly, only density fluctuations of wavenumbers k *smaller* than 1/R (i.e., wavelengths *larger* than R) contribute to the above integral.

As an example let us consider a cold dark matter (CDM) dominated universe. The expected peculiar velocity of the Virgo Cluster and of a spherical region of radius of 50 h^{-1} Mpc are respectively (Vittorio and Turner, 1986):

$$v_{VC} = 322 kms^{-1} \, \Omega_0^{+0.03} h^{-0.57} \nu^{-1} \tag{2a}$$

and

$$v_{50} = 83 kms^{-1} \, \Omega_0^{-0.33} h^{-0.92} \nu^{-1} \tag{2b}$$

where ν is the biasing factor, and usually it is assumed that $\nu \sim 3$ (Kaiser, 1986). If the galaxy formation was not biased towards high density regions, ν=1. For inflation-produced fluctuations the primordial density field is gaussian, and each component of the peculiar velocity is normally distributed. The probability of measuring a peculiar velocity with magnitude in the interval $v_1 \rightarrow v_2$ is given by

$$P = \sqrt{\frac{54}{\pi}} \int_{v_1/v_{rms}}^{v_2/v_{rms}} x^2 exp(-1.5x^2) dx \tag{3}$$

Hence, at the 90% confidence level, we have:

$$110 < \frac{v_{VC} \, h^{0.57} \nu}{km \, s^{-1}} < 520 \tag{4a}$$

$$30 < \frac{v_{50} \, \Omega_0^{0.33} \, h^{0.92} \nu}{km \, s^{-1}} < 135 \tag{4b}$$

It is clear that for $\nu \sim 3$ the velocity on scales $\sim 50 h^{-1} Mpc$ predicted in a flat CDM dominated universe falls short of the observed values (Vittorio, Juszkiewicz, Davis, 1986). Moreover, for the same value of ν, the theoretical prediction for v_{VC} is too low to explain the observed velocity $\sim 400 kms^{-1}$ (Vittorio and Turner, 1986).

We want to conclude this section by emphasizing the following point. Measurements of our peculiar velocity relative to a galaxy sample do not necessarily give cosmologically relevant information. Selecting a sample implies ignoring all the universe outside the sample itself. Fortunately, we have at our disposal an independent measure of our peculiar velocity relative to the comoving frame, the CMB dipole anisotropy. It is by subtracting the two velocities that we can evaluate the drift of the considered sample relative to the comoving frame.

Peculiar Acceleration

The analysis of the IRAS Point Source Catalogue has proven to be a powerful probe of the Hubble flow field. It has been shown that the distribution of galaxies in the catalogue exhibits a dipole anisotropy in reasonable agreement with the direction of the CMB dipole anisotropy (to within $\sim 20° - 30°$; Yahil et al., 1986; Meiksin and Davis, 1986). For small perturbations, the relation between the peculiar velocity **v** and acceleration **g** is given by (see,e.g., Peebles 1980):

$$\mathbf{v} = \frac{1}{3}\Omega_0^{0.6} H_0 \mathbf{g} \tag{5}$$

We interpret **g** as the peculiar acceleration of the Local Group induced by all the galaxies, clusters, and voids in the Universe. By selecting a sample, say a spherical shell of galaxies with an inner and outer radii r and R, we ignore all the material outside of the shell. Since gravity is a long range force we have to be sure that all the ignored material is not contributing to **g**. The standard way of dealing with this problem in the existing literature (e.g., Yahil et al., 1986) is to calculate **g** for subsamples of increasing outer radius R. If **g** remains constant within the measurement errors, it is concluded that the "convergence" criterion is satisfied and that the material outside the sample does not contribute to the total acceleration. Then, the inhomogeneities responsible for the dipole anisotropy should be contained in the sample. The sample itself should be at rest relative to the comoving frame and in Eq.(5) we can set $\mathbf{v} = \mathbf{V}_{LG}$. Since both **g** and $\mathbf{V}_{LG}$ are measured, Ω_0 can be evaluated. We want to show that this is a very dangerous procedure. Indeed, the peculiar acceleration can be written as

$$\mathbf{g} = \frac{3}{4\pi} \int_{shell} \delta(\mathbf{r}) \frac{\mathbf{r}}{r^3} d^3r \tag{6}$$

By Fourier transforming Eq.(6) we can calculate the expected peculiar acceleration induced by the material inside the shell:

$$\begin{aligned} <|\mathbf{g}|^2> &= \frac{9}{2\pi^2} \int_0^\infty dk \; |\delta_k|^2 \; (\frac{sinkr}{kr} - \frac{sinkR}{kR})^2 \\ &\simeq \frac{9}{2\pi^2} \int_{R^{-1}}^{r^{-1}} dk \; |\delta_k|^2 \end{aligned} \tag{7}$$

Note that only perturbations of wavenumber $R^{-1} < k < r^{-1}$ contribute to the above integral [in contrast, only perturbations of wavenumber $k < R^{-1}$ contribute to the rms bulk motion of the shell, see Eq.(1)]. Let us take a single power law spectrum with an effective spectral index n: $|\delta_k|^2 \propto k^n$. If $n \geq -1$, we have

$$\begin{aligned} <|\mathbf{g}|^2> &\propto ln\, \frac{R}{r}; & n = -1 \\ &\propto \frac{1}{r^{n+1}} - \frac{1}{R^{n+1}}; & n > -1 \end{aligned} \tag{8}$$

First let us consider the case n=-1. In this case $< |g|^2 >^{1/2}$ depends logarithmically on R and it may easily deceive the observers as "convergent", while the peculiar velocity of the whole sample, given by Eq.(1), is formally infinite! This problem does not exist in models where mass inhomogeneities have little power on large scales (e.g., n=1). In this case, however, the inferred value of Ω_0 is extremely sensitive to the observationally uncertain lower cutoff r. We illustrate this in Fig.1 showing $< |g|^2 >^{1/2}$ predicted for the CDM model with $\Omega_0 = 1$, h=0.5 and a scale- invariant adiabatic perturbations. The three different curves correspond to r = 1, 5, 10 Mpc respectively. The differences in the asymptotic values of $|g|$ for $R >> r$ reflects the fact that the acceleration in this case is dominated by the contribution from nearby inhomogeneities.

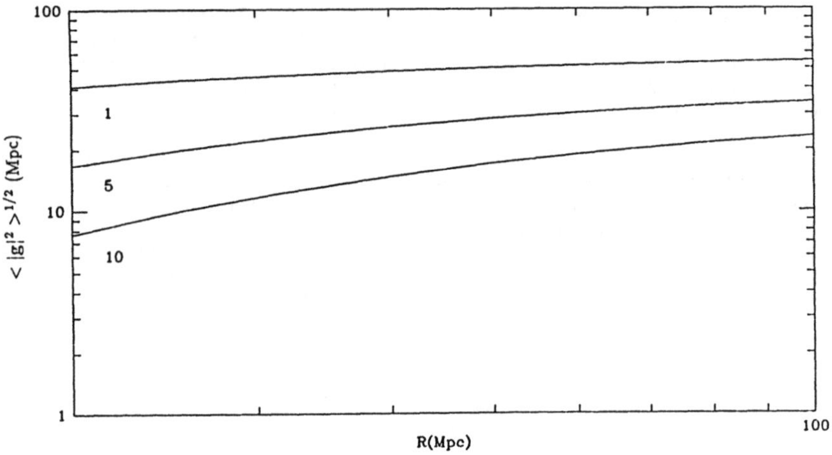

We conclude that it is impossible to measure Ω_0 using the IRAS sample unless we have an independent measurement of the peculiar velocity of the sample itself. If the large scale drift of galaxies is confirmed, we doubt that the IRAS sample can define the comoving frame. The details of this calculation will be published elsewhere (Juszkiewicz and Vittorio, 1986).

References

Burstein, D., Davies, R.L., Dressler, A., Faber, S.M., Lynden-Bell, D., Terlevich, R., Wegner, G. 1986, in *"Galaxy distances and Deviations from the universal expansion"*, B.F. Madore and R.B.Tully Eds. (Reidel)

Collins, A., Joseph, R.D., and Robertson 1986, N.A., *Nature*, **320**, 506

Juszkiewicz, R. and Vittorio, N. 1986, *Ap.J. (Letters)*, submitted

Kaiser, N. 1986, in *Inner Space/Outer Space*, eds. E.W. Kolb et al. (Chicago: University of Chicago Press).

Lubin, P., and Villela, T. , 1986 , in *"Galaxy distances and Deviations from the universal expansion"*, B.F. Madore and R.B.Tully Eds. (Reidel)

Meiksin A. and Davis, M. 1986, *A.J.*, **91**, 191

Peebles, P.J.E. 1980, *Large-Scale Structure of the Universe* (Pri nceton: Princeton University Press).

Rubin, V., Ford, W.K., Thonnard, N., and Roberts, M.S. 1976, *A.J.* , **81**, 687.

Vittorio, N., Juszkiewicz, R., and Davis, M. 1986, *Nature*, **323**, 132

Vittorio, N. and Turner, M.S. 1986, *Ap.J.*, , in press

Yahil, A., Walker, D., and Rowan-Robinson, M. 1986, *Ap.J. (Letters)*, **301**, L1.

Conference Summary

JAMES E. GUNN

So vast was the subject of this conference that summarizing it in a small space is very difficult, and I will be able to do no more than outline a few memorable results and perhaps synthesize a point of view or two from the overwhelming amount of material we have heard in the past two weeks. One problem was that the subject matter was never very well defined, and the workshop eventually covered essentially all of extragalactic astronomy with the possibly notable exception of active nuclei. What is a "nearly normal" galaxy? Virginia Trimble remarked a few days ago that a nearly normal person is anyone one does not know very well.

I. Stellar Populations

A long-standing problem in extragalactic astronomy has been the one of synthesis of stellar populations; a primary difficulty has been the virtual certainty that one's library of stars, compiled from the occupants of the solar neighborhood, is not at all representative chemically of many of the populations one attempts to synthesize, notably the ones in giant ellipticals. Since the equations one attempts to solve in doing synthesis are mathematically notoriously ill-posed, the lack of a proper library is potentially disastrous, and at the very least renders interpretations of population ages and metallicities highly suspect. The recent work of Frogel, Whitford, and Rich, reported here by Jay Frogel, may at last solve the problem. There appear to exist in the central regions of the Galaxy old stars of very high metallicity, precisely the (known) missing ingredient in our libraries for ellipticals. There remains a large amount of work to do, most notably reliably to assign metallicities in a uniform way from early K giants to late M stars, so that one can assign complete giant branches to some (known) metallicity, and to assemble a large spectral library of stars spanning temperature and metallicity. Work on the main sequence in the central regions must await Space Telescope, but we may understand metallicity effects on the main sequence well enough already, even though we do not have any dwarfs with [Fe/H] of +1 to study.

A central, usually neglected aspect of the library problem is that not only is there a lack of high-metallicity objects, there is not nearly enough diversity to mock up even a simple closed-box model spread in metallicity. Since large changes in metallicity grossly affect the temperature of the giant branch, proper inclusion of the variance in metallicity must have a significant effect on the synthesis. The Frogel-Whitford-Rich sample should fix this problem as well.

There are still, of course, good reasons to believe that the problems will not magically go away even when a good galactic center sample is developed. "Metallicity" is clearly much too gross a concept in many applications involving low-metallicity populations, and there are disturbing signs that that may be the case in high-metallicity ones as well. The mysterious gradients seen by Davies and Sadler in ellipticals, in which the Fe I remains constant while magnesium declines precipitously with radius *may* simply reflect compensatory effects of giant-branch morphology with metallicity, or may be in fact due to gradients in [Mg/Fe], which would be very difficult to understand in themselves, and incidentally would render our nice universal library useless. Other evidence was presented by Burstein: the relatively metal-rich globular cluster population in M31 has enhanced CN relative to galactic globulars of the same metallicity. This may be "only" a mixing effect, but is in any case profoundly disturbing, or should be, to those who synthesize populations. Neither large variations in early-synthesized nitrogen nor large variations in atmospheric nitrogen on account of mixing variations on the scale of a whole galaxy are easy to incorporate in any easy synthesis scheme (nor, incidentally, are they possible to understand on any simple model of galactic evolution.)

Another obvious problem which any set of observations, however complete, of a 10^{10} year-old population cannot address, is the evolution of the asymptotic giant branch at a given luminosity as the progenitors come from less and less massive stars on the main sequence. To the extent that AGB stars contribute to the light in some band, this difficulty makes evolutionary calculations suspect. We will return to this point later.

Closely connected with the observational problem of population synthesis — indeed, if the theory of stellar evolution and death were more perfectly understood, almost equivalent to it — is the question of the initial mass function. Is there a universal one? Or are there two universal ones, which can occur in various mixes in different circumstances? Three? Four? We heard arguments from Larson and from Dopita in support of a bimodal IMF. It seems to me that the situation is extremely unclear, and that there is indeed no clear evidence at the present time which strongly suggests that the IMF is anything but universal...That, I think, is a comment on the quality of the data and the depth of our understanding rather than any conviction concerning the universality of the solar neighborhood IMF. There seem to be no easy, clean tests, but clearly we need badly to clean up the situation in the Clouds and in galactic globular clusters, both of which can probably be done from the ground with a great deal of work and will be easy with ST.

A classical problem which may or may not have something to do with the constancy of the IMF with time is what one might call the extended G-dwarf problem, represented by the absence of stars of extremely low metallicity in the halo and the paucity of stars of metallicity of a tenth to a third solar in the disk. Beers, reporting on the work of Preston, Shectman, and himself, showed that very low metallicity stars are probably present in the (very small) expected numbers in the halo; the work of Kuijken and Gilmore on the thick disk indicates that the intermediate metallicity stars may be present there in sufficient numbers to explain the disk metallicity distribution. In any case, possible complications such as infall of low-metallicity material make the connection with changing IMF tenuous at best.

II. Evolution of Galaxies

One of the most exciting developments of recent years and, I think, one of the most exciting promises of the HST, is the availability of detectors of sufficient sensitivity to allow the direct exploration of the evolution of galaxies, spectroscopically from the ground and, (soon?) morphologically from space. The pioneering work of Oemler and Butcher and the more recent spectroscopic work of Dressler and the author has demonstrated quite securely that galaxies do in fact evolve—that is, their stellar populations evolve in some statistical sense. The seemingly simpler question of whether the populations in ellipticals evolve as one would expect them to simply from the aging of their stellar populations is yet a little unclear. The fact that very high redshift objects of the sort observed by Spinrad and Djorgovsky are all peculiar rather muddies the question, and all one can say is that there seems to be an envelope of red objects which evolves as Bruzual's "c-models" are supposed to—modulo the considerable uncertainty which attends our utter ignorance of the IMF. Whether there are objects too red at high redshifts for even this minimal evolution is not clear. There seem to be lots of objects which are brighter and bluer at high redshifts than c-model evolution would suggest, but it is not clear what present-day descendents of these galaxies are, and that would seem to be a crucial point if one wishes to argue about evolution, as Gustavo Bruzual has often stressed.

In any case, the danger of using effectively flux-limited samples (in any band) to study evolution seems as obvious as it seems to be ignored. What one must do, it would seem, is to study a typical population of galaxies whose analogues today are well studied, if not understood. Even the study of clusters, however, is suspect, because of the likelihood that clusters themselves evolve strongly, in form at least, and not unlikely in total population as well. Since the morphological mix of the galaxies in clusters is closely correlated with the dynamical state of the cluster itself, one must attempt to understand that evolution. Fortunately, the observed evolution of cluster galaxies cannot be explained by cluster evolution, since that evolution (to more "relaxed", almost certainly more massive, systems) exaggerates the necessity for population evolution from the blue galaxies seen at high redshift to the old, red ones which populate clusters today.

Another avenue to the understanding of evolution might be the wonderfully deep counts that are being produced with CCD's by Tyson and MacKay, and the "counts" which come from the damped Lyman-alpha absorption lines being studied by Wolfe. The picture presented by the ensemble of data at present is a little bewildering, and it is not clear that any simple picture will explain it all. It may be, for example, that Wolfe's absorption lines at high redshift may not be the disks of galaxies but may be protogalactic lumps of the sort envisaged by Searle and Zinn to have made up the metal-poor spheroid of the galaxy; and that the faint counts, which seem to contain too many objects, may be related. Such objects, if they exist, must not convert too much of their mass into stars, as we shall see later.

The stress in this workshop on evolution has been on evolution connected with star formation history, but it is worth remembering that even in the absence of star formation ("c-models") the evolution is not simple. It is directly affected by the logarithmic slope x of the mass function, as given by the Tinsley relation

$d \ln L / d \ln t = -1.3 + 0.3x.$

This relation assumes that the giant branch does not evolve significantly with progenitor mass, and that assumption is sensitive to the band in which the observations are made. In particular, at K, where we have heard statements about the wonderful simplicity of the observational evolutionary problem, the contribution of asymptotic giant branch stars is large; the present situation regarding the understanding of the dependence of the AGB on metallicity and progenitor mass is hardly satisfactory. In summary, there is the tantalizing certainty that one has observed the evolution of galaxies, but the ability to make quantitative statements about the evolution is woefully lacking ... and will be so until a great deal more work is done.

III. Systematics in the Structure and Dynamics of Galaxies

It is conventional these days to discuss the systematics of galaxies within the framework of some sort of hierarchy: in one's imagination either a relic of the original perturbation spectrum or one imposed by some other formation process. That there exist well-defined relations between the dynamical properties of galaxies as functions of mass (luminosity, really) is clear. My favorite picture for looking at these relations is the mass, velocity dispersion diagram I drew for the Vatican conference five years ago, in which one can place any structure one observes if one is willing to make a few simplifying assumptions. Broadly speaking, the bound systems in the universe that one is willing to believe come from primordial structure (reasonably big galaxies, groups, clusters, say) define a relation in which $\sigma \propto M^{0.16}$, steepening to $M^{0.25}$ or so on the scale of large galaxies and to perhaps 0.35 or more for small galaxies. This diagram contains the essence of the Tully-Fisher and Faber-Jackson relations, and shows that there is continuity in these relations up to scales of 10^{15} $M_{\odot}$ or more. An important question for galaxies is the relation of morphology to this heirarchy, a topic still hotly debated for the Tully-Fisher relation; how does one cut the heirarchy to get E's? Sa's? Sc's? The plot would seem to indicate strongly that the early-to-late progression defines a series of sequences essentially parallel to the main heirarchy, with early-type systems having short collapse times and high binding energies, later types longer and lower, but systematic errors in mass-to-light ratio would have the same effect. In any case, if the M/L is reasonably behaved (i.e. reasonably monotonic) along a given morphological sequence, the virial theorem clearly demands that systems should lie on a surface in the space which is augmented by some size coordinate. This behavior has been found by Burstein et al (hereinafter the Magnificent Seven, or M7 for short) and by Djorgovsky, for elliptical galaxies, and has been used by M7 to define a new, seemingly very accurate distance scale, described here by Faber—which, however, as soon as applied to our neighborhood, yields a bewildering result, about which you all know and to which I will return later.

It would be surprising if spiral galaxies did not partake of such a relation, which could be used to augment Tully-Fisher, but we heard from Pierce that no convincing relation can be found.

Kormendy and Lauer told us about the systematics of core properties, and the former made clear yet another datum in the growing list which connects the enigmatic dwarf spheroidals with the spiral-elliptical sequence: they have the same core properties. It now seems overwhelmingly probable that they are simply very low-mass irregulars which have, either by some chance interaction with their parent or by hydrodynamic processes of their own, shed their interstellar gas. Their core properties are so strikingly different from real ellipticals, of course, because they are all core, and the cores of ellipticals are tiny fractions of their total mass and luminosity.

Among the cores of ellipticals there are also strong relationships among surface brightness, velocity dispersion, and core radius. I was struck by the possibility that all of these things can be understood as variations in the radius at which an essentially r^{-2} density distribution whose properties are consistent with the augmented Faber-Jackson relation for the global properties of the galaxy, chooses to round itself off into a regular core. Certainly the gross features can be so understood, but the details need to be studied a bit more.

Aaronson and Freeman both showed convincing evidence of quite high mass-to-light ratios among dwarf galaxies. It is not clear whether this is to be attributed to the presence of the same kind of dark matter which makes up most of the mass in clusters and groups and the halos of spiral galaxies, or simply a dramatic shift in the mass function toward the manufacture of very low-mass stars. The latter would help explain the low metallicity of these systems by decreasing the yield. It is, on the other hand, possible that dissipation does not play as significant a role for these systems as for their bigger siblings. After all, their velocity dispersions are not much bigger than the 10 km/s or so to which primordial material can happily cool, so perhaps their hydrogen did not efficiently separate from the associated dark matter in the perturbation from which they formed.

Binggeli discussed the luminosity function of galaxies, and gave quantitative support to the suspicion that different kinds of galaxies have different luminosity functions, which moreover, change in relative amplitude as one goes from dense environments to the field. Thus the density dependence of morphology which Dressler first quantified some years ago can perhaps be understood as the relative changes of type-dependent luminosity functions which retain their approximate shape within a given morphological type but change amplitude as a function of density. Such a scheme is more complicated, to be sure, than a universal LF, but much simpler than any one of a number of other possibilities that suggest themselves. It would appear to me that to explain this might be difficult for blossoming theories of galaxy formation—one has to cut the heirarchy to make Sc's, say, in such a way that in a dense region the relative number of bright and faint Sc's does not change, even though the number of Sc's relative to earlier types is suppressed markedly. Any takers ?

IV. Spheroids, Mergers, Disks, and all That

My good friend and colleague Alar Toomre attempted to convice us that everything we did not understand was the result of mergers. Summarizers are, I suppose, supposed to be dispassionate, but I will here abuse my privilege of having the last word. I do not think you can make rocks by merging clouds.

In my favorite mass-dispersion diagram, it is clear that ellipticals define the high specific binding energy edge of the heirarchy. It still seems to me remarkable that two loosely-bound things can merge and make a tightly bound one. Carlberg has recently demonstrated that the phase density in the central parts of ellipticals is much higher than in any part of candidate victim spirals, and unless one invokes gas dynamics (which, of course, one can) that makes for difficulty.

To make the bulges of big spirals by mergers seems to me silly, but it is a case well worth considering in view of other benefits we may accrue from studying it. The evidence grows that the bulges of big spirals are just like ellipticals, even to their angular momentum-mass relationship. If, therefore, I can convince you that making bulges by mergers is silly, I will perhaps not have too hard a time convincing you that making ellipticals by mergers is likewise. (I still believe that cDs are merger products, which for some reason is almost as unpopular these days as the view that ellipticals are not.)

Quinn discussed the nice but not-surprising result that disks are dynamically very fragile. They are very cold, of course, and it takes very little in the way of disturbance by an invading meal to heat them substantially. This is especially true in their outer parts, where the ratio of thermal to binding energy is quite low. It is precisely these outer parts which are supposed to survive cold while the galaxy gobbles up one or several (low-density) satellites to puff up its (high-density) core. It seems very unlikely that the average spiral could have suffered even a minor merger during the period after the disk has formed, and seventy percent of the big galaxies in the universe are spirals. (Needless to say, a merger between nearly equals can never result in a galaxy with a majority of its mass in a cold disk unless both parents are mostly gaseous at the time of the merger.) If one does not subscribe to the making of bulges by mergers, one is left with the remarkable need for the other thirty percent to be the products of many mergers. To be fair, the densities from elliptical-rich regions to the field vary over several orders of magnitude, and one needs to look carefully at the statistics.

One needs, to be sure, very efficient star formation to make ellipticals and spheroids. Why, I ask, invoke the pokey processes we observe in the quiescent interstellar medium, when we can appeal to the spectacular star-formation rates *observed* in the starbursts in colliding and interacting galaxies in the energetic shocks produced in the interaction. In the present-day universe, these shocks happen for us only in interacting systems. Where in the early universe might one have several-hundred km/s shocks in relatively high-density gas? One might think about the collapse of protogalaxies? Yes?

All of this might pose another challenge for the current ideas about the formation of galaxies. The favorite spectra nowadays are very flat at small scales, and mergers of protogalactic lumps is rampant at early epochs. One could reasonably ask whether any of these scenarios is consistent with the existence of as much mass as is observed in cold spiral disks. One might be forced to the view that disk formation occurred rather late, a subject about which I have muttered words on occasion.

In summary and apology, it seems to me very reasonable that there was much merging and general chaotic activity early among very gas-rich structures which resembled galaxies not at all, and in these violent events were created bulges and elliptical galaxies. Disks must have been formed after all the chaos was over, out of the remaining gas–on what timescale is debatable, but later.

As an important footnote, past passion about spiral structure was notable at this workshop by its absence. It (the subject, not the passion) was discussed very nicely by Carlberg; it seems very likely that Toomre's swing amplifier, acting either on the wakes of a lumpy disk without any aid except mild refrigeration caused by either the fresh supply of gas from stellar mass loss, perhaps helped by a little infall, or a bit of feedback through the center of the galaxy, can adequately explain the observed spirality. The bars and companions doubtless help, but are not really necessary to explain the bulk of the phenomenon, and real galaxies probably do not care in the slightest about grand-design modes.

V. Dark Matter

It seems clear this week that we have to have it. One thing which has emerged beautifully in the last year, and was here discussed by Barnes and by Oh, is that there is no longer any mystery about flat rotation curves in systems in which the inner parts are dominated by ordinary matter and the outer parts by dark matter. The near-coincidence between f, the ratio of "luminous" to dark matter and the mean of the Λ angular-momentum parameter, a bit of natural cooperation with core radii, and adiabatic compression of the dark halo by the dissipating baryons, very naturally make flat or gently rising rotation curves.

Galaxy formation by gravitational collapse with a wide variety of spectra produces very nearly (within a factor of about two) "maximal" disks for f about 0.07, rotation curves which rise too gently for smaller f, and falling rotation curves and strong central baryon domination for larger f. Thus the question of whether disks are maximal or not is a very important one to the immensely important question of the value of f. Faber and I independently came to the value of 0.07 some years ago, I from the local group and she from other small groups, but it must be considered very uncertain indeed. We have heard values at the workshop from 1/40 (Quinn) to 1/4 (Henriksen), based on very different considerations.

Scott Tremaine dropped what was perhaps the biggest bombshell of the workshop with the announcement of an absurdly low mass for the Galaxy, about 3×10^{11} $M_{\odot}$, which, if representative, gives a value of f of about 1/4. It seems to me that the timing mass for the local group, of order 3×10^{12} $M_{\odot}$, is difficult to reconcile with such a low mass for the galaxy unless the major part of the dark matter is sufficiently hot that it cannot fall freely into the halos of big galaxies. The arguments, however, for the low mass seem persuasive, unless the orbits of the satellites are preferentially very circular (they may plausibly be.)

It is worth saying again that it is quite clear that if one wishes Ω to be unity, one must have bias, which means that the **effective** value of f, in the sense not of "baryonic" to dark, but "luminous" to dark, must decrease in low-density regions–i.e., there must be a lot of dark baryonic matter, in failed galaxies or whatever, in regions of low galaxy density. It is perhaps disturbing that there is no obvious trend whatever in the inferred value of f from small aggregates like the local group to great clusters like Coma to the whole of the Local Supercluster–a convenient measure is the M/L value referred to the population in spiral disks, which seems to be about $150h$ in all cases. This is especially puzzling for the local supercluster, where the mean density is only perhaps three times the global mean.

There was much discussion of biasing mechanisms; the only one one can be reasonably sure of is the automatic statistical one (the "Kaiser effect") which is simply that a little lump on top of a bigger lump need not be as big as a little lump in a big hole in order to form a bound system of a given mass and binding. What is not clear and needs very much to be made clear is whether that is enough. Other schemes, even as presented by Martin Rees, seem almost impossibly difficult to arrange. The Frenk-White-Davis-Efstathiou (hereinafter the Four Horsemen, 4H) simulations are almost convincing. Almost.

VI. Large-scale Structure: the Beauty Contest

I have now happily forgotten who made the remark (I think it was either Simon White or Adrian Melott) that the N-body simulations of large-scale structure should be pitted against each other in a beauty contest (better the simulations than the simulators), since it was clear that the real universe with bubbles, filaments, and warps, was much prettier and more complex than any simulation to date. The question of interest, perhaps, is who judges the contest. The Huchra-Geller-Lapparent data seem to suggest a regularity which is notably absent in the larger-volume sample of Haynes and Giovanelli. Both data sets have structure which is remarkably like in character, if perhaps not quite in scale, the 4H cold dark matter simulations discussed by Simon White. Nothing else I have seen even comes close. The CDM picture may run aground on other obstacles, but when it comes to making galaxies and distributing them, it seems to be remarkably successful, unless structure on very large scales and very large peculiar velocities are demanded.

Are they? We heard a great deal about large-scale structure, from Neta Bahcall, who with her collaborators has wrung an enormous amount of statistical information out of the Abell catalog; however much those results may fly in the face of what current models are able to describe (I except strings from "current models"—certainly the naive scale-invariance they predict is promising in the light of the Bahcall-Soniera results, but the string models are not quite, I think, at the level of development required to make meaningful comparison with the observed correlation data. We may look forward to that development in the coming short while.), they pale before the structure that Tully attempted to convince us that he sees. The matter seems to me now to be in a very unsatisfactory state. The Abell catalog is itself a terribly inhomogeneous collection—not, I think, in the sense that any substantial number of nearby rich clusters were missed in those regions where they are reasonably easy to find, but the classification into distance and richness classes is not very precise and probably not very homogeneous. There are demonstrable correlations with galactic latitude (which probably reflects the difficulty of seeing clusters in a field full of stars), declination (which reflects Palomar zenith distance and hence background sky brightness) and funny right ascension dependences which might reflect large-scale structure and might reflect systematic plate quality variations with season—it is very much more difficult to get good plates at Palomar in the spring than it is in the fall. I think there is little doubt that the clusters are strongly clustered, but how much to trust the data into the large-distance, weak-signal region I do not know. I think we need machine-counted catalogs either from CCD data or from calibrated plates before we will know. I think it is a bit premature to abandon cold dark matter (or any other pet theory) on the basis of the existence of a positive cluster-cluster correlation function to untenably large distances.

The other embarrassment to CDM (and, probably, to any other picture which has gaussian noise initially and has a Zel'dovich spectrum on large scales and does not conflict with the background fluctuations, as Davis and collaborators have recently argued) is the possible existence of very large random velocities (~700 km/s) on scales of $60h^{-1}$ Mpc as seen in the M7 data. If the present interpretation results is correct, the problem seems very serious indeed. We may be driven to very non-gaussian initial conditions, or to non-gravitational formation scenarios; strings seem at this point the primary hope, either as seeds for gravitational growth in the usual picture or as electromagnetic bombs in the recent Ostriker-Witten-Thompson explosive picture, but much more needs to be done. Alternatively, one might seek to relax one of the constraints—reionization might reduce the background fluctuation embarrassment, for instance, in which case one of the not-so-dark-matter-dominated models discussed by Bond, with low Ω *and* low h, might work, the problem of reconciling inflation with curvature aside.

The data certainly suggest that the object doing the pulling is not too far outside the surveyed sphere, and work to clarify that needs to be done. It is worth noting also that if one chooses not to believe that there is an object pulling, the severity of the problem decreases markedly if one dismisses the bulk motion inferred from our motion with respect to the background. It does not seem impossible to imagine that the microwave dipole arises in some other fashion than through local gravitationally induced velocities. Most global gravitational influences produce quadrupole but no dipole (basically because of the equivalence principle), but one might imagine seeing residual gradients of *something* from incomplete inflation which might cause a "global" dipole.

On the subject of gaussian initial fluctuations, the topological investigations of Gott and collaborators and discussed here by Mellott, seem a particularly beautiful new way to investigate the initial conditions. Whether Nature has been kind enough to supply enough galaxies to trace the structure with the required accuracy is not clear, and to apply the technique adequately will require both bigger and more accurate (*i.e.* objective) catalogs than we have now, but the scheme should be pursued as vigorously as possible.

VII. Last Thoughts

We live in a universe in which some unknown but probably large fraction of the mass is in some unknown form. Many interesting and crucial quantities in that universe depend upon its dimensional scale to some high power, and we persuade ourselves, probably incorrectly, that we know that scale to within a factor of *two.* The GUT (if there is one) may or may not determine everything about the initial fluctuations and hence the present structure, but there is no satisfactory GUT, or perhaps there are too many. We trace the structure of the universe with nearly normal galaxies made mostly of nearly normal stars whose structure and evolution we understand not very well. There seems more than enough to do for the forseeable future, but the probability of our learning the answers to any of the big questions raised here anytime soon appears small.

There is a short list of large concrete tasks facing us, and my pet list of them will make no difference to what any of us does, but I will write it down anyway:

1. Determine the Hubble Constant to of order ten percent. The present situation is intolerable, and has been for twenty years.

2. Understand the giant and asymptotic giant branches in the central bulge of the Galaxy.

3. Make an objective galaxy catalog (from which can be constructed an objective cluster catalog), in which the limits and selection effects are understood and the photometry is accurate. Obtain redshifts for a large enough subsample that the motions in a cube of order 200 h^{-1} Mpc can be understood, and the median/n-tile isodensity surfaces can be drawn.

4. Understand the velocity fields and two-dimensional surface photometry and HI maps of a handful of nearby spiral galaxies well enough that detailed comparison can be made with n-body models and theory to settle finally the halo-to-disk ratio question.

5. Understand all the evidence from "test-particle structures" in elliptical galaxies (polar rings, ripples, HI disks ...) well enough to get the run of mass with radius.

6. Use 4.) and 5.) and group catalogs on the real sky from 3.) and ones constructed from the most realistic numerical simulations available with the same selection effects to get reliable group M/Ls and galaxy M/Ls as functions of radius to attempt to understand whether f is constant or variable on the scale of groups and larger.

7. *Find* the microwave background fluctuations.

And a couple of gauntlets for theorists:

8. Determine whether any reasonable gaussian-noise fluctuation model is consistent with all the constraints. It is worth reminding the protagonists in this regard that "half-and-half" baryon and CDM models throw away at least some of the beauties of the highly CDM dominated models, including the naturalness of the observed rotation curves and the easy understanding of the angular momentum of galaxies.

9. Make a complete string model, complete enough to predict the luminosity function, correlation function, and angular momentum distribution in galaxies and groups.

And we can all come back in twenty years for another go ... though I confidently predict that nearly normal galaxies will by then be such a vanishing breed that the workshop will have to have another title.

MIX
Papier aus verantwortungsvollen Quellen
Paper from responsible sources
FSC® C105338

If you have any concerns about our products,
you can contact us on
ProductSafety@springernature.com

In case Publisher is established outside the EU,
the EU authorized representative is:
Springer Nature Customer Service Center GmbH
Europaplatz 3, 69115 Heidelberg, Germany

Printed by Libri Plureos GmbH
in Hamburg, Germany